Human Embryonic Stem Cells

Human Embryonic Stem Cells

Edited by

J. Odorico

*Department of Surgery,
University of Wisconsin, Madison,
Wisconsin, USA*

S.-C. Zhang

*Department of Anatomy and Neurology,
University of Wisconsin, Madison,
Wisconsin, USA*

R. Pedersen

*Addenbrooke's Hospital,
University of Cambridge,
Cambridge, UK*

BIOS Scientific Publishers
Taylor & Francis Group

© Garland Science/BIOS Scientific Publishers, 2005

First published 2005

All rights reserved. No part of this book may be reprinted or reproduced or utilised in any form or by any electronic, mechanical, or other means, now known or hereafter invented, including photocopying and recording, or in any information storage or retrieval system without permission in writing from the publishers.

A CIP catalogue record for this book is available from the British Library.

ISBN 1 85996 278 5

Garland Science/BIOS Scientific Publishers
4 Park Square, Milton Park, Abingdon, Oxon, OX14 4RN, UK and
270 Madison Avenue, New York, NY 10016, USA
World Wide Web home page: www.garlandscience.com

Garland Science/BIOS Scientific Publishers is a member of the Taylor & Francis Group.

Distributed in the USA by
Fulfilment Center
Taylor & Francis
10650 Toebben Drive
Independence, KY 41051, USA
Toll Free Tel.: +1 800 634 7064; E-mail: taylorandfrancis@thomsonlearning.com

Distributed in Canada by
Taylor & Francis
74 Rolark Drive
Scarborough, Ontario M1R 4G2, Canada
Toll Free Tel: +1877 226 2237; E-mail: tal_fran@istar.ca

Distributed in the rest of the world by
Thomson Publishing Services
Cheriton house
North Way
Andover, Hampshire SP10 5BE, UK
Tel: +44(0)1264 332424; E-mail: salesorder.tandf@thompsonpublishingservices.co.uk

Production Editor: Andrew Watts
Typeset by Phoenix Photosetting, Chatham, UK
Printed by Cromwell Press, Trowbridge, UK

Table of Contents

Contributors		xi
Abbreviations		xiii
Foreword		xvii
Preface		xx

1. Biology of embryonic stem cells — 1
M.B. Morris, J. Rathjen, R.A. Keough and P.D. Rathjen

Introduction	1
Derivation and definition of mouse ES cells	1
The molecular basis of ES cell pluripotency	3
Differentiation of ES cells	10
Applications of ES cell technology	15

2. Characteristics of human embryonic stem cells, embryonal carcinoma cells and embryonic germ cells — 29
M.J. Shamblott and J.L. Sterneckert

Introduction	29
Sources of stem cells	29
Embryoid body-derived (EBD) cells	37
Markers of pluripotency	39

3. Adult stem cell plasticity — 45
R.E. Schwartz and C.M. Verfaillie

Introduction	45
Stem cells – definition	45

Adult stem cells	46
Adult stem cells – plasticity	46
Plasticity of hematopoietic bone marrow cells	47
Plasticity of mesenchymal stem cells	49
Mulitpotent adult progenitor cells	50
Plasticity of skeletal muscle cells	51
Plasticity of neural cells	52
Mechanisms of plasticity	52
Potential uses of adult stem cells	55

4. Human and murine embryonic stem cell lines: windows to early mammalian development   **61**
J.S. Odorico and S.-C. Zhang

Introduction	61
Derivation, growth and morphology of murine and human ES cells	62
ES cells as an *in vitro* model of early mammalian development	69
ES cells: a renewable source of functional cells	73
Summary	75

5. Human mesenchymal stem cells and multilineage differentiation to mesoderm lineages   **83**
V. Sottile and J. McWhir

Human mesenchymal stem cells (hMSCs) in culture	83
Differentiation towards mesenchymal lineages: lessons from hMSCs	88
Synergy of hMSC and hES research	91
Conclusion	92

6. Trophoblast differentiation from embryonic stem cells   **101**
T.G. Golos and R.-H. Xu

Origin and development of the placenta: introduction	101
Bridging the mouse–human gap in placental biology	108
Summary and future prospects	114

7. Current and future prospects for hematopoiesis studies using human embryonic stem cells   **121**
D.S. Kaufman

Introduction	121
Lessons from mouse ES cell-based hematopoiesis	123
Hematopoiesis from human ES cells: studies to date	124
Hematopoiesis from human ES cells: the next stage	126

	Human embryonic stem cells, preimplantation genetic diagnosis, and hematopoiesis	129
	Summary	132

8. Derivation of endothelial cells from human embryonic stem cells 137
S. Levenberg, N.F. Huang and R. Langer

Introduction	137
Development of endothelial cell progenitors	137
Isolation of endothelial cells and their progenitors	140
Characterization techniques for isolated endothelial cells	142
Therapeutic applications of endothelial cell progenitors	144
Challenges today and hopes for tomorrow	148

9. Neural specification from human embryonic stem cells 153
S.-C. Zhang

Introduction	153
Neural induction in vertebrates	154
Embryonic stem cells as a window to mammalian neural development	155
Neural differentiation from mouse ES cells	156
Neural differentiation from human ES cells	159
Outstanding questions	166

10. Modeling islet development through embryonic stem cell differentiation 173
J.S. Odorico, B. Kahan, D.A. Hullett, L.M. Jacobson and V.L. Browning

Introduction	173
Development of the pancreas and islets of Langerhans in vertebrates	174
Islet differentiation from embryonic stem cells	182
Recapitulating developmental pathways of islet differentiation in ES cells	187
Remaining questions	188
Summary	189

11. Cardiomyocyte differentiation in human embryonic stem cell progeny 199
I. Kehat, J. Itskovitz-Eldor and L. Gepstein

Introduction	199
Early signals in cardiac development	200

	In vitro differentiation of mouse and human ES cells to cardiomyocytes	201
	Prospects for myocardial regeneration	204
	Summary	209

12. Genetic engineering of human embryonic stem cells — 215
M. Drukker, S.K. Dhara and N. Benvenisty

Introduction	215
Methods for introduction of DNA into human ES cells	216
Alteration of gene expression in human ES cells	220
The potential clinical applications of genetically modified human ES cells	224
Conclusions	226

13. ES cells for transplantation: coping with immunity — 231
J.A. Bradley, E.M. Bolton and R.A. Pedersen

Introduction	231
Immune profile	232
Strategies for matching donor and recipient	237
Strategies for preventing allograft rejection	245
Concluding comments	250

14. Clinical applications for human ES cells — 257
T.J. Kamp and J.S. Odorico

Introduction	257
Goals for bringing hES cell-based therapy to clinical practice	258
Tissue engineering with hES cells	269
Cell-based therapy for delivery of bioactive molecules	270
Somatic cell nuclear transfer and hES cells	273
Progress and promise in disease-specific cell therapies	275
Future	280

15. Production of human embryonic stem cell-derived cellular product for therapeutic use — 289
R. Mandalam, Y. Li, S. Powell, E. Brunette and J. Lebkowski

Introduction	289
Required properties for a hESC-based cell therapy	290
Qualification of hESCs and raw materials	290
Cell production	292
Conclusions	297

16. Ethical and policy considerations in embryonic stem cell research — 301
R. Alta Charo

Federal regulation of embryo research — 301
The origins of the *de facto* ban on federal funding for embryo research — 302
Origins of the *de jure* ban on federal funding for embryo research — 303
Origins of the decision to permit general federal funding of research on embryonic stem cell lines — 304
The decision to narrow the eligibility requirements for federal funding of research with human embryonic stem cells — 306
The intersection of embryo research funding and the abortion debate — 307
Summary — 312

17. Legal framework pertaining to research creating or using human embryonic stem cells — 315
C.E. Gulbrandsen, M. Falk, E. Donley, D. Kettner and L. Koop

Introduction — 315
Federal statute — 315
Patent rights, licensing programs and agreements — 320
State regulation of research involving embryos — 326
International legal framework — 328
Summary — 329

18. Genomic approaches to stem cell biology — 339
T.S. Tanaka, M.G. Carter, K. Aiba, S.A. Jaradat, and Minoru S.H. Ko

Introduction — 339
Large-scale isolation of new genes from early embryos and stem cells — 340
Methods for gene expression profiling — 341
Data analysis and bioinformatics — 344
Expression profiling of stem cells — 346
Follow-up study of cDNA microarrays — 350
cDNA microarray analysis of cloned animals — 351
Large-scale functional studies of genes — 353
Future perspectives — 354

19. Proteomics and embryonic stem cells — 363
M.R. Sussman, A.D. Hegeman, A.C. Harms and C.J. Nelson

Introduction — 363
Making elephants fly — 364

Mass spectrometry instrumentation	367
Protein and peptide chemistry	368
Better proteomics through chemistry	371
Baby steps	373
Appendix: Human embryonic stem cell resources	379
Index	383

Color plates can be found between p. 136 and p. 137.

Contributors

Aiba, K., Developmental Genomics and Aging Section, Laboratory of Genetics, National Institute on Aging, NIH, Baltimore, Maryland, USA
Benvenisty, N., Department of Genetics, The Hebrew University of Jerusalem, Jerusalem, Israel
Bolton, E.M., Addenbrooke's Hospital, University of Cambridge, Cambridge, UK
Bradley, J.A., Addenbrooke's Hospital, University of Cambridge, Cambridge, UK
Browning, V.L., Department of Surgery, University of Wisconsin, Madison, Wisconsin, USA
Brunette, E., Geron Corporation, Menlo Park, California, USA
Carter, M.G., Developmental Genomics and Aging Section, Laboratory of Genetics, National Institute on Aging, NIH, Baltimore, Maryland, USA
Charo, R.A., Law School, University of Wisconsin, Madison, Wisconsin, USA
Dhara, S.K., Department of Genetics, The Hebrew University of Jerusalem, Jerusalem, Israel
Donley, E., Wisconsin Alumni Research Foundation, Madison, Wisconsin, USA
Drukker, M., Department of Genetics, The Hebrew University of Jerusalem, Jerusalem, Israel
Falk, M., Wisconsin Alumni Research Foundation, Madison, Wisconsin, USA
Gepstein, L., Technion – Israel Institute of Technology, Haifa, Israel
Golos, T.G., Department of Obstetrics and Gynecology and National Primate Research Center, University of Wisconsin, Madison, Wisconsin, USA
Gulbrandsen, C.E., Wisconsin Alumni Research Foundation, Madison, Wisconsin, USA
Harms, A.C., Biotechnology Center, University of Wisconsin, Madison, Wisconsin, USA
Hegeman, A.D., Biotechnology Center, University of Wisconsin, Madison, Wisconsin, USA
Huang, N.F., University of California–Berkeley and University of California–San Francisco Joint Graduate Group in Bioengineering, Berkeley, California, USA
Hullett, D.A., Department of Surgery, University of Wisconsin, Madison, Wisconsin, USA
Itskovitz-Eldor, J., Rambam Medical Center, Haifa, Israel
Jacobson, L.M., Department of Surgery, University of Wisconsin, Madison, Wisconsin, USA
Jaradat, S.A., Biotechnology Center, Jordan University of Science and Technology, Irbid, Jordan

Kahan, B., Department of Surgery, University of Wisconsin, Madison, Wisconsin, USA
Kamp, T., Clinical Sciences Center, University of Wisconsin, Madison, Wisconsin, USA
Kaufman, D.S., Stem Cell Institute, University of Minnesota, Minneapolis, Minnesota, USA
Kehat, I., Technion – Israel Institute of Technology, Haifa, Israel
Kettner, D.M., Wisconsin Alumni Research Foundation, Madison, Wisconsin, USA
Keough, R.A., ARC Special Research Centre for Molecular Genetics of Development, School of Molecular and Biomedical Science, University of Adelaide, Adelaide, Australia
Ko, M.S.H., Developmental Genomics and Aging Section, Laboratory of Genetics, National Institute on Aging, NIH, Baltimore, Maryland, USA
Koop, L., Wisconsin Alumni Research Foundation, Madison, Wisconsin, USA
Langer, R., Chemical Engineering Department, Massachusetts Institute of Technology, Cambridge, Massachusetts, USA
Lebkowski, J., Geron Corporation, Menlo Park, California, USA
Levenberg, S., Biomedical Engineering Department, Technion – Israel Institute of Technology, Haifa, Israel
Li, Y., Geron Corporation, Menlo Park, California, USA
Mandalam, R., Geron Corporation, Menlo Park, California, USA
McWhir, J., Roslin Institute, Midlothian, Scotland, UK
Morris, M.B., Australian Stem Cell Centre, and Network in Genes and Environment in Development, School of Molecular and Biomedical Science, University of Adelaide, Adelaide, Australia
Nelson, C.J., Biotechnology Center, University of Wisconsin, Madison, Wisconsin, USA
Odorico, J., Department of Surgery, University of Wisconsin, Madison, Wisconsin, USA
Pedersen, R.A., Addenbrooke's Hospital, University of Cambridge, Cambridge, UK
Powell, S., Geron Corporation, Menlo Park, California, USA
Rathjen, P.D., Australian Stem Cell Centre, ARC Special Research Centre for Molecular Genetics of Development, and Network in Genes and Environment in Development, Faculty of Science, University of Adelaide, Adelaide, Australia
Rathjen, J., Australian Stem Cell Centre, and ARC Special Research Centre for Molecular Genetics of Development, School of Molecular and Biomedical Science, University of Adelaide, Adelaide, Australia
Scadlock, C., Wisconsin Alumni Research Foundation, Madison, Wisconsin, USA
Schwartz, R.E., Stem Cell Institute, University of Minnesota, Minneapolis, Minnesota, USA
Shamblott, M.J., Johns Hopkins University School of Medicine, Institute for Cellular Engineering, Baltimore, Maryland, USA
Sottile, V., Institute of Genetics, University of Nottingham Medical School, Nottingham, UK
Sterneckert, J., Johns Hopkins University School of Medicine, Institute for Cellular Engineering, Baltimore, Maryland, USA
Sussman, M.R., Biotechnology Center, University of Wisconsin, Madison, Wisconsin, USA
Tanaka, T.S., Laboratory of Stem Cell Biology and Functional Genomics, Institute of Medical Science, University of Toronto, Toronto, Ontario, Canada
Thomson, J.A., Department of Anatomy, University of Wisconsin, Madison, Wisconsin, USA
Verfaillie, C.M., Stem Cell Institute, University of Minnesota, Minneapolis, Minnesota, USA
Xu, R.-H., Wisconsin Alumni Research Foundation, Madison, Wisconsin, USA
Zhang, S.-C., Department of Anatomy and Neurology, University of Wisconsin, Madison, Wisconsin, USA

Abbreviations

2-DGE	two-dimensional gel electrophoresis
ANF	atrial naturetic factor
ANP	atrial natriuretic peptide
AP	alkaline phosphatase
AVE	anterior visceral endoderm
bFGF	basic fibroblast growth factor
bHLH	basic helix-loop-helix
BMP	bone morphogenetic protein
CAFC	cobblestone area-forming cell
cDNA	complementary DNA
CFC	colony-forming cell
CFU-f	colony-forming unit-fibroblastic
CG	chorionic gonadotrophin
cGMP	current Good Manufacturing Practice
CIITA	class II transactivator
CM	conditioned medium
CML	chronic myelogenous leukemia
CMV	cytomegalovirus
COA	Certificate of Analysis
CTB	cytotrophoblast
dsRNA	double-stranded RNA
EB	embryoid body
EBD	embryoid body-derived
EC	embryonal carcinoma
ECM	extracellular matrix
EG	embryonic germ
EGC	embryonic germ cell
EGF	epidermal growth factor

eGFP	enhanced green fluorescent protein
EPC	endothelial progenitor cell
EPL	primitive ectoderm-like
EPLEBs	differentiation of EPL cells as EB
ERRβ	estrogen related receptor beta
ES	embryonic stem
ESC	embryonic stem cell
ESRF	ES cell renewal factor
EST	expressed sequence tag
EVT	extravillous cytotrophoblast
FAA	fumarylacetoacetate
FACS	fluorescence-activated cell sorting
FAH	fumarylacetoacetate hydrolase
FBS	fetal bovine serum
FCFC	fibroblast colony-forming cell
FCS	fetal calf serum
FDA	Food and Drug Administration
FGF	fibroblast growth factor
FGFR	fibroblast growth factor receptor
FKBP	FK-binding protein
FMEA	failure mode and effect analysis
FT-ICR	Fourier transform ion cyclotron resonance
GCSFR	GCSF receptor
GDF5	growth and differentiation factor-5
GDNF	glial cell line-derived neurotrophic factor
GM-CSF	granulocyte/macrophage colony stimulating factor
HCT	hematopoietic cell transplantation
HEF	human embryonic fibroblast
hES	human embryonic stem
HLA	human leukocyte antigens
hMSC	human mesenchymal stem cell
HPLC	high performance liquid chromatography
HPRT	hypoxanthine phosphoribosyltransferase
HSC	hematopoietic stem cell
IBMX	isobutylmethylxanthine
ICAT	isotope coded affinity tag
ICM	inner cell mass
IGF	insulin-like growth factor
IHH	Indian hedgehog
IL	interleukin
IMAC	immobilized metal affinity chromatography
IVF	*in vitro* fertilization
KSR	knockout serum replacement
LIF	leukemia inhibitory factor
LIFR	LIF receptor
LKLF	lung Kruppel-like factor

LPM	lateral plate mesoderm
LTR	long terminal repeat
MACS	magnetic column separation
MALDI-TOF	matrix-assisted laser desorption ionization/time of flight
MAPC	multipotent adult progenitor cell
M-CSF	macrophage colony stimulating factor
MEA	microelectrode array
MEF	murine (mouse) embryonic fibroblast
MEF2C	myocyte enhancer binding factor 2C
MHC	major histocompatibility complex; myosin heavy chain
mHC	minor histocompatibility complex
ML-IC	myeloid-lymphoid initiating cell
MPSS	massively parallel signature sequencing
MS	mass spectrometry
MSC	mesenchymal stem cell
NCAM	neural cell adhesion molecule
NFAT	nuclear factor of activated T cells
NSC	neural stem cell
NT	nuclear transplantation
NTN	neurturin
OCT	octomer-binding transcription factor
PDGF	platelet-derived growth factor
PECAM	platelet/endothelial cell adhesion molecule
PEI	polyethylenimine
PGA	polyglycolic acid
PGC	primordial germ cell
PGD	pre-implantation genetic diagnosis
PhIAT	phosphoprotein isotope-coded affinity tag
PNS	positive negative selection
PSA-NCAM	poly-sialylated neural cell adhesion molecule
PTLD	post-transplant lymphoproliferative disease
PTM	post-translational modification
Q-PCR	quantitative-PCR
RA	retinoic acid
RESC	rat ES-cell like
RF	radiofrequency
RNAi	RNA interference
RT-PCR	reverse transcriptase polymerase chain reaction
SAGE	serial analysis of gene expression
SCF	stem cell factor
SCID	severe combined immunodeficiency
SCNT	somatic cell nuclear transfer
SCX	strong cation exchange
SDIA	stromal cell-derived neural inducing activity
SDS-PAGE	sodium-dodecyl sulfate polyacrylamide gel electrophoresis
siRNA	short inhibitory (or small interfering) RNA

SOM	self-organizing map
SRC	SCID reconstituting cell
SSEA	stage-specific embryonic antigens
STB	syncytiotrophoblast
TDGF1	teratocarcinoma-derived growth factor 1
TGFβ	transforming growth factor beta
TOF	time-of-flight
TPO	thrombopoietin
TSC	trophoblast stem cell
VEGF	vascular endothelial growth factor
vWF	von Willebrand factor

Foreword
The health of human ES cell research

James A. Thomson

The reports of the derivation of human Embryonic Stem (ES) cells (Thomson et al., 1998) and of human embryonic germ (EG) cells (Shamblott et al., 1998) in late 1998 sparked both a wave of scientific enthusiasm and a political controversy that remains incompletely resolved. A partial resolution of that controversy in the United States was made by President George W. Bush when he restricted federal funding to human ES cell lines derived before August 9, 2001. As of this writing (April 8, 2003), only 11 human ES cell lines are listed as available by the National Institutes of Health Embryonic Stem Cell Registry. How damaging to human ES cell research have this and other compromises been over the last four years? A comparison to the early years of mouse ES cell work is useful. Two groups first reported the derivation of mouse ES cells in 1981 (Evans and Kaufman, 1981; Martin, 1981). An informal search of PubMed from July 1981 through November 1985 revealed 14 citations (excluding reviews) involving mouse ES cells. A search covering a similar period (November 1998 through March 2003) revealed 35 non-review articles involving human ES cells. By this superficial measure, at least, human ES cell research appears to be progressing at a reasonable rate.

However, this enumeration of citations ignores significant differences between the initial derivation of mouse and human ES cells. At the time of the initial derivation of mouse ES cells, the mouse experimental embryology community was a small, tightly knit group with only a handful of laboratories having the required expertise to work with mouse embryos or ES cells. To the few members of that small community, the idea of making knock-out mice with ES cells was just a dream, little appreciated by outside researchers. The development of homologous recombination for mouse ES cells and the resulting ability to make knock-out mice spawned an intense interest in mouse ES cells that was no longer restricted to the mouse embryology community. What started with a handful of mouse embryologists now involves most universities and institutes with significant biomedical research programs.

Against this backdrop, the current progress of human ES cell research is somewhat disappointing. Unlike the initial mouse ES cell derivations, there was an almost immediate appreciation that human ES cell research would be broadly important across biomedical research disciplines. This appreciation was due, in part, to the two decades of previous experience by the scientific community with mouse ES cells and to the intense media coverage that made the cells widely known to the scientific community. Yet the rate at which investigators have joined the field has been slower than might have been predicted given the level of interest. There are multiple contributing explanations for this. In the USA, there was no federally funded human ES cell research prior to President Bush's August 9, 2001 announcement restricting federal funding to those ES cell lines derived prior to that date. The initial lack of federal funding and the political uncertainty surrounding the work made investigators hesitant to enter the field. Ignoring the dubious public policy merit of President Bush's compromise, it did have the effect of giving investigators confidence that this work would go forward and be supported. However, the restricted number of existing cell lines created a bottleneck as the investigators involved with the initial derivation scrambled to set up the necessary infrastructure to meet the demand. An even more significant bottleneck was the limited number of groups with the expertise to use human ES cells effectively. Although a number of training courses for human ES cell culture have now been set up to address this need, there remains a significant, inherent time lag between this initial training and the emergence of quality publications. Indeed, the inherent cycle times of graduate and postdoctoral studies are likely to have the most significant long-term effects on the growth curve of the field.

The diversity of investigators contributing to the chapters in this volume suggests that the initial lag phase for the human ES cell field is already coming to an end and that an exponential growth phase is beginning. During the next year or two, it is likely that therapeutically useful human ES cell derivatives will be purified, and that defined culture conditions eliminating the need both for feeder layers and for non-human proteins will be developed. When these events occur, there will be intense pressure for public policy to go beyond President Bush's compromise and for multiple groups to derive new cell lines. Although the political controversy has certainly increased the time lag, the growth curve of the human ES cell field ultimately will be driven primarily by the scientific and medical merit of the cells.

References

Evans M, Kaufman M (1981) Establishment in culture of pluripotential cells from mouse embryos. *Nature* 292, 154–156.

Martin GR (1981). Isolation of a pluripotent cell line from early mouse embryos cultured in medium conditioned by teratocarcinoma stem cells. *Proc. Natl Acad. Sci. USA* 78, 7634–7638.

Shamblott MD, Axelman J, Wang S *et al.* (1998). Derivation of pluripotent stem cells from cultured human primordial germ cells. *Proc. Natl Acad. Sci. USA* 95, 13726–13731.

Thomson JA, Itskovitz-Eldor J, Shapiro SS *et al.* (1998). Embryonic stem cell lines derived from human blastocysts. *Science* **282**, 1145–1147.

James A. Thomson, VMD, PhD, Diplomate ACVP,
John D. MacArthur Professor, Department of Anatomy,
University of Wisconsin – Madison Medical School,
The Genome Center of Wisconsin, and The Wisconsin
National Primate Research Center, Madison, Wisconsin

Preface

The derivation of human embryonic stem cells in 1998 was a landmark discovery that will ultimately allow us to more profoundly comprehend human developmental processes, and which in the future could provide medical therapies for diseases characterized by the failure or destruction of specialized cells. Human embryonic stem cell research crosses many disciplines, including stem cell biology, reproductive biology, molecular biology, immunology, ethics, policy, embryology, neurobiology, oncology, and transplantation. Several chapters in this monograph illustrate the potential for cross-fertilization of ideas and technologies, such as proteomics, gene expression profiling, gene therapy, somatic cell nuclear transfer, prenatal genetic diagnosis, and tissue engineering. It was our goal as editors to stimulate a possible bridging of these disciplines in a fruitful way in future human embryonic stem cell research. In planning this text as a resource for scientists and students with a basic understanding of the principles of cell biology, we felt we should cover the unique biological properties of the cells and present this material in the greater context of other stem cell populations that are present in fetuses and adult organisms. These topics are discussed in the first five chapters. Much has been made of the possibility of generating medically useful cell-based therapies from human embryonic stem cells and the editors felt that an essential part of this text was a discussion of the current status of research towards generating cell therapies for treatment of diseases such as diabetes, Parkinson's disease and heart failure, among others. In Chapter 14 and several chapters that deal with differentiation into specific lineages, experts present recent achievements, current controversies, and future challenges as scientists develop and refine strategies to produce purified populations of functional cells for transplantation. Given the intense public and ethical debate surrounding human embryonic stem cell research, important social, moral, ethical and policy issues as they pertain to research in this field and therapeutic cloning are also presented. However, most would admit we are presently in the 'morula' phase of development of such therapies, and much work still needs to be done. It is clear that we are only at the end of the beginning, but already human embryonic stem cells have catalyzed the emerging field of regenerative medicine and will likely impact it for many years.

There are two underlying themes of human embryonic stem cell research that deserve mention and which the editors feel do not receive enough attention in the public debate about current research and the merits of these cells. In presenting their respective topics, the editors asked each of the contributors to address both of these themes. First, human embryonic stem cells provide a unique and important window to study early human development. Second, and in turn, a better understanding of developmental mechanisms and how tissues form will help achieve directed differentiation of desired lineages and facilitate effective transplantation therapies (*Figure 1*).

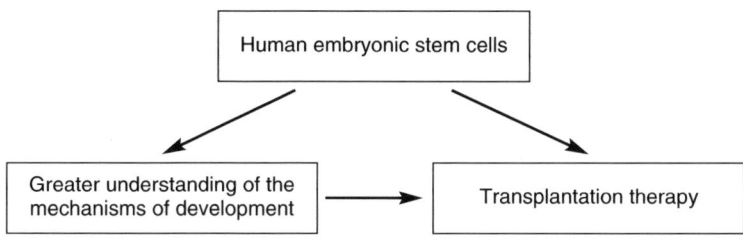

Figure 1: The impact of human embryonic stem cells on biology and medicine.

Although we have learned a great deal about the genetic control of development in many lower species such as frogs, chickens, zebrafish, and mice, we know very little about how and if this knowledge holds true for human development. Despite our understanding of the morphology of organogenesis in human development, there is a major gap in our knowledge of the molecular events that control these processes. Consequently, there is a significant 'species gap' in developmental biology and we are profoundly ignorant about our own, particularly the very earliest stages. For the first time, we have a means to study how the diversity of cell types that make up the human body form emerge, that is, which transcription factors are involved, what other cell types or tissues are inductive, and what signaling pathways regulate these processes in a human context. In areas of study such as placentogenesis, where there are no good small animal models, human embryonic stem cells will provide an insightful supplemental model system. Even if human embryonic stem cells were to fall short of their therapeutic promise, we believe future studies using human embryonic stem cells to study developmental mechanisms will have lasting impact on our understanding of both normal and abnormal human development. In turn, a better understanding of developmental mechanisms and signaling pathways will facilitate the development of effective cell-culture protocols for differentiating human embryonic stem cells into the desired specific cell lineages and will most likely drive the establishment of effective transplantation therapies. The editors hope that the reader will appreciate these two important and unique aspects of this text and that these themes will stimulate many new and exciting experiments.

We apologize to our many colleagues whose work was not included. Because this is a rapidly moving field, some aspects of the science, technology and policy may have evolved significantly since the chapters were initially written; for this we also apologize to the reader. However, while some details may evolve, we believe that the overarching themes of this book will remain important principles as the science moves forward.

The expertise required to generate this text far exceeds that of its editors. For the superb contributions of each of the authors we owe our sincerest gratitude. Moreover, his book would not have been possible without the assistance of our support staff. We are indebted to Janet Fox, Karen Heim, Kathy Worrall and Liz Cadman. We would also like to express our appreciation for the staff at the Taylor & Francis Group, including Nigel Farrar and Andrew Watts, for their outstanding technical support. Finally, we are grateful for the patience and support of our families during this project.

We hope that this book will provide students and scientists with a greater appreciation of the truly unique properties of human embryonic stem cells. We also hope it will entice new outstanding scientists to enter the field, that it will engender many new experiments among existing stem-cell researchers, and will stimulate focused and redoubled efforts towards generating stem cell-based therapies. At the minimum, we hope that it conveys the important impact this rapidly emerging, but young field can have on our understanding of human development.

Jon S. Odorico, M.D.
Su-Chun Zhang, Ph.D.
Roger A. Pedersen, Ph.D.

1. Biology of embryonic stem cells

Michael B. Morris, Joy Rathjen, Rebecca A. Keough and Peter D. Rathjen

1.1 Introduction

The first reports of cultured pluripotent mouse embryonic stem (ES) cell lines appeared more than 20 years ago (Evans and Kaufman, 1981; Martin, 1981). Early optimism that equivalent cell lines could be established from other mammalian species was not realized and until recently the mouse remained the only mammalian species from which stable pluripotent cell lines could be isolated. Accordingly, mouse ES cells have become the paradigm for the study and exploitation of mammalian stem cells. The greatest impact of ES cells over the last 20 years has been the development of gene targeting technology, in which ES cells provide a vector for the creation and analysis of precise alterations to the mouse genome. Experimental analysis and manipulation of ES cells has also increased our knowledge of the biochemical basis of pluripotence and extended our understanding of embryogenesis, principally in the processes of differentiation and cell fate determination. The lessons learned from mouse ES cells will inform the characterization of the recently established human ES cell lines (Thomson *et al.*, 1998; Reubinoff *et al.*, 2000), and the application of these cells to the study of human development and treatment of human disease.

1.2 Derivation and definition of mouse ES cells

The inner cell mass of the mouse blastocyst at about 4 days post-coitum is normally fated to form all cells and tissues of the embryo and extraembryonic yolk sac. Mouse ES cells were originally derived from this population of 20–40 pluripotent cells by culturing whole blastocysts or the surgically removed inner cell mass on a feeder layer of mitotically inactivated mouse embryonic fibroblasts (Brook and Gardner, 1997; Evans and Kaufman, 1981; Hogan *et al.*, 1994; Martin, 1981). More recently, successful culture has been performed in medium supplemented with one of a number of cytokines from the interleukin 6 (IL-6) family (Nichols *et al.*, 1990,

1994; Pease *et al.*, 1990). After several days, proliferating cells are disaggregated and replated. Colonies with the undifferentiated morphology characteristic of pluripotent ES cells (*Figure 1.1*) can be selected and propagated clonally by disaggregating to a single-cell suspension and reseeding. Successful derivation of ES cells has been achieved using blastocysts obtained from particular inbred strains of mice, principally 129 and C57BL/6 (Brook and Gardner, 1997; Evans and Kaufman, 1981; Hogan *et al.*, 1994; Kaufman *et al.*, 1983; Martin, 1981), but derivation from other strains has proven problematic.

1.2.1 Pluripotency of ES cells

A number of different approaches show that ES cells are pluripotent and can differentiate to form cell populations derived from all three primary germ layers, endoderm, ectoderm, and mesoderm. These include (1) directed and random differentiation of ES cells in culture (Bain *et al.*, 1995; Doetschman *et al.*, 1985; Guan *et al.*, 1999; Lake *et al.*, 2000; Nakano *et al.*, 1994; Wobus *et al.*, 1984), (2) implantation under the kidney capsule of adult mice resulting in the formation of teratomas (Damjanov *et al.*, 1987; Kaufman *et al.*, 1983), and (3) introduction into the mouse morula or blastocyst resulting in chimeras in which the ES cells contribute to all fetal and adult tissues including germ-line cells (Beddington and Robertson, 1989; Bradley *et al.*, 1984; Lallemand and Brulet, 1990; Smith, 1992; Wood *et al.*, 1993).

Although usually considered pluripotent, ES cells may be totipotent since in chimeras they sometimes contribute to extraembryonic visceral endoderm, the parietal endoderm of the yolk sac and also, albeit rarely, to the trophoblast-derived placenta (Beddington and Robertson, 1989). Consistent with this, ES cells in culture readily form parietal and visceral endoderm (Lake *et al.*, 2000), and a 50% reduction in the expression of the POU transcription factor Oct4 is sufficient to convert ES cells to trophoblast cells (Niwa *et al.*, 2000). These data suggest that in the appropriate environment ES cells are capable of forming all tissues of the embryo and adult, including all extraembryonic tissue.

1.2.2 Self-renewal of ES cells

In vitro, ES cells grow as domed colonies (*Figure 1.1*) which can proliferate without differentiation in medium supplemented with either leukemia inhibitory factor (LIF) (Pease and Williams, 1990; Williams *et al.*, 1988) or one of a number of cytokines from the IL-6 family (Conover *et al.*, 1993; Nichols *et al.*, 1994; Pennica *et al.*, 1995; Rose *et al.*, 1994; Yoshida *et al.*, 1994). These cytokines signal through the gp130 receptor subunit (Chow *et al.*, 2002) and are both necessary and sufficient for the isolation and maintenance of ES cells (Nichols *et al.*, 1990, 1994; Pease *et al.*, 1990). Support of ES cells by more complex culture environments, such as medium supplemented with Buffalo Rat Liver conditioned medium or by co-culture with embryonic fibroblasts, has been shown to be dependent on the paracrine supply of LIF (Rathjen *et al.*, 1990a,b; Smith *et al.*, 1988).

In the presence of LIF, ES cells retain an indefinite capacity for self-renewal without transformation, and their growth is not restricted by contact inhibition or proliferative senescence. Compared with cancer cells, ES cells remain substantially

Chapter 1 – Biology of embryonic stem cells

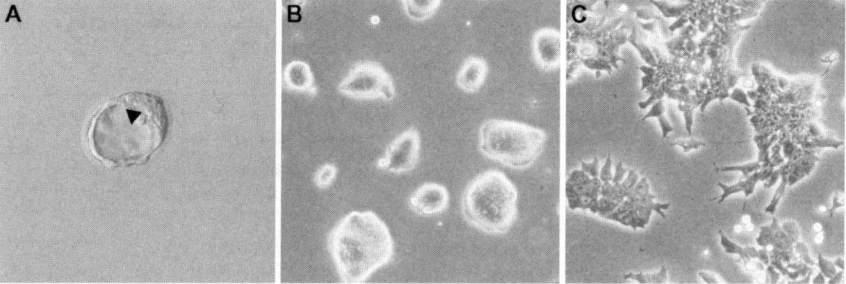

Figure 1.1: (A) 3.5-day post-coitum mouse blastocyst. The arrow indicates the inner cell mass (ICM), a population of pluripotent cells localized at one pole of the blastocelic cavity. The image was taken using Hoffman interference contrast and a 20× objective. (B) Mouse ES cells in culture, growing in characteristic domed, three-dimensional colonies in which individual cells cannot be discerned. (C) ES cells cultured in HepG2 conditioned medium differentiate to form a homogeneous population of EPL cells. Differentiation is accompanied by alterations in colony morphology, with cells growing in two-dimensional sheets in which individual cells can be easily observed. (B) and (C) were photographed using phase contrast and a 20× objective.

normal karyotypically over extended culture. However, many ES cell lines contain subpopulations of cells with chromosomal abnormalities. Trisomy-8 ES cells have a selective growth advantage over euploid cells and rarely contribute to germ-line transmission in chimeras (Liu *et al.*, 1997b). Increasing passage number can result in aneuploidy, which correlates with reduced efficiency of both chimera formation and germ-line transmission (Longo *et al.*, 1997).

1.2.3 Markers of ES cells

Aside from their pluripotency and self-renewing properties, undifferentiated mouse ES cells can be characterized by a number of markers (*Table 1.1*), some of which are known to be essential for ES cell survival and self-renewal (see below) and almost all of which are known to be expressed in inner cell mass cells of the embryo. Within ES and inner cell mass cells, alkaline phosphatase activity is high (Johnson *et al.*, 1977; Matsui *et al.*, 1992; Mulnard and Huygens, 1978; Wobus *et al.*, 1984) as is telomerase activity (Armstrong *et al.*, 2000; Liu *et al.*, 2000b), the latter being consistent with the self-renewal properties and genomic stability of pluripotent cells. The expression of most of these markers and activities is not exclusive to pluripotent cells but in combination their expression appears to define mouse ES cells uniquely.

1.3 The molecular basis of ES cell pluripotency

Several genes and signaling pathways have been identified as important for survival and maintenance of pluripotence of ES cells in culture and inner cell mass cells in the embryo.

1.3.1 Transcriptional regulators

A key regulator of pluripotence is the POU transcription factor Oct4, also referred to as *POU5f1*, Oct3 or Oct3/4 (Niwa *et al.*, 2000, 2002). *Oct4* is expressed in ES cells

Table 1.1: Markers of ES cells.

Marker	Type of protein	Expressed in ICM?	References
SSEA-1 [a]	Cell-surface antigen	Yes	(Matsui et al., 1992; Solter and Knowles, 1978)
Fgf4	Growth factor	Yes	(Niswander and Martin, 1992; Rappolee et al., 1994; Yuan et al., 1995)
Tpo [a]	Growth factor	Not known	(Xie et al., 2002)
Esg-1 [a]	RNA binding?	Yes	(Bierbaum et al., 1994; Tanaka et al., 2002)
Psc1	mRNA regulation?	Yes	(Pelton et al., 2002)
CRTR-1	CP2 TF [a]	Yes	(Pelton et al., 2002)
Foxd3 [b]	Forkhead TF	Yes [d]	(Hanna et al., 2002)
Gbx2	Homeobox TF	Yes	(Chapman et al., 1997)
Oct4 [c]	POU TF	Yes	(Nichols et al., 2001; Niwa et al., 2000; Pelton et al., 2002)
Rex1	Zinc-finger TF	Yes	(Pelton et al., 2002; Rogers et al., 1991)
Sox2	SRY-related HMG Box TF	Yes	(Avilion et al., 2003; Tomioka et al., 2002)
Taube nuss	Transcription coactivator?	Yes	(Voss et al., 2000)

[a] Abbreviations: SSEA-1, stage-specific embryonic antigen-1; Tpo, thrombopoietin; TF, transcription factor; Esg-1, embryonal stem cell-specific gene-1.
[b] Hfh2 or genesis.
[c] POU5f1, Oct3 or Oct3/4.
[d] Detected in whole blastocysts and 6.5 days post-coitum epiblast.

(Nichols et al., 1998) and continues to be expressed as these cells differentiate to early primitive ectoderm-like (EPL) cells, an *in vitro* equivalent of primitive ectoderm in the embryo at ~5.5 days post-coitum (*Figure 1.1*; Pelton et al., 2002). Oct4 expression is downregulated as ES cells differentiate further to cells representative of the three germ layers (Palmieri et al., 1994; Rathjen et al., 1999).

The pluripotency of mouse ES cells appears to depend on tightly regulated Oct4 expression. Using a tetracycline-regulated *Oct4* transgene, Niwa et al. (2000) showed that a reduction of 50% or more in expression induced differentiation of ES cells to extraembryonic trophoblast cells, whilst a 50% increase in *Oct4* expression triggered differentiation to endoderm and mesoderm. Production of primitive endoderm in F9 embryonal carcinoma cells stimulated by retinoic acid treatment is also accompanied by a transient increase in *Oct4* expression (Botquin et al., 1998).

The expression and function of *Oct4* in the early embryo is also consistent with a requirement for tightly regulated *Oct4* expression to maintain pluripotency and self-renewal. (1) *Oct4* in the embryo is expressed in the inner cell mass cells of the blastocyst, continues to be expressed as these cells differentiate to the pluripotent but developmentally restricted primitive ectoderm, and thereafter is downregulated in almost every cell type (Pelton et al., 2002; Saijoh et al., 1996). (2) *Oct4* expression is downregulated as the trophectoderm surrounding the inner cell mass differentiates (Palmieri et al., 1994). (3) *Oct4* expression is increased in the primitive endoderm of late blastocysts compared with its expression in the inner cell mass cells of the early blastocyst (Palmieri et al., 1994). (4) $Oct4^{-/-}$ embryos fail just after implantation and cannot be used to establish ES cell lines because the inner cell mass cells do not survive. Instead, inner cell mass cells form trophoblast (Nichols et al., 1998).

Oct4 binds a variety of DNA sequences including the consensus octamer motif, the AT-rich sequence (Okamoto et al., 1990; Saijoh et al., 1996), and various Oct factor recognition elements such as PORE and MORE (Tomilin et al., 2000). The survival of ES cells and their self-renewal requires the Oct4 POU domain and at least one of its proline-rich transactivation domains, or a similar transactivation domain supplied, for example, from Oct2 (Niwa et al., 2002).

Oct4 can homodimerize or heterodimerize with other transcriptional regulators to activate or repress gene expression (Pesce and Scholer, 2001). For example, Oct4 represses the expression of α and β human chorionic gonadotrophin genes in choriocarcinoma cells by binding octamer motifs in the promoters (Liu and Roberts, 1996; Liu et al., 1997a). Oct4 and Sox2 act synergistically to activate the expression of *Fgf4* (Yuan et al., 1995) via formation of a complex in which each component binds adjacent sites on an *Fgf4* enhancer element located in the 3' untranslated region (Ambrosetti et al., 1997; Yuan et al., 1995).

As with $Oct4^{-/-}$ embryos (Nichols et al., 1998), $Sox2^{-/-}$ (Avilion et al., 2003) and $Fgf4^{-/-}$ (Feldman et al., 1995) embryos die shortly after implantation because the pluripotent cells fail to survive. Consistent with this, ES cells cannot be derived from $Sox2^{-/-}$ embryos. Cultured $Ffg4^{-/-}$ embryos can be rescued by the addition of Fgf4 (Feldman et al., 1995; Wilder et al., 1997) but the addition of Fgf4 does not rescue either $Oct4^{-/-}$ or $Sox2^{-/-}$ embryos (Avilion et al., 2003) nor is Fgf4 required for the survival of $Fgf4^{-/-}$ ES cells (Wilder et al., 1997). Collectively, these data confirm that Fgf4 lies downstream of both Oct4 and Sox2 and demonstrate that both Oct4 and Sox2 regulate the expression of other genes important for survival and maintenance of pluripotent cells.

More recently, the forkhead (winged helix) transcription factor Foxd3 (also known as *Hfh2* and genesis) has been implicated in the maintenance of early pluripotent cells in the embryo and in the establishment of ES cells in culture (Hanna et al., 2002). $Foxd3^{-/-}$ embryos show a similar phenotype to $Oct4^{-/-}$, $Sox2^{-/-}$ and $Fgf4^{-/-}$ embryos and die shortly after implantation because of a failure of the pluripotent cells to survive. $Foxd3^{-/-}$ inner cell mass cells cannot be used to establish ES cells in culture and neither the embryos nor the ES cells can be rescued by the addition of Fgf4. $Foxd3^{-/-}$ cells show apparently normal levels of Oct4, Sox2 and Fgf4, indicating that Foxd3 does not regulate the expression of these genes. However, Oct4 and Foxd3 have been shown to interact to regulate gene expression (Guo et al., 2002) and this interaction may be required for pluripotent cells to respond appropriately to Fgf4 signals (Hanna et al., 2002). As with Oct4, close control of the expression of Foxd3 may provide a means of controling pluripotency of ES cells and preventing spontaneous differentiation, since it has been shown that continued expression of Foxd3 in migrating neural crest cells interferes with their differentiation (Dottori et al., 2001). It would be of interest to determine if LIF-dependent signaling in ES cells regulates the expression and activity of Foxd3.

1.3.2 LIF signaling

LIF signals through a plasma membrane receptor consisting of the LIF receptor β (LIFRβ) oligomerized with the ubiquitous cytokine receptor component gp130 (Chow *et al.*, 2001, 2002; Davis *et al.*, 1993; Smith *et al.*, 1988; Yoshida *et al.*, 1994). Other IL-6 cytokines, which signal through gp130, including oncostatin M (Rose *et al.*, 1994), ciliary neurotrophic factor (Conover *et al.*, 1993) and cardiotrophin-1 (Pennica *et al.*, 1995), can replace LIF. LIFRβ has been shown to be dispensable using a combination of IL-6 and soluble IL-6 receptor, which together activate gp130 homodimers (Nichols *et al.*, 1994). Similarly, ES cells expressing chimeric receptors consisting of the GCSF receptor (GCSFR) fused with the intracellular domain of either LIFRβ or gp130 have been used to show that gp130 but not LIFRβ is required for self-renewal (Niwa *et al.*, 1998).

In the embryo, a clear dependence on LIF or IL-6 family cytokines for stimulation or maintenance of the inner cell mass during normal development has not been demonstrated. LIF signaling does not appear to be important for development of the early embryo. Pluripotent cells of the inner cell mass and primitive ectoderm behave normally in embryos disrupted for the LIF, LIFRβ or gp130 genes (Stewart *et al.*, 1992; Ware *et al.*, 1995; Yoshida *et al.*, 1996), indicating that alternative pathways can regulate the maintenance of pluripotent cells in the early embryo. Instead, LIF appears to be important in diapause, in which a developing embryo is arrested and implantation delayed, while, for example, the feeding mother weans the previous litter (Nichols *et al.*, 2001; Smith, 2001). Unlike wild-type embryos, pluripotent cells in $gp130^{-/-}$ embryos die soon after the onset of diapause. Thus, for an embryo in diapause, LIF signaling appears to prevent pluripotent cell death and keeps the pluripotent cells primed for self-renewal in the absence of cell growth. ES cell lines derived from embryos, including diapaused embryos, may have acquired artifactually a dependence on LIF signaling which is not required for pluripotent cells during normal embryogenesis (Smith, 2001).

The LIF–LIFRβ/gp130 complex regulates the activity of two parallel signaling pathways, the JAK/STAT3 pathway and the Shp2/Ras-dependent pathway signaling via ERK1 and ERK2 (*Figure 1.2*; Burdon *et al.*, 1999; Niwa *et al.*, 1998). The relative activities of these two pathways may provide the 'switch' between pluripotence and differentiation.

1.3.3 LIF signaling via JAK/STAT promotes self-renewal of ES cells

Binding of LIF to the receptor complex results in conformational changes to the receptor which leads to autophosphorylation of quiescent JAK molecules previously recruited to the C-terminal intracellular domains of the receptor components (*Figure 1.2*). Phosphorylated JAKs phosphorylate tyrosine residues on the receptor, which act as docking sites for STAT3 molecules via their SH2 domains (Hou *et al.*, 2002). Bound STAT3 molecules are tyrosine phosphorylated by JAK, released from the receptor, dimerize via their SH2 domains, and are translocated to the nucleus where the dimer binds promoter consensus sequences and regulates gene expression (Schindler and Darnell, 1995).

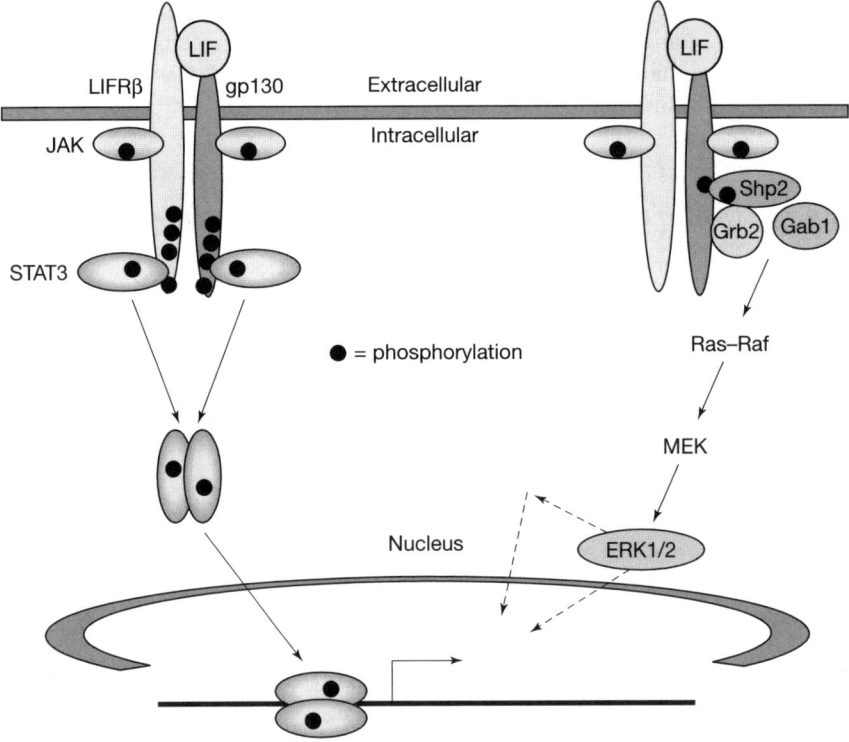

Figure 1.2: LIF signaling in ES cells. Extracellular LIF binds the LIFRβ–gp130 receptor complex in the plasma membrane and triggers the JAK/STAT and Shp2/Ras pathways. The receptor undergoes a conformational change resulting in autophosphorylation of tyrosines on JAK molecules bound to the intracellular domains of the receptor. The activated JAKs phosphorylate the receptor subunits promoting the recruitment and phosphorylation of STAT3. Activated STAT3 leaves the receptor, dimerizes, and enters the nucleus where it acts as a transcription factor regulating the expression of genes required to maintain the pluripotence of ES cells. For the Shp2/Ras pathway, LIF binding induces JAK-dependent phosphorylation of the receptor, resulting in the recruitment, phosphorylation and activation of Shp2. Other components, including Grb2 and Gab1, are also recruited resulting in the activation of ERK1 and ERK2 via MEK phosphorylation. Activated ERKs translocate to the nucleus and regulate transcription or phosphorylate cytoplasmic components, which can eventually result in transcriptional regulation (dotted arrows). This pathway appears to induce differentiation.

In many cell types, activation of STAT3 results in differentiation but in ES cells STAT3 activation results in self-renewal and maintenance of pluripotence for reasons that are not clear. Several results highlight the importance of activated STAT3 for self-renewal of ES cells. (1) Mutation of the intracellular tyrosines of gp130, which act as docking sites for STAT3, induces differentiation of ES cells (Matsuda et al., 1999; Niwa et al., 1998). (2) Expression in ES cells of the dominant negative STAT3F mutant, in which the tyrosine required for dimerization and nuclear translocation is replaced with a phenylalanine, also results in differentiation (Niwa et al., 1998). (3) STAT3–estradiol receptor fusion protein heterologously expressed in ES cells dimerizes in the presence of 4-hydroxy tamoxifen and is translocated to the nucleus with the outcome that pluripotence is maintained, and differentiation inhibited, even in the absence of LIF (Matsuda et al., 1999). (4) Consistent with this, cultured $STAT3^{-/-}$ blastocysts show reduced proliferation of the inner cell mass (Takeda et al., 1997). However, failure of the embryo is not observed until later in development following egg cylinder formation.

1.3.4 LIF signaling via ERKs appears to promote ES cell differentiation

LIF-dependent signaling also activates the Shp2/Ras-dependent ERK pathway in ES cells (*Figure 1.2*; Burdon et al., 2002). Recruitment and phosphorylation of Shp2 to the LIF-activated receptor complex induces recruitment of Grb2 and other proteins to the receptor resulting in activation of sequential elements in the pathway including phosphorylation and activation of ERK1 and ERK2. Activated ERKs are translocated to the nucleus and modulate the expression of various genes, including those for transcription factors.

Various results show that the Shp2/ERK pathway does not promote self-renewal of ES cells but instead inhibits it and promotes differentiation. (1) Expression of a mutated gp130 in which the Shp2 binding site has been removed from the intracellular domain promotes LIF-dependent self-renewal (Burdon et al., 1999). (2) Knockouts of Grb2 (Cheng et al., 1998) and Shp2 (Qu and Feng, 1998) or elimination of the gp130 binding site from Shp2 (Qu and Feng, 1998) also reduce differentiation and promote self-renewal. (3) Chemical inhibition of MEK, the kinase which phosphorylates and activates the ERKs, or reducing the phosphorylation of ERKs with ERK phosphatases produces similar results (Burdon et al., 1999).

It would appear that under conditions that promote self-renewal (i.e., in the presence of LIF) the balance of the 'competing' STAT3 and ERK pathways in ES cells lies in favor of STAT3 and pluripotence. With a reduction (or complete removal) of LIF, the system switches to differentiation.

1.3.5 LIF-independent signaling

Other signaling pathways appear to contribute to ES cell self-renewal. ES cells express thrombopoietin and its receptor c-Mpl and formation of the ligand–receptor complex results in recruitment and activation of STAT3 and suppression of differentiation (Xie et al., 2002). The ligand–receptor complex also recruits and activates Shp2 but, similar to results with LIF-dependent signaling, removal of the recruitment domain from Shp2 further suppresses differentiation (Xie et al., 2002). An ES cell renewal factor (ESRF) in the conditioned medium of a parietal endoderm-like cell line has been postulated to promote both ES cell renewal in the absence of LIF and LIF signaling via STAT3 (Dani et al., 1998). Self-renewal of pluripotent EPL cells has also been shown to occur in HepG2 conditioned medium MedII even in the presence of a neutralizing antibody to human LIF (Rathjen et al., 1999). The existence of LIF-independent signaling pathways for pluripotent cell maintenance may provide an explanation for the normal progression of early embryogenesis in the absence of LIF/gp130 signaling (Stewart et al., 1992; Ware et al., 1995; Yoshida et al., 1996).

1.3.6 Gene profiling of pluripotent cells

Gene expression profiling has been used to identify sets of 'signature' candidate genes common to embryonic and adult stem cells (Ivanova et al., 2002; Ramalho-Santos et al., 2002; Tanaka et al., 2002). These sets contain a small percentage of genes scattered over the mouse genome and are enriched in genes, such as *Oct4* (Tanaka et al., 2002) and components of the JAK/STAT signaling pathway (Ramalho-Santos et al., 2002), whose expression controls critical stem cell properties like self-renewal and pluripotency. Many of these 'stem cell' genes are known to be involved in transcriptional regulation and chromatin remodeling, cell cycle regulation, DNA repair, RNA processing, and the fidelity of protein folding and degradation, consistent with the dual requirements for rapid self-renewal and maintenance of genome integrity (Ramalho-Santos et al., 2002). The sets are also characterized by a higher proportion of novel genes in the stem cell sets than is found in differentiated cell types (Ramalho-Santos et al., 2002; Tanaka et al., 2002), and by a greater overlap between ES cells and neural stem cells than between ES cells and hematopoietic stem cells, indicating that ES cells are more similar to neural stem cells than hematopoietic stem cells (Ivanova et al., 2002; Ramalho-Santos et al., 2002). This last point may explain why ES cells appear primed to adopt a neural fate by default since many of the genes expressed in neural stem cells are already expressed in ES cells (Ramalho-Santos et al., 2002).

Tanaka et al. (2002) compared the expression profiles of mouse ES cells, trophoblast stem cells and embryonic fibroblasts and in this way defined a unique set of largely novel genes whose expression is upregulated exclusively in ES cells. From this profiling study, *embryonal stem cell-specific gene 1 (Esg-1)* appears to be a promising candidate for involvement in ES cell pluripotence. Like *Oct4*, *Esg-1* is expressed in the early embryo and expression is downregulated upon differentiation in both the embryo and in ES cells in culture. Furthermore, *Esg-1* expression (but not *Oct4* expression) is strongly downregulated in $STAT3^{-/-}$ cells and also strongly downregulated by forced reduction in *Oct4* expression. These results indicate that *Esg-1* is placed downstream of both STAT3 signaling and Oct4 transcriptional regulation, consistent with the identification of potential binding sites for STAT3 and Oct4 within 10 kb of *Esg-1*. The *Esg-1* protein contains a KH domain indicating a role for the protein in RNA binding. Such a role is consistent with the fact that 'stemness', as defined by other array studies, includes a number of genes implicated in RNA processing (Ramalho-Santos et al., 2002; Rodda et al., 2002).

1.3.7 The ES cell cycle

The pluripotent cells of the early mouse embryo have an unusually rapid growth rate with a doubling time of ~10 h between days 4.5 and 6.0 post-coitum reducing to 4.4 h around the time of gastrulation at day 6.5 (Hogan et al., 1994). Similarly, the generation time of mouse ES cells is rapid (~11 h) and is reduced to ~8 h following conversion of ES cells to EPL cells (Stead et al., 2002). The cell cycle of ES cells and pluripotent cells *in vivo* is characterized by greatly reduced G phases and a high proportion of time (~50%) spent in S phase (Stead et al., 2002). Unlike somatic cell cycles, ES cells constitutively express high levels of cdk2, cyclin A and cyclin E kinase activities throughout the cell cycle, whilst the activity of cyclin D kinase is undetectable (Savatier et al., 1996; Stead et al., 2002). High cyclin A and E activities result in hyperphosphorylation and inactivation of the pocket protein p107, and retinoblastoma tumor suppressor protein (pRb). In somatic cells, reduced phosphorylation of pocket proteins results in binding to E2F, which represses the expression of E2F target genes such as cyclin E, B-myb and cdc2 thereby preventing cell-cycle progression at the G1 checkpoint (Stead et al., 2002). Instead, in ES cells constitutively hyperphosphorylated pocket proteins fail to bind E2F resulting in cell cycle-independent expression of E2F target genes and progression through G1/S (Stead et al., 2002). Thus, ES cells do not appear to sense or receive signals from their environment that would impose the G1 checkpoint, slow the cell cycle, and regulate cyclin activities in a cell cycle-dependent manner.

It is unclear if cell signaling contributes to the establishment and maintenance of this unusual cell cycle structure and regulation in ES cells. One hypothesis is that cell cycle regulation is mediated by LIF signaling, and perhaps other signaling pathways, which activate STAT3 (Burdon et al., 2002; Hirano et al., 2000). In other cells, gp130-dependent proliferation depends on STAT3 activation of c-myc and Pim-1, which cooperate to overcome cell cycle arrest at the G1/S transition by driving hyperphosphorylation of pRb (Shirogane et al., 1999). When ES cells differentiate upon LIF withdrawal, downregulation of STAT3 activity may result in induction of a normal somatic cell cycle structure via activation of the pRb-dependent restriction point that regulates the G1/S transition (Burdon et al., 2002).

1.4 Differentiation of ES cells

The differentiation potential of ES cells can be exploited experimentally in culture for a number of purposes, primarily production of specific cell populations for experimental and therapeutic outcomes, understanding pathways and signaling environments that regulate differentiation, and defining the role of extracellular and intracellular regulators of development.

1.4.1 Spontaneous differentiation: the broad potential of ES cells in culture

The ability to differentiate ES cells to representative cell populations of the three primary germ layers provides a readily accessible source of normal, non-transformed progenitor and differentiated cells for experimental manipulations and functional analysis *in vitro* and *in vivo*. A number of methodologies are available for the differentiation of ES cells, including withdrawal of factors required for maintenance of pluripotence and addition of chemical inducers of differentiation, such as retinoic acid, dimethyl sulphoxide and 3-methoxybenzamide (Smith, 1991, 1992).

Differentiation of ES cells as embryoid bodies (EBs), ES cell aggregates cultured in suspension in the absence of gp130 agonists, has been used extensively. EBs undergo a program of differentiation which recapitulates the early events of mammalian embryogenesis and results in the formation of multiple cell populations representative of the three primary germ lineages (Doetschman et al.,

1985; Rathjen and Rathjen, 2001). Although a wide variety of cell populations can be formed in EBs, including contracting cardiomyocytes, hematopoietic progenitors, erythrocytes, dendritic cells, neural progenitors, neurons and gut epithelial cells (reviewed in Rathjen and Rathjen, 2001), the abundance of each cell type within a body is small. The experimental tractability of the system is restricted by a high degree of temporal and spatial heterogeneity within and between bodies in a single population, a consequence of the lack of positional information or organization within EBs.

Differentiation within EBs initiates with the formation of extraembryonic endoderm from the 'outer' cells of the aggregate, which further differentiates to form parietal, or yolk sac, endoderm and visceral endoderm. Analogous with the proposed role of visceral endoderm in the embryo (Beddington and Robertson, 1999), this tissue is thought to act as a source of signaling molecules that regulate differentiation. While these signals presumably program the subsequent differentiation of remaining pluripotent cells, the presence of endogenous signaling impedes the ability of exogenous cytokines and growth factors to direct differentiation within EBs.

1.4.2 Enrichment for differentiated cell populations

Purification from EBs, and recently from differentiating cells in adherent culture (Ying *et al.*, 2003), has been used to produce populations enriched for specific cell types. Generic approaches to purification have included engineering ES cells to express selectable markers under the control of lineage specific promoters and the use of selective culture conditions that favor the growth of particular lineages. Genetically marked ES cells, expressing the green fluorescent protein under the control of the myosin light chain-2v promoter, have enabled efficient purification of cardiac muscle progenitors from EBs by FACS (Müller *et al.*, 2000). Similarly, expression of neomycin resistance under the control of the *Sox2* promoter coupled with application of selective conditions has been used for the formation of populations highly enriched in neural progenitor cells (Li *et al.*, 1998). Most of these protocols initiate ES cell differentiation within EBs and demonstrate the power of this system for the generation of a variety of cells. Further, functional characterization of the differentiated and progenitor cells produced from EBs has demonstrated equivalence with analogous populations formed during embryogenesis (Doevendans *et al.*, 2000; Guan *et al.*, 2001; Kolossov *et al.*, 1998). However, the dysregulated and disorganized formation of cells within EBs leads to exposure of cells to inappropriate signaling molecules and potentially adverse effects of this on cell identity and function cannot be excluded (Rathjen and Rathjen, 2001).

Alternatively, the addition of growth factors to the EB differentiation environment can be used to manipulate the differentiation outcome and enrich specific cell populations. For example, addition of granulocyte/macrophage colony stimulating factor (GM-CSF) and IL-3 supports the formation of immature dendritic cells from EBs (Fairchild *et al.*, 2000). Similarly, the addition of growth factors, including vascular endothelial growth factor (VEGF) and bone morphogenetic protein 4 (BMP4), enhances the formation of hematopoietic lineages (Adelman *et al.*, 2002; Kennedy *et al.*, 1997; Nakayama *et al.*, 2000).

Lastly, differentiation of ES cells genetically modified to express constitutively gene(s) implicated in determination of cell fate has been used to produce differentiated cell populations enriched in specific target cells. Constitutive expression in ES cells of *Nurr1*, a transcription factor critical for the development of midbrain dopaminergic neurons, coupled with *in vitro* differentiation and neural enrichment resulted in a 4–5 fold increase in the formation of dopaminergic neurons, without altering the levels of other neural populations (Chung *et al.*, 2002a). Likewise, forced expression in ES cells of *Pax4*, a protein required for β cell production (Sosa-Pineda *et al.*, 1997), significantly promoted the production of insulin-producing cells (Blyszczuk *et al.*, 2003), and the expression of *HoxB4* in ES cells resulted in a significant increase in the formation of progenitors for erythroid/myeloid and definitive erythroid colonies from EBs (Helgason *et al.*, 1996) and the generation of progenitors capable of reconstituting the hematopoietic system in mice (Kyba *et al.*, 2002).

1.4.3 Directed differentiation of ES cells in culture

Directed differentiation of ES cells to homogeneous populations of progenitor or differentiated cells in culture would circumvent many of the limitations of EBs. Several reports suggest that this approach to ES cell differentiation is feasible.

ES cells can be differentiated to a second pluripotent cell population, early primitive ectoderm-like (EPL) cells (*Figure 1.1C*), by culture in conditioned medium from the human hepatocellular carcinoma cell line, HepG2. EPL cell formation is homogeneous and accompanied by alterations in morphology, gene expression, cytokine responsiveness and differentiation potential consistent with the formation of a population analogous to primitive ectoderm (Rathjen *et al.*, 1999; Lake *et al.*, 2000). The formation of EPL cells *in vitro* occurs without formation of the extraembryonic visceral endoderm lineage. ES cells differentiate to EPL cells in response to a combination of factors within the conditioned medium; a component of the extracellular matrix and a low molecular weight activity of less than 3 kDa (Rathjen *et al.*, 1999; Bettess, 2001). The signals regulating primitive ectoderm formation in the embryo are thought to emanate from the primitive/visceral endoderm (reviewed in Rodda *et al.*, 2002). Several lines of evidence suggest that signaling from HepG2 cells/hepatic cells recapitulates such signaling (Rathjen *et al.*, 2001).

The absence of visceral endoderm in populations of EPL cells provides a means to direct formation of somatic lineages from pluripotent cells in the absence of endogenous signals. Differentiation of EPL cells as EBs (EPLEBs) results in populations of cells highly enriched in mesodermal progenitors followed by terminally differentiated mesodermal lineages. Mesodermal progenitors form earlier and more efficiently in EPLEBs compared with EBs, with an absence of embryonic ectodermal and endodermal lineages (Lake *et al.*, 2000). Addition of the cytokines interleukin 3 (IL-3) and macrophage colony stimulating factor (M-CSF) to differentiating EPLEBs results in enrichment for macrophages (Lake *et al.*, 2000), suggesting that the mesodermal progenitors are multipotent and capable of responding to exogenous factors. The preferential formation of mesodermal progenitors in EPLEBs has been hypothesized to result from recapitulation of the epithelial to mesenchyme transition that occurs at the site of mesoderm formation during embryonic gastrulation (Rodda et al., 2002).

EPL cells can also be formed and maintained as cellular aggregates in the presence of HepG2 conditioned medium, conditions that result in the synchronous and homogeneous formation of neurectoderm, or neural precursors (Rathjen et al., 2002). Greater than 95% of the population comprises cells expressing neural markers, while markers and cell types characteristic of mesoderm or extra-embryonic endoderm are not observed. The differentiation of EPL cells to neurectoderm in response to HepG2 conditioned medium recapitulates neurectoderm formation *in vivo*, which follows the sequential elaboration of primitive ectoderm and definitive ectoderm intermediates. EPL cell-derived neurectoderm is multipotent and can be further differentiated to neurons, glia and neural crest. Both glia and neural crest can be derived as near homogeneous populations in response to biologically relevant signaling environments (Rathjen et al., 2002). Mechanistically, the determination of ectodermal cell fate from EPL cells is thought to result from maintenance of cell:cell and cell:ECM contact during differentiation and continued exposure of the cells to visceral endoderm signaling, supplied by the HepG2 conditioned medium (Rodda et al., 2002).

The differentiation of ES cells to neural progenitors and neurons in culture has been approached by several other methodologies, all of which give substantially enriched populations of neural progenitors and are, in some capacity, directed. Formation and culture of ES cell aggregates in serum-free medium supplemented with LIF results in formation of nestin positive sphere colonies (Tropepe et al., 2001). This approach exploits a potential 'default differentiation pathway' inherent in ES cells and leads to a population enriched in neural cells and deficient in the expression of mesodermal markers, although genes characteristic of early endoderm lineages were expressed, perhaps indicating that homogeneity was not achieved. The frequency of neural sphere formation was low (~0.2%), suggesting that extensive cell death had occurred, and questions the relevance of this pathway to pluripotent cell differentiation *in vivo*.

Alternatively, ES cells have been co-cultured with the stromal cell line PA6, conditions which result in efficient neural differentiation, with 92% of colonies expressing the neural marker NCAM, and <2% of colonies containing cells expressing mesodermal markers (Kawasaki et al., 2000). ES cells are seeded directly into contact with the stromal cells, thereby eliminating EB formation. Neural induction was shown to require a PA6-associated activity named stromal cell-derived neural inducing activity (SDIA), which was in part associated with the surface of the PA6 cells, and in part soluble. Although neural induction appears to be directed, synchrony within this system is lacking, with cells within individual colonies heterogeneous with respect to expression of stage-specific markers.

1.4.4 Formation of specialized cell types: combining lineage induction with positional information

Obtaining populations of cells committed to specialized cell types, like dopaminergic neurons or ventricular cardiomyocytes, is a major goal of ES cell differentiation, both as a route to a more sophisticated appreciation of developmental pathways and for the generation of cell populations suited to use as cell therapeutics. Development *in vivo* involves initial induction of the lineage, such as formation of neurectoderm, followed by patterning of the cell population in response to positional information (Bally-Cuif and Hammerschmidt, 2003). A number of the signals that impart positional information in naïve lineages emanate from the visceral endoderm (Beddington and Robertson, 1999). The absence of visceral endoderm during differentiation, achieved either by directed differentiation or purification, is required to limit endogenous patterning of the cells and provide a naïve substrate for the coordinated and homogeneous specification of positional information. The addition to neural precursors selected from differentiating EBs of sonic hedgehog, fibroblast growth factor 8 (FGF8), dibutyryl-cAMP and ascorbic acid, or 'survival promoting factors', a mixture containing IL-1β, glial cell line-derived neurotrophic factor (GDNF), neurturin (NTN), TGF-$β_3$ and dibutyryl-cAMP, resulted in enrichment for dopaminergic neurons (34% and 43% respectively; Lee *et al.*, 2000; Rolletschek *et al.*, 2001). Similarly, neurectoderm generated by co-culture of ES cells with PA6 cells has been used as a substrate to generate progenitors committed to the motor neurons of the spinal cord (Wichterle *et al.*, 2002). Caudalization of the progenitors by treatment with retinoic acid and ventralization with an agonist of the sonic hedgehog receptor resulted in a population of cells enriched for motor neurons (20–30% compared with 0% in untreated populations; Wichterle *et al.*, 2002). Although neither of these approaches yielded homogeneous populations of specified cells, these data demonstrate the ability of ES cell-derived lineage progenitors to acquire positional information in response to biologically relevant signaling.

1.4.5 Understanding embryogenesis through ES cell modeling

Understanding mammalian embryogenesis through analysis of the early embryo is complicated by a number of factors including size, availability and the complexity of the embryo and uterine environment. Characterization of knockout phenotypes has been informative in many processes, but can be limited by early stage lethality or requirement for the gene product in the extraembryonic support tissues. Systems for differentiation of ES cells *in vitro*, combined with the availability of genetically modified ES cell lines, provide experimental models that can be used to augment *in vivo* studies of mammalian embryogenesis, promoting a greater understanding of genes and signaling pathways regulating developmental decisions.

Although ES cell differentiation is potentially applicable to all developmental systems, the use of ES cells is most advanced in characterization of the earliest events in hematopoietic lineage formation. Hematopoiesis initiates early in EBs with the formation of a 'transitional colony', a progenitor cell population characterized by expression of the nascent mesoderm marker *brachyury* and the early hemopoietic marker *Flk1*, and a developmental potential to form hematopoietic, endothelial and other terminally differentiated mesodermal lineages (Robertson *et al.*, 2000). ES cell-derived transitional cells can give rise to two populations of $Flk1^+$ cells distinguished by expression of *Scl* (*Tal1*) (Chung *et al.*, 2002b; Ema *et al.*, 2003). $Flk1^+$ Scl^+ cells, or hemangioblasts, have the potential to differentiate into hematopoietic, endothelial and smooth muscle cells (Choi *et al.*,

1998; Chung et al., 2002b; Ema et al., 2003; Kennedy et al., 1997; Nishikawa et al., 1998; Robertson et al., 2000). Furthermore, analysis of *Scl* in ES cell differentiation, using ES cells engineered to express *Scl* under the control of the *Flk1* promoter or null for *Scl*, defined *Scl* as a determinant of hematopoietic potential in hemangioblasts (Ema et al., 2003). $Flk1^+$ Scl^- cells, or angioblasts, have a more restricted developmental potential, being able to form endothelial and smooth muscle lineages but not the hematopoietic lineages (Chung et al., 2002b; Ema et al., 2003). From the analysis of ES cell differentiation it is proposed that these two populations remain spatially distinct in the embryo, with the hemangioblast arising in the yolk sac from migratory transitional cells and the angioblast arising from transitional cells that remain in the embryo proper, effectively restricting primitive hematopoietic development to the yolk sac (Ema et al., 2003).

In vitro studies have defined several signaling pathways involved in establishment of the hematopoietic and vascular lineages. BMP4 signaling is required for the initial formation of ventral, or hematopoietic-competent, mesoderm, a population encompassing the transitional cell (Johansson and Wiles, 1995; Nakayama et al., 2000; Adelman et al., 2002). Further induction of hematopoietic lineages in EBs was enhanced by addition of the Flk1 ligand VEGF, suggesting that formation of blast cells from the transitional cells requires activation of Flk1 (Kennedy et al., 1997; Nakayama et al., 2000; Robertson et al., 2000). Although not required for specification of cell fate, a role for FGF signaling in proliferation of the blast cells has been demonstrated (Faloon et al., 2000). The analysis of EBs *in vitro* has led to an understanding of the sequential action of signaling pathways and transcription factors in the early stages of the hematopoietic lineage, from pluripotent cells to the committed progenitor populations.

Formation of highly enriched populations of hematopoietic progenitors in culture, both intermediary states and blast cell colonies, enables correlation of gene expression and differentiation potential, and allows construction of cellular phylogenies. Characterization of the hemangioblast and angioblast *in vitro* demonstrated that the differentiation potential of these cells encompassed both the endothelial and mural, or smooth muscle cell, populations of the vasculature, formed in response to VEGF and platelet-derived growth factor BB (PDGF-BB) respectively (Ema et al., 2003; Yamashita et al., 2000). This broad potential is contrary to the conclusions of conventional embryology, which state that the smooth muscle cell component of the vasculature derives from neural crest.

1.5 Applications of ES cell technology

1.5.1 Animal models of development and disease

Genetic modification of ES cells or gene targeting has been used extensively to study gene function (Bedell et al., 1997a,b; Brandon et al., 1995a,b,c). Homologous recombination in ES cells enables predetermined alterations to be made in specific genetic loci such that the expression of the targeted gene is abolished ('knocked out') or mutations introduced ('knocked in'). Mutations can range from single nucleotide changes to gross deletions, insertions, inversions or translocations.

Homologous recombination is often a rare event, therefore the success of gene targeting in ES cells relies on efficient transfection, and selection and propagation of clonal, targeted cell lines coupled with retention of full pluripotentiality following genetic manipulation. The effect of genetic modification can then be assessed during embryonic development and in adult mice produced from the ES cells (Thomas and Capecchi, 1987; Zijlstra et al., 1989).

Gene targeting continues to provide a powerful mechanism for analyzing gene function and genetic pathways in the context of the whole animal and in cell lines derived from such mice. This technology has yielded significant discoveries in diverse areas of biology, including neurobiology, hematology, immunology, reproductive biology and developmental biology (reviewed in Brandon et al., 1995a,b,c). With the sequencing of the human and mouse genomes, this technology may also provide a means for rapid functional analysis of the large number of genes of unknown function.

A further application of gene targeting is the creation of mouse models for the study of human diseases involving single gene disorders (Bedell et al., 1997b), including cystic fibrosis (Colledge et al., 1995; Zeiher et al., 1995), hemophilia B (Lin et al., 1997), skin disorders such as epidermolytic hyperkeratosis (Porter et al., 1996), and numerous metabolic diseases (Bedell et al., 1997b). Whilst the generation of null mutations is common, modifications that mimic those commonly found in the human disease (e.g., point mutations) generally recapitulate the disease phenotype more closely. However, not every model has been found to mimic the pathology of the equivalent human disease well (Bedell et al., 1997b). Mouse models of more complex polygenic or multifactorial disorders such as obesity (Brockmann and Bevova, 2002) and psychiatric disorders (Seong et al., 2002) are being generated using combinations of genetic manipulation, mouse strain differences, and environmental factors. These models should provide an understanding of the genetic, biochemical and pathological basis of disease, and might be used as vehicles for gene therapy and pharmacological testing.

1.5.2 Cell-based therapies and tissue engineering

The effectively unlimited capacity for ES cell expansion provides an opportunity for production of large quantities of differentiated cells *in vitro*, which can be transplanted into animal models of disease to assess therapeutic efficacy. Cells can be transplanted directly or seeded onto structural scaffolds to generate three-dimensional tissues for transplantation. Successful trials will require purified populations of defined and validated differentiated cells that are karyotypically normal. These cells must then be delivered in appropriate numbers to the appropriate site, engraft and survive, and correct the condition without aberrant cell growth such as teratoma formation. Although most of these problems have been anticipated for some time, largely through work with mouse ES cells, solving them has proved very difficult in most instances.

Several types of mature and progenitor cells produced by differentiation of mouse ES cells have been transplanted into rodents and shown to engraft and survive (Blyszczuk et al., 2003; Chinzei et al., 2002; Klug et al., 1996; Kyba et al., 2002; Liu et al., 2000a; Morizane et al., 2002; Yin et al., 2002). In some cases, amelioration of symptoms in animal models of disease has been observed. For example, using injected ES-derived dopaminergic neurons, an improvement of symptoms in a rat model of Parkinson's disease has been demonstrated (Kim *et al.*, 2002), as has the normalization of hyperglycemia by ES cell-derived insulin secreting cells injected into a mouse model of diabetes (Blyszczuk *et al.*, 2003; Soria *et al.*, 2000). Transplantation of more differentiated cell types, such as ES cell-derived hepatocytes and dopaminergic neurons, appears to decrease the risk of teratoma formation (Chinzei *et al.*, 2002; Kim *et al.*, 2002; Yin *et al.*, 2002).

Genetic modification of ES cells may provide a means of reducing immunological rejection of grafted cells through modification of the major histocompatibility complex. A second approach to overcoming rejection of transplanted cells is to use nuclear transfer to establish ES cell lines syngeneic with diseased individuals (Munsie *et al.*, 2000; Wakayama *et al.*, 2001). An elegant 'proof of principle' in mice has been demonstrated recently (Rideout *et al.*, 2002) in which an ES cell line was generated by transfer of the nucleus from a tail-tip cell of an immunodeficient $Rag2^{-/-}$ mouse into an enucleated oocyte. The specific gene defect was repaired using homologous recombination and the genetically modified ES cells were then expanded and differentiated *in vitro* into hematopoietic cells and engrafted back into syngeneic $Rag2^{-/-}$ immunodeficient mice to restore immune function. This work establishes the potential for combined nuclear transfer, gene therapy and cellular therapy for the treatment of genetic diseases.

1.5.3 Evaluation of drugs and toxins

ES cells and their differentiated derivatives can be used in screening assays for potential pharmaceuticals and toxic or mutagenic compounds. While primary cell cultures or established cell lines are commonly used for both purposes, ES cells offer several advantages. In contrast to most permanent, transformed cell lines and cell lines established from specialized somatic cells, ES cells are non-transformed, karyotypically normal and have the ability to differentiate and effectively produce unlimited numbers of cells representative of the three germ layers of the embryo. The developmental equivalence of ES cell-derived and embryo populations provides a more rigorous system for evaluating the teratogenic and embryotoxic effects of a substance, in addition to general mutagenic and cytotoxic effects (Rohwedel *et al.*, 2001). A protocol based on ES cell differentiation, the 'embryonic stem cell test', has been established and validated for use in toxicity testing (Bremer, 2002; Scholz *et al.*, 1999). Additionally, genetic modification enables the tailoring of ES cell lines for specific purposes. For example, specific genes can be altered to increase sensitivity to mutagens (Ogi *et al.*, 2002) or drugs (Lorico *et al.*, 1996), or tissue-specific reporter genes can be introduced to detect changes in gene expression induced by toxic chemicals or therapeutic agents (Li *et al.*, 2002).

Acknowledgments

We gratefully acknowledge support from the Australian Research Council (ARC), the ARC Special Research Centre for Molecular Genetics of Development, the

National Health and Medical Research Council of Australia, Raymond Ryce, and from past and present members of the laboratory who have contributed to our understanding of mouse ES cells.

References

Adelman CA, Chattopadhyay S, Bieker J (2002) The BMP/BMPR/Smad pathway directs expression of the erythroid-specific EKLF and GATA1 transcription factors during embryoid body differentiation in serum-free media. *Development* **129**, 539–549.

Ambrosetti DC, Basilico C, Dailey L (1997) Synergistic activation of the fibroblast growth factor 4 enhancer by Sox2 and Oct-3 depends on protein–protein interactions facilitated by a specific spatial arrangement of factor binding sites. *Molec. Cell. Biol.* **17**, 6321–6329.

Armstrong L, Lako M, Lincoln J, Cairns PM, Hole N (2000) mTert expression correlates with telomerase activity during the differentiation of murine embryonic stem cells. *Mech. Develop.* **97**, 109–116.

Avilion AA, Nicolis SK, Pevny LH, Perez L, Vivian N, Lovell-Badge R (2003) Multipotent cell lineages in early mouse development depend on SOX2 function. *Genes Develop.* **17**, 126–140.

Bain G, Kitchens D, Yao M, Huettner JE, Gottlieb DI (1995) Embryonic stem cells express neuronal properties *in vitro*. *Develop. Biol.* **168**, 342–357.

Bally-Cuif L, Hammerschmidt M (2003) Induction and patterning of neuronal development and its connection to cell cycle control. *Curr. Opin. Neurobiol.* **13**, 16–25.

Beddington RS, Robertson EJ (1989) An assessment of the developmental potential of embryonic stem cells in the midgestation mouse embryo. *Development* **105**, 733–737.

Beddington RS, Robertson EJ (1999) Axis development and early asymmetry in mammals. *Cell* **96**, 195–209.

Bedell MA, Jenkins NA, Copeland NG (1997a) Mouse models of human disease. Part I: techniques and resources for genetic analysis in mice. *Genes Develop.* **11**, 1–10.

Bedell MA, Largaespada DA, Jenkins NA, Copeland NG (1997b) Mouse models of human disease. Part II: recent progress and future directions. *Genes Develop.* **11**, 11–43.

Bettess MD (2001) Purification, identification and characterisation of signals directing embryonic stem (ES) cell differentiation. In PhD thesis, Department of Molecular Biosciences, University of Adelaide.

Bierbaum P, MacLean-Hunter S, Ehlert F, Moroy T, Muller R (1994) Cloning of embryonal stem cell-specific genes: characterization of the transcriptionally controlled gene esg-1. *Cell Growth Different.* **5**, 37–46.

Blyszczuk P, Czyz J, Kania G, Wagner M, Roll U, St-Onge L, Wobus AM (2003) Expression of Pax4 in embryonic stem cells promotes differentiation of nestin-positive progenitor and insulin-producing cells. *Proc. Natl Acad. Sci. USA* **100**, 998–1003.

Botquin V, Hess H, Fuhrmann G, Anastassiadis C, Gross MK, Vriend G, Scholer HR (1998) New POU dimer configuration mediates antagonistic control of an osteopontin preimplantation enhancer by Oct-4 and Sox-2. *Genes Develop.* 12, 2073–2090.

Bradley A, Evans M, Kaufman MH, Robertson E (1984) Formation of germ-line chimaeras from embryo-derived teratocarcinoma cell lines. *Nature* 309, 255–256.

Brandon EP, Idzerda RL, McKnight GS (1995a) Knockouts. Targeting the mouse genome: a compendium of knockouts (Part I). *Curr. Biol.* 5, 625–634.

Brandon EP, Idzerda RL, McKnight GS (1995b) Targeting the mouse genome: a compendium of knockouts (Part II). *Curr. Biol.* 5, 758–765.

Brandon EP, Idzerda RL, McKnight GS (1995c) Targeting the mouse genome: a compendium of knockouts (Part III). *Curr. Biol.* 5, 873–881.

Bremer S (2002) Development of a testing strategy for detecting embryotoxic hazards of chemicals *in vitro* by using embryonic stem cell models. *Altern. Lab. Anim.* 30, 107–109.

Brockmann GA, Bevova MR (2002) Using mouse models to dissect the genetics of obesity. *Trends Genet.* 18, 367–376.

Brook FA, Gardner RL (1997) The origin and efficient derivation of embryonic stem cells in the mouse. *Proc. Natl Acad. Sci. USA* 94, 5709–5712.

Burdon T, Stracey C, Chambers I, Nichols J, Smith A (1999) Suppression of SHP-2 and ERK signalling promotes self-renewal of mouse embryonic stem cells. *Develop. Biol.* 210, 30–43.

Burdon T, Smith A, Savatier P (2002) Signalling, cell cycle and pluripotency in embryonic stem cells. *Trends Cell Biol.* 12, 432–438.

Chapman G, Remiszewski JL, Webb GC, Schulz TC, Bottema CD, Rathjen PD (1997) The mouse homeobox gene, Gbx2: genomic organization and expression in pluripotent cells *in vitro* and *in vivo*. *Genomics* 46, 223–233.

Cheng AM, Saxton TM, Sakai R, Kulkarni S, Mbamalu G, Vogel W, Tortorice CG, Cardiff RD, Cross JC, Muller WJ *et al.* (1998) Mammalian Grb2 regulates multiple steps in embryonic development and malignant transformation. *Cell* 95, 793–803.

Chinzei R, Tanaka Y, Shimizu-Saito K, Hara Y, Kakinuma S, Watanabe M, Teramoto K, Arii S, Takase K, Sato C *et al.* (2002) Embryoid-body cells derived from a mouse embryonic stem cell line show differentiation into functional hepatocytes. *Hepatology* 36, 22–29.

Choi K, Kennedy M, Kazarov A, Papadimitriou JC, Keller G (1998) A common precursor for hematopoietic and endothelial cells. *Development* 125, 725–732.

Chow D, He X, Snow AL, Rose-John S, Garcia KC (2001) Structure of an extracellular gp130 cytokine receptor signaling complex. *Science* 291, 2150–2155.

Chow D, Brevnova L, He X, Martick MM, Bankovich A, Garcia KC (2002) A structural template for gp130-cytokine signaling assemblies. *Biochim. Biophys. Acta* 1592, 225–235.

Chung S, Sonntag KC, Andersson T, Bjorklund LM, Park JJ, Kim DW, Kang UJ, Isacson O, Kim KS (2002a) Genetic engineering of mouse embryonic stem cells by Nurr1 enhances differentiation and maturation into dopaminergic neurons. *Eur. J. Neurosci.* 16, 1829–1838.

Chung YS, Zhang WJ, Arentson E, Kingsley PD, Palis J, Choi K (2002b) Lineage analysis of the hemangioblast as defined by FLK1 and SCL expression. *Development* 129, 5511–5520.

Colledge WH, Abella BS, Southern KW, Ratcliff R, Jiang C, Cheng SH, MacVinish LJ, Anderson JR, Cuthbert AW, Evans MJ (1995) Generation and characterization of a Δ F508 cystic fibrosis mouse model. *Nature Genet.* 10, 445–452.

Conover JC, Ip NY, Poueymirou WT, Bates B, Goldfarb MP, DeChiara TM, Yancopoulos GD (1993) Ciliary neurotrophic factor maintains the pluripotentiality of embryonic stem cells. *Development* 119, 559–565.

Damjanov I, Damjanov A, Solter D (1987) Production of teratocarcinomas from embryos transplanted to extra-uterine sites. In: *Teratocarcinomas and Embryonic Stem Cells: A Practical Approach* (ed. EJ Robertson) IRL Press, Oxford, pp. 1–18.

Dani C, Chambers I, Johnstone S, Robertson M, Ebrahimi B, Saito M *et al.* (1998) Paracrine induction of stem cell renewal by LIF-deficient cells: a new ES cell regulatory pathway. *Develop. Biol.* 203, 149–162.

Davis S, Aldrich TH, Stahl N, Pan L, Taga T, Kishimoto T, Ip NY, Yancopoulos GD (1993) LIFR beta and gp130 as heterodimerizing signal transducers of the tripartite CNTF receptor. *Science* 260, 1805–1808.

Doetschman TC, Eistetter H, Katz M, Schmidt W, Kemler R (1985) The *in vitro* development of blastocyst-derived embryonic stem cell lines: formation of visceral yolk sac, blood islands and myocardium. *J. Embryol. Exp. Morphol.* 87, 27–45.

Doevendans PA, Kubalak SW, An RH, Becker DK, Chien KR, Kass RS (2000) Differentiation of cardiomyocytes in floating embryoid bodies is comparable to fetal cardiomyocytes. *J. Molec. Cell. Cardiol.* 32, 839–851.

Dottori M, Gross MK, Labosky P, Goulding M (2001) The winged-helix transcription factor Foxd3 suppresses interneuron differentiation and promotes neural crest cell fate. *Development* 128, 4127–4138.

Ema M, Faloon P, Zhang WJ, Hirashima M, Reid T, Stanford WL, Orkin S, Choi K, Rossant J (2003) Combinatorial effects of Flk1 and Tal1 on vascular and hematopoietic development in the mouse. *Genes Develop.* 17, 380–393.

Evans MJ, Kaufman MH (1981) Establishment in culture of pluripotential cells from mouse embryos. *Nature* 292, 154–156.

Fairchild PJ, Brook FA, Gardner RL, Graca L, Strong V, Tone Y, Tone M, Nolan KF, Waldman H (2000) Directed differentiation of dendritic cells from mouse embryonic stem cells. *Curr. Biol.* 10, 1515–1518.

Faloon P, Arentson E, Kazarov A, Deng CX, Porcher C, Orkin S, Choi K (2000) Basic fibroblast growth factor positively regulates hematopoietic development. *Development* 127, 1931–1941.

Feldman B, Poueymirou W, Papaioannou VE, DeChiara TM, Goldfarb M (1995) Requirement of FGF-4 for postimplantation mouse development. *Science* 267, 246–249.

Guan K, Rohwedel J, Wobus AM (1999) Embryonic stem cell differentiation models: cardiogenesis, myogenesis, neurogenesis, epithelial and vascular smooth muscle cell differentiation *in vitro*. *Cytotechnology* 30, 211–226.

Guan K, Chang H, Rolletschek A, Wobus AM (2001) Embryonic stem cell-derived neurogenesis. Retinoic acid induction and lineage selection of neuronal cells. *Cell Tissue Res.* **305**, 71–76.

Guo Y, Costa R, Ramsey H, Starnes T, Vance G, Robertson K, Kelley M, Reinbold R, Scholer H, Hromas R (2002) The embryonic stem cell transcription factors Oct-4 and FoxD3 interact to regulate endodermal-specific promoter expression. *Proc. Natl Acad. Sci. USA* **99**, 3663–3667.

Hanna LA, Foreman RK, Tarasenko IA, Kessler DS, Labosky PA (2002) Requirement for Foxd3 in maintaining pluripotent cells of the early mouse embryo. *Genes Develop.* **16**, 2650–2661.

Helgason CD, Sauvageau G, Lawrence HJ, Largman C, Humphries RK (1996) Overexpression of HOXB4 enhances the hematopoietic potential of embryonic stem cells differentiated *in vitro*. *Blood* **87**, 2740–2749.

Hirano T, Ishihara K, Hibi M (2000) Roles of STAT3 in mediating the cell growth, differentiation and survival signals relayed through the IL-6 family of cytokine receptors. *Oncogene* **19**, 2548–2556.

Hogan B, Beddington R, Constantini F, Lacy E (1994) *Manipulating the Mouse Embryo: A Laboratory Manual*. Cold Spring Harbor Laboratory Press, New York.

Hou SX, Zheng Z, Chen X, Perrimon N (2002) The JAK/STAT pathway in model organisms: Emerging roles in cell movement. *Develop. Cell* **3**, 765–778.

Ivanova NB, Dimos JT, Schaniel C, Hackney JA, Moore KA, Lemischka IR (2002) A stem cell molecular signature. *Science* **298**, 601–604.

Johansson BM, Wiles MV (1995) Evidence for involvement of activin A and bone morphogenetic protein 4 in mammalian mesoderm and hematopoietic development. *Molec. Cell. Biol.* **15**, 141–151.

Johnson LV, Calarco PG, Siebert ML (1977) Alkaline phosphatase activity in the preimplantation mouse embryo. *J. Embryol. Exp. Morphol.* **40**, 83–89.

Kaufman MH, Robertson EJ, Handyside AH, Evans MJ (1983) Establishment of pluripotential cell lines from haploid mouse embryos. *J. Embryol. Exp. Morphol.* **73**, 249–261.

Kawasaki H, Mizuseki K, Nishikawa S, Kaneko S, Kuwana Y, Nakanishi S, Nishikawa S, Sasai Y (2000) Induction of midbrain dopaminergic neurons from ES cells by stromal cell-derived inducing activity. *Neuron* **28**, 31–40.

Kennedy M, Firpo M, Choi K, Wall C, Robertson S, Kabrun N, Keller G (1997) A common precursor for primitive erythropoiesis and definitive haematopoiesis. *Nature* **386**, 488–493.

Kim JH, Auerbach JM, Rodriguez-Gomez JA, Velasco I, Gavin D, Lumelsky N, Lee SH, Nguyen J, Sanchez-Pernaute R, Bankiewicz K *et al.* (2002) Dopamine neurons derived from embryonic stem cells function in an animal model of Parkinson's disease. *Nature* **418**, 50–56.

Klug MG, Soonpaa MH, Koh GY, Field LJ (1996) Genetically selected cardiomyocytes from differentiating embryonic stem cells form stable intracardiac grafts. *J. Clin. Invest.* **98**, 216–224.

Kolossov E, Fleischmann BK, Liu Q, Bloch W, Viatchenko-Karpinski S, Manzke O, Ji GJ, Bohlen H, Addicks K, Hescheler J (1998) Functional characteristics of

ES cell-derived cardiac precursor cells identified by tissue-specific expression of the green fluorescent protein. *J. Cell Biol.* 143, 2045–2056.

Kyba M, Perlingeiro RC, Daley GQ (2002) HoxB4 confers definitive lymphoid-myeloid engraftment potential on embryonic stem cell and yolk sac hematopoietic progenitors. *Cell* 109, 29–37.

Lake J, Rathjen J, Remiszewski J, Rathjen PD (2000) Reversible programming of pluripotent cell differentiation. *J. Cell Sci.* 113, 555–566.

Lallemand Y, Brulet P (1990) An in situ assessment of the routes and extents of colonisation of the mouse embryo by embryonic stem cells and their descendants. *Development* 110, 1241–1248.

Lee SH, Lumelsky N, Studer L, Auerbach JM, McKay RD (2000) Efficient generation of midbrain and hindbrain neurons from mouse embryonic stem cells. *Nature Biotechnol.* 18, 675–679.

Li M, Pevny L, Lovell-Badge R, Smith A (1998) Generation of purified neural precursors from embryonic stem cells by lineage selection. *Curr. Biol.* 8, 971–974.

Li K, Ramirez MA, Rose E, Beaudet AL (2002) A gene fusion method to screen for regulatory effects on gene expression: application to the LDL receptor. *Human Molec. Genet.* 11, 3257–3265.

Lin HF, Maeda N, Smithies O, Straight DL, Stafford DW (1997) A coagulation factor IX-deficient mouse model for human hemophilia B. *Blood* 90, 3962–3966.

Liu L, Roberts RM (1996) Silencing of the gene for the β subunit of human chorionic gonadotropin by the embryonic transcription factor Oct-3/4. *J. Biol. Chem.* 271, 16683–16689.

Liu L, Leaman D, Villalta M, Roberts RM (1997a) Silencing of the gene for the α-subunit of human chorionic gonadotropin by the embryonic transcription factor Oct-3/4. *Molec. Endocrinol.* 11, 1651–1658.

Liu S, Qu Y, Stewart TJ, Howard MJ, Chakrabortty S, Holekamp TF, McDonald JW (2000a) Embryonic stem cells differentiate into oligodendrocytes and myelinate in culture and after spinal cord transplantation. *Proc. Natl Acad. Sci. USA* 97, 6126–6131.

Liu X, Wu H, Loring J, Hormuzdi S, Disteche CM, Bornstein P, Jaenisch R (1997b) Trisomy eight in ES cells is a common potential problem in gene targeting and interferes with germ line transmission. *Develop. Dyn.* 209, 85–91.

Liu Y, Snow BE, Hande MP, Yeung D, Erdmann NJ, Wakeham A, Itie A, Siderovski DP, Lansdorp PM, Robinson MO *et al.* (2000b) The telomerase reverse transcriptase is limiting and necessary for telomerase function *in vivo*. *Curr. Biol.* 10, 1459–1462.

Longo L, Bygrave A, Grosveld FG, Pandolfi PP (1997) The chromosome make-up of mouse embryonic stem cells is predictive of somatic and germ cell chimaerism. *Transgenic Res.* 6, 321–328.

Lorico A, Rappa G, Flavell RA, Sartorelli AC (1996) Double knockout of the MRP gene leads to increased drug sensitivity *in vitro*. *Cancer Res.* 56, 5351–5355.

Martin GR (1981) Isolation of a pluripotent cell line from early mouse embryos cultured in medium conditioned by teratocarcinoma stem cells. *Proc. Natl Acad. Sci. USA* **78**, 7634–7638.

Matsuda T, Nakamura T, Nakao K, Arai T, Katsuki M, Heike T, Yokota T (1999) STAT3 activation is sufficient to maintain an undifferentiated state of mouse embryonic stem cells. *EMBO J.* **18**, 4261–4269.

Matsui Y, Zsebo K, Hogan BL (1992) Derivation of pluripotential embryonic stem cells from murine primordial germ cells in culture. *Cell* **70**, 841–847.

Morizane A, Takahashi J, Takagi Y, Sasai Y, Hashimoto N (2002) Optimal conditions for *in vivo* induction of dopaminergic neurons from embryonic stem cells through stromal cell-derived inducing activity. *J. Neurosci. Res.* **69**, 934–939.

Müller M, Fleischmann BK, Selbert S, Ji GJ, Endl E, Middeler G *et al.* (2000) Selection of ventricular-like cardiomyocytes from ES cells *in vitro*. *FASEB J.* **14**, 2540–2548.

Mulnard J, Huygens R (1978) Ultrastructural localization of non-specific alkaline phosphatase during cleavage and blastocyst formation in the mouse. *J. Embryol. Exp. Morphol.* **44**, 121–131.

Munsie MJ, Michalska AE, O'Brien CM, Trounson AO, Pera MF, Mountford PS (2000) Isolation of pluripotent embryonic stem cells from reprogrammed adult mouse somatic cell nuclei. *Curr. Biol.* **10**, 989–992.

Nakano T, Kodama H, Honjo T (1994) Generation of lymphohematopoietic cells from embryonic stem cells in culture. *Science* **265**, 1098–1101.

Nakayama N, Lee J, Chiu L (2000) Vascular endothelial growth factor synergistically enhances bone morphogenetic protein-4-dependent lymphohematopoietic cell generation from embryonic stem cells *in vitro*. *Blood* **95**, 2275–2283.

Nichols J, Evans EP, Smith AG (1990) Establishment of germ-line-competent embryonic stem (ES) cells using differentiation inhibiting activity. *Development* **110**, 1341–1348.

Nichols J, Chambers I, Smith A (1994) Derivation of germline competent embryonic stem cells with a combination of interleukin-6 and soluble interleukin-6 receptor. *Exp. Cell Res.* **215**, 237–239.

Nichols J, Zevnik B, Anastassiadis K, Niwa H, Klewe-Nebenius D, Chambers I, Scholer H, Smith A (1998) Formation of pluripotent stem cells in the mammalian embryo depends on the POU transcription factor Oct4. *Cell* **95**, 379–391.

Nichols J, Chambers I, Taga T, Smith A (2001) Physiological rationale for responsiveness of mouse embryonic stem cells to gp130 cytokines. *Development* **128**, 2333–2339.

Nishikawa SI, Nishikawa S, Hirashima M, Matsuyoshi N, Kodama H (1998) Progressive lineage analysis by cell sorting and culture identifies FLK1$^+$ VE-cadherin$^+$ cells at a diverging point of endothelial and hemopoietic lineages. *Development* **125**, 1747–1757.

Niswander L, Martin GR (1992) Fgf-4 expression during gastrulation, myogenesis, limb and tooth development in the mouse. *Development* **114**, 755–768.

Niwa H, Burdon T, Chambers I, Smith A (1998) Self-renewal of pluripotent embryonic stem cells is mediated via activation of STAT3. *Genes Develop.* **12**, 2048–2060.

Niwa H, Miyazaki J, Smith AG (2000) Quantitative expression of Oct-3/4 defines differentiation, dedifferentiation or self-renewal of ES cells. *Nature Genet.* **24**, 372–376.

Niwa H, Masui S, Chambers I, Smith AG, Miyazaki J (2002) Phenotypic complementation establishes requirements for specific POU domain and generic transactivation function of Oct-3/4 in embryonic stem cells. *Molec. Cell. Biol.* **22**, 1526–1536.

Ogi T, Shinkai Y, Tanaka K, Ohmori H (2002) Polκ protects mammalian cells against the lethal and mutagenic effects of benzo[*a*]pyrene. *Proc. Natl Acad. Sci. USA* **99**, 15548–15553.

Okamoto K, Okazawa H, Okuda A, Sakai M, Muramatsu M, Hamada H (1990) A novel octamer binding transcription factor is differentially expressed in mouse embryonic cells. *Cell* **60**, 461–472.

Palmieri SL, Peter W, Hess H, Scholer HR (1994) Oct-4 transcription factor is differentially expressed in the mouse embryo during establishment of the first two extraembryonic cell lineages involved in implantation. *Develop. Biol.* **166**, 259–267.

Pease S, Williams RL (1990) Formation of germ-line chimeras from embryonic stem cells maintained with recombinant leukemia inhibitory factor. *Exp. Cell Res.* **190**, 209–211.

Pease S, Braghetta P, Gearing D, Grail D, Williams RL (1990) Isolation of embryonic stem (ES) cells in media supplemented with recombinant leukemia inhibitory factor (LIF). *Develop. Biol.* **141**, 344–352.

Pelton TA, Sharma S, Schulz TC, Rathjen J, Rathjen PD (2002) Transient pluripotent cell populations during primitive ectoderm formation: correlation of *in vivo* and *in vitro* pluripotent cell development. *J. Cell Sci.* **115**, 329–339.

Pennica D, Shaw KJ, Swanson TA, Moore MW, Shelton DL, Zioncheck KA, Rosenthal A, Taga T, Paoni NF, Wood WI (1995) Cardiotrophin-1. Biological activities and binding to the leukemia inhibitory factor receptor/gp130 signaling complex. *J. Biol. Chem.* **270**, 10915–10922.

Pesce M, Scholer HR (2001) Oct-4: gatekeeper in the beginnings of mammalian development. *Stem Cells* **19**, 271–278.

Porter RM, Leitgeb S, Melton DW, Swensson O, Eady RA, Magin TM (1996) Gene targeting at the mouse cytokeratin 10 locus: severe skin fragility and changes of cytokeratin expression in the epidermis. *J. Cell Biol.* **132**, 925–936.

Qu CK, Feng GS (1998) Shp-2 has a positive regulatory role in ES cell differentiation and proliferation. *Oncogene* **17**, 433–439.

Ramalho-Santos M, Yoon S, Matsuzaki Y, Mulligan RC, Melton DA (2002) 'Stemness': transcriptional profiling of embryonic and adult stem cells. *Science* **298**, 597–600.

Rappolee DA, Basilico C, Patel Y, Werb Z (1994) Expression and function of FGF-4 in peri-implantation development in mouse embryos. *Development* **120**, 2259–2269.

Rathjen J, Rathjen PD (2001) Mouse ES cells: Experimental exploitation of pluripotent differentiation potential. *Curr. Opin. Genet. Develop.* **11**, 589–595.

Rathjen J, Lake JA, Bettess MD, Washington JM, Chapman G, Rathjen PD (1999) Formation of a primitive ectoderm like cell population, EPL cells, from ES cells in response to biologically derived factors. *J. Cell Sci.* **112**, 601–612.

Rathjen J, Dunn S, Bettess MD, Rathjen PD (2001) Lineage specific differentiation of pluripotent cells *in vitro*: a role for extraembryonic cell types. *Reprod. Fert. Develop.* **13**, 15–22.

Rathjen J, Haines BP, Hudson KM, Nesci A, Dunn S, Rathjen PD (2002) Directed differentiation of pluripotent cells to neural lineages: homogeneous formation and differentiation of a neurectoderm population. *Development* **129**, 2649–2661.

Rathjen PD, Nichols J, Toth S, Edwards DR, Heath JK, Smith AG (1990a) Developmentally programmed induction of differentiation inhibiting activity and the control of stem cell populations. *Genes Develop.* **4**, 2308–2318.

Rathjen PD, Toth S, Willis A, Heath JK, Smith AG (1990b) Differentiation inhibiting activity is produced in matrix-associated and diffusible forms that are generated by alternate promoter usage. *Cell* **62**, 1105–1114.

Reubinoff BE, Pera MF, Fong CY, Trounson A, Bongso A (2000) Embryonic stem cell lines from human blastocysts: somatic differentiation *in vitro*. *Nature Biotechnol.* **18**, 399–400.

Rideout WM, 3rd, Hochedlinger K, Kyba M, Daley GQ, Jaenisch R (2002) Correction of a genetic defect by nuclear transplantation and combined cell and gene therapy. *Cell* **109**, 17–27.

Robertson SM, Kennedy M, Shannon JM, Keller G (2000) A transitional stage in the commitment of mesoderm to hematopoiesis requiring the transcription factor SCL/tal-1. *Development* **127**, 2447–2459.

Rodda SJ, Kavanagh SJ, Rathjen J, Rathjen PD (2002) Embryonic stem cell differentiation and the analysis of mammalian development. *Int. J. Develop. Biol.* **46**, 449–458

Rogers MB, Hosler BA, Gudas LJ (1991) Specific expression of a retinoic acid-regulated, zinc-finger gene, Rex-1, in preimplantation embryos, trophoblast and spermatocytes. *Development* **113**, 815–824.

Rohwedel J, Guan K, Hegert C, Wobus AM (2001) Embryonic stem cells as an *in vitro* model for mutagenicity, cytotoxicity and embryotoxicity studies: present state and future prospects. *Toxicol. In Vitro* **15**, 741–753.

Rolletschek A, Chang H, Guan K, Czyz J, Meyer M, Wobus AM (2001) Differentiation of embryonic stem cell-derived dopaminergic neurons is enhanced by survival-promoting factors. *Mech. Develop.* **105**, 93–104.

Rose TM, Weiford DM, Gunderson NL, Bruce AG (1994) Oncostatin M (OSM) inhibits the differentiation of pluripotent embryonic stem cells *in vitro*. *Cytokine* **6**, 48–54.

Saijoh Y, Fujii H, Meno C, Sato M, Hirota Y, Nagamatsu S, Ikeda M, Hamada H (1996) Identification of putative downstream genes of Oct-3, a pluripotent cell-specific transcription factor. *Genes Cells* **1**, 239–252.

Savatier P, Lapillonne H, van Grunsven LA, Rudkin BB, Samarut J (1996) Withdrawal of differentiation inhibitory activity/leukemia inhibitory factor up-regulates D-type cyclins and cyclin-dependent kinase inhibitors in mouse embryonic stem cells. *Oncogene* 12, 309–322.

Schindler C, Darnell JE (1995) Transcriptional responses to polypeptide ligands: the JAK-STAT pathway. *Ann. Rev. Biochem.* 64, 621–651.

Scholz G, Pohl I, Genschow E, Klemm M, Spielmann H (1999) Embryotoxicity screening using embryonic stem cells *in vitro*: correlation to *in vivo* teratogenicity. *Cells Tiss. Organs* 165, 203–211.

Seong E, Seasholtz AF, Burmeister M (2002) Mouse models for psychiatric disorders. *Trends Genet.* 18, 643–650.

Shirogane T, Fukada T, Muller JM, Shima DT, Hibi M, Hirano T (1999) Synergistic roles for Pim-1 and c-Myc in STAT3-mediated cell cycle progression and antiapoptosis. *Immunity* 11, 709–719.

Smith AG (1991) Culture and differentiation of embryonic stem cells. *J. Tiss. Culture Meth.* 13, 89–94.

Smith AG (1992) Mouse embryo stem cells: their identification, propagation and manipulation. *Semin. Cell Biol.* 3, 385–399.

Smith AG (2001) Embryo-derived stem cells: of mice and men. *Ann. Rev. Cell Develop. Biol.* 17, 435–462.

Smith AG, Heath JK, Donaldson DD, Wong GG, Moreau J, Stahl M, Rogers D (1988) Inhibition of pluripotential embryonic stem cell differentiation by purified polypeptides. *Nature* 336, 688–690.

Solter D, Knowles BB (1978) Monoclonal antibody defining a stage-specific mouse embryonic antigen (SSEA-1). *Proc. Natl Acad. Sci. USA* 75, 5565–5569.

Soria B, Roche E, Berna G, Leon-Quinto T, Reig JA, Martin F (2000) Insulin-secreting cells derived from embryonic stem cells normalize glycemia in streptozotocin-induced diabetic mice. *Diabetes* 49, 157–162.

Sosa-Pineda B, Chowdhury K, Torres M, Oliver G, Gruss P (1997) The Pax4 gene is essential for differentiation of insulin-producing beta cells in the mammalian pancreas. *Nature* 386, 399–402.

Stead E, White J, Faast R, Conn S, Goldstone S, Rathjen J, Dhingra U, Rathjen P, Walker D, Dalton S (2002) Pluripotent cell division cycles are driven by ectopic Cdk2, cyclin A/E and E2F activities. *Oncogene* 21, 8320–8333.

Stewart CL, Kaspar P, Brunet LJ, Bhatt H, Gadi I, Kontgen F, Abbondanzo SJ (1992) Blastocyst implantation depends on maternal expression of leukaemia inhibitory factor. *Nature* 359, 76–79.

Takeda K, Noguchi K, Shi W, Tanaka T, Matsumoto M, Yoshida N, Kishimoto T, Akira S (1997) Targeted disruption of the mouse Stat3 gene leads to early embryonic lethality. *Proc. Natl Acad. Sci. USA* 94, 3801–3804.

Tanaka TS, Kunath T, Kimber WL, Jaradat SA, Stagg CA, Usuda M, Yokota T, Niwa H, Rossant J, Ko MSH (2002) Gene expression profiling of embryo-derived stem cells reveals candidate genes associated with pluripotency and lineage specificity. *Genome Res.* 12, 1921–1928.

Thomas KR, Capecchi MR (1987) Site-directed mutagenesis by gene targeting in mouse embryo-derived stem cells. *Cell* 51, 503–512.

Thomson JA, Itskovitz-Eldor J, Shapiro SS, Waknitz MA, Swiergiel JJ, Marshall VS, Jones JM (1998) Embryonic stem cell lines derived from human blastocysts. *Science* **282**, 1145–1147.

Tomilin A, Remenyi A, Lins K, Bak H, Leidel S, Vriend G, Wilmanns M, Scholer HR (2000) Synergism with the coactivator OBF-1 (OCA-B, BOB-1) is mediated by a specific POU dimer configuration. *Cell* **103**, 853–864.

Tomioka M, Nishimoto M, Miyagi S, Katayanagi T, Fukui N, Niwa H, Muramatsu M, Okuda A (2002) Identification of Sox-2 regulatory region which is under the control of Oct-3/4-Sox-2 complex. *Nucleic Acids Res.* **30**, 3202–3213.

Tropepe V, Hitoshi S, Sirard C, Mak TW, Rossant J, van der Kooy D (2001) Direct neural fate specification from embryonic stem cells: A primitive mammalian neural stem cell stage acquired through a default mechanism. *Neuron* **30**, 65–78.

Voss AK, Thomas T, Petrou P, Anastassiadis K, Scholer H, Gruss P (2000) Taube nuss is a novel gene essential for the survival of pluripotent cells of early mouse embryos. *Development* **127**, 5449–5461.

Wakayama T, Tabar V, Rodriguez I, Perry AC, Studer L, Mombaerts P (2001) Differentiation of embryonic stem cell lines generated from adult somatic cells by nuclear transfer. *Science* **292**, 740–743.

Ware CB, Horowitz MC, Renshaw BR, Hunt JS, Liggitt D, Koblar SA, Gliniak BC, McKenna HJ, Papayannopoulou T, Thoma B (1995) Targeted disruption of the low-affinity leukemia inhibitory factor receptor gene causes placental, skeletal, neural and metabolic defects and results in perinatal death. *Development* **121**, 1283–1299.

Wichterle H, Lieberam I, Porter JA, Jessell TM (2002) Directed differentiation of embryonic stem cells into motor neurons. *Cell* **110**, 385–397.

Wilder PJ, Kelly D, Brigman K, Peterson CL, Nowling T, Gao QS, McComb RD, Capecchi MR, Rizzino A (1997) Inactivation of the FGF-4 gene in embryonic stem cells alters the growth and/or the survival of their early differentiated progeny. *Develop. Biol.* **192**, 614–629.

Williams RL, Hilton DJ, Pease S, Willson TA, Stewart CL, Gearing DP, Wagner EF, Metcalf D, Nicola NA, Gough NM (1988) Myeloid leukaemia inhibitory factor maintains the developmental potential of embryonic stem cells. *Nature* **336**, 684–687.

Wobus AM, Holzhausen H, Jakel P, Schoneich J (1984) Characterization of a pluripotent stem cell line derived from a mouse embryo. *Exp. Cell Res.* **152**, 212–219.

Wood SA, Allen ND, Rossant J, Auerbach A, Nagy A (1993) Non-injection methods for the production of embryonic stem cell-embryo chimaeras. *Nature* **365**, 87–89.

Xie X, Chan R, Yoder M (2002) Thrombopoietin acts synergistically with LIF to maintain an undifferentiated state of embryonic stem cells homozygous for a Shp-2 deletion mutation. *FEBS Letts* **529**, 361.

Yamashita J, Itoh H, Hirashima M, Ogawa M, Nishikawa S, Yurugi T, Naito M, Nakao K, Nishikawa SI (2000) Flk1-positive cells derived from embryonic stem cells serve as vascular progenitors. *Nature* **408**, 92–93.

Yin Y, Lim YK, Salto-Tellez M, Ng SC, Lin CS, Lim SK (2002) AFP(+), ESC-derived cells engraft and differentiate into hepatocytes *in vivo. Stem Cells* 20, 338–346.

Ying QL, Stavridis M, Griffiths D, Li M, Smith A (2003) Conversion of embryonic stem cells into neuroectodermal precursors in adherent monoculture. *Nature Biotechnol.* 21, 183–186.

Yoshida K, Chambers I, Nichols J, Smith A, Saito M, Yasukawa K, Shoyab M, Taga T, Kishimoto T (1994) Maintenance of the pluripotential phenotype of embryonic stem cells through direct activation of gp130 signalling pathways. *Mech. Develop.* 45, 163–171.

Yoshida K, Taga T, Saito M, Suematsu S, Kumanogoh A, Tanaka T, Fujiwara H, Hirata M, Yamagami T, Nakahata T *et al.* (1996) Targeted disruption of gp130, a common signal transducer for the interleukin 6 family of cytokines, leads to myocardial and hematological disorders. *Proc. Natl Acad. Sci. USA* 93, 407–411.

Yuan H, Corbi N, Basilico C, Dailey L (1995) Developmental-specific activity of the FGF-4 enhancer requires the synergistic action of Sox2 and Oct-3. *Genes Develop.* 9, 2635–2645.

Zeiher BG, Eichwald E, Zabner J, Smith JJ, Puga AP, McCray PB, Capecchi MR, Welsh MJ, Thomas KR (1995) A mouse model for the Δ F508 allele of cystic fibrosis. *J. Clin. Invest.* 96, 2051–2064.

Zijlstra M, Li E, Sajjadi F, Subramani S, Jaenisch R (1989) Germ-line transmission of a disrupted beta 2-microglobulin gene produced by homologous recombination in embryonic stem cells. *Nature* 342, 435–438.

2. Characteristics of human embryonic stem cells, embryonal carcinoma cells and embryonic germ cells

Michael J. Shamblott and Jared L. Sterneckert

2.1 Introduction

2.1.1 Stem cells and developmental potential

A stem cell can replicate itself and produce cells that take on more specialized functions. The breadth of function adopted by the more differentiated daughter cells and their progeny is commonly referred to as the developmental potential or potency of the stem cell. Stem cells are classified by this potential and by the source of tissue from which they are derived. Those stem cells that give rise to only one type of differentiated cell are termed unipotent. In common usage, the terms oligopotent, multipotent, and pluripotent are used to represent different classes of stem cells that are able to give rise to an increasing number of differentiated cell types, from few to many or most cells of the adult body. On one hand, some stem cells may normally generate cells within a particular lineage, such as neural stem cells that give rise to neurons and glia, or hematopoietic stem cells that give rise to particular subsets of immune cells. The term totipotent in contrast describes a cell that can generate the totality of cell types that comprise the organism, including the placenta, and is therefore often restricted in use to describe the potential of fertilized eggs and blastomeres of the early embryo.

2.2 Sources of stem cells

2.2.1 Endogenous stem cells

Stem cells can be found in many of the tissues of our adult body. They play critical roles in wound healing and the processes of regeneration that are a normal part of survival. In some instances, cells from these endogenous stem cell pools have been isolated, proliferated and/or manipulated *in vitro*. These cells are often termed 'adult' stem cells since they are derived from non-embryonic sources such as bone marrow, peripheral blood, umbilical cord blood, neural tissues, liver, gastrointestinal tract, skin, muscle, and other fetal and adult tissues. Early work with these cell types led to the belief that they were limited in their capacity to proliferate and in their potential to differentiate. However, these generalizations have been challenged by recent observations of extensive proliferative capacity and developmental potency exhibited by stem cells from some non-embryonic sources (Jiang et al., 2002). It remains to be seen, however, whether the observed plasticity following cellular administration into animal models is a result of cellular differentiation or other mechanism, such as cell fusion (Liu and Rao, 2003).

2.2.2 Stem cells derived from the embryo

In order to understand the various embryonic sources of stem cells, the process must be considered by which a fertilized egg develops into a complex multicellular organism in mammals. Within several days following fertilization, the processes of cell cleavage, compaction, and cavitation lead to the formation of the blastocyst. The blastocyst consists of an outer layer of cells surrounding a fluid-filled cavity and a mass of cells within the cavity. The outer cells (trophectoderm) contribute to the placenta but not to the embryo proper. The inner cell mass (ICM) gives rise to two groups of cells. The hypoblast, which contributes to extraembryonic endoderm, and the epiblast, which further differentiates during gastrulation to form the three germ layers of the embryo: ectoderm, mesoderm, and endoderm. Through a complex series of interactions and migrations, cells of the germ layers generate all the tissues of the embryo. During gastrulation, cells that will go on to form the germ cells (eggs and sperm) are allocated.

There are two types of stem cells that are generally referred to as 'embryonic' (*Figure 2.1*). Embryonic stem (ES) cells and embryonic germ (EG) cells are derived from the blastocysts and embryonic primordial germ cells, respectively. Embryonal carcinoma (EC) cells are stem cells derived from tumors, but share the characteristic presence of specific cell surface markers, and a broad capacity to differentiate *in vitro* and *in vivo*.

2.2.3 Embryonal carcinoma (EC) cells

Some of the first clues to the existence of cells capable of differentiating into a wide variety of cell types came from observations of spontaneously occurring tumors termed teratocarcinomas that occur at high frequency in the 129 strain of mice (Stevens, 1958). Within these malignant tumors, or the benign teratomas derived from ES cells (*Figure 2.2*) can be identified a surprising variety of cell types and partially formed tissues such as teeth, bone, hair follicles, neural elements, respiratory epithelia, glandular structures, and layered skin and gut elements. Although the cellular origin of the tumor was not fully understood at the time, it was demonstrated that teratocarcinomas contain a relatively undifferentiated stem cell population that could be clonally isolated, propagated in culture and used to form new teratocarcinomas following transplantation. Importantly, cells that are normally derived from each of the three germ layers were present in these newly formed tumors, demonstrating the pluripotency of the stem cells used to generate them. The cultured stem cells were termed embryonal carcinoma (EC) cells. Later, it was observed that transplantation of preimplantation mouse embryos,

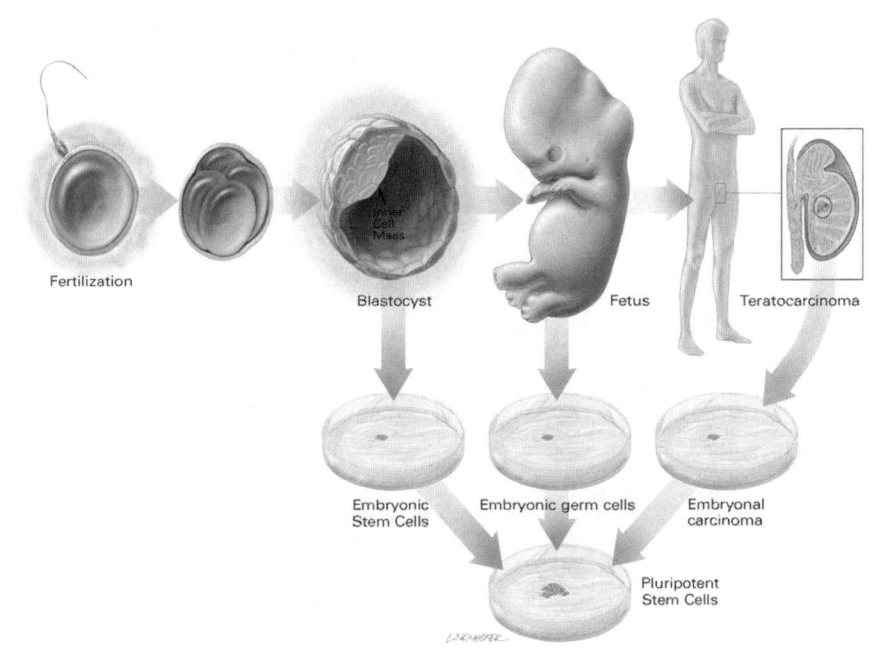

Figure 2.1: Three sources of human pluripotent stem cells. Stem cells can be derived from the epiblast cells of the blastocyst inner cell mass, embryonic and fetal genital ridge, and from teratocarcinomas.

embryonic gonads, and other embryonic tissues also resulted in this type of tumor (Solter *et al.*, 1970; Stevens, 1970a,b).

The ultimate test of stem cell potential is to reintroduce it into a developing blastocyst and to observe the contribution it makes to the tissues of the resultant chimeric fetus. A few mouse EC lines were found to be able to contribute to embryogenesis; however, they were abnormal in both chromosome number and structure, and therefore unlikely to proceed through meiosis to form mature gametes (Smith, 2001).

EC cell cultures have also been derived from human teratocarcinomas, which are classified as germ cell tumors due to their origin from human primordial germ cells (Peyron, 1939). Most mouse and human EC cell cultures cannot be differentiated extensively *in vitro*, however there are several notable exceptions such as the Tera-2 culture and its neural-biased subculture NTera-2 (Andrews, 1984). The status of human EC cells as chromosomally abnormal and neoplastic make their differentiated post-mitotic cell products an unlikely candidate for human cell therapy. However, the strong desire to treat certain diseases has led to their use in clinical trials (Nelson *et al.*, 2002).

Several aspects of EC work served to move the entire field of embryonic stem cell biology forward. One early observation of EC cells is that they often grew better when adjacent to differentiated cells. This finding led to our current use of mouse embryonic fibroblasts as 'feeder' cells. The search for factors responsible for the

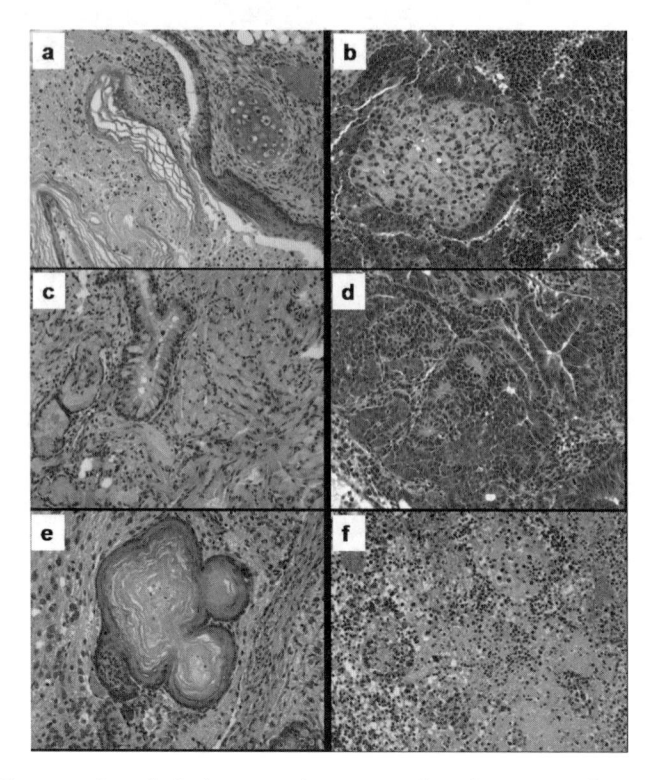

Figure 2.2: Teratomas formed after intramuscular injection of 5×10^5 mouse ES cells. Possible identification of elements are as follows: (a) keratinizing stratified, squamous epithelium, stroma, and a small focus of cartilage formation below the basal layer of the squamous epithelium; (b) morphologically poorly defined structure lined by transitional-like epithelium and filled with cells containing a large amount of pale cytoplasm. Surrounding this structure are numerous small, dark undifferentiated cells; (c) numerous well-formed glandular structures within a stromal background; (d) numerous small, dark undifferentiated cells forming primitive neural tube-like structures; (e) a focus of keratinizing stratified, squamous epithelium within a cellular stromal background; (f) early neural differentiation with areas of high cellularity alternating with areas containing amorphous neuropil.

observed growth enhancement resulted in the identification of leukemia inhibitory factor (LIF), a cytokine that interacts with the gp130 receptor to promote proliferation of stem cells and to maintain their pluripotent status (Smith and Hooper, 1983; Smith et al., 1992).

The EC component of teratocarcinomas has a characteristic blue haematoxylin staining pattern, but in many ways is not histologically distinct. One type of morphologically distinct structure found in EC-derived teratocarcinomas is spherical structures termed embryoid bodies (EBs), which are often found in cysts and cavities within the primary tumor or in the ascites of secondary tumors passaged by intraperitoneal injection. These structures can be simple solid cellular aggregates or, in mouse EBs, can take on the appearance of post-implantation embryos with an ICM-like core of EC cells and an outer layer of visceral endoderm (Damjanov et al., 1987). Embryoid bodies also formed in suspension cultures of some feeder cell-dependent EC cell lines (Martin and Evans, 1974, 1975), and were shown to represent cellular differentiation. The formation of these structures remains an important method of embryonic stem cell differentiation.

The fact that PGCs can form teratocarcinomas *in situ* provided early evidence that, like cells of the epiblast, germ cells can give rise to embryonic stem cell cultures. This fact was borne out several years later, with the description of embryonic germ (EG) cell cultures (Matsui *et al.*, 1992; Resnick *et al.*, 1992).

The tools used to characterize EC cells remain important to this day. These consist of the glycolipid stage-specific embryonic antigens (SSEA-1, 3, 4) and the antibodies TRA-1-60 and TRA-1-81 that recognize glycoprotein antigens (Solter and Knowles, 1978; Kannagi *et al.*, 1983).

2.2.4 Embryonic stem (ES) cells

ES cells were first derived from the epiblast of delayed-implantation or pre-implantation mouse blastocysts in 1981 (Evans and Kaufman, 1981; Martin, 1981). The methods used to derive and grow mouse ES cells have changed very little since then. First, expanded blastocyst-stage embryos are either directly plated or plated after the ICM is immunosurgically isolated to remove trophectoderm (Solter and Knowles, 1975). In either case, the cells are plated on a feeder layer of mouse embryonic fibroblasts that have been mitotically inactivated with gamma radiation or mitomycin-C. LIF is almost universally added to the culture medium to retain pluripotency. Some mouse ES cell lines can be grown in the absence of a feeder layer. It is difficult to overstate the impact mouse ES cells have had on cell biology. This work encompasses *in vitro* studies of cellular development and *in vivo* studies involving targeted mutagenesis.

Mouse ES cells are characterized by high level expression of alkaline phosphatase (AP) activity and the expression of the embryonic cell surface antigens including SSEA-1. Morphologically, mouse ES cells are small (~10 μm diameter) and grow as tightly adherent multicellular colonies on top of the feeder layer. They can be continuously passaged while retaining a normal karyotype, with the exceptions that XX ES cells usually lose an X chromosome and many mouse ES lines can become aneuploid unless great care is taken to monitor cultures and subclone when required.

The most important characteristic of mouse ES cells, however, are their capacity to differentiate. This is demonstrated in several ways. Cells that derive from all three embryonic germ layers can be identified in experimentally induced teratomas following transplantation into immunocompromised or isogenic mice or within EBs formed *in vitro*. They also can efficiently participate in embryogenesis when introduced into mouse blastocysts, contributing to every tissue (other than the placenta) including the germ line. Additionally, ES cells can be differentiated directly (without EB formation) *in vitro*. Embryonic stem cell lines that share some of these characteristics have also been reported for chicken (Pain *et al.*, 1996), mink (Sukoyan *et al.*, 1993), hamster (Doetschman *et al.*, 1988), pig (Shim *et al.*, 1997; Wheeler, 1994), rhesus monkey (Thomson *et al.*, 1995), and common marmoset (Thomson *et al.*, 1996). Recently, human ES cells have been described (Thomson *et al.*, 1998) and will be discussed in greater detail elsewhere. Human ES cells differ from mouse ES cells in several important ways. Most fundamentally, they do not seem to require LIF or gp130 signaling for maintenance of pluripotency. Other differences, such as a significantly slower cellular proliferation rate and differences in cell culture methodology may be species-specific, or due to relatively limited experience and access.

2.2.5 Embryonic germ (EG) cells

Embryonic germ (EG) cells are a lesser-known type of embryonic pluripotent stem cell, as compared with EC and ES cells. EG cells are derived from primordial germ cells (PGCs) (Matsui *et al.*, 1991, 1992; Resnick *et al.*, 1992). During normal mouse development, PGCs are first observed, by lineage tracking and strong AP activity, at E6.5 in the proximal half of the epiblast (Lawson and Hage, 1994). These cells migrate to the allantois then along the mesenteries of the invaginating hindgut into the developing gonad by day E10.5. PGCs undergo significant proliferation during this migration and gonad colonization, from ~150 cells at E8.5 to ~26 000 at E12.5 (Tam and Snow, 1981). PGCs stop dividing by E13.5 in the now sexually differentiated gonad. In females PGCs directly enter meiosis and arrest at meiotic prophase until a few days after birth (Siracusa *et al.*, 1985). In males, PGCs undergo mitotic arrest until about 5 days after birth (Bellve, 1997; Ginsburg *et al.*, 1990). Developing germ cells then re-enter mitosis, forming spermatogonia that eventually give rise to functional sperm. To derive mouse EGs, embryos are harvested at E8.5 to E12.5. At E8.5 the embryo proper is isolated, disaggregated with trypsin and plated onto a feeder layer of certain mouse fibroblast lines that have been mitotically inactivated. At E12.5, the genital ridges are disaggregated and plated similarly. The frequency of EG cell derivation from these later stage embryos is much lower than that of E8.5 (Labosky *et al.*, 1994).

Feeder layers alone cannot support the conversion of solitary PGCs to multicellular EG cultures. PGCs die after 7–14 days unless the growth media is supplemented with basic fibroblast growth factor (bFGF, FGF2) and LIF. Unlike mouse ES cells, EG cell derivation relies on the expression of the transmembrane form of stem cell factor (SCF, c-kit ligand, steel factor) by the feeder layer (Dolci *et al.*, 1991; Matsui *et al.*, 1991). During the derivation process the requirements for SCF and bFGF are usually lost, and EG cells can be routinely passaged under the same conditions as mouse ES cells.

Mouse EG cells share many characteristics with mouse ES cells such as high level of AP activity, the presence of certain embryonic cell surface antigens and growth as tightly adherent multicellular colonies. They can be continuously passaged while retaining a normal karyotype, but unlike mouse ES cells, stable XX EG cell lines can be derived and propagated.

Cells from mouse EG cell lines can participate in embryogenesis when introduced into a blastocyst and can contribute to all tissues including the germ line (Labosky *et al.*, 1994; Stewart *et al.*, 1994). However, imprinting patterns are erased during germ cell development. This can compromise the developmental potential of EG cultures if they are established from late stage PGCs (Tada *et al.*, 1998). Detailed examination of the methylation status of imprinting in the insulin-like growth factor 2 receptor (*Igf2r*) gene in several mouse ES and EG lines demonstrated that although the methylation state of most EG cell lines is different from ES and somatic cell lines, there was no correlation between the methylation pattern and the ability to contribute to the germ line of chimeric mice. It is not clear whether the methylation differences noted between EG lines and as compared with ES and somatic lines were due to differences inherent to PGCs or to their response to EG derivation and culture (Labosky *et al.*, 1994).

However, two experiments suggest that EG cells may have difficulty in reacquiring a completely normal imprinting pattern upon differentiation. When EG cell nuclei were transplanted into enucleated oocytes, the developing placenta was abnormal and reminiscent of a mouse achaete-scute complex homolog 2 (*Mash2*) -/- conceptus (Kato *et al.*, 1999). Further analysis confirmed that *Mash2* imprinting was indeed abnormal. This result is reminiscent of the results of recent ES nuclear transfer experiments in which the inherent instability of H19 imprinting of the parent ES cells was reflected in their cloned offspring (Humpherys *et al.*, 2001). Also, in chimeras of 25–50% EG contribution, abnormally heavy weight and gross skeletal abnormalities were observed (Tada *et al.*, 1998).

Alkaline phosphatase positive human PGCs are observed in the yolk sac and migrate through the embryo to the developing gonads (Witschi, 1948). This information, and well developed protocols for the derivation of mouse EGs (De Felici *et al.*, 1993; Resnick *et al.*, 1992), led to the derivation of human XX and XY EG cultures from 5–11 week post-fertilization gonadal tissue (Shamblott *et al.*, 1998). Human EG cultures are derived in the presence of a mitotically inactivated mouse embryonic fibroblast cell line (STO) feeder layer, hrbFGF, forskolin, and hrLIF. Cells within human EG colonies are chromosomally normal, are AP positive, express SSEA-1, SSEA-3 (weak), and SSEA-4 antigens, and are immunoreactive for TRA-1-60 and TRA-1-81. Undifferentiated human EG cells are also Oct4 positive by RT-PCR and have elevated levels of telomerase. These markers are rapidly lost during the differentiation that accompanies routine culture. Unlike mouse EGs, human EGs do not readily lose their dependency on bFGF and factors provided by the feeder layer such as transmembrane SCF.

Like mouse ES and EG cells, human EG cells grow as large tightly compacted multicellular colonies (*Figure 2.3a,b*). Unlike mouse ES and EG cells, however, human EG cells are relatively resistant to enzymatic disaggregation. A comparison of the cellular junctions of human EG cells with those of mouse EG and ES cells indicates one possible explanation (*Figure 2.3c–e*). Human EG cells appear to be more tightly adherent to each other, with less interstitial space. This might impact the permeability of the EG cell colony to any agent carried in the media, including enzymes, nutrients and growth factors.

Some undesirable consequences of incomplete disaggregation are that cultures cannot be expanded rapidly or grown robustly. These are obvious limitations to cryostorage, manipulation of any kind, and collaboration. When mouse ES and EG cell colonies are not efficiently disaggregated (either improperly or purposefully) they become very large and begin to differentiate. This is the status quo for human EG culture, with 10% or more of the marker positive EG cell colonies converting to differentiated EBs each week in the presence of LIF. Difficulty with disaggregation

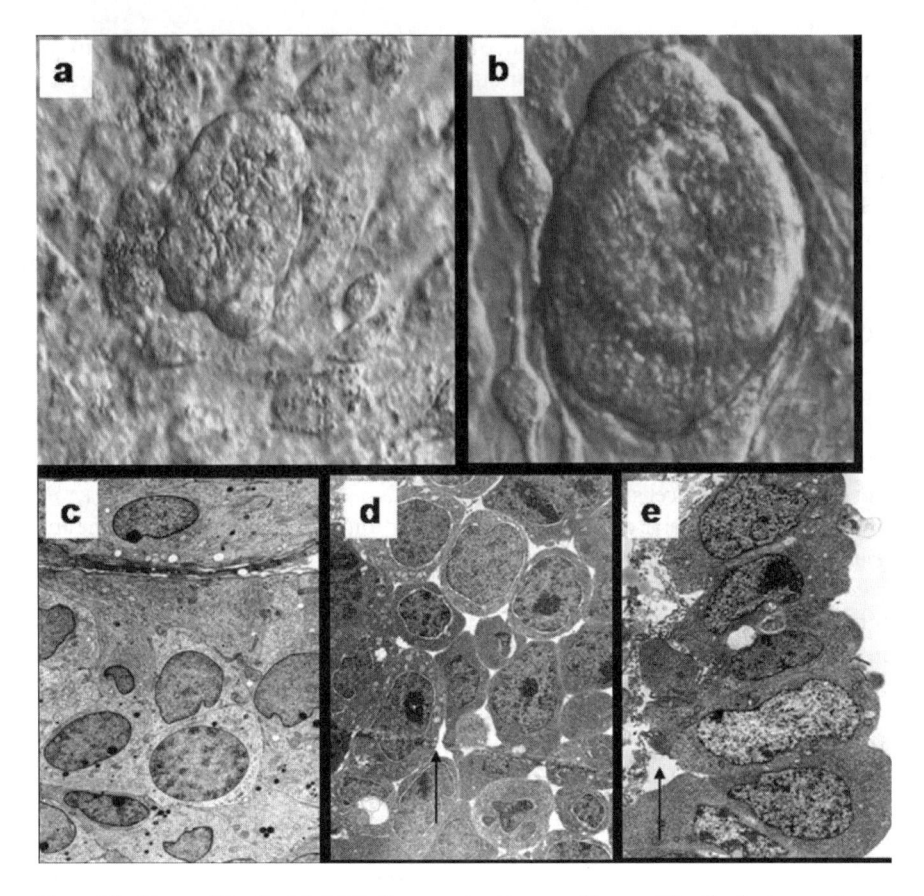

Figure 2.3: Human EG cells. (a)–(b) Hoffman contrast images of human EG cell colonies growing on a feeder layer of mouse STO fibroblasts; (c)–(e) Electron microscopic images of (c) Human EG; (d) Mouse EG; and (e) Mouse ES cells. Interstitial spaces indicated by arrows in **d** and **e**.

is a trait shared with many non-mouse ES and EG cell cultures, including some primate ES cell cultures.

Perhaps the only advantage afforded by inefficient disaggregation is the ready supply of differentiated cells. When human EBs are harvested from the culture, they can be analyzed immunohistochemically to reveal the presence of cells that normally derive from all three germ layers. This provided the only evidence directly to support a pluripotent status, as every attempt to form teratocarcinomas following injection into immunocompromised mice failed.

Embryoid bodies can be disaggregated to some extent with a mixture of collagenase and dispase or other enzymes. Based on immunohistochemical evidence, a wide variety of differentiated cell types from this process could be expected, and it is at least theoretically possible to isolate and expand these populations. Unfortunately (for the cell biologist), many fully differentiated cell types do not proliferate robustly in culture. This is due to inherent limitation such as tight cell cycle control and telomere length, as well as to suboptimal culture conditions. Isolation of mature, possibly post-mitotic cell types can result in compelling proof-of-principle but is unlikely to generate sufficient numbers of cells to allow careful study or affect some future cell-based transplantation therapy.

Progenitor and precursor cells play a central role in many well-established cellular differentiation pathways such as those occurring during neural and hematopoietic differentiation. As such, they are seen as less developmentally potent (lineage-restricted) than the stem cell pool from which they were derived, yet they are still capable of further differentiation into multiple cell types. Progenitor and precursor cells often retain some capacity to proliferate *in vivo* and *in vitro*.

2.3 Embryoid body-derived (EBD) cells

EBs formed from EC, ES and EG cells contain collections of cells that represent a continuum of differentiation. Several theoretical outcomes are possible when selecting for cells within EBs that can proliferate extensively in culture. One possibility is that rapidly dividing stem cells such as ECs will grow out of these cultures. Another possibility is that a rapidly dividing stromal cell type, such as fibroblasts, will predominate. Lastly, it is possible that progenitor and precursor cells that give rise to the more terminally-differentiated cell types found in EBs will be produced.

To test these possibilities and to obtain human cells capable of proliferation and subsequent differentiation, EBs from four genetically distinct human EG cultures (2 XX and 2 XY) were picked, disaggregated and plated into one of six growth environments. The media components consisted of either RPMI 1640 supplemented with 15% fetal calf serum (FCS) or EGM2mv, a commercially available medium (Clonetics) containing 5% FCS supplemented with bFGF, epidermal growth factor (EGF), insulin-like growth factor I (IGF I) and vascular endothelial growth factor (VEGF). Three different cell attachment surfaces were used: tissue culture plastic, bovine collagen type I, and human placental extracellular matrix extract, a commercially available mixture of laminin, collagen IV and heparin sulfate proteoglycan. All six growth environments supported cell proliferation and the resultant cells were termed EB-derived (EBD) cell cultures (*Figure 2.4*) (Shamblott *et al.*, 2001). This process has been repeated on more than 30 human EG cultures, resulting in more than 100 EBD cultures and clonal lines. EBD cell cultures have a normal karyotype and senesce after 70 to 80 population doublings. Under the conditions of derivation, EBD cells grow as a monolayer and are amenable to enzymatic disaggregation and genetic manipulation using chemical and viral methods. Many EBD cultures are also clonogenic.

Extensive analyses of mRNA and protein expression have been carried out on EBD cultures and clonal lines. This was done initially to test whether derivation conditions alone could be used to derive lineage-restricted cell types. With few exceptions, the results of these analyses suggested that EBD cell cultures are a heterogeneous cell population containing cells capable of simultaneously expressing markers of multiple distinct lineages including neural, muscle, vascular/hematopoietic and endodermal lineages (Shamblott *et al.*, 2001). This was demonstrated by immunocytochemical staining and mRNA expression analysis of clonal

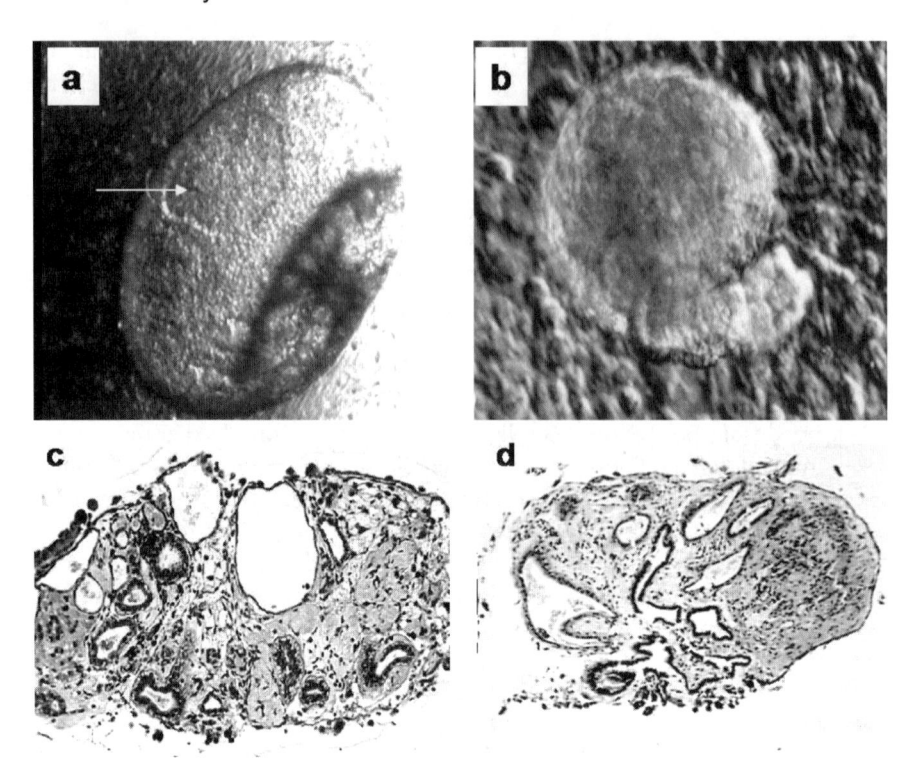

Figure 2.4: Human EG-derived embryoid bodies. (a) Low magnification image of large cyctic EB (~5 mm diameter) arising from an EG colony; Arrow indicates an EG cell colony; (b)Solid human EB; and (c)–(d) Haematoxylin & eosin stained paraffin sections of a human EBs.

lines. The biological significance and mechanisms of multilineage gene expression pattern are not clear, however it is a phenomenon shared by some stem and progenitor cell populations (Colucci-D'Amato et al., 1999; Piper et al., 2000; Hu et al., 1997). It is also interesting to note that human ES cells differentiated in culture in the presence of different growth factors demonstrated a prominently heterogeneous mRNA expression profile regardless of the growth factor used (Schuldiner et al., 2000).

EBD cells bear little resemblance to EC cells. They are euploid, senesce and so far are non-tumorigenic to the many hundreds of rodents and primates that have received these cells experimentally. Cellular function *in vitro* or following engraftment into animal models has not yet been demonstrated in peer-reviewed publication, but many of these experiments suggest that cells from at least two EBD cultures can vigorously engraft into various locations, and can provide relevant biological functions. It has also been demonstrated that EBD cells have a normal imprinting pattern (Onyango et al., 2002).

2.4 Markers of pluripotency

Oct-3/4 is the only transcription factor known to be specifically expressed in early embryos, the germ line and the pluripotent stem cells from which they are derived. Encoded by the *Pou5f1* locus, Oct-3/4 is necessary for pluripotency, as defined by transgenesis experiments (Nichols et al., 1998). Mice in which the *Pou5f1* locus has been inactivated fail to develop an inner cell mass. Quantitative studies revealed that ES cells either maintain pluripotency or differentiate depending upon the levels of Oct-3/4 (Niwa et al., 2000). When Oct-3/4 expression in ES cells is eliminated, trophoblastic differentiation ensues. However, maintenance of Oct-3/4 expression is insufficient to prevent differentiation of ES cells when LIF is withdrawn. Several studies have defined a few target genes regulated by Oct-3/4. Genes dependent on Oct-3/4 activity for expression in ES cells include: *Fgf-4* (Yuan et al., 1995), *Rex-1* (Ben-Shushan et al., 1998), *Utf-1* (Nishimoto et al., 1999), platelet-derived growth factor receptor alpha (*Pdgf-R*) (Kraft et al., 1996), *Opn* (Botquin et al., 1998), *Lefty-1* (Niwa et al., 2000), *Upp* (Niwa et al., 2000), and *Tera* (Niwa et al., 2000). Human chorionic gonadotropin (*HCG*) is repressed by Oct-3/4 activity (Liu and Roberts, 1996).

Several other genetic markers of pluripotency have been described. The homeoprotein Nanog is expressed in ES cells and preimplantation embryos, and is capable of maintaining ES cell pluripotency and self renewal independently of the LIF/Stat3 pathway (Chambers et al., 2003; Mitsui et al., 2003). *FoxD3* (also called *Genesis*) and *Sox-2* are both expressed in ES, EG and EC cells as well as in several other non-pluripotent cell types (Yuan et al., 1995; Sutton et al., 1996). Both of these transcription factors have been demonstrated to regulate downstream gene expression through an interaction with Oct-3/4 (Yuan et al., 1995; Guo et al., 2002). Though not specific to pluripotent lineages, *Pem* has been identified as a gene whose expression is sufficient to interfere with normal ES cell differentiation (Fan et al., 1999). However, when *Pem* was inactivated through transgenesis, the ES cells remained pluripotent.

Finally, there are two activities specific to totipotent and pluripotent cells, yet the genes involved remain unknown. High expression levels of Hsp70 in EC cells were explained by an E1A-like activity (Imperiale et al., 1984). This activity was further demonstrated through E1A independent activation of the adenoviral E2A promoter. This activity has also been found in oocytes and pre-implantation embryos and is lost upon differentiation (Dooley et al., 1989). It is possible that Oct-3/4 may activate downstream gene expression through protein(s) involved in this E1A-like activity, but this remains to be definitively elucidated (Brehm et al., 1999). Also, cellular fusion experiments with ES, EG, and EC cells have demonstrated dominant trans-activating factors that are capable of reprogramming a somatic nucleus (Miller and Ruddle, 1976; Tada et al., 1997, 2001; Takagi et al., 1983). The mechanisms of reprogramming may include DNA demethylation, X-chromosome activation, and/or Oct-3/4 expression, and may be related to reprogramming activities found in oocytes. Though the extent of this pluripotent reprogramming is unknown, characterization of the factors involved will no doubt shed light upon the nature of pluripotency of embryo-derived stem cells.

In an effort to identify pathways essential for 'stemness', two groups simultaneously reported the results of an expression array analysis comparing ES cells, neural stem cells and hematopoietic stem cells (Ivanova *et al.*, 2002; Ramalho-Santos *et al.*, 2002). The results were a group of approximately 230–280 genes; only a fraction of which have known function.

Although the derivation, culture requirements and cellular properties of human ES, EG and EC cells differ widely, they share the capacity to differentiate into a wide variety of cell types. Present and future efforts strive to direct this differentiation in order to formulate safe and effective cellular therapies and model some aspects of human development and cell biology.

References

Andrews PW (1984) Retinoic acid induces neuronal differentiation of a cloned human embryonal carcinoma cell line *in vitro*. *Dev. Biol.* **103**, 285–293.

Bellve A (1997) The molecular biology of spermatogenesis. In: *Oxford Reviews of Reproductive Biology* (ed. C. Finn). Clarendon Press, Oxford, pp. 159–261.

Ben-Shushan E, Thompson JR, Gudas LJ, Bergman Y (1998) Rex-1, a gene encoding a transcription factor expressed in the early embryo, is regulated via Oct-3/4 and Oct-6 binding to an octamer site and a novel protein, Rox-1, binding to an adjacent site. *Mol. Cell. Biol.* **18**, 1866–1878.

Botquin V, Hess H, Fuhrmann G, Anastassiadis C, Gross M.K, Vriend G, Scholer HR (1998) New POU dimer configuration mediates antagonistic control of an osteopontin preimplantation enhancer by Oct-4 and Sox-2. *Genes Dev.* **12**, 2073–2090.

Brehm A, Ohbo K, Zwerschke W, Botquin V, Jansen-Durr P, Scholer HR (1999) Synergism with germ line transcription factor Oct-4: Viral oncoproteins share the ability to mimic a stem cell-specific activity. *Mol. Cell Biol.* **19**, 2635–2643.

Chambers I, Colby D, Robertson M, Nichols J, Lee S, Tweedie S, Smith A (2003) Functional expression cloning of Nanog, a pluripotency sustaining factor in embryonic stem cells. *Cell* **113**, 643–655.

Colucci-D'Amato GL, Tino A, Pernas-Alonso R, ffrench-Mullen JM, di Porzio U (1999) Neuronal and glial properties coexist in a novel mouse CNS immortalized cell line. *Exp. Cell Res.* **252**, 383–391.

Damjanov I, Damjanov A, Solter D (1987) Production of teratocarcinomas from embryos transplanted to extra-uterine sites. In: *Teratocarcinomas and Embryonic Stem Cells: A Practical Approach* (ed. E. Roberson). IRL Press, Oxford, pp. 1–18.

De Felici M, Dolci S, Pesce M (1993) Proliferation of mouse primordial germ cells *in vitro*: a key role for camp. *Dev. Biol.* **157**, 277–280.

Doetschman T, Williams P, Maeda N (1988) Establishment of hamster blastocyst-derived embryonic stem (ES) cells. *Dev. Biol.* **127**, 224–227.

Dolci S, Williams DE, Ernst MK, Resnick JL, Brannan CI, Lock LF, Lyman SD, Boswell HS, Donovan PJ (1991) Requirement for mast cell growth factor for primordial germ cell survival in culture. *Nature* **352**, 809–811.

Dooley TP, Miranda M, Jones NC, DePamphilis ML (1989) Transactivation of the adenovirus EIIA promoter in the absence of adenovirus E1A protein is restricted to mouse oocytes and preimplantation embryos. *Development* **107**, 945–956.

Evans MJ, Kaufman MH (1981) Establishment in culture of pluripotential cells from mouse embryos. *Nature* **292**, 154–156.

Fan Y, Melhem MF, Chaillet JR (1999) Forced expression of the homeobox-containing gene Pem blocks differentiation of embryonic stem cells. *Dev. Biol.* **210**, 481–496.

Ginsburg M, Snow MH, McLaren A (1990) Primordial germ cells in the mouse embryo during gastrulation. *Development* **110**, 521–528.

Guo Y, Costa R, Ramsey H, Starnes T, Vance G, Robertson K, Kelley M, Reinbold R, Scholer H, Hromas R (2002) The embryonic stem cell transcription factors Oct-4 and FoxD3 interact to regulate endodermal-specific promoter expression. *Proc. Natl Acad. Sci. USA* **99**, 3663–3667.

Hu M, Krause D, Greaves M, Sharkis S, Dexter M, Heyworth C, Enver T (1997) Multilineage gene expression precedes commitment in the hemopoietic system. *Genes Dev.* **11**, 774–785.

Humpherys D, Eggan K, Akutsu H, Hochedlinger K, Rideout WM 3rd, Biniszkiewicz D, Yanagimachi R, Jaenisch R (2001) Epigenetic instability in ES cells and cloned mice. *Science* **293**, 95–97.

Imperiale MJ, Kao HT, Feldman LT, Nevins JR, Strickland S (1984) Common control of the heat shock gene and early adenovirus genes: Evidence for a cellular E1A-like activity. *Mol. Cell Biol.* **4**, 867–874.

Ivanova NB, Dimos JT, Schaniel C, Hackney JA, Moore KA, Lemischka IR (2002) A stem cell molecular signature. *Science* **298**, 601–604.

Jiang Y, Jahagirdar BN, Reinhardt RL, Schwartz RE, Keene CD, Ortiz-Gonzalez XR *et al.* (2002) Pluripotency of mesenchymal stem cells derived from adult marrow. *Nature* **418**, 41–49.

Kannagi R, Cochran NA, Ishigami F, Hakomori S, Andrews PW, Knowles BB, Solter D (1983) Stage-specific embryonic antigens (SSEA-3 and -4) are epitopes of a unique globo-series ganglioside isolated from human teratocarcinoma cells. *EMBO J.* **2**, 2355–2361.

Kato Y, Rideout WM 3rd, Hilton K, Barton SC, Tsunoda Y, Surani MA (1999) Developmental potential of mouse primordial germ cells. *Development* **126**, 1823–1832.

Kraft HJ, Mosselman S, Smits HA, Hohenstein P, Piek E, Chen Q, Artzt K, van Zoelen EJ (1996) Oct-4 regulates alternative platelet-derived growth factor alpha receptor gene promoter in human embryonal carcinoma cells. *J. Biol. Chem.* **271**, 12873–12878.

Labosky P, Barlow D, Hogan B (1994) Mouse embryonic germ (EG) cell lines: Transmission through the germline and differences in the methylation imprint of insulin-like growth factor 2 receptor (igf2r) gene compared with embryonic stem (ES) cell lines. *Development* **120**, 3197–3204.

Lawson K, Hage W (1994) Clonal analysis of the origin of primordial germ cells in the mouse. In: *Germline Development* (eds J. Marsh and J. Goode). John Wiley & Sons, New York, pp. 68–91.

Liu L, Roberts RM (1996) Silencing of the gene for the beta subunit of human chorionic gonadotropin by the embryonic transcription factor Oct-3/4. *J. Biol. Chem.* **271**, 16683–16689.

Liu Y, Rao MS (2003) Transdifferentiation-fact or artifact. *J. Cell Biochem.* **88**, 29–40.

Martin GR (1981) Isolation of a pluripotent cell line from early mouse embryos cultured in media conditioned by teratocarcinoma stem cells. *Proc. Natl Acad. Sci. USA* **78**, 7634–7638.

Martin GR, Evans MJ (1974) The morphology and growth of a pluripotent teratocarcinoma cell line and its derivatives in tissue culture. *Cell* **2**, 163–172.

Martin GR, Evans MJ (1975) Differentiation of clonal lines of teratocarcinoma cells: Formation of embryoid bodies *in vitro*. *Proc. Natl Acad. Sci. USA* **72**, 1441–1445.

Matsui Y, Toksoz D, Nishikawa S, Nishikawa S, Williams D, Zsebo K, Hogan BL (1991) Effect of steel factor and leukaemia inhibitory factor on murine primordial germ cells in culture. *Nature* **353**, 750–752.

Matsui Y, Zsebo K, Hogan BL (1992) Derivation of pluripotential embryonic stem cells from murine primordial germ cells in culture. *Cell* **70**, 841–847.

Miller RA, Ruddle FH (1976) Pluripotent teratocarcinoma-thymus somatic cell hybrids. *Cell* **9**, 45–55.

Mitsui K, Tokuzawa Y, Hoh H, Segawa K, Murakami M, Kazutoshi T, Maruyama M, Maeda M, Yamanaka S (2003) The homeoprotein nanog is required for maintenance of pluripotency in mouse epiblast and ES cells. *Cells* **113**, 631–642.

Nelson PT, Kondziolka D, Wechsler L, Goldstein S, Gebel J, DeCesare S *et al.* (2002) Clonal human (hNT) neuron grafts for stroke therapy: Neuropathology in a patient 27 months after implantation. *Am. J. Pathol.* **160**, 1201–1206.

Nichols J, Zevnik B, Anastassiadis K, Niwa H, Klewe-Nebenius D, Chambers I, Scholer H, Smith A (1998) Formation of pluripotent stem cells in the mammalian embryo depends on the pou transcription factor Oct4. *Cell* **95**, 379–391.

Nishimoto M, Fukushima A, Okuda A, Muramatsu M (1999) The gene for the embryonic stem cell coactivator Utf1 carries a regulatory element which selectively interacts with a complex composed of Oct-3/4 and Sox-2. *Mol. Cell Biol.* **19**, 5453–5465.

Niwa H, Miyazaki J, Smith AG (2000) Quantitative expression of Oct-3/4 defines differentiation, dedifferentiation or self-renewal of ES cells. *Nat. Genet.* **24**, 372–376.

Onyango P, Jiang, S, Uejima, H, Shamblott, M.J, Gearhart, JD, Cui, H, Feinberg, AP (2002) Monoallelic expression and methylation of imprinted genes in human and mouse embryonic germ cell lineages. *Proc. Natl Acad. Sci. USA* **99**, 10599-10604.

Pain B, Clark ME, Shen M, Nakazawa H, Sakurai M, Samarut J, Etches RJ (1996) Long-term *in vitro* culture and characterisation of avian embryonic stem cells with multiple morphogenetic potentialities. *Development* **122**, 2339–2348.

Peyron A (1939) Faits nouveaux relatifs à l'origine et à l'histogénèse des embryomes. *Bull. Assn Franc. Étude Cancer* **28**, 658–681.

Piper DR, Mujtaba T, Rao MS, Lucero MT (2000) Immunocytochemical and physiological characterization of a population of cultured human neural precursors. *J. Neurophysiol.* **84**, 534–548.

Ramalho-Santos M, Yoon S, Matsuzaki Y, Mulligan RC, Melton DA (2002) 'Stemness': Transcriptional profiling of embryonic and adult stem cells. *Science* **298**, 597–600.

Resnick JL, Bixler LS, Cheng L, Donovan PJ (1992) Long-term proliferation of mouse primordial germ cells in culture. *Nature* **359**, 550–551.

Schuldiner M, Yanuka O, Itskovitz-Eldor J, Melton DA, Benvenisty N (2000) Effects of eight growth factors on the differentiation of cells derived from human embryonic stem cells. *Proc. Natl Acad. Sci. USA* **97**, 11307–11312.

Shamblott MJ, Axelman J, Wang S, Bugg EM, Littlefield JW, Donovan PJ, Blumenthal PD, Huggins GR, Gearhart JD (1998) Derivation of pluripotent stem cells from cultured human primordial germ cells. *Proc. Natl Acad. Sci. USA* **95**, 13726–13731.

Shamblott M, Axelman J, Littlefield JW, Blumenthal PD, Huggins GR, Cui Y, Cheng L, Gearhart JD (2001) Human embryonic germ cell derivatives express a broad range of developmentally distinct markers and proliferate extensively *in vitro*. *Proc. Natl Acad. Sci. USA* **98**, 113–118.

Shim H, Gutierrez-Adan A, Chen LR, BonDurant RH, Behboodi E, Anderson GB (1997) Isolation of pluripotent stem cells from cultured porcine primordial germ cells. *Biol. Repro.* **57**, 1089–1095.

Siracusa G, Defelici M, Salustri A (1985) The proliferative and meiotic history of mammalian female germ cells. *Biol. Fertil.* **1**, 253–297.

Smith AG, Nichols J, Robertson M, Rathjen PD (1992) Differentiation inhibiting activity (DIA/LIF) and mouse development. *Dev. Biol.* **151**, 339–351.

Smith AG (2001) Embryo-derived stem cells: of mice and men. *Annu. Rev. Cell Dev. Biol.* **17**, 435–462.

Smith TA, Hooper ML (1983) Medium conditioned by feeder cells inhibits the differentiation of embryonal carcinoma cultures. *Exp. Cell Res.* **145**, 458–462.

Solter D, Knowles BB (1975) Immunosurgery of mouse blastocyst. *Proc. Natl Acad. Sci. USA* **72**, 5099–5102.

Solter D, Knowles B (1978) Monoclonal antibody defining a stage-specific mouse embryonic antigen (SSEA-1). *Proc. Natl Acad. Sci. USA* **75**, 5565–5569.

Solter D, Skreb N, Damjanov I (1970) Extrauterine growth of mouse egg-cylinders results in malignant teratoma. *Nature* **227**, 503–504.

Stevens LC (1958) Studies on transplantable testicular teratomas of strain 129 mice. *J. Natl Cancer Inst.* **20**, 1257–1270.

Stevens LC (1970a) Experimental production of testicular teratomas in mice of strains 129, A/He, and their F1 hybrids. *J. Natl Cancer Inst.* **44**, 923–929.

Stevens LC (1970b) The development of transplantable teratocarcinomas from intratesticular grafts of pre- and postimplantation mouse embryos. *Dev. Biol.* **21**, 364–382.

Stewart C, Gadi I, Bhatt H (1994) Stem cells from primordial germ cells can reenter the germ line. *Dev. Biol.* **161**, 626–628.

Sukoyan MA, Vatolin SY, Golubitsa AN, Zhelezova AI, Semenova LA, Serov OL (1993) Embryonic stem cells derived from morulae, inner cell mass, and blastocysts of mink: Comparisons of their pluripotencies. *Mol. Reprod. Dev.* **36**, 148–158.

Sutton J, Costa R, Klug M, Field L, Xu D, Largaespada DA *et al.* (1996) Genesis, a winged helix transcriptional repressor with expression restricted to embryonic stem cells. *J. Biol. Chem.* **271**, 23126–23133.

Tada M, Tada T, Lefebvre L, Barton SC, Surani MA (1997) Embryonic germ cells induce epigenetic reprogramming of somatic nucleus in hybrid cells. *EMBO J.* **16**, 6510–6520.

Tada M, Takahama Y, Abe K, Nakatsuji N, Tada T (2001) Nuclear reprogramming of somatic cells by *in vitro* hybridization with ES cells. *Curr. Biol.* **11**, 1553–1558.

Tada T, Tada M, Hilton K, Barton SC, Sado T, Takagi N, Surani MA (1998) Epigenotype switching of imprintable loci in embryonic germ cells. *Dev. Genes Evol.* **207**, 551–561.

Takagi N, Yoshida MA, Sugawara O, Sasaki M (1983) Reversal of x-inactivation in female mouse somatic cells hybridized with murine teratocarcinoma stem cells *in vitro*. *Cell* **34**, 1053–1062.

Tam PP, Snow MH (1981) Proliferation and migration of primordial germ cells during compensatory growth in mouse embryos. *J. Embryol. Exp. Morphol.* **64**, 133–147.

Thomson JA, Kalishman J, Golos TG, Durning M, Harris CP, Becker RA, Hearn JP (1995) Isolation of a primate embryonic stem cell line. *Proc. Natl Acad. Sci. USA* **92**, 7844–7848.

Thomson JA, Kalishman J, Golos TG, Durning M, Harris CP, Hearn JP (1996) Pluripotent cell lines derived from common marmoset (*Callithrix jacchus*) blastocysts. *Biol. Reprod.* **55**, 254–259.

Thomson JA, Itskovitz-Eldor J, Shapiro SS, Waknitz MA, Swiergiel JJ, Marshall VS, Jones JM (1998) Embryonic stem cell lines derived from human blastocysts. *Science* **282**, 1145–1147.

Wheeler MB (1994) Development and validation of swine embryonic stem cells: A review. *Reprod. Fertil. Dev.* **6**, 563–568.

Witschi E (1948) Migration of the germ cells of human embryos from the yolk sac to the primitive gonadal folds. In: *Contributions to Embryology.* Carnegie Institution of Washington, pp. 76–79.

Yuan H, Corbi N, Basilico C, Dailey L (1995) Developmental-specific activity of the FGF-4 enhancer requires the synergistic action of Sox2 and Oct-3. *Genes Dev.* **9**, 2635–2645.

3. Adult stem cell plasticity

Robert E. Schwartz and Catherine M. Verfaillie

3.1 Introduction

Historically each organ and tissue was thought to possess cells capable of self-renewal and of giving rise to a large number of differentiated descendents. These cells, termed adult stem cells, were believed to be restricted in their potential. This tissue specific restriction was accompanied by the belief that as cells differentiate, they lose their ability to make fate choices. This is exemplified by many well-characterized cell types, such as hematopoietic stem cells and neural stem cells. In the past few years several reports have suggested that such stem cells may possess developmental capabilities resembling those of more immature and potent cells such as embryonic stem cells. These findings are raising fundamental questions about the traditional hierarchical view of developmental biology and challenging the established beliefs and dogmas developed in biology over the past century.

In this chapter, we will review several reports on adult stem cell plasticity and address the strengths and concerns of each report. We will discuss the possible explanations for stem cell plasticity and its potential consequences.

3.2 Stem cells – definition

Through the years, scientists have defined stem cells in many ways but the consensus definition would encompass three main principles. First, a stem cell must be capable of self-renewal, i.e., undergoing symmetric or asymmetric divisions through which the stem cell population is maintained. Secondly, a single cell must be capable of multilineage differentiation. The third principle is *in vivo* functional reconstitution of a given tissue.

Human and mouse embryonic stem (ES) cells are the quintessential pluripotent stem cells and fulfil all of these criteria (Evans and Kaufman 1981; Martin, 1981; Thomson *et al.*, 1995, 1998). Both mouse and human ES cells are capable of 300–400 cell doublings in culture while maintaining a stable karyotype and phenotype. ES cells are able to form all the somatic tissues (i.e., all three lineages, endoderm, ectoderm, and mesoderm) as well as the germ cells of the mouse as demonstrated by injection into blastocysts. Furthermore, human, as well as mouse, embryonic stem cells can form mature cells as shown through embryoid body and teratoma formation. More importantly, these cells are functional as whole mice have been derived from embryonic stem cells. Taken together, ES cells fulfil all three criteria that define stem cells.

3.3 Adult stem cells

Adult stem cells also fulfil these criteria. However, the degree of self-renewal and differentiation potential is far more restricted when compared with embryonic stem cells. Over the past 50 years the most extensively studied adult stem cell is without question the hematopoietic stem cell (HSC) (Bhatia et al., 1997; Spangrude et al., 1988). The HSC undergoes self-renewal, differentiates into many different blood-forming units at the single cell level, and, when transplanted, functionally repopulates the hematopoietic system of an ablated recipient. More recently, other adult stem cells have been studied. For example, neural stem cells (NSC) give rise to neurons, astrocytes, and oligodendrocytes (Gage, 2000). Another example is mesenchymal stem cells (MSC) that differentiate into fibroblasts, osteoblasts, chondroblasts, adipocytes, and skeletal muscle (Fridenshtein, 1982; Pittenger et al., 1999; Prockop, 1997). Some cells also termed adult stem cells, such as corneal stem cells (Daniels et al., 2001) and angioblasts (Rafii et al., 1994), fulfil all these principles although they only differentiate into a single mature cell type.

3.4 Adult stem cells – plasticity

Many reports over the past six years have been published that suggest that cells from a given tissue might be capable of differentiating into cells of a different tissue (LaBarge and Blau, 2002; Gussoni et al., 1999; Mezey et al., 2000; Theise et al., 2000a,b; Wang et al., 2001, 2002). If true this would suggest that our previous understanding of tissue specificity of stem cells may not be correct. The ability of a tissue-specific stem cell to acquire the fate of a cell type different from the original tissue has been termed adult stem cell plasticity, although no consensus exists as to what the exact definition should be. To many, 'stem cell plasticity' may be a new concept. However the idea is almost a century old. In the late 19th and early 20th century it was recognized that there are epithelial changes in tissues in response to different stresses (Cotran, 1999a). These changes, in which one adult cell type is replaced by another cell type, was termed metaplasia. An example includes the change from columnar epithelium to squamous epithelium in the respiratory tract of smokers in response to chronic irritation caused by smoking (Cotran, 1999b). Another example is the change from squamous epithelium to columnar epithelium due to gastric reflux that occurs in Barrett's esophagus (Cotran, 1999c). The possible mechanisms for this plasticity will be described later.

More recently, reports on stem cell plasticity have brought much excitement within the lay and scientific communities (Verfaillie et al., 2002). In addition, they have also generated great skepticism (Hawley and Sobieski, 2002; Holden and Vogel, 2002). This is a consequence of the fact that most studies still await independent confirmation, show a very low frequency of 'plasticity', and do not show that the demonstrated plasticity results from a single stem cell. Most importantly such studies' conclusions conflict with the established dogma of stem cell hierarchy and role in developmental biology.

The majority of studies that have shown plasticity were based on sex-mismatched bone marrow transplants of either marked cells (rodents) or unmarked cells (human and rodent). It is important to note that bone marrow contains MSC, endothelial progenitors, and possibly even hepatic progenitors in addition to HSC (Asahara *et al.*, 1997; Avital *et al.*, 2001; Fiegel *et al.*, 2003; Fridenshtein, 1982; Lin *et al.*, 2000). Although this could theoretically explain bone, cartilage, fat, endothelial (Crosby *et al.*, 2000; Grant *et al.*, 2002) and hepatic cell differentiation (Krause *et al.*, 2001; Lagasse *et al.*, 2000; Theise *et al.*, 2000a,b) of donor origin, reports have shown donor-derived cells of ectodermal origin including skin epithelium (Krause *et al.*, 2001) and neuronal cells (Brazelton *et al.*, 2000; Mezey *et al.*, 2000), albeit at low frequency.

Further criticism stems from studies that have identified and characterized cells via phenotypic and morphologic characteristics to define 'differentiation' of one cell type into another without examining the functional characteristics. In addition most studies fail to address the issue of cell fusion as a potential mechanism (Terada *et al.*, 2002; Ying *et al.*, 2002).

3.5 Plasticity of hematopoietic bone marrow cells

The majority of studies demonstrating adult stem cell plasticity involve hematopoietic stem cells either from bone marrow, peripheral blood, or from samples enriched for HSC. HSC traditionally have been believed to form the various components of the hematopoietic system, i.e.. lymphoid, myeloid, erythroid, and megakaryocytic cells. However, more recent studies have claimed that HSC appear to differentiate into skeletal muscle, cardiac muscle, smooth muscle, neuroectoderm, endodermal lineages such as hepatocytes and pancreatic duct, endothelium, and lung epithelium.

A common thread through all these experiments is the transplantation of marked (LacZ or GFP) or sex-mismatched whole or partially purified bone marrow. Different tissues are then examined to determine whether bone marrow-derived cells have undergone a switch in cell morphology and phenotype. Almost all studies defined differentiation, and thus plasticity, based on morphologic and phenotypic characteristics of the differentiated cells and not by their functional characteristics. We will address this point later.

A series of studies have suggested that bone marrow cells can give rise to skeletal muscle cells. Ferrari *et al.* (1998) were the first to demonstrate that a subpopulation of bone marrow cells was capable of migrating into areas of induced muscle damage, undergoing myogenic differentiation, and participating in muscle regeneration. Likewise Bittner *et al.* (1999) demonstrated that bone marrow cells were capable of myogenic differentiation and could partially restore dystrophin expression in a few muscle fibers of the mdx mouse (a mouse model of muscular dystrophy in which mice are deficient in dystrophin). Taking this one step further, Gussoni and colleagues (1999) demonstrated that not only transplantation of whole bone marrow but transplantation of enriched hematopoietic stem cells into irradiated mdx mice resulted in the reconstitution of the hematopoietic compartment, the incorporation of donor-derived male nuclei into skeletal muscle, and the partial restoration of dystrophin expression in the affected muscle. LaBarge et al. demonstrate that GFP^+ whole bone marrow was capable of becoming satellite cells in lethally irradiated animals (LaBarge and Blau, 2002). Following exercise induced muscle injury, GFP-labeled multinucleated muscle fibers were detected. However, neither of these studies demonstrates that a single cell gave rise to hematopoietic and muscle cells. In addition, all these studies rely on morphologic and phenotypic characteristics to demonstrate differentiation and thus plasticity. Although some studies demonstrated differentiation of these cells in a temporal manner similar to normal physiologic processes, functional assessment of the marrow-derived muscle fibers was not performed.

A second example is the perceived lineage switch from hematopoietic to endodermal epithelial cells, including hepatocytes, gastrointestinal epithelial cells and lung epithelium. In livers from females who received a sex-mismatched bone marrow transplantation, 5% to 40% (depending on recipient) of the liver parenchyma appeared to be derived from the donor bone marrow (Theise et al., 2000b). When the lineage-switched hepatocytes were examined by cytogenetic analysis, they were shown to bear only one X and one Y chromosome (Korbling et al., 2002). In cases of graft versus host disease levels of engraftment were found to be even higher among cells of the liver and gastrointestinal tract. In all studies except the study by Krause et al. (2001), mixed cell populations were transplanted. Consequently, the demonstration of engraftment into multiple tissues does not truly demonstrate adult stem cell plasticity, as it is possible that progenitors for each of these tissues reside in the bone marrow (Avital et al., 2001; Fiegel et al., 2003; Krause et al., 2001; Lagasse et al., 2000). Krause et al. (2001) demonstrated that 'homed' $CD34^+$ $Sca1^+$ mouse bone marrow cells were capable of differentiation into epithelium of liver and lung along with the hematopoietic stem cells. However, in a similar single cell transplantation study, Wagers et al., 2002 found that transplantation of fresh sorted $cKit^+Thy1^+Lin^-Sca1^+$ cells gave rise to considerably less 'lineage-switch' (only X hepatocytes). Whether the different phenotype of the transplanted cells plays a role in these differing results is not known. However, none of these studies suggesting bone marrow to endoderm differentiation proved that the bone marrow-derived endodermal epithelial cells were functional. However, in a landmark study by Lagasse et al. (2000) it was demonstrated that bone marrow-derived cells can successfully rescue mice lacking the enzyme fumarylacetoacetate hydrolase (FAH), a key enzyme in the tyrosine metabolism pathway. Mice lacking this enzyme develop acute liver failure. This results from the accumulation of the upstream metabolite, fumarylacetoacetate (FAA), which is broken down into toxic metabolites through other pathways (Grompe et al., 1993). FAA production is prevented by the drug NTBC, which acts on an enzyme upstream of FAA (Grompe et al., 1995). Therefore, liver failure can be controlled through the administration of NTBC. Lagasse *et al.* (2000) showed that FAH mutant animals transplanted with normal BM or normal BM enriched for HSC could be taken off NTBC. These animals quickly developed acute liver failure but a majority of animals recovered and, when examined more closely, had evidence of donor-derived hepatocytes, whereas animals that received no transplant died. This study demonstrated functional hepatic repopulation derived from donor HSC. One criticism that can be leveled at this study is that a minimum of 50 purified HSC was necessary for animal survival and hepatic repopulation. Therefore the possibility remains that one cell in this fraction was capable of differentiating into hepatocytes while the other cells were capable of reconstituting the hematopoietic system.

A third set of studies suggests bone marrow differentiation to neuroectoderm. Mezey *et al.* (2000) showed that prior to transplantation, bone marrow does not contain cells expressing neuronal markers. However, several months after transplantation, donor-derived cells that expressed neuronal specific markers were found throughout the brain. Kabos *et al.* (2002) suggested that neural stem cells could be derived from whole BM *in vitro*. As has plagued many such studies, the authors of these studies rely solely on morphology and immunofluorescence to illustrate their point, but do not address a fundamental criterion of stem cells, functionality. Moreover, none of these studies demonstrate that a single bone marrow cell gave rise to hematopoietic and neural progenies.

One other noteworthy study in which single cells were shown to give rise to two different differentiated cells is the study by Grant *et al.* (2002) They transplanted single HSC into lethally irradiated murine recipients, and showed that hematopoietic stem cells and retinal endothelial cells were derived from a single common progenitor.

3.6 Plasticity of mesenchymal stem cells

First reported in 1976, Fridenshtein showed that bone marrow contains mesenchymal stem cells in addition to hematopoietic stem cells. MSCs were initially isolated as the plastic adherent fraction of bone marrow that can be cultured *ex vivo* for several passages, and differentiate into limb bud mesodermal cell lineages such as adipocytes, chondroblasts, fibroblasts, osteoblasts, and skeletal myoblasts both *in vitro* and *in vivo* (Fridenshtein, 1982; Gronthos and Simmons, 1996; Haynesworth *et al.*, 1992; Prockop, 1997). More recent studies have suggested that MSCs may give rise to cells outside the limb bud mesodermal lineages including endoderm, endothelium, and neuroectoderm (Black and Woodbury, 2001; Deng *et al.*, 2001; Woodbury *et al.*, 2000, 2002). A study by Kopen *et al.* suggests that MSCs are capable of becoming neurons and glia upon transplantation (Kopen *et al.*, 1999). This study was unable to examine the function of such cells or, more importantly, the mechanism by which these neurons formed. Kim *et al.* (2002) and Woodbury *et al.* (2002) tried to address this in their *in vitro* studies. Many different culture conditions were examined, and differentiation was defined as presence of mRNAs for neural genes and detection of proteins found in neural cells. However, no functional analysis of the presumed neuroectodermal cells was performed. These studies documented the somewhat surprising observation that activation and loss of neuronal gene and protein expression occurs within a matter of hours, not days or weeks, as one would expect from studies initiated with NSC and developmental biology. In another study, mouse recipients of MSCs isolated from a EGFP-transgenic mouse were found to have a large number of EGFP+ cells in the brain. FACS analysis demonstrated that over 20% of the EGFP+ cells were CD45 and CD11b negative, suggesting that these cells were a non-hematopoietic population. These cells expressed the astrocyte marker GFAP, or the neuron-specific markers, NeuN or Neurofilament-H, suggesting that MSCs may adopt a neuroectodermal phenotype *in vivo* (Brazelton *et al.*, 2000). Again, no functional analysis was carried out, and in none of the studies was it shown that the same cell that gives rise to limb-bud mesodermal cell types also give rise to cells with neural markers.

Makino and colleagues (Makino *et al.*, 1999) showed that mouse MSCs can produce spontaneously beating cardiomyocytes. However this work has not been reproduced *in vivo* as Pereira found no evidence of MSC-derived cells in the hearts of mice 2–3 months after intraperitoneal injection (Pereira *et al.*, 1998).

3.7 Mulitpotent adult progenitor cells

The initial studies that led to the discovery of Multipotent Adult Progenitor Cells or MAPC began with the attempt to isolate mesenchymal stem cells (Reyes and Verfaillie, 2001). Using modified conditions, it was shown that cells with unexpected self-renewal ability and lineage differentiation ability could be isolated from postnatal human bone marrow (Jiang *et al.*, 2002a; Reyes and Verfaillie, 2001; Reyes *et al.*, 2001). At the clonal level, shown by retroviral marking studies, MAPC appeared to differentiate into the classical mesenchymal lineages; adipocytes, chondroblasts, osteoblasts, and skeletal myoblasts and the non-mesenchymal mesodermal lineages, endothelium and hepatocyte-like cells. Using similar conditions with the addition of leukemia inhibitory factor, MAPC have been isolated from both mouse and rat (Jiang *et al.*, 2002a) (*Color Plate 1*).

Human MAPC have maintained telomere length for over 80 doublings while mouse and rat MAPC have maintained telomere length for over 120 doublings (Jiang *et al.*, 2002a; Reyes *et al.*, 2001). All three cells express telomerase and express the transcription factors OCT-4 and Rex-1 similar to ES cells. MAPCs are CD34, c-kit, CD45, MHC1, and MHC2 negative, Flk1, Sca1, CD44, and Thy-1 low, and CD13 positive. Mouse MAPC express SSEA1, a marker normally associated with ES cells. Like human MAPC, mouse MAPC have been shown to differentiate at the clonal level into cells of all three germ layers; neurons (ectoderm), hepatocytes (endoderm), and endothelium (mesoderm). Most surprising, when single MAPC were injected into a mouse blastocyst, up to one-third of animals born were chimeric with chimerism in two animals as high as 45%. When tissues of these animals were examined, MAPC contributed to all examined tissues which included brain, cardiac and skeletal muscle, kidney, small intestine, liver, spleen and blood as demonstrated by donor marker expression through immunofluorescence and quantitative real time PCR.

In another study, MAPCs were injected intravenously into 6–8 week old NOD/SCID mice. MAPC engrafted and differentiated into several tissues including small intestine, liver, lung, and bone marrow. Furthermore, tissue engraftment and differentiation occurred in a manner consistent with our understanding of organ formation. For example, in the epithelium of the small intestine, one half of a villus from the crypt to the tip was derived from donor MAPC, consistent with the notion that MAPC may have engrafted in the stem cell compartment of the intestinal epithelium and contributed subsequently to small intestine epithelial turnover. In contrast to the blastocyst injection, MAPC did not engraft and differentiate into skeletal or cardiac muscle, brain tissue or skin in adult immuno-incompetent mice. Engraftment and differentiation occurred in tissues that generally have higher rates of cell turnover and have little barrier to engraftment. Interestingly we did not see engraftment in the body's most proliferative tissue, skin epithelium, despite contribution of MAPC to skin in the chimeric animals. MAPC did engraft and repopulate the bone marrow, which was shown to persist beyond the initial transplant into secondary transplants demonstrating their self-renewal capability. However, in adult recipients of MAPCs, CD3+ T-cells were not found despite repeated examinations (though they were found in the chimeric animals generated from the blastocyst experiment).

More recently, Jiang *et al.* (2002b) demonstrated that MAPC could not only be isolated from postnatal bone marrow but also from brain and skeletal muscle. Microarray analysis of these three populations of cells (bone marrow-, brain-, and skeletal muscle-derived MAPCs) shows a significant similarity or 'MAPC' signature in the transcriptional programs of MAPCs despite their differing tissue origins.

When taken together, question arises as to what are MAPC: cells that require culture conditions similar to ES cells, express at least some of the genetic markers of ES cells (Oct-4, Rex1, SSEA-1), have extensive proliferation and clonal multi-lineage differentiation potential, contribute to all organs when injected in a blastocyst, and engraft and differentiate into tissue specific cells in response to organ specific cues. Later in this chapter we will address the different possibilities and explanations for stem cell plasticity, which apply to MAPC and the field in general.

3.8 Plasticity of skeletal muscle cells

Gussoni and colleagues showed that muscle SP cells could reconstitute the hematopoietic compartment (Gussoni *et al.*, 1999). Their findings were corroborated by Jackson *et al.* (1999) who reported that transplantation of side population muscle cells could give rise to hematopoietic cells following transplantation into lethally irradiated hosts and could compete with bone marrow-derived HSCs. Not answered in both studies was the question of whether the same cell that gave rise to HSCs also gave rise to myoblasts and thus skeletal muscle cells. However, in subsequent studies, Kawada and Ogawa (2001) and McKinney-Freeman *et al.* (2002) demonstrated that this plasticity may not be caused by transdifferentiation, but by HSCs of bone marrow origin that are resident in substantial numbers within the skeletal muscle compartment.

3.9 Plasticity of neural cells

Bjornson et al. (1999) showed that murine neural stem cells (NSCs) could differentiate into hematopoietic cells *in vivo* and was found in 48 of 60 transplanted mice. These results were confirmed for human NSCs by Shih et al. (2001) who saw engraftment and contribution in 100% of animals. In both studies, isolated NSCs were cultured *ex vivo* and expanded for several population doublings, thereby fulfilling the first requirement of stem cells, i.e., self-renewal. While both studies used cells derived from neurospheres that they believed were derived from a single neural stem cell, neither report documents this with retroviral marking or single cell sorting to illustrate this point clearly. Without such evidence, it is not clear whether the cells used in the experiment were truly derived from a single cell. Furthermore, in both studies animals received sublethal doses of irradiation, thereby decreasing one's ability to discern whether the hematopoietic stem cells derived from the neural stem cells *in vivo* are truly functional. A more recent study casts doubt on the ability of NSC to contribute to hematopoiesis when transplanted in postnatal animals. Morsehead et al. (2002) examined the ability of early passage and late passage NSCs to contribute to hematopoiesis. In both instances no detectable contribution was found. However, changes in adherence, altered proliferation kinetics, loss of growth factor dependence, and changes in gene expression were noted in comparing late passage with early passage NSCs. The authors suggest that hematopoietic contribution must be rare and may be due to genetic and/or epigenetic changes believed to cause the changes found in the longer passage cultures.

To assess further the differentiation potential of adult NSCs *in vivo*, Clarke et al. (2000) examined their contribution to other tissues after inoculation into the chick embryo or the mouse blastocyst. In both cases, it was shown that NSCs contribute to multiple tissues but most surprisingly did not contribute to the hematopoietic system. Like Bjornson et al. (1999) and Shih et al. (2001), Clarke et al. (2000) used *ex vivo* cultured cells that had undergone several cell doublings but also did not fully prove that such cells were derived from a single progenitor. Furthermore, none of the chimeric mice partially derived from neural stem cells were born, precluding the ability to test whether donor-derived cells were functional *in vivo*.

3.10 Mechanisms of plasticity

There are several possible explanations for the perceived stem cell plasticity: (1) multiple tissue-specific stem cells are present in each organ; (2) stem cells are capable of dedifferentiation and re-differentiation into another cell type; (3) fusion of the donor cell with the host cells; (4) multipotent or pluripotent stem cells actually exist in adults. Present data support all four explanations with examples in nature (*Figure 3.1*).

The first mechanism, namely that stem cells for a given tissue may reside in unrelated tissues, has now been demonstrated in several studies. It has been long established that HSC exit the bone marrow and either specifically home to or are resident in various different organs. This appears to be the case for skeletal muscle

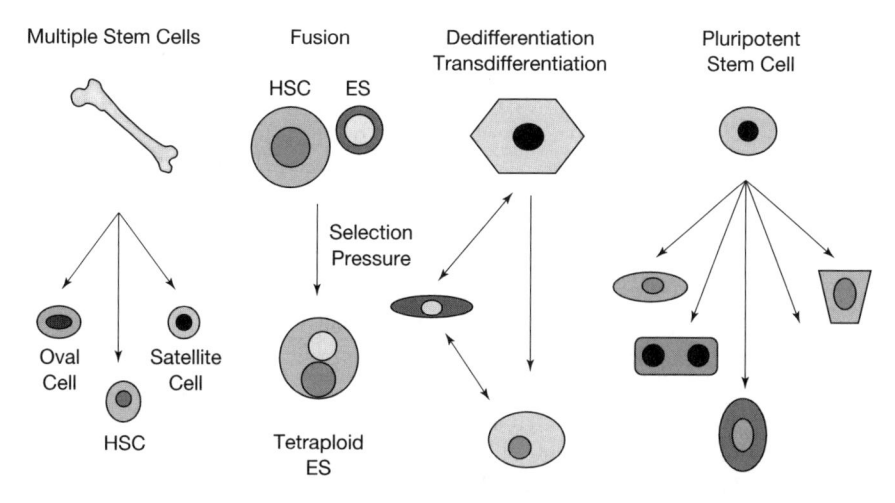

Figure 3.1: Proposed mechanisms of plasticity. Plasticity as a biological phenomenon has been widely reported over the past few years. Four proposed mechanisms may underlie such reports: (1) multiple tissue-specific stem cells are present in each organ; (2) fusion of the donor cell with the host cells; (3) stem cells are capable of dedifferentiation and re-differentiation into another cell type; (4) multipotent or pluripotent stem cells actually exist in adults.

as it has been shown that HSC can be isolated from skeletal muscle. Several experiments have shown that sex-mismatched bone marrow transplants in humans or rodents result in the appearance of a small number of donor-derived cells with the phenotype of hepatocytes suggesting transdifferentiation of HSC into hepatocytes. However at least two studies showed that liver progenitors may be present in the bone marrow. Avital *et al.* (2001) showed that in the rat, a population of hepatocyte progenitors characterized as Thy-1 positive and Beta2-microglobulin negative may exist. These cells can be induced to express mature hepatocyte markers and produce urea when cultured *in vitro*. Likewise, Fiegel *et al.* (2003) showed that in cultures of human bone marrow, cells with hepatocyte markers can be found, even though they did not examine functional activity of such hepatocyte-like cells. Therefore, these studies suggest the possibility that the bone marrow contains hepatic progenitors. Consequently when transplanted these may be the cells that contribute to the host liver. In both instances, the apparent lineage switch would then not be caused by transdifferentiation of a single stem cell but rather caused by the presence of multiple different types of stem cells, giving the perception of plasticity. These studies and the fact that most studies suggesting plasticity have not proven single cell derivation of both hematopoietic cells and a second lineage has created skepticism. Many studies tried to address clonal origin of differentiated progeny using cloning rings. This approach is not without pitfalls, as cells are very motile in culture and therefore one cannot fully demonstrate that single cells give rise to multiple lineages. Other studies have relied on better and more reliable methods such as single cell sorting, or retroviral marking strategies to demonstrate single cell derivation of multiple lineage differentiation.

A second possible explanation for plasticity is that fusion of the transplanted cells with a host cell of a different lineage may occur. This would lead to the transfer of the cell contents, including proteins, DNA, and RNA from the transplanted cells to the host cell. This idea is decades old and has been studied since the early 20th century, and was then known as the heterokaryon technique. For example, myoblast fusion with fibroblasts induces expression of muscle proteins in the fibroblasts. This indicates that the cell cytoplasm contains factors, which induce specification, and differentiation, which is not surprising. The cloning of 'Dolly' and 'Copy-cat' is a clear example of this. Nuclear cloning involves the transfer of a nucleus from a somatic cell into an oocyte. It is widely known that some factors in the cytoplasm are involved in the dedifferentiation of the somatic nucleus though the specific factors and mechanisms are still not known. Two independent studies clearly demonstrated that though rare ($\sim$1/100 000–1/1 000 000), co-culture of adult cells with embryonic stem cells leads to cell fusion (Terada *et al.*, 2002; Ying *et al.*, 2002). When co-cultured with ES cells, bone marrow cells or NSC acquired ES cell-like characteristics and appeared to have transdifferentiated. On closer examination however, karyotyping and cell marking clearly demonstrated fusion. In both studies this phenomenon required strong selection pressure. These *in vitro* studies suggest that apparent lineage switching may be caused by donor–host cell fusion and that this phenomenon, which is likely inefficient, might explain rare events of transdifferentiation. Correlating with the prevalence of studies claiming transdifferentiation phenomenon would likely occur more frequently when strong selection pressure exists *in vivo*, such as in acute organ failure or tissue death. One might also speculate that this phenomenon may also be more likely in organs that normally exhibit polyploidy such as muscle, hepatocytes, and cerebellar Purkinje cells.

That this may occur *in vivo* was recently demonstrated by Wang *et al.* (2003). They showed that the rescue of FAH mice with bone marrow-derived cells may not be the result of the transdifferentiation of HSC to hepatocytes but the result of fusion of HSC or their hematopoietic progeny with hepatocytes. The transfer of genetic material from the normal HSC to the hepatocyte with the genetic defect resulted in hepatocytes that were able to produce the missing enzyme and consequently rescue the mice. The importance of these studies cannot be underscored though these results cannot be generalized to all studies suggesting plasticity and their interpretation should be limited.

A third explanation is that cells can undergo dedifferentiation or transdifferentiation. A somatic cell from many mammalian species can be reprogrammed to dedifferentiate into pluripotent cells, as demonstrated by cloning experiments. There are many other examples of this phenomenon in nature such as that which occurs during regeneration in *Drosophila*, planaria, and newts. For example, experiments with newts have shown that during limb regeneration, postmitotic cells undergo 'dedifferentiation' at the site of injury, and form a cap-like structure known as a blastema (Lo *et al.*, 1993; Simon *et al.*, 1995). These 'dedifferentiated' cells now can enter mitosis and are capable of reforming the limb. Extracts from blastema cells were capable of 'dedifferentiating' fully formed newt myotubes and murine myotubes (McGann *et al.*, 2001). When the extract was removed, these 'dedifferentiated' cells expressed markers consistent with cells undergoing adipogenesis, chondrogenesis, osteogenesis, and myogenesis. It has been suggested that the nuclear protein msx1 may play a role in blastema formation and for the observed dedifferentiation process (Simon *et al.*, 1995). In developing mouse limbs, the expression or lack of expression of msx1 demarcates the boundary between differentiated and undifferentiated cells (Simon *et al.*, 1995). Over-expression of msx1 in myotubes derived from C2C12 (a mouse skeletal muscle cell line) resulted in loss of expression of skeletal muscle markers and transcription factors and the regression of myotubes into smaller multinucleated myotubules or mononucleated myoblasts (Odelberg *et al.*, 2000). It has not been determined whether these pathways examined *in vitro* or *in vivo* in newts and mice plays a role in higher mammals. It is possible that this process is similar to the cloning mechanism described earlier. In both cases differentiated cells adopt a more immature and less differentiated state. In nuclear cloning this has been shown to result from epigenetic changes, i.e., decreased DNA methylation and histone acetylation (Rideout *et al.*, 2001). A similar process may partially explain the dedifferentiation and re-differentiation of myotubules. The exact mechanism involved in dedifferentiation and re-differentiation is not yet known but closer examination of these pathways may reveal a key role for these processes in stem cell plasticity.

The fourth explanation is the persistence of truly multipotent or pluripotent stem cells into postnatal life. Pluripotent stem cells are exemplified by embryonic stem cells. Embryonic stem cells have been characterized based on cell surface markers including stage specific embryonic antigens (SSEA 1–4), expression of the transcription factors Oct4 and Rex1. Oct4 is a transcription factor expressed in the pre-gastrulation embryo, the inner cell mass, germ cells, and in embryonic carcinoma cells. When ES cells differentiate, Oct4 is downregulated. Oct4 expression is required for maintenance of the undifferentiated phenotype of ES cells and plays a major role in early embryogenesis. Another property of ES cells is the expression of telomerase, which prevents shortening of telomeres and thus allows ES cells to undergo virtually unlimited cell divisions. As ES cells are the quintessential pluripotent stem cells, it is notable that many characteristics of these cells are shared by MAPCs and adult marrow-derived stem cells.

3.11 Potential uses of adult stem cells

Often in the controversy of adult stem cells and stem cell plasticity, the potential opportunities for therapeutic purposes is lost. Millions of people suffer from diseases that could benefit from a cellular therapy. Tissues and organs that actively regenerate themselves from stem cells have been the best targets for cellular therapy. This maxim explains the present successes in cellular therapy. For example, adult stem cells such as hematopoietic stem cells have been used clinically to reestablish the hematopoietic system following radiation and/or chemotherapy for over 30 years. More recently, keratinocyte stem cells are being used as a source of artificial skin and the use of corneal and neural stem cells are now being evaluated in a number of studies. This principle underlies both the success and failure of cellular therapy. Diseases resulting from a intrinsic stem cell defect (i.e., the problem lies in the stem cells themselves) would be an ideal target for cellular therapy. This explains the success of bone marrow transplant in diseases resulting in bone marrow failure such as Fanconi anemia. In contrast, diseases that result from an extrinsic stem cell failure (i.e., the problem lies outside the stem cells) or results from remodeling of the stem cell environment may be a more intractable problem. For instance, transplantation of hepatic stem cells in the setting of hepatitis C-mediated liver cirrhosis may have little positive impact.

Despite these concerns it is important to examine the use of stem cells and mature cells in cellular therapy. Our appreciation of the possible capabilities of adult cells is not being realized. As early as the 1980s several groups reported on whether bone marrow or peripheral blood cells contribute to tissues outside the hematopoietic system. Results from these and present studies were inconclusive and mixed. For instance, many studies strongly argue that stromal cells were of donor origin while others asserted the opposite. However, more recent reports such as that by Horwitz *et al.* indicate that bone marrow transplantation may ameliorate symptoms of osteogenesis imperfecta such as bone brittleness and fractures. It is thought that this results from MSC engraftment and donor-derived osteoblast formation, although characterization of the originating cell or the precise mechanism remains unclear .

The notion that HSC may for instance engraft and differentiate into satellite and skeletal muscle cells is exciting as it opens the possibility for treatment of diseases such as muscular dystrophy. Clinical usefulness will however require that the degree of 'plasticity' is higher than the commonly reported 0.1–1%. Nevertheless, if the current published studies are correct, they would serve as the proof of principle experiment for a technology and method still in its infancy. Initial experiments with bone marrow transplantation in animals yielded poor results. Only after better appreciation for the overall mechanisms was the success rate increased. Over the next few years it will be important that the field address these fundamental questions and criticisms in order to gain a better understanding of the mechanisms that underlie the 'plasticity' of stem cells so that these mechanisms can be applied to clinical treatment.

References

Asahara T, Murohara T, Sullivan A, Silver M, van der Zee R, Li T, Witzenbichler B, Schatteman G, Isner JM (1997) Isolation of putative progenitor endothelial cells for angiogenesis. *Science* 275, 964–967.

Avital I, Inderbitzin D, Aoki T, Tyan DB, Cohen AH, Ferraresso C, Rozga J, Arnaout WS, Demetriou AA (2001) Isolation, characterization, and transplantation of bone marrow-derived hepatocyte stem cells. *Biochem. Biophys. Res. Commun.* 288, 156–164.

Bhatia M, Wang JC, Kapp U, Bonnet D, Dick JE (1997) Purification of primitive human hematopoietic cells capable of repopulating immune-deficient mice. *Proc. Natl Acad. Sci. USA* 94, 5320–5325.

Bittner RE, Schofer C, Weipoltshammer K, Ivanova S, Streubel B, Hauser E,

Freilinger M, Hoger H, Elbe-Burger A, Wachtler F (1999) Recruitment of bone-marrow-derived cells by skeletal and cardiac muscle in adult dystrophic mdx mice. *Anat. Embryol.* (Berl) **199**, 391–396.

Bjornson CR, Rietze RL, Reynolds BA, Magli MC, Vescovi AL (1999) Turning brain into blood: a hematopoietic fate adopted by adult neural stem cells *in vivo*. *Science* **283**, 534–537.

Black IB, Woodbury D (2001) Adult rat and human bone marrow stromal stem cells differentiate into neurons. *Blood Cells Mol. Dis.* **27**, 632–636.

Brazelton TR, Rossi FM, Keshet GI, Blau HM (2000) From marrow to brain: expression of neuronal phenotypes in adult mice. *Science* **290**, 1775–1779.

Clarke DL, Johansson CB, Wilbertz J, Veress B, Nilsson E, Karlstrom H, Lendahl U, Frisen J (2000) Generalized potential of adult neural stem cells. *Science* **288**, 1660–1663.

Cotran RS (1999a) *Pathologic Basis of Disease.* W.B. Saunders Company, Philadelphia, pp. 31–38, 266–268, 466.

Cotran RS (1999b) *Pathologic Basis of Disease.* W.B. Saunders Company, Philadelphia, pp. 710–712.

Cotran RS (1999c) *Pathologic Basis of Disease.* W.B. Saunders Company, Philadelphia, pp. 781–782.

Crosby JR, Kaminski WE, Schatteman G, Martin PJ, Raines EW, Seifert RA, Bowen-Pope DF (2000) Endothelial cells of hematopoietic origin make a significant contribution to adult blood vessel formation. *Circ. Res.* **87**, 728–730.

Daniels JT, Dart JK, Tuft SJ, Khaw PT (2001) Corneal stem cells in review. *Wound Repair Regen.* **9**, 483–494.

Deng W, Obrocka M, Fischer I, Prockop DJ (2001) *In vitro* differentiation of human marrow stromal cells into early progenitors of neural cells by conditions that increase intracellular cyclic AMP. *Biochem. Biophys. Res. Commun.* **282**, 148–152.

Evans MJ, Kaufman MH (1981) Establishment in culture of pluripotential cells from mouse embryos. *Nature* **292**, 154–156.

Ferrari G, Cusella-De Angelis G, Coletta M, Paolucci E, Stornaiuolo A, Cossu G, Mavilio F (1998) Muscle regeneration by bone marrow-derived myogenic progenitors. *Science* **279**, 1528–1530.

Fiegel HC, Lioznov MV, Cortes-Dericks L, Lange C, Kluth D, Fehse B, Zander AR (2003) Liver-specific gene expression in cultured human hematopoietic stem cells. *Stem Cells* **21**, 98–104.

Fridenshtein A (1982) [Stromal bone marrow cells and the hematopoietic microenvironment]. *Arkh. Patol.* **44**, 3–11.

Gage FH (2000) Mammalian neural stem cells. *Science* **287**, 1433–1438.

Grant MB, May WS, Caballero S, Brown GA, Guthrie SM, Mames RN *et al.* (2002) Adult hematopoietic stem cells provide functional hemangioblast activity during retinal neovascularization. *Nat. Med.* **8**, 607–612.

Grompe M, al-Dhalimy M, Finegold M, Ou CN, Burlingame T, Kennaway NG, Soriano P (1993) Loss of fumarylacetoacetate hydrolase is responsible for the neonatal hepatic dysfunction phenotype of lethal albino mice. *Genes Dev.* **7**, 2298–2307.

Grompe M, Lindstedt S, al-Dhalimy M, Kennaway NG, Papaconstantinou J, Torres-Ramos CA, Ou CN, Finegold M (1995) Pharmacological correction of neonatal lethal hepatic dysfunction in a murine model of hereditary tyrosinaemia type I. *Nat. Genet.* **10**, 453–460.

Gronthos S, Simmons PJ (1996) The biology and application of human bone marrow stromal cell precursors. *J. Hematother.* **5**, 15–23.

Gussoni E, Soneoka Y, Strickland CD, Buzney EA, Khan MK, Flint AF, Kunkel LM, Mulligan RC (1999) Dystrophin expression in the mdx mouse restored by stem cell transplantation. *Nature* **401**, 390–394.

Hawley RG, Sobieski DA (2002) Somatic stem cell plasticity: to be or not to be. *Stem Cells* **20**, 195–197.

Haynesworth SE, Baber MA, Caplan AI (1992) Cell surface antigens on human marrow-derived mesenchymal cells are detected by monoclonal antibodies. *Bone* **13**, 69–80.

Holden C, Vogel G. Stem cells (2002) Plasticity: time for a reappraisal? *Science* **296**, 2126–2129.

Horwitz EM, Gordon PL, Koo WK, Marx JC, Neel MD, McNall RY, Muul L, Hofmann T (2002) Isolated allogeneic bone marrow-derived mesenchymal cells engraft and stimulate growth in children with osteogenesis imperfecta: Implications for cell therapy of bone. *Proc. Natl Acad. Sci. USA* **99**(13), 8932–8937.

Horwitz EM, Prockop DJ, Fitzpatrick LA, Koo WW, Gordon PL, Neel M et al. (1999) Transplantability and therapeutic effects of bone marrow-derived mesenchymal cells in children with osteogenesis imperfecta. *Nat. Med.* **5**, 309–313.

Horwitz EM, Prockop DJ, Gordon PL, Koo WW, Fitzpatrick LA, Neel MD, McCarville ME, Orchard PJ, Pyeritz RE, Brenner MK (2001). Clinical responses to bone marrow transplantation in children with severe osteogenesis imperfecta. *Blood* **97**, 1227–1231.

Jackson KA, Mi T, Goodell MA (1999) Hematopoietic potential of stem cells isolated from murine skeletal muscle. *Proc. Natl Acad. Sci. USA* **96**, 14482–14486.

Jiang Y, Jahagirdar BN, Reinhardt RL, Schwartz RE, Keene CD, Ortiz-Gonzalez XR et al. (2002a) Pluripotency of mesenchymal stem cells derived from adult marrow. *Nature* **418**, 41–49.

Jiang Y, Vaessen B, Lenvik T, Blackstad M, Reyes M, Verfaillie CM (2002b) Multipotent progenitor cells can be isolated from postnatal murine bone marrow, muscle, and brain. *Exp. Hematol.* **30**, 896–904.

Kabos P, Ehtesham M, Kabosova A, Black KL, Yu JS (2002) Generation of neural progenitor cells from whole adult bone marrow. *Exp. Neurol.* **178**, 288–293.

Kawada H, Ogawa M (2001) Bone marrow origin of hematopoietic progenitors and stem cells in murine muscle. *Blood* **98**, 2008–2013.

Kim BJ, Seo JH, Bubien JK, Oh YS (2002) Differentiation of adult bone marrow stem cells into neuroprogenitor cells *in vitro*. *Neuroreport* **13**, 1185–1188.

Kopen GC, Prockop DJ, Phinney DG (1999) Marrow stromal cells migrate throughout forebrain and cerebellum, and they differentiate into astrocytes after injection into neonatal mouse brains. *Proc. Natl Acad. Sci. USA* **96**, 10711–10716.

Korbling M, Katz RL, Khanna A, Ruifrok AC, Rondon G, Albitar M, Champlin RE, Estrov Z (2002) Hepatocytes and epithelial cells of donor origin in recipients of peripheral-blood stem cells. *N. Engl. J. Med.* **346**, 738–746.

Krause DS, Theise ND, Collector MI, Henegariu O, Hwang S, Gardner R, Neutzel S, Sharkis SJ (2001) Multi-organ, multi-lineage engraftment by a single bone marrow-derived stem cell. *Cell* **105**, 369–377.

LaBarge MA, Blau HM (2002) Biological progression from adult bone marrow to mononucleate muscle stem cell to multinucleate muscle fiber in response to injury. *Cell* 111, 589–601.

Lagasse E, Connors H, Al-Dhalimy M, Reitsma M, Dohse M, Osborne L, Wang X, Finegold M, Weissman IL, Grompe M (2000) Purified hematopoietic stem cells can differentiate into hepatocytes *in vivo*. *Nat. Med.* 6, 1229–1234.

Lin Y, Weisdorf DJ, Solovey A, Hebbel RP (2000) Origins of circulating endothelial cells and endothelial outgrowth from blood. *J. Clin. Invest.* 105, 71–77.

Lo DC, Allen F, Brockes JP (1993) Reversal of muscle differentiation during urodele limb regeneration. *Proc. Natl Acad. Sci. USA* 90, 7230–7234.

Makino S, Fukuda K, Miyoshi S, Konishi F, Kodama H, Pan J et al. (1999) Cardiomyocytes can be generated from marrow stromal cells *in vitro*. *J. Clin. Invest.* 103, 697–705.

Martin GR (1981) Isolation of a pluripotent cell line from early mouse embryos cultured in medium conditioned by teratocarcinoma stem cells. *Proc. Natl Acad. Sci. USA* 78, 7634–7638.

McGann CJ, Odelberg SJ, Keating MT (2001) Mammalian myotube dedifferentiation induced by newt regeneration extract. *Proc. Natl Acad. Sci. USA* 98, 13699–13704.

McKinney-Freeman SL, Jackson KA, Camargo FD, Ferrari G, Mavilio F, Goodell MA (2002) Muscle-derived hematopoietic stem cells are hematopoietic in origin. *Proc. Natl Acad. Sci. USA* 99, 1341–1346.

Mezey E, Chandross KJ, Harta G, Maki RA, McKercher SR (2000) Turning blood into brain: cells bearing neuronal antigens generated *in vivo* from bone marrow. *Science* 290, 1779–1782.

Morshead CM, Benveniste P, Iscove NN, van der Kooy D (2002) Hematopoietic competence is a rare property of neural stem cells that may depend on genetic and epigenetic alterations. *Nat. Med.* 8, 268–273.

Odelberg SJ, Kollhoff A, Keating MT (2000) Dedifferentiation of mammalian myotubes induced by msx1. *Cell* 103, 1099–1109.

Pereira RF, O'Hara MD, Laptev AV, Halford KW, Pollard MD, Class R, Simon D, Livezey K, Prockop DJ (1998) Marrow stromal cells as a source of progenitor cells for nonhematopoietic tissues in transgenic mice with a phenotype of osteogenesis imperfecta. *Proc. Natl Acad. Sci. USA* 95, 1142–1147.

Pittenger MF, Mackay AM, Beck SC, Jaiswal RK, Douglas R, Mosca JD, Moorman MA, Simonetti DW, Craig S, Marshak DR (1999) Multilineage potential of adult human mesenchymal stem cells. *Science* 284, 143–147.

Prockop DJ (1997) Marrow stromal cells as stem cells for nonhematopoietic tissues. *Science* 276, 71–74.

Rafii S, Shapiro F, Rimarachin J, Nachman RL, Ferris B, Weksler B, Moore MA, Asch AS (1994) Isolation and characterization of human bone marrow microvascular endothelial cells: hematopoietic progenitor cell adhesion. *Blood* 84, 10–19.

Reyes M, Verfaillie CM (2001) Characterization of multipotent adult progenitor cells, a subpopulation of mesenchymal stem cells. *Ann. NY Acad. Sci.* 938, 231–235.

Reyes, M, Lund, T, Lenvik, T, Aguiar, D, Koodie, L, Verfaillie, C.M (2001) Purification and *ex vivo* expansion of postnatal human marrow mesodermal progenitor cells. *Blood* 98, 2615–2625.

Rideout WM 3rd, Eggan K, Jaenisch R (2001) Nuclear cloning and epigenetic reprogramming of the genome. *Science* **293**, 1093–1098

Shih CC, Weng Y, Mamelak A, LeBon T, Hu MC, Forman SJ (2001) Identification of a candidate human neurohematopoietic stem-cell population. *Blood* **98**, 2412–2422.

Simon HG, Nelson C, Goff D, Laufer E, Morgan BA, Tabin C (1995) Differential expression of myogenic regulatory genes and Msx-1 during dedifferentiation and redifferentiation of regenerating amphibian limbs. *Dev. Dyn.* **202**, 1–12.

Spangrude GJ, Heimfeld S, Weissman IL (1988) Purification and characterization of mouse hematopoietic stem cells. *Science* **241**, 58–62.

Terada N, Hamazaki T, Oka M, Hoki M, Mastalerz DM, Nakano Y, Meyer EM, Morel L, Petersen BE, Scott EW (2002) Bone marrow cells adopt the phenotype of other cells by spontaneous cell fusion. *Nature* **416**, 542–545.

Theise ND, Badve S, Saxena R, Henegariu O, Sell S, Crawford JM, Krause DS (2000a) Derivation of hepatocytes from bone marrow cells in mice after radiation-induced myeloablation. *Hepatology* **31**, 235–240.

Theise ND, Nimmakayalu M, Gardner R, Illei PB, Morgan G, Teperman L, Henegariu O, Krause DS (2000b) Liver from bone marrow in humans. *Hepatology* **32**, 11–16.

Thomson JA, Kalishman J, Golos TG (1995) Durning M, Harris CP, Becker RA, Hearn JP. Isolation of a primate embryonic stem cell line. *Proc. Natl Acad. Sci. USA* **92**, 7844–7848.

Thomson JA, Itskovitz-Eldor J, Shapiro SS, Waknitz MA, Swiergiel JJ, Marshall VS, Jones JM (1998) Embryonic stem cell lines derived from human blastocysts. *Science* **282**, 1145–1147.

Verfaillie CM, Pera MF, Lansdorp PM (2002) Stem cells: hype and reality. *Hematology (Am. Soc. Hematol. Educ. Program)*, **2002**, 369–391.

Wagers AJ, Sherwood RI, Christensen JL, Weissman IL (2002) Little evidence for developmental plasticity of adult hematopoietic stem cells. *Science* **297**, 2256–2259.

Wang X, Al-Dhalimy M, Lagasse E, Finegold M, Grompe M (2001) Liver repopulation and correction of metabolic liver disease by transplanted adult mouse pancreatic cells. *Am. J. Pathol.* **158**, 571–579.

Wang X, Montini E, Al-Dhalimy M, Lagasse E, Finegold M, Grompe M (2002) Kinetics of liver repopulation after bone marrow transplantation. *Am. J. Pathol.* **161**, 565–574.

Wang X, Willenbring H, Akkari Y, Torimaru Y, Foster M, Al-Dhalimy M, Lagasse E, Feingold M, Olson S, Grompe M (2003) Cell fusion is the principal source of bone marrow-derived hepatocytes. *Nature* **422**, 897–901.

Woodbury D, Schwarz EJ, Prockop DJ, Black IB (2000) Adult rat and human bone marrow stromal cells differentiate into neurons. *J. Neurosci. Res.* **61**, 364–370.

Woodbury D, Reynolds K, Black IB (2002) Adult bone marrow stromal stem cells express germline, ectodermal, endodermal, and mesodermal genes prior to neurogenesis. *J. Neurosci. Res.* **69**, 908–917.

Ying QL, Nichols J, Evans EP, Smith AG (2002) Changing potency by spontaneous fusion. *Nature* **416**, 545–548.

4. Human and murine embryonic stem cell lines: windows to early mammalian development

Jon S. Odorico and Su-Chun Zhang

4.1 Introduction

Mice, chickens, and zebrafish have long been considered model organisms for the study of vertebrate development. Studies of these organisms have provided many insights into the molecular mechanisms underlying normal development, and are beginning to suggest potential pathophysiological mechanisms of some important developmental/congenital abnormalities in humans. However, in our ultimate quest to understand the mechanisms of human development with the goal of preventing and/or treating developmental defects in humans, these studies fall short. Important questions arise, such as how do the findings of these studies reflect mechanisms and events that occur specifically in human development, and what unique molecular events dictate formation of morphologically and functionally distinct organisms?

The direct study of human embryos is severely restricted by the inability to obtain adequate amounts of tissue at all developmental stages. Furthermore, ethical considerations and laws in many countries prohibit the manipulation of human embryos. In this context of limited experimentation on human embryos in the USA and in many cultures of the world, human embryonic stem (ES) cells provide nearly the only viable opportunity for direct study of human development. Because some developmental processes are different between mice and humans (for example, placental development and early embryogenesis), direct comparative studies with murine and human ES cells may ultimately reveal important differences in the molecular mechanisms controlling these and other aspects of development.

We will review here the differences and similarities in the derivation and properties of human and murine ES cells and outline how an *in vitro* ES cell differentiation system can be used to compare fundamental developmental mechanisms between these two species. This will lead, in time, to a better understanding of human development.

4.2 Derivation, growth and morphology of murine and human ES cells

Experiments over the last 20 years have determined that ES cells can be derived from murine, non-human primate, and human intact blastocysts using essentially similar techniques (Rossant and Papaioannou, 1984; Thomson et al., 1995, 1996,1998; Umeda et al., 2004). From either freshly recovered or cultured embryos, the inner cell mass (ICM) is isolated, using antisera or microdissection, and plated onto mitotically-inactivated murine embryonic fibroblast (MEF) feeder layers in tissue culture (Solter and Knowles, 1975; Brook and Gardner, 1997). The ICM cell outgrowths are passaged in the presence of serum, and colonies with the appropriate undifferentiated morphology are subsequently selected and expanded.

Most available murine ES cell lines have been derived from hybrid 129 mouse strains (D3, R1, etc.), but lines have been derived successfully from several more-resistant inbred strains (Brook and Gardner, 1997; Brook et al., 2003). Since the initial report of the derivation of human ES cells in 1998 (Thomson et al., 1998), more than 100 human ES cell lines have been derived in different laboratories around the world (www.isscr.org/science/sclines.htm). The derivation of human ES cell lines using human feeder layers and human serum, avoiding potential contamination with xenoproteins and xenogeneic tissues has been reported (Richards et al., 2002), but reports of successful derivation of pluripotent stem cells from human embryos without the use of feeder layers have not yet emerged.

Human and non-human primate ES cell colonies share a similar morphology that is distinct from both mouse ES cells and human embryonal germ (EG) cells, which are derived not from blastocyst ICMs but from primordial germ cells (Evans and Kaufman, 1981; Shamblott et al., 1998; Thomson et al., 1995, 1998). Undifferentiated colonies of most human ES and mouse cell lines show a compact morphology with a high nucleus-cytoplasm ratio (*Figure 4.1*). Whereas human ES cells form relatively flat, compact colonies that easily dissociate into single cells in trypsin or in Ca2+- and Mg2+- free medium, human EG cells form tight, more spherical colonies that are refractory to standard dissociation methods, but which more closely resemble the morphology of mouse ES cell colonies. Electron microscopy demonstrates gap junctions and microvillae in both mouse and human ES cells (Carpenter et al., 2004; Ginis et al., 2004). Cytoplasmic organelles, such as autophagosomes, which are associated with non-apoptotic cell death, may be more prevalent in mouse ES cells than human ES cells (Ginis et al., 2004). Under conditions of reduced cell density, such as during the establishment of clonal cell lines, human ES cell lines are more difficult to propagate than mouse ES cells, with a cloning efficiency of approximately 0.25% (Amit et al., 2000). Because of this property, it was found to be more effective to electroporate human ES cells as colonies rather than single cells (Zwaka and Thomson, 2003), which is different from the typical protocol for electroporating mouse ES cells. Human ES cells grow more slowly than mouse ES cells: the population doubling time of mouse ES cells is ~12 h, whereas the population doubling time of human ES cells is 35–40 h (Amit et al., 2000).

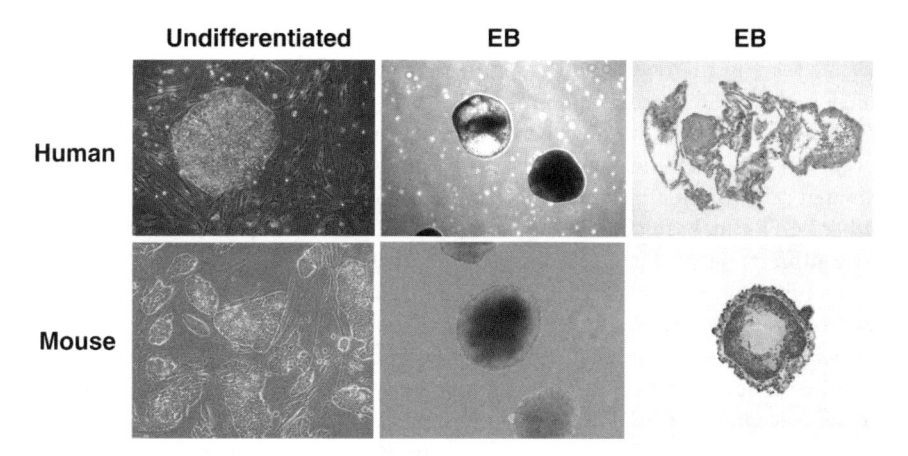

Figure 4.1: Morphology of commonly employed human (H9;WA09) and mouse (D3,R1) ES cell lines. Left panel: undifferentiated ES cell colonies on fibroblast feeder layers. Human ES cell colonies show a rounded appearance. (original magnification-50×). Center panel: Phase contrast images of 14 day human embryoid bodies (EBs) and 7 day murine EBs. ES cells from both species are able to generate simple, cavitating, and cystic EBs as shown. (original magnification – 50× human, 100× mouse). Right panel: Hematoxylin and eosin stained paraffin sections of similar stage EBs. (original magnification – 100×).

4.2.1 Self-renewal factors

Whereas the derivation of mouse ES cells and their propagation in an undifferentiated state requires leukemia inhibitory factor (LIF) (Niwa *et al.*, 1998; Sato *et al.*, 2004), human ES cell self-renewal requires basic fibroblast growth factor (bFGF), and LIF alone is not sufficient to prevent differentiation of human ES cells *in vitro* (Reubinoff *et al.*, 2000; Sato *et al.*, 2004; Thomson *et al.*, 1998). One possible reason for this difference stems from the growth factor receptor expression profile of mouse and human ES cells. Mouse ES cells express cell surface LIF receptor (LIFR)/CD130 (*Interleukin 6 signal transducer [IL6ST]* gene product, gp130) complexes, which bind LIF and mediate pluripotency through downstream activation of STAT3 (Carpenter *et al.*, 2004). In contrast, human ES cells do not express LIF receptors or gp130 receptors to any great degree and no ESTs for *gp130* were found in an undifferentiated human ES cell cDNA library (Brandenberger *et al.*, 2004; Carpenter *et al.*, 2004). Likewise, few or no transcripts were detected in undifferentiated human ES cell cDNA libraries for *Janus kinases 1/2/3*, *LIFR*, and LIF pathway inhibitory molecules such as *suppressor of cytokine signaling (SOCS)* genes (Brandenberger *et al.*, 2004). Interestingly, however, these and other studies demonstrate the presence of *STAT1 and STAT3* transcripts in undifferentiated human ES cells. In addition, *LIF* and *IL6ST* transcripts are detected in embryoid bodies during the early differentiation phases. The precise roles of these genes in human ES cell self-renewal and differentiation remain to be elucidated.

Whereas mouse ES cell self-renewal is primarily controlled through LIF/STAT pathway signaling, FGF signaling is thought currently to be the predominant mechanism by which human ES cell pluripotency is maintained in culture. A common feature in the derivation of numerous human ES cell lines over the last several years in many different laboratories worldwide has been the inclusion of bFGF in the culture medium to promote expansion and self-renewal. This practice is consistent with the finding that undifferentiated human ES cells express FGF receptor 1 (FGFR1), the cognate receptor for bFGF, more abundantly than differentiated cells (Brandenberger et al., 2004; Carpenter et al., 2004; Sato et al., 2003). Other FGFRs, including FGFR2, FGFR3, and FGFR4 also appear to be enriched in the undifferentiated state in some of these studies.

As bFGF alone is insufficient to prevent differentiation in feeder layer-free or feeder layer conditioned media-free cultures, other soluble molecules must participate in regulating self-renewal in human ES cells. Potential candidates were revealed in several recent studies. In one study, human ES cells were found to express receptors for Flt3L (CD135) and stem cell factor (SCF, CD117), cytokines linked with growth control of hematopoietic stem cells (Carpenter et al., 2004). Whether the ligands for these receptors will promote undifferentiated growth of human ES cells should be investigated.

Recent gene expression profiling studies have suggested a role for Wnt and TGFβ1/BMP signaling pathways in self-renewal mechanisms of ES cells (Brandenberger et al., 2004; Sato et al., 2003, 2004; Ying et al., 2003a). The role of Wnt signaling has been investigated further by a study in which mouse and human ES cells were able to be maintained short term in an undifferentiated state *in vitro* in the absence of feeder layers or conditioned medium. Both *in vitro* and *in vivo* pluripotency were retained through Wnt pathway activation by exposing cells to inhibitors of glycogen synthase kinase-3 or recombinant Wnt3a protein (Sato et al., 2004). In their response to Wnt pathway activation in this study, human and mouse ES cells appeared to be similar; however, it was not clear whether human or mouse ES cells could be maintained in the undifferentiated state over a prolonged period in culture under these conditions. Correlating with a potential role of TGFβ, Amit et al. demonstrated the efficacy of a combination of TGFβ, bFGF, and fibronectin in supporting proliferation of undifferentiated human ES cells (Amit et al., 2004). Proteome analyses are beginning to delineate proteins expressed by feeder layers and in serum that might mediate the maintenance of self-renewal of human ES cells (Lim and Bodnar, 2002). Ying et al. (2003a) found in mouse ES cells that BMP and LIF are sufficient to sustain self-renewal and maintain cells in an undifferentiated state without the need for serum or feeder layers. Whether BMP plays such a vital role in human ES cell self-renewal remains to be investigated. It is hoped that as the different mechanisms underlying self-renewal in murine and human ES cells are unraveled, essential growth factors, cytokines, and signaling molecules will be discovered that could be used to maintain human ES cells in simplified culture systems for prolonged periods of time. This would permit feasible scale-up for clinical applications while cells maintain the phenotypic and functional features of pluripotency.

4.2.2 Novel feeder layers

Typically, after the initial derivation on feeder layers and in the presence of serum, human ES cell lines can be maintained and propagated on feeder layers in medium containing serum alone, or serum replacement plus bFGF. Recently, reports of other means of propagating human ES cells in an undifferentiated state have been described and include using media conditioned by MEFs and growth factor-reduced Matrigel (Carpenter et al., 2004; Rosler et al., 2004; Xu et al., 2001) and adult and fetal fibroblast feeder layers derived from human tissue (Richards et al., 2002). Advantages of feeder-free cultures are their simplicity, scalability, and the lack of concern of fibroblast carryover to initial differentiation cultures. Whereas mouse ES cells grow readily under feeder-free conditions in the presence of LIF, human ES cells may show some signs of differentiation. Indeed, when grown in MEF-conditioned media, undifferentiated human ES cell colonies are often surrounded by differentiated, fibroblast-like stroma, which gradually disappear with longer passage in feeder-free conditions (Rosler et al., 2004). Moreover, it has been reported that some cell lines may grow well in MEF-conditioned media, whereas others may have a tendency to differentiate (Richards et al., 2002). This propensity to differentiate may be of practical interest as the presence of some differentiated cells in so-called 'undifferentiated' cell cultures may confound interpretation of gene expression studies or lead to inadvertent selection of clones having abnormal proliferative or differentiative capacities.

As efforts are made to move towards transplantation applications of human ES cell-derived tissues, a potential concern exists for zoonotic transmission of pathogenic agents, such as endogenous retroviruses, that might have occurred as a result of derivation and culture on murine fibroblast feeder layers. In this light, feeder layers from human tissues, although still allogeneic to a future recipient, might provide an advantage. Recent studies have tested whether human feeders support human ES cell growth. It appears that feeder layers derived from human fetal muscle, fetal skin, adult fallopian tube (Richards et al., 2002), adult truncal skin (Richards et al., 2003) and human neonatal foreskin fibroblast cell lines (unpublished observations, JSO) are all capable of supporting undifferentiated growth of human ES cells. In addition, it was shown that human ES cell lines could be newly derived using human feeder layers (Richards et al., 2002). In order to remove both the feeder layer and all animal proteins, recent efforts have been made to propagate human ES cells in feeder-free and serum-free conditions (Amit et al., 2004). Various combinations of TGFβ1, bFGF, LIF, bovine fibronectin and human fibronectin in a serum-free medium (notably without MEF-conditioned media) were compared with MEF feeder layers. Amit et al. (2004) found that a medium comprised of 15% serum replacement, 0.12 ng/ml TGFβ1, 4 ng/ml bFGF, and human fibronectin matrix supported undifferentiated growth of human ES cells to an equivalent degree to MEFs. These promising studies demonstrate the potential for large-scale development of a well-defined culture system without feeder layers or serum, thereby reducing the exposure to animal pathogens. For these culture systems to be worthwhile, it will be necessary to derive new lines under similar conditions. It is important to note, however, that it is not clear to what extent there is an actual risk of animal pathogen exposure. As human ES cell-based therapies are being developed, further studies should be undertaken to determine precisely whether there is a need to avoid all exposure to non-human-animal-derived materials.

4.2.3 Markers of the undifferentiated state in mouse and human ES cells

Paralleling differences in cellular morphology, human ES cells differ from their murine counterparts with regard to cell-surface antigen phenotype. Like undifferentiated primate ES cells and human embryonal carcinoma (EC) cells, human ES cells express stage-specific embryonic antigens 3 and 4 (SSEA-3 and SSEA-4), TRA-1-60, TRA-1-81, and alkaline phosphatase (Thomson et al., 1995, 1998; Thomson and Marshall, 1998). Pluripotent stem cells derived from the primordial germ cells of human fetuses, also termed embryonal germ (EG) cells, display a similar profile (Table 4.1). Mouse ES cells, on the other hand, do not express SSEA-3 or SSEA-4, but do express the lactoseries glycolipid SSEA-1, which in turn is not expressed in human ES cells, rhesus ES cells, or human EC cells (Table 4.1). Also, mouse ES cells do not express the TRA antigens to a significant degree. Although the functional significance of these antigens is unknown, both preimplantation human embryos (inner cell mass cells) and human ES cells demonstrate similar expression profiles of these antigens (Henderson et al., 2002). When compared in one laboratory, different hES cell lines expressed comparable levels of these key stem cell markers (Abeyta et al., 2004; Carpenter et al., 2004).

Other proteins and genes that are expressed prominently in both human ES cells and mouse ES cells and appear to be enriched or exclusively expressed in undifferentiated ES cells include the POU domain transcription factor Oct 3/4 (*POU5F1*); the teratocarcinoma-derived growth factor 1 (TDGF1/Cripto); the zinc finger protein REX-1; the homeobox domain transcription factor SOX-2; the transcriptional activator UTF-1; the divergent homeodomain protein Nanog; the telomerase associated genes TERT and TERFs; Thy-1 (CD90); CD133; CD9; and gap junction proteins connexin 43, among others (Bhattacharya et al., 2004; Brandenberger et al., 2004; Carpenter et al., 2004; Chambers et al., 2003; Ginis et al., 2004; Mitsui et al., 2003; Rosler et al., 2004; Sato et al., 2003). In contrast, the forkhead box gene, *Foxd3*, formerly called *Genesis*, is expressed in murine ES cells where it is required for the derivation of ES cell lines from the ICM, but does not appear to be expressed in the human ES cell line H1 (Ginis et al., 2004; Hanna et al., 2002). Our knowledge of the molecular markers of undifferentiated human and mouse ES cells has dramatically increased over the last several years; yet, the biological functions of many of these molecules remain to be defined.

4.2.4 Differentiation in vitro and in vivo

Do pluripotent stem cells from mice and humans have the same broad differentiative capacity? Whereas EC cell lines may have differing developmental potentials (Andrews et al., 1980), this has yet to be described convincingly for any human ES cell lines. A defining characteristic of ES cells is the ability to form teratomas comprised of tissue structures derived from all three embryonic germ layers upon injection into immunoincompetent murine hosts. Indeed, there appear to be great similarities in the types of tissues formed in teratomas grown from human ES cell lines derived from different embryo donors, and maintained in different laboratoriess under a variety of feeder layer or feeder-free conditions (Cowan et al., 2004; Odorico et al., 2001; Richards et al., 2002, 2003; Rosler et al.,

Table 4.1: Cell surface antigen expression by pluripotent cells from mice, rhesus monkeys and humans And by mouse and human blastocyst stage embryos.*

	SSEA1	SSEA3	SSEA4	TRA1-60	TRA1-81	AP	CD9	CD133	Thy 1 (CD90)	CD130
Murine ES	+	–	–	–	–	+	+	?	?	+
Mouse Embryo*,**,+	++	–	–	–	–	+	?	?	?	+
Primate ES	–	+	+	+	+	+	?	?	?	?
Human EC	–	+	+	+	+	+	?	?	?	?
Human EG	+	+	+	+	+	+	?	?	?	?
Human ES	–	+	+	+	+	+	+	+	+	–
Human Embryo*,+	–	++	++	++	++	+	?	?	?	?

* Staining specifically on the inner cell mass of preimplantation blastocyst stage embryos.
** Earlier stage mouse embryos also express SSEA3 and SSEA4.
+ Trophectoderm of both human and mouse embryos also express SSEA1.
Data compiled from references (Henderson et al., 2002; Carpenter et al., 2004; Thomson et al., 1998; Thomson and Marshall, 1998; Ginis et al., 2004).

2004; Thomson *et al.*, 1998; Xu *et al.*, 2001). Mouse ES cell lines also seem to be consistent in the tissue types that develop *in vitro* and *in vivo*. If human ES cell sublines are generated or if karyotypically abnormal cell lines are studied, it is possible that specific differentiative abilities may evolve. Few direct comparisons of *in vitro* or *in vivo* differentiation capacity of mouse and human ES cells have been carried out (Hay *et al.*, 2004).

Human ES cells and non-human primate ES cells are able to differentiate into trophoblast/trophectoderm in culture (Gerami-Naini *et al.*, 2004; Odorico *et al.*, 2001; Thomson *et al.*, 1995, 1998; Xu *et al.*, 2002b). On the other hand, mouse ES cells typically are not capable of differentiating into trophoblast (Rossant and Papaioannou, 1984). Based on the ability of human ES cells to differentiate into trophoblast and the significant differences that exist between primate and murine placental development, human ES cells may serve as a valuable tool to study the molecular and cellular mechanisms of human placentogenesis (Gerami-Naini *et al.*, 2004; Xu *et al.*, 2002b; also reviewed below and in Chapter 6). So far, many of the specific cell lineages that have been derived from mouse ES cells have been found in differentiated human ES cell cultures. With regard to *in vivo* differentiation, teratomas derived from murine and human ES cells are strikingly similar (personal observations, JSO).

4.2.5 Chromosomal alterations

It is widely known that mouse ES cell lines can develop karyotypic changes that are associated with loss of germ-line competence (e.g., D3 ES cells are known to harbor several mutations that limit their utility for generating chimeras) (Rossant and Papaioannou, 1984). Like their mouse counterparts, human ES cells may develop karyotypic abnormalities with time in culture (Cowan *et al.*, 2004; Draper *et al.*, 2004). The development of cytogenetic abnormalities may be promoted by culture in feeder-free conditions, if cloning is attempted, or in other suboptimal culture conditions. Genetic variants having a selective growth advantage may arise spontaneously and survive eventually to dominate a population. Gain of chromosome 17q or chromosome 12 are the most commonly reported abnormalities to date in human ES cell cultures (Cowan *et al.*, 2004; Draper *et al.*, 2004). Whether these changes affect the differentiative potential of particular lineages *in vitro* or *in vivo* is unknown. It is known that these gross chromosomal alterations do not affect expression of canonical ES cell markers, such as SSEA4. Aneuploidy may be only the 'tip of the iceberg' in these abnormal cells. Minor rearrangements and point mutations would go undetected by chromosomal banding and FISH; other methods yielding finer resolution are necessary to identify such changes. Detailed gene expression profiling or genetic analyses comparing karyotypically abnormal cell lines and normal cell lines have not been carried out. It is possible that these alterations may contribute to subtle or overt changes in differentiation capacity to a specific lineage. These new observations of karyotypic abnormalities in human ES cells are not unexpected given the plethora of culture conditions being tested and they emphasize the critical importance of periodically testing cell lines for chromosomal changes, as well as the need for careful maintenance of normal hES cell stocks. It is important to keep in mind that, when comparing cell lines, karyotypic abnormalities in one or both of the lines could account for some or all of the observed differences. This is true even when the cell lines are derived from different species.

4.3 ES cells as an *in vitro* model of early mammalian development

Our understanding of embryonic human development is largely based on an assumption that the process is similar to that in other animals. Although undoubtedly similar in many aspects, critical differences are clearly present. The limited available information relating specifically to human development comes only from observations made on tissue sections of human embryos. Even these scarce embryo sections are obtained from embryos that have passed the critical period of embryonic induction, which occurs within 3–4 weeks after the start of gestation. Thus, human ES cells appear to be the only viable tool at the present time to examine cell lineage segregation and organogenesis in early human development. Direct observations of cellular development and functional consequence can be made when human ES cells are differentiated along a particular pathway under defined conditions and/or when the ES cells are genetically modified. Human ES cells will likely play major roles in confirming the biological principles learned from studies of other animals and identifying novel rules governing the development of human primates.

4.3.1 *Early post-implantation embryonic development and placentogenesis*

Despite the similarities between mouse and human development, of which there are many, early post-implantation embryogenesis differs greatly between lower and higher vertebrate species. For example, mice and humans differ in the overall morphology of the early embryo (egg cylinder in mice vs. disc in humans) and in the process by which the embryo proceeds through gastrulation (complex rotation and inversion in the mouse) (Hogan *et al.*, 1994; Kaufman, 1992; O'Rahilly and Muller, 1987). Moreover, the formation of the placenta and extraembryonic membranes (allantois, chorion, etc.) are significantly different in human and mouse embryos. Whereas mice have a labyrinthine placenta in which maternal blood cells come into direct contact with fetal trophoblast, the human placenta is villous in nature (Cross, 1998). Consequently, mice provide an inadequate model of human placental development, which is more closely approximated by the non-human primate species (reviewed in Chapter 6 in this volume). Because human ES cells are capable of differentiating into trophoblast (Gerami-Naini *et al.*, 2004; Xu *et al.*, 2002b), they provide an important model system for studying gene and cell functions in placentogenesis that are unique to humans, and may ultimately give insight into human congenital diseases mediated by placental dysfunction.

4.3.2 *Neurogenesis*

Development of the nervous system is one of the earliest events in embryonic germ layer induction, and it has long been thought of as a step following formation of the embryonic ectoderm. This view has been modified recently by observations showing that neuroectodermal cells may be specified as early as late blastocyst stage in chickens, and before segregation of the three basic embryonic germ layers during gastrulation (Streit et al., 2000; Wilson et al., 2000; Wilson and Edlund, 2001). Confirmation of this phenomenon in mammals is hindered by inaccessibility to embryos at this particular stage for cell lineage tracing.

The study of ES cells, however, is an acceptable alternative to the direct use of human embryos. Using human ES cells as a model, we have found that, by morphology, neural tube-like rosettes develop from ES cells by approximately two weeks of differentiation culture (Zhang et al., 2001). This corresponds to day 19–20 of a human embryo, roughly the stage when the neural tube appears in a human embryo by the end of the third gestational week (O'Rahilly and Muller, 1994). A closer look at the gene expression pattern along with the corresponding cellular changes reveals that neuroectodermal cells are specified after the ES cells are differentiated for about 10 days, which translates to the beginning of the third week in an intact human embryo. This would confirm observations made in lower vertebrate animals that neuroectodermal induction takes place before gastrulation. However, the sequence and pattern of some specific neuroectoderm-associated genes may be very different between humans and animals including mice (Pankratz et al., 2003). Would this mean that neuroectodermal specification in human embryos employs different signaling pathways? Neuroectoderm is thought to be induced by active FGF signaling and/or inhibition of BMP signaling although the degree of effect of each signaling pathway may vary in different species (Tropepe et al., 2001; Wilson and Edlund, 2001; Ying et al., 2003b). The most controversial notion is perhaps the role of Wnt signaling in neuroectoderm induction. In *Xenopus* embryos, activation of Wnt signaling promotes neural induction (Baker et al., 1999) whereas in chickens, as well as mice, Wnt signaling appears to be inhibitory to neural induction (Aubert et al., 2002; Wilson et al., 2000). While these phenomena could be attributed to the possible variants of Wnt proteins, they also illustrate possible key mechanistic difference among species. Is the Wnt pathway involved in control of neural differentiation in humans? ES cells will likely play a major role in finding answers to questions like these.

4.3.3 Pancreaticogenesis

A better understanding of islet ontogeny and the phenotype of putative islet stem cells in humans would further our ability to generate a cell replacement therapy for treating diabetes. Furthermore, whether islet progenitor cells in mice and humans have the same phenotype and/or developmental potential will need to be determined prospectively. In fact, a comprehensive comparison of pancreatic differentiation between mice and humans has not been undertaken. Although some parallels can be drawn between mouse and human pancreaticogenesis, certain observations predict critical differences. For example, the anatomy is very different: the mouse pancreas consists of a thin film of acinar tissue between two folds of peritoneal membrane whereas the human organ is entirely retroperitoneal and is composed of a fibrous stroma. Moreover, human β cells appear to be relatively resistant to the β cell toxin streptozotocin, compared with fetal or adult murine β cells (Tuch and Chen, 1993; Yang and Wright, 2002). In the adult murine islet, the endocrine cells have a stereotypical arrangement where the β cells make up the core of the islet with α cells distributed in the periphery. In the primate islet including those in man, the distribution of α and β cells is more random (Andersson et al., 1996). The complement of integrins expressed in islets and the pancreas may also differ among species (Wang et al., 1999). Complementing these morphological observations, human and mouse embryos differ in the timing and/or nature of embryonic gene expression (e.g., insulin-like growth factor family genes, pancreatic polypeptide, insulin, tissue factor, and LIF among others) (Gregor et al., 1996; Liu et al., 1997; Luther et al., 1996). Thus, there are essential dissimilarities between human and mouse development that establish the fact that, *a priori*, one cannot assume all aspects of development, and in particular pancreaticogenesis, are identical in both species. An *in vitro* differentiation culture system in which pancreatic islet endocrine cells can be generated from human ES cells could be used to study the mechanisms regulating pancreatic islet development (Kahan et al., 2003; also reviewed in Chapter 10 in this volume).

4.3.4 Hematopoietic development

Blood stem cells are perhaps the most thoroughly studied subject in the stem cell field, with a number of standardized assay systems and a list of cell surface molecules that may be used for defining and isolating progenitors at various developmental stages. Blood cells are normally produced in different organs at different developmental stages thus demonstrating a carefully orchestrated temporal and spatial pattern of development. First, blood formation begins in the yolk sac at an early embryonic stage, followed by production of blood cells in the fetal liver and ultimately in the bone marrow microenvironment. Differences exist between mouse and human embryogenesis including the development of the yolk sac in which blood cells are first produced (Palis and Yoder, 2001; Yoder, 2001). Moreover, whereas the fetal liver dominates as the site of hematopoiesis in mice through late gestation and into the first week of post-natal life, in humans, the transition from fetal liver hematopoiesis to bone marrow hematopoiesis occurs in mid gestation. The expression of fetal globin genes also differs between the species. Thus, to gain insights into some blood-related disorders in humans, it will be important to examine early embryonic hematopoiesis in a human context (Ciavatta et al., 1995). Furthermore, most studies on hematopoietic stem cells use stem cells taken from the bone marrow or the cord blood, which is not representative of early hematopoiesis. Consequently, ES cells provide a means specifically to examine early embryonic hematopoiesis.

Studies to date have employed similar approaches for hematopoietic differentiation from both mouse and human ES cells (Chen et al., 2003; Kaufman et al., 2001). ES cells are aggregated and differentiated in suspension for several days during which time characteristic blood stem cells begin to develop. The aggregates are then dissociated and individual cells are placed in semi-solid methylcellulose-based media, in which colonies of blood stem/progenitors form in the presence of serum, conditioned media of cell lines, and/or cytokines (Kaufman et al., 2001). This approach represents spontaneous differentiation with selective expansion of certain populations of blood progenitors, which may explain why ES-derived blood stem/progenitor cells, either of mouse or human origin, generally demonstrate limited stable long-term multilineage hematopoietic engraftment (Hole *et al.*, 1996). A recent study by Kyba *et al.* (2002) takes a new approach to direct mouse ES cells to primitive blood progenitors under the influence of a homeobox gene, *HoxB4*. Progenitors generated in this manner appear capable of reconstituting lymphoid and myeloid lineages to some degree. The findings of this study should be instrumental in human ES cell studies to drive the full range of hematopoietic development and to generate engraftable blood stem cells for therapeutic purposes. In summary, human ES cells provide a unique window to early hematopoiesis in humans that is not achievable through the study of human embryos.

4.3.5 Cardiac development

Like hematopoietic differentiation from ES cells, cardiac differentiation is achieved by spontaneous differentiation of ES cell aggregates into beating cardiomyocytes, generally carried out under non-selective culture conditions (Boheler *et al.*, 2002). Unlike the hematopoietic system, studies in the development of the cardiovascular system appear to be hindered by the lack of defined cell surface markers that could be used to isolate putative cardiac stem/progenitor cells. Although both mouse and human ES cells can differentiate into the various sub-specialized cardiomyocytic cell types such as cells of the atria, ventricles, and conduction system, efficient strategies for directed differentiation and isolation of specific populations of human cardiac cells are needed for therapeutic trials (Boheler *et al.*, 2002; He *et al.*, 2003; Klug *et al.*, 1996; Mummery *et al.*, 2003; Xu *et al.*, 2001). Functionally distinct populations of cardiac cells in different parts of the heart are likely differentiated from multipotent precursors in response to unique sets of morphogens. An understanding of the molecular mechanism of heart morphogenesis should provide important clues to enable the directed differentiation of human ES cells into specific populations of cardiac muscle cells.

4.3.6 Human ES cells as a model to study human development

The ES cell *in vitro* differentiation system has been used to study the molecular and cellular mechanisms regulating normal development. In the mouse ES cell system, many investigations have studied the genetic regulation of cell lineage differentiation by comparing 'knockout' or transgenic ES cells and wild-type ES cells differentiating in culture. Others have rapidly identified novel developmentally-regulated genes by screening gene-trapped ES cell colonies after differentiation *in vitro*. Still others have examined the effects of growth factors on cell lineage differentiation pathways. Tissue-restricted progenitor cells or specific functional cell types could be isolated using ES cells that have been engineered to express a fluorescent protein under control of a tissue-specific promoter. By this means, tissue-restricted human progenitor cells from a variety of cell lineages could be prospectively isolated for study of their developmental potential and putative plasticity (Baker and Lyons, 1996; Duncan *et al.*, 1998; Evans, 1998; Forrester *et al.*, 1996; Keller, 1995; Meyer *et al.*, 2000; Robertson *et al.*, 2000; Roy *et al.*, 1999; Vallier *et al.*, 2001; Zambrowicz *et al.*, 1998). Apart from the ability to ask interesting biological questions about the differences in developmental pathways between humans and mice, human ES cells are a tool that can be used to isolate differentiated human tissues for transplantation medicine.

4.4 ES cells: a renewable source of functional cells

Scientific and public interest in stem cells is generated, at least partly, from the expectation that ES cells may become a long-lasting source for the production of functional cells for replacement therapy. Mouse ES cells can be preferentially directed to a wide variety of differentiated progenies and the function of some of these *in vitro*-generated cell types has been demonstrated in animal models of diseases such as Parkinson's disease, among others (Kim *et al.*, 2002; McDonald *et al.*, 1999). Human ES cells entered the stem cell stage only recently but have already shown promise in producing multiple functional cell types (Assady *et al.*, 2001; He *et al.*, 2003; Kaufman *et al.*, 2001; Xu *et al.*, 2002a; Zhang *et al.*, 2001). Future investigations will likely refine the culture conditions for directed differentiation of key lineages and eventually demonstrate a capacity for differentiation of human ES cells into less common and progressively more restricted lineages.

4.4.1 Similar principles governing differentiation of mouse and human ES cells

From the standpoint of developmental biology, principles of cell lineage specification in animals are likely to carry over to human development. Thus, differentiation protocols developed for mouse ES cells may be applicable to human ES cells as well. In the past few years, strategies for directed differentiation of specific cell types from ES cells have begun to emerge. Most of these *in vitro* directed differentiation protocols are designed based on principles of embryonic cell lineage development, which in turn serve as a model system for dissecting molecular mechanisms underlying the developmental processes. In practice, directed differentiation leads to a much more efficient production of a defined cell population than the empirical method of EB formation (Lee *et al.*, 2000; Wichterle *et al.*, 2002; Ying *et al.*, 2003b). Lessons learned from studies in specific cell lineage differentiation in mouse ES cells should provide a rich framework to begin directed differentiation studies in human ES cells.

Early embryogenesis is essentially a process of cell lineage segregation accompanied by morphogenesis. Precursors at each stage differentiate into more-specialized cells and occupy unique locations in response to a set of morphogens. This positional information may be essential for cellular function in later life as is the case for neurons in the brain and spinal cord. Hence, developmental insights gained from embryogenesis studies will be critical to achieve directed differentiation of specific cell lineages from ES cells. The most obvious example of this is neural cell lineage development. Mouse ES cell-derived neuroepithelial cells are caudalized by retinoic acid to generate cells with a spinal cord progenitor identity and, in the presence of the ventralizing signal sonic hedgehog, become ventral neuronal cell types including motor neurons (Wichterle *et al.*, 2002). Similarly, mouse ES cell-produced neuroepithelial cells take on a midbrain identity in response to FGF8 and generate dopaminergic neurons in the presence of sonic hedgehog (Lee *et al.*, 2000). Subtype neuronal differentiation from human ES cells appears to follow the same principles (Zhang, unpublished). Cell lineage development of other organs and tissues is likely to occur in a similar manner.

4.4.2 Mechanistic versus technical differences in mouse and human ES cell differentiation

In practice, translation of cell differentiation protocols from mouse to human ES cells may not be straightforward due to both mechanistic and technical differences. Cell lineage development depends on the interplay of intrinsic cell programs and extrinsic signals. Precursor cells at a particular developmental stage respond to environmental signals, often sets of morphogens and/or cytokines, in order to make fate choices. Understanding the interactions among morphogens and precursor cells at a particular developmental stage will be instrumental in induction of ES cells to a specific fate. In rodents, organogenesis begins within a week of fertilization. Consequently, ES cells may choose one fate or another by this time in culture. An example is the differentiation of neuroepithelial cells from mouse ES cells in 2–3 days of differentiation culture (Ying *et al.*, 2003b). In contrast, neuroepithelial cells appear in human ES cell differentiation cultures by 2 weeks (Zhang *et al.*, 2001). Also, in the case of pancreatic islet cell differentiation, the appearance of hormone-producing cells requires a considerably longer period of time in human ES cell cultures than in their mouse counterparts (Kahan *et al.*, 2003) (reviewed in Chapter 10). These time-related determinants must be considered when designing directed differentiation protocols for human ES cells. Due to involvement of multiple factors, such as the presence or absence of different morphogens and/or cytokines at particular precursor stages in long-term cultures, it may become a technically complicated process to differentiate human ES cells to subtypes of functional cells derived from the endoderm, such as pancreatic or liver cells.

An important method for evaluating the function of *in vitro*-generated cells is by transplantation into the animal models of development and/or disease. Implantation of mouse ES cell-generated motoneurons into chick neural tube results in the maturation of motoneurons and connection of axons with muscles (Wichterle *et al.*, 2002). Similarly, transplantation of mouse ES-produced dopaminergic neurons into the striatum of Parkinson's rats leads to the restoration of motor function deficits (Kim *et al.*, 2002). These seemingly simple experiments may not be as straightforward for testing the function of human ES cell-derived cells because of time-related factors and species differences, which may come into play to confound the interpretation of such experiments. For example, while the transplantation of mouse ES cells or ES cell-derived dopaminergic neurons into denervated rodent striatum restores motor functional deficit in 2–3 weeks (Kim *et al.*, 2002), it requires several months for dopaminergic neurons isolated from fetal human midbrain to contribute to functional restoration in a rat model of Parkinson's disease (Svendsen *et al.*, 1997). Also, human ES cell-generated neural precursors slowly differentiate (several weeks) into neurons and glia after transplantation despite the developmental environment of neonatal mice (Zhang *et al.*, 2001). These observations may indicate that human cells may mature largely in a cell autonomous manner partly or alternatively, may simply reflect a mismatch of signals between transplanted cell and recipient. In addition, technical issues, such as limited lifespan of certain rodent models, requirement of long-term immunosuppression for xenografts, and hostile disease environments, may pose restraints to success and validity of such analyses. Consequently, non-human primate models will likely play a vital role in examination of the function of human ES cell derivatives.

4.4.3 ES cell differentiation as a tool for discovery

Aside from the perspective of clinical application of a cell replacement therapy, terminally differentiated functional cells from ES cells provide a useful tool for examining gene function and screening pharmaceuticals and toxic reagents. The testing of pharmaceuticals on human ES cell derivatives will have direct relevance to application in patients. Genetic manipulation in human ES cells with genes that are involved in certain pathological processes and/or the derivation of novel human ES cell lines from diseased human embryos would provide unprecedented *in vitro* models of human diseases.

Human ES cells are an important tool for understanding the identity and nature of tissue-restricted multipotent progenitor cells that could be used in human cell-based therapies. Progenitors readily respond to signals in their microenvironment enabling them to differentiate, and potentially functionally integrate into tissues in a recipient. As a cell replacement therapy, immature progenitors have a theoretical advantage as donor cells over terminally differentiated mature cells in that they may exhibit reduced immunogenicity and significantly greater proliferative potential. The ability to identify and isolate progenitor cells will be a critical step towards the effective and safe clinical application of ES cell progenies.

Cell surface antigens that can be used for purification of progenitors at different developmental stages are not available or known for most lineages such as neural, cardiac, and pancreatic cells. Only in hematopoietic precursor cells are unique cell surface markers readily available. These are commonly used in the clinical arena, thus providing proof-of-concept. Despite lack of adequate cell surface markers for progenitor cell isolation, transcriptional regulation of cell lineage development is well delineated and can be employed to isolate progenitor cells at various stages. This means of genetic selection of lineage restricted ES cell progeny has been used in several instances. By knocking reporters into genes that transcriptionally control the specification of a cell lineage, for example *olig2* in a subset of neural progenitors (Xian *et al.*, 2003) or *Brachyury* in mesodermal precursors (Kubo *et al.*, 2004), specific progenitor-stage populations can be isolated. These purified populations, which can be isolated in large numbers, will form the foundation for identifying specific cell surface molecules unique to these populations for FACS through genomic and proteomic analyses. Such a strategy is likely to have increasing impact when combined with human ES cells in the future.

4.5 Summary

Apart from their promise for generating cell-based therapies in transplantation medicine, human ES cells will play an increasingly important role in understanding gene function in human development and lineage specification. Mutant human ES cells, which can be obtained through targeted mutagenesis or gene trap screens, or perhaps even somatic cell nuclear transfer, may serve as novel models for human pathological conditions. Even if human ES cells fall short of their therapeutic promise, they will remain a valuable tool to explore the genetic and epigenetic regulation of human development.

References

Abeyta MJ, Clark AT, Rodriguez RT, Bodnar MS, Pera RA, Firpo MT (2004) Unique gene expression signatures of independently-derived human embryonic stem cell lines. *Hum. Mol. Genet.* **13**, 601–608.

Amit M, Carpenter MK, Inokuma MS, Chiu CP, Harris CP, Waknitz MA, Itskovitz-Eldor J, Thomson JA (2000) Clonally derived human embryonic stem cell lines maintain pluripotency and proliferative potential for prolonged periods of culture. *Dev. Biol.* **227**, 271–278.

Amit M, Shariki C, Margulets V, Itskovitz-Eldor J (2004) Feeder layer- and serum-free culture of human embryonic stem cells. *Biol. Reprod.* **70**, 837–845.

Andersson A, Eizirik DL, Bremer C, Johnson RC, Pipeleers DG, Hellerstrom C (1996) Structure and function of macroencapsulated human and rodent pancreatic islets transplanted into nude mice. *Horm. Metab. Res.* **28**, 306–309.

Andrews PW, Bronson DL, Benham F, Strickland S, Knowles BB (1980) A comparative study of eight cell lines derived from human testicular teratocarcinoma. *Int. J. Cancer* **26**, 269–280.

Assady S, Maor G, Amit M, Itskovitz-Eldor J, Skorecki KL, Tzukerman M (2001) Insulin production by human embryonic stem cells. *Diabetes* **50**, 1691–1697.

Aubert J, Dunstan H, Chambers I, Smith A (2002) Functional gene screening in embryonic stem cells implicates Wnt antagonism in neural differentiation. *Nat. Biotechnol.* **20**, 1240–1245.

Baker JC, Beddington RS, Harland RM (1999) Wnt signaling in Xenopus embryos inhibits bmp4 expression and activates neural development. *Genes Dev.* **13**, 3149–3159.

Baker RK, Lyons GE (1996) Embryonic stem cells and in vitro muscle development. *Curr. Top. Dev. Biol.* **33**, 263–279.

Bhattacharya B, Miura T, Brandenberger R, Mejido J, Luo Y, Yang AX *et al.* (2004) Gene expression in human embryonic stem cell lines: unique molecular signature. *Blood* **103**, 2956–2964.

Boheler KR, Czyz J, Tweedie D, Yang HT, Anisimov SV, Wobus AM (2002) Differentiation of pluripotent embryonic stem cells into cardiomyocytes. *Circ. Res.* **91**, 189–201.

Brandenberger R, Wei H, Zhang S, Lei S, Murage J, Fisk GJ *et al.* (2004) Transcriptome characterization elucidates signaling networks that control human ES cell growth and differentiation. *Nat. Biotechnol.* **22**, 707–716.

Brook FA, Gardner RL (1997) The origin and efficient derivation of embryonic stem cells in the mouse. *Proc. Natl Acad. Sci. USA* **94**, 5709–5712.

Brook FA, Evans EP, Lord CJ, Lyons PA, Rainbow DB, Howlett SK, Wicker LS,

Todd JA, Gardner RL (2003) The derivation of highly germline-competent embryonic stem cells containing NOD-derived genome. *Diabetes* 52, 205–208.

Carpenter MK, Rosler ES, Fisk GJ, Brandenberger R, Ares X, Miura T, Lucero M, Rao MS (2004) Properties of four human embryonic stem cell lines maintained in a feeder-free culture system. *Dev. Dyn.* 229, 243–258.

Chambers I, Colby D, Robertson M, Nichols J, Lee S, Tweedie S, Smith A (2003) Functional expression cloning of Nanog, a pluripotency sustaining factor in embryonic stem cells. *Cell* 113, 643–655.

Chen D, Lewis RL, Kaufman DS (2003) Mouse and human embryonic stem cell models of hematopoiesis: past, present, and future. *Biotechniques* 35, 1253–1261.

Ciavatta DJ, Ryan TM, Farmer SC, Townes TM (1995) Mouse model of human beta zero thalassemia: targeted deletion of the mouse beta maj- and beta min-globin genes in embryonic stem cells. *Proc. Natl Acad. Sci. USA* 92, 9259–9263.

Cowan CA, Klimanskaya I, McMahon J, Atienza J, Witmyer J, Zucker JP *et al.* (2004) Derivation of embryonic stem-cell lines from human blastocysts. *N. Engl. J. Med.* 350, 1353–1356.

Cross JC (1998) Formation of the placenta and extraembryonic membranes. *Ann. NY Acad. Sci.* 857, 23–32.

Draper JS, Smith K, Gokhale P, Moore HD, Maltby E, Johnson J, Meisner L, Zwaka TP, Thomson JA, Andrews PW (2004) Recurrent gain of chromosomes 17q and 12 in cultured human embryonic stem cells. *Nat. Biotechnol.* 22, 53–54.

Duncan SA, Navas MA, Dufort D, Rossant J, Stoffel M (1998) Regulation of a transcription factor network required for differentiation and metabolism. *Science* 281, 692–695.

Evans M, Kaufman M (1981) Establishment in culture of pluripotent cells from mouse embryos. *Nature* 292, 154–156.

Evans MJ (1998) Gene trapping–a preface. *Dev. Dyn.* 212, 167–169.

Forrester LM, Nagy A, Sam M, Watt A, Stevenson L, Bernstein A, Joyner AL, Wurst W (1996) An induction gene trap screen in embryonic stem cells: Identification of genes that respond to retinoic acid in vitro. *Proc. Natl Acad. Sci. USA* 93, 1677–1682.

Gerami-Naini B, Dovzhenko OV, Durning M, Wegner FH, Thomson JA, Golos TG (2004) Trophoblast differentiation in embryoid bodies derived from human embryonic stem cells. *Endocrinology* 145, 1517–1524.

Ginis I, Luo Y, Miura T, Thies S, Brandenberger R, Gerecht-Nir S *et al.* (2004) Differences between human and mouse embryonic stem cells. *Dev. Biol.* 269, 360–380.

Gregor P, Feng Y, DeCarr LB, Cornfield LJ, McCaleb ML (1996) Molecular characterization of a second mouse pancreatic polypeptide receptor and its inactivated human homologue. *J. Biol. Chem.* 271, 27776–27781.

Hanna LA, Foreman RK, Tarasenko IA, Kessler DS, Labosky PA (2002) Requirement for Foxd3 in maintaining pluripotent cells of the early mouse embryo. *Genes Dev.* 16, 2650–2661.

Hay DC, Sutherland L, Clark J, Burdon T (2004) Oct-4 knockdown induces similar patterns of endoderm and trophoblast differentiation markers in human and mouse embryonic stem cells. *Stem Cells* 22, 225–235.

He JQ, Ma Y, Lee Y, Thomson JA, Kamp TJ (2003) Human embryonic stem cells develop into multiple types of cardiac myocytes: action potential characterization. *Circ. Res.* **93**, 32–39.

Henderson JK, Draper JS, Baillie HS, Fishel S, Thomson JA, Moore H, Andrews PW (2002) Preimplantation human embryos and embryonic stem cells show comparable expression of stage-specific embryonic antigens. *Stem Cells* **20**, 329–337.

Hogan B, Beddington R, Costantini F, Lacy E (1994) *Manipulating the Mouse Embryo: A Laboratory Manual* (2nd edn).Cold Spring Harbor Laboratory Press, Plainview, NY.

Hole N, Graham GJ, Menzel U, Ansell JD (1996) A limited temporal window for the derivation of multilineage repopulating hematopoietic progenitors during embryonal stem cell differentiation in vitro. *Blood* **88**, 1266–1276.

Kahan BW, Jacobson LM, Hullett DA, Oberley TD, Odorico JS (2003) Pancreatic precursors and differentiated islet cell types from murine embryonic stem cells: an in vitro model to study islet differentiation. *Diabetes* **52**, 2016–2024.

Kaufman DS, Hanson ET, Lewis RL, Auerbach R, Thomson JA (2001) Hematopoietic colony-forming cells derived from human embryonic stem cells. *Proc. Natl Acad. Sci. USA* **98**, 10716–10721.

Kaufman MH (1992) *The Atlas of Mouse Development*.Academic Press, San Diego.

Keller GM (1995) In vitro differentiation of embryonic stem cells. *Curr. Opin. Cell Biol.* **7**, 862–869.

Kim JH, Auerbach JM, Rodriguez-Gomez JA, Velasco I, Gavin D, Lumelsky N *et al.* (2002) Dopamine neurons derived from embryonic stem cells function in an animal model of Parkinson's disease. *Nature* **418**, 50–56.

Klug MG, Soonpaa MH, Koh GY, Field LJ (1996) Genetically selected cardiomyocytes from differentiating embryonic stem cells form stable intracardiac grafts. *J. Clin. Invest.* **98**, 216–224.

Kubo A, Shinozaki K, Shannon JM, Kouskoff V, Kennedy M, Woo S, Fehling HJ, Keller G (2004) Development of definitive endoderm from embryonic stem cells in culture. *Development* **131**, 1651–1662.

Kyba M, Perlingeiro RC, Daley GQ (2002) HoxB4 confers definitive lymphoid-myeloid engraftment potential on embryonic stem cell and yolk sac hematopoietic progenitors. *Cell* **109**, 29–37.

Lee SH, Lumelsky N, Studer L, Auerbach JM, McKay RD (2000) Efficient generation of midbrain and hindbrain neurons from mouse embryonic stem cells. *Nat. Biotechnol.* **18**, 675–679.

Lim JW, Bodnar A (2002) Proteome analysis of conditioned medium from mouse embryonic fibroblast feeder layers which support the growth of human embryonic stem cells. *Proteomics* **2**, 1187–1203.

Liu HC, He ZY, Tang YX, Mele CA, Veeck LL, Davis O, Rosenwaks Z (1997) Simultaneous detection of multiple gene expression in mouse and human individual preimplantation embryos. *Fertility & Sterility* **67**, 733–741.

Luther T, Flossel C, Mackman N, Bierhaus A, Kasper M, Albrecht S *et al.* (1996) Tissue factor expression during human and mouse development. *Am. J. Pathol.* **149**, 101–113.

McDonald JW, Liu XZ, Qu Y, Liu S, Mickey SK, Turetsky, D, Gottlieb DI, Choi DW (1999) Transplanted embryonic stem cells survive, differentiate and promote recovery in injured rat spinal cord. *Nat. Med.* 5, 1410–1412.

Meyer N, Jaconi M, Landopoulou A, Fort P, Puceat M (2000) A fluorescent reporter gene as a marker for ventricular specification in ES-derived cardiac cells. *FEBS Lett* 478, 151–158.

Mitsui K, Tokuzawa Y, Itoh H, Segawa K, Murakami M, Takahashi K, Maruyama M, Maeda M, Yamanaka S (2003) The homeoprotein Nanog is required for maintenance of pluripotency in mouse epiblast and ES cells. *Cell* 113, 631–642.

Mummery C, Ward-van Oostwaard D, Doevendans P, Spijker R, van den Brink S, Hassink, R *et al.* (2003) Differentiation of human embryonic stem cells to cardiomyocytes: role of coculture with visceral endoderm-like cells. *Circulation* 107, 2733–2740.

Niwa H, Burdon T, Chambers I, Smith A (1998) Self-renewal of pluripotent embryonic stem cells is mediated via activation of STAT3. *Genes Dev.* 12, 2048–2060.

O'Rahilly R, Muller F (1987) *Developmental Stages in Human Embryos.* Carnegie Institution of Washington, Washington, DC.

O'Rahilly RR, Muller F (1994) *The Embryonic Human Brain.* Wiley-Liss, New York.

Odorico J S, Kaufman DS, Thomson JA (2001) Multilineage differentiation from human embryonic stem cell lines. *Stem Cells* 19, 193–204.

Palis J, Yoder MC (2001) Yolk-sac hematopoiesis: the first blood cells of mouse and man. *Exp. Hematol.* 29, 927–936.

Pankratz MT, Lyons EA, Moreno P, Zhang SC (2003) Gene expression during human embryonic stem cell differentiation into neuroepithelia. *Soc. Neurosci.* (Abstr), 673.5.

Reubinoff BE, Pera MF, Fong CY, Trounson A, Bongso A (2000) Embryonic stem cell lines from human blastocysts: somatic differentiation in vitro. *Nat. Biotechnol.* 18, 399–404.

Richards M, Fong CY, Chan WK, Wong PC, Bongso A (2002) Human feeders support prolonged undifferentiated growth of human inner cell masses and embryonic stem cells. *Nat. Biotechnol.* 20, 933–936.

Richards M, Tan S, Fong CY, Biswas A, Chan WK, Bongso A (2003) Comparative evaluation of various human feeders for prolonged undifferentiated growth of human embryonic stem cells. *Stem Cells* 21, 546–556.

Robertson SM, Kennedy M, Shannon JM, Keller G (2000) A transitional stage in the commitment of mesoderm to hematopoiesis requiring the transcription factor SCL/tal-1. *Development* 127, 2447–2459.

Rosler ES, Fisk GJ, Ares X, Irving J, Miura T, Rao MS, Carpenter MK (2004) Long-term culture of human embryonic stem cells in feeder-free conditions. *Dev. Dyn.* 229, 259–274.

Rossant J, Papaioannou VE (1984) The relationship between embryonic, embryonal carcinoma and embryo-derived stem cells. *Cell Differ.* 15, 155–161.

Roy NS, Wang S, Harrison-Restelli C, Benraiss A, Fraser RA, Gravel M, Braun PE, Goldman SA (1999) Identification, isolation, and promoter-defined separation of mitotic oligodendrocyte progenitor cells from the adult human subcortical white matter. *J. Neurosci.* 19, 9986–9995.

Sato N, Sanjuan IM, Heke M, Uchida M, Naef F, Brivanlou AH (2003) Molecular signature of human embryonic stem cells and its comparison with the mouse. *Dev. Biol.* 260, 404–413.

Sato N, Meijer L, Skaltsounis L, Greengard P, Brivanlou AH (2004) Maintenance of pluripotency in human and mouse embryonic stem cells through activation of Wnt signaling by a pharmacological GSK-3-specific inhibitor. *Nat. Med.* 10, 55–63.

Shamblott MJ, Axelman J, Wang S, Bugg EM, Littlefield JW, Donovan PJ, Blumenthal PD, Huggins GR, Gearhart JD (1998) Derivation of pluripotent stem cells from cultured human primordial germ cells. *Proc. Natl Acad. Sci. USA* 95, 13726–13731.

Solter D, Knowles BB (1975) Immunosurgery of mouse blastocyst. *Proc. Natl Acad. Sci. USA* 72, 5099–5102.

Streit A, Berliner AJ, Papanayotou C, Sirulnik A, Stern CD (2000) Initiation of neural induction by FGF signalling before gastrulation. *Nature* 406, 74–78.

Svendsen CN, Caldwell MA, Shen J, ter Borg MG, Rosser AE, Tyers P, Karmiol S, Dunnett SB (1997) Long-term survival of human central nervous system progenitor cells transplanted into a rat model of Parkinson's disease. *Exp. Neurol.* 148, 135–146.

Thomson JA, Marshall VS (1998) Primate embryonic stem cells. *Curr. Top. Dev. Biol.* 38, 133–165.

Thomson JA, Kalishman J, Golos TG, Durning M, Harris CP, Becker RA, Hearn JP (1995) Isolation of a primate embryonic stem cell line. *Proc. Natl Acad. Sci. USA* 92, 7844–7848.

Thomson JA, Kalishman J, Golos TG, Durning M, Harris CP, Hearn JP (1996) Pluripotent cell lines derived from common marmoset (*Callithrix jacchus*) blastocysts. *Biol. Reprod.* 55, 254–259.

Thomson JA, Itskovitz-Eldor J, Shapiro SS, Waknitz MA, Swiergiel JJ, Marshall VS, Jones J M (1998) Embryonic stem cell lines derived from human blastocysts. *Science* 282, 1145–1147.

Tropepe V, Hitoshi S, Sirard C, Mak TW, Rossant J, van der Kooy D (2001) Direct neural fate specification from embryonic stem cells: a primitive mammalian neural stem cell stage acquired through a default mechanism. *Neuron* 30, 65–78.

Tuch BE, Chen J (1993) Resistance of the human fetal beta-cell to the toxic effect of multiple low-dose streptozotocin. *Pancreas* 8, 305–311.

Umeda K, Heike T, Yoshimoto M, Shiota M, Suemori H, Luo HY *et al.* (2004) Development of primitive and definitive hematopoiesis from nonhuman primate embryonic stem cells in vitro. *Development* 131, 1869–1879.

Vallier L, Mancip J, Markossian S, Lukaszewicz A, Dehay C, Metzger D, Chambon P, Samarut J, Savatier P (2001) An efficient system for conditional gene expression in embryonic stem cells and in their in vitro and in vivo differentiated derivatives. *Proc. Natl Acad. Sci. USA* 98, 2467–2472.

Wang RN, Paraskevas S, Rosenberg L (1999) Characterization of integrin expression in islets isolated from hamster, canine, porcine, and human pancreas. *J. Histochem. Cytochem.* **47**, 499–506.

Wichterle H, Lieberam I, Porter JA, Jessell TM (2002) Directed differentiation of embryonic stem cells into motor neurons. *Cell* **110**, 385–397.

Wilson SI, Edlund T (2001) Neural induction: toward a unifying mechanism. *Nat. Neurosci.* **4 Suppl**, 1161–1168.

Wilson SI, Graziano E, Harland R, Jessell TM, Edlund T (2000) An early requirement for FGF signalling in the acquisition of neural cell fate in the chick embryo. *Curr. Biol.* **10**, 421–429.

Xian HQ, McNichols E, St Clair A, Gottlieb DI (2003) A subset of ES-cell-derived neural cells marked by gene targeting. *Stem Cells* **21**, 41–49.

Xu C, Inokuma MS, Denham J, Golds K, Kundu P, Gold JD, Carpenter MK (2001) Feeder-free growth of undifferentiated human embryonic stem cells. *Nat. Biotechnol.* **19**, 971–974.

Xu C, Police S, Rao N, Carpenter MK (2002a) Characterization and enrichment of cardiomyocytes derived from human embryonic stem cells. *Circ. Res.* **91**, 501–508.

Xu RH, Chen X, Li DS, Li R, Addicks GC, Glennon C, Zwaka TP, Thomson JA (2002b) BMP4 initiates human embryonic stem cell differentiation to trophoblast. *Nat. Biotechnol.* **20**, 1261–1264.

Yang H, Wright JR Jr (2002) Human beta cells are exceedingly resistant to streptozotocin in vivo. *Endocrinology* **143**, 2491–2495.

Ying QL, Nichols J, Chambers I, Smith A (2003a) BMP induction of Id proteins suppresses differentiation and sustains embryonic stem cell self-renewal in collaboration with STAT3. *Cell* **115**, 281–292.

Ying QL, Stavridis M, Griffiths D, Li M, Smith A (2003b) Conversion of embryonic stem cells into neuroectodermal precursors in adherent monoculture. *Nat. Biotechnol.* **21**, 183–186.

Yoder MC (2001) Introduction: spatial origin of murine hematopoietic stem cells. *Blood* **98**, 3–5.

Zambrowicz BP, Friedrich GA, Buxton EC, Lilleberg SL, Person C, Sands AT (1998) Disruption and sequence identification of 2,000 genes in mouse embryonic stem cells. *Nature* **392**, 608–611.

Zhang SC, Wernig M, Duncan ID, Brustle O, Thomson JA (2001) In vitro differentiation of transplantable neural precursors from human embryonic stem cells. *Nat. Biotechnol.* **19**, 1129–1133.

Zwaka TP, Thomson JA (2003) Homologous recombination in human embryonic stem cells. *Nat. Biotechnol.* **21**, 319–321.

5. Human mesenchymal stem cells and multilineage differentiation to mesoderm lineages

Virginie Sottile and Jim McWhir

5.1 Human mesenchymal stem cells (hMSCs) in culture

5.1.1 Introduction: from bone marrow stromal cells to mesenchymal stem cells

The work of Friedenstein and Owen on the osteogenic potential of bone marrow cells is widely acknowledged as the foundation of the present research on adult stem cells. Early transplantation studies in rodents 30 years ago established that fragments of marrow had osteogenic as well as chondrogenic potential when implanted *in vivo* (Friedenstein *et al.*, 1966). Subsequently a heterogeneous population of non-hematopoietic cells with fibroblastic morphology were isolated *in vitro*: the stromal cells from the bone marrow. Because these cells adhere to tissue culture plastic, they could be easily separated from the more prevalent hematopoietic cellular elements in crude bone marrow cell preparations. In addition, they are able to grow by giving rise to FCFC (fibroblast colony-forming cells) or CFU-f (colony-forming unit-fibroblastic), thereby permitting expansion *in vitro*. Upon re-implantation *in vivo* using diffusion chambers, these cells gave rise to osteochondrogenic and connective tissue (Friedenstein *et al.*, 1968; Friedenstein, 1976; Ashton *et al.*, 1980; Owen, 1988), even after more than 20 cell divisions in culture (Friedenstein *et al.*, 1987). This result suggested that marrow stroma contained osteogenic progenitors as well as precursors of other cell types. Comparing tissues formed *in vivo* from cultures derived from individual CFU-f cultures indicated that stromal cells are a heterogeneous population with regard to their differentiation capacity (Kuznetsov *et al.*, 1997). But the fact that at least some samples generated from a single initial CFU-f contained several cell types (osteoblasts, chondrocytes, adipocytes, fibroblasts) supported the hypothesis that these various lineages formed in diffusion chambers may arise from a common precursor present in the stromal population (Friedenstein, 1980; Friedenstein *et al.*, 1987; Owen, 1988). Initial animal model studies were later extended to stromal cells derived from human bone marrow (Gundle *et al.*, 1995; Mankani *et al.*, 2001), leading to similar observations.

There was then a legitimate temptation to apply to this stromal compartment the hematopoietic model, where a hematopoietic stem cell (HSC) gives rise to all cell types of the immune system (Prockop, 1997). Caplan first introduced the notion of a mesenchymal stem cell with similar properties in the early 1990s (Caplan, 1991). According to this model, MSCs give rise to osteochondrogenic, adipogenic, myogenic, fibroblastic and stromal lineages in the bone marrow (Pittenger and Marshak, 2001).

The concept of the MSC has modified the way in which many researchers think about tissue differentiation. Traditionally, organs have been understood to harbor specialized precursors with restricted potential. Confusion about the multipotency of precursors has led to the appearance of a rich nomenclature to refer to these cells: bone marrow stromal cells, bone marrow precursor cells, bone marrow mesenchymal cells, marrow stromal stem cells, mesenchymal stem cells (Owen, 1988; Caplan, 1991; Kuznetsov et al., 1997; Krebsbach et al., 1999). This semantic ambiguity reveals the lack of obvious and rigorous criteria to characterize this cell population, and it is this latest term of mesenchymal stem cells that persists despite the controversy. Despite this widely acknowledged ambiguity (Pittenger and Marshak, 2001), the term human mesenchymal stem cell (hMSC) will be used in this chapter.

The proportion of MSCs in bone marrow is relatively low. Various groups report 1 to 10 MSCs per 10^5 mononucleated marrow cells (Friedenstein, 1976, 1980; Jaiswal et al., 1997; Zuk et al., 2001). Interestingly, the frequency of HSCs is estimated to be 1 to 50 out of 10^5 nucleated cells (Micklem et al., 1987; Abkowitz et al., 2000), indicating comparable frequencies for these two stem cell populations. However, a major difference between HSCs and MSCs resides in the remarkable plasticity among MSC descendants (Park et al., 1999; Bianco et al., 1999; Bianco and Gehron Robey, 2000), whereas HSC-derived cell types are terminally differentiated lineages (Lawman et al., 1992).

5.1.2 A controversial cell system

Progressive refinement of culture conditions has enabled a closer *in vitro* study of marrow stromal cells, which started being publicized when the word 'stem' became incorporated into their common name. According to the minimalist definition by Morrison et al., (1997), 'stemness' can be defined by two primitive criteria: a proliferative potential enabling the maintenance of the stem cell subset, and the capacity to generate several cell types (Mezey and Chandross, 2000). Many studies have shown that stromal cells mineralize under osteogenic treatment, generate fat cells under adipogenic treatment, and form chondrogenic pellets exposed to an appropriate protocol (Mackay et al., 1998; Pittenger et al., 1999; Muraglia et al., 2000). This property was mainly reported using cultures derived from distinct CFU-f, suggesting that MSCs might contain a mixture of progenitors having pre-existing monopotential accounting for the observed multipotency. However, stromal cultures expanded from a single stromal cell have been demonstrated to be multipotent *in vitro* (Halleux et al., 2001), thus ruling out this possibility.

Recent evidence that spontaneous cell fusion can occur in an *in vitro* co-culture setting (Ying et al., 2002; Terada et al., 2002; Pells et al., 2002) restarted the scientific debate about the adult stem cell concept, and made re-examination of previous data supporting multipotency a priority. Without entering the debate, which needs facts rather than arguments, it is obvious that the controversy benefits from the widespread lack of systematic karyotypic analysis of primary cultures studied as adult stem cell lines. However, at least one of the early reports on MSCs provided evidence of a normal karyotype for the culture studied (Pittenger et al., 1999; supplementary data), thus supporting the concept that the multipotency observed would not be due to aneuploidy or massive chromosomal re-arrangements.

5.1.3 hMSC isolation and characterization: a poorly defined cell type

Isolation techniques

Murine MSCs can be readily extracted from long bones (femur, tibiae) by marrow flushing (Friedenstein et al., 1968; Beresford et al., 1992; Hanada et al., 1997). The relatively simple purification protocol relies on the capacity of stromal cells to adhere in culture. Deprived of this property, hematopoietic material is progressively eliminated through medium changes. MSC cultures have been established from various animal species (Jessop et al., 1994; Kuznetsov and Gehron Robey, 1996; Kadiyala et al., 1997; Awad et al., 1999) including non-human primates (Bartholomew et al., 2001). Because of their therapeutic relevance, human MSCs have rapidly emerged as the focus of this research. Human samples are increasingly accessible to research laboratories, generally obtained by the invasive protocol of marrow aspiration from the iliac crest (Jaiswal et al., 1997; Conget et al., 1999). The panel of recommended intermediate steps of purification is quite large and undergoes constant updating through emerging publications.

Adherence of MSCs to tissue culture plastic has historically been the classic method of purification of marrow stromal cells (Phinney et al., 1999), and despite its 'low tech' connotation, this method is still used in many laboratories, sometimes combined with a preliminary step of Percoll or Ficoll-based gradient centrifugation to enrich for mononuclear cells (Colter et al., 2000; Hung et al., 2002a,b; Quirici et al., 2002). Because of the complex composition of bone marrow, a further purification step was devised to sort the MSC subset using empirical physical criteria such as size and granularity (Zohar et al., 1997; Colter et al., 2001; Hung et al., 2002a,b). It is however possible that these somewhat subjective criteria are influenced by the specific culture regimen used. If so, it may prove difficult to rely on such steps in the absence of standardized culture conditions.

Early attempts to identify stromal precursors from the bone marrow aimed at purifying osteoprogenitors from other cell populations. This approach provided some candidate antibodies such as SB-10 (Bruder et al., 1998), which were subsequently shown to have a variable specificity according to the maturity status of the precursor (Caplan, 1991; Bruder et al., 1997). Tracking down the osteoprogenitor eventually led to the development of monoclonal antibodies against human marrow stromal cells (Haynesworth et al., 1992a). Antibodies such as SH-2, SH-3 (Pittenger et al., 1999) or Stro-1 (Gronthos et al., 1994; Walsh et al., 2000) recognize a stromal cell population containing MSCs, but solid evidence is still lacking to support their strict specificity for stem cells as opposed to more committed precursors from mesenchymal lineages (Haynesworth et al., 1992b). Similarly, progressive detection of surface receptors like nerve growth factor receptor (NGF-R), although non-specific, drives some affinity-based selection attempts (Jones et al., 2002; Quirici et al., 2002).

Acknowledging this absence of reliable specific antibodies to select unequivocally MSCs, several groups have developed a more pragmatic approach involving negative selection to subtract cells from the marrow population belonging to defined lineages. Practically, cells are selected for failing to be recognized by markers of known lineages, which in effect represents more enrichment than sorting *per se*. Whereas some protocols rely on basic hematopoietic cell elimination by excluding CD45+ (broad hematopoietic marker) and Glycophorin A+ (erythrocyte marker) cells (Lodie et al., 2002; Zhao et al., 2002), others report the use of increasingly complex combinations of multiple antibodies (Krause et al., 2001; Orlic et al., 2001), which gradually become commercially available as preparation kits (Merino et al., 2003). The diversity of isolation methods makes direct comparison of reported observations difficult, although some evidence suggests similarities between the various cell populations independently described (Lodie et al., 2002). Heterogeneity may therefore partly arise from the analysis criteria rather than the analyzed material itself.

Standard culture conditions

The maintenance of hMSCs in culture commonly involves a high serum content, typically 10% (Pittenger et al., 1999; Muraglia et al., 2000; Jones et al., 2002) and sometimes up to 20% FCS (Colter et al., 2000), which is in some cases batch-tested to promote a better growth in culture. Serum-free conditions have, however, been investigated (Gronthos and Simmons, 1995), in order better to define the factors necessary to sustain MSCs in culture. The emergence of the work on MAPCs, grown in 2% serum (Jiang et al., 2002, see Chapter 3), further questions the need for culture in the presence of high serum concentrations. Unlike ES cells, MSCs are relatively slowly cycling cells. The addition of basic FGF has been shown to improve culture expansion rate, while preserving their differentiation potential (Locklin et al., 1999; Banfi et al., 2000; Bianchi et al., 2001; Halleux et al., 2001; Tsutsumi et al., 2001). Other growth factors, mainly platelet-derived growth factor (PDGF) and epidermal-derived growth factor (EGF), have been reported to support MSC growth (Gronthos and Simmons, 1995). MSC cultures are generally passaged in the same way as fibroblasts, using trypsin/EDTA (Pittenger et al., 1999; Colter et al., 2000).

Compiling the literature on MSC cell surface markers (markers expressed and not expressed) is quite complex since the markers assessed, the isolation protocols, culture media and coating substrates all vary from report to report, and these are factors known to influence the expression profile (Zhao et al., 2002, supplementary data). However, six different publications present consistent data for the positive expression of CD90 (Thy1), CD49b, CD105 (endoglin, recognized by the aforementioned SH2 antibody), and CD44, and the absence of expression of c-kit, CD45, and CD34 in hMSCs (Pittenger et al., 1999; Colter et al., 2000; Hung et al.,

2002a; Jones et al., 2002; Lodie et al., 2002; Zhao et al., 2002). Expression of receptors for NGF, FGF, EGF, PDGF is also reported.

One of the inherent difficulties of MSC culture is the absence of objective markers for 'normality'. Unlike hES cells, for which Oct4 and coupled SSEA-1/ SSEA-4 analysis enable close monitoring of the undifferentiated state (Draper et al., 2002), no such criteria are available to ensure consistency of MSC cultures over time. As illustrated by some approaches focusing on shape and surface area parameters (Sekiya et al., 2002), research on adult stem cells is still handicapped by a lack of references which make cell culture a blind exercise.

5.1.4 Multiple sources

One interesting feature of this biological field is that it is rapidly expanding, as different laboratories investigate the presence of cells with similar characteristics in their respective tissues of interest. Several reports present the isolation of mesenchymal cells with stem cell characteristics from various sources. Historically, the skeleton has been the source of MSCs solely through its marrow, but thorough analysis of bone tissue itself reveals cells with similar potential (Sottile et al., 2002; Noth et al., 2002). Non-invasive sources are being intensely investigated. Hence, there is enthusiasm surrounding the publication of cord blood-derived 'stromal-like' cells (Zvaifler et al., 2000; Erices et al., 2000), which although different from marrow MSCs (Mareschi et al., 2001) exhibit some similarities. In the 'same vein', circulating skeletal stem cells are reported to be present at very low frequencies in human and mouse blood samples, and to possess MSC characteristics both *in vitro* and after *in vivo* implantation (Kuznetsov et al., 2001). Satellite cells isolated from skeletal muscle have recently been shown to exhibit similar characteristics (Asakura et al., 2001). The prospect of purifying MSCs from adipose tissue is suggested by reports describing multipotent mesenchymal progenitors isolated from liposuction aspirates (Zuk et al., 2001). This particular source of multipotent cells is particularly exciting for the relative ease with which they can be recovered. The increasing number of stem cell sources raises the question of whether purification of adult stem cells from mesenchymal origin may be limited by the culture conditions currently applied for this purpose. However, rigorous caution is required to ensure the consistency of adult stem cell research. With the increasing flow of publications describing MSC-like cell isolation, it appears critical to define stringent common criteria and rigorously assess, as has been done for bone marrow cells, whether their properties are due to a single stem cell or multiple pre-existing progenitors having a more restricted potential.

5.1.5 Recent development: MAPCs

The field of adult stem cell biology has been revitalized by the discovery of MAPCs, and the remarkable work by C. Verfaillie and colleagues establishing the broad multipotency of such cells isolated from adult bone marrow, at a single cell level. These findings, discussed by the discoverer herself in a separate chapter, ought to be acknowledged as a decisive step, which gives adult stem cells, and among them mesenchymal stem cells, a new dynamism in terms of scientific questioning and clinical investigations. Part of the renewed curiosity is stimulated by the MAPC specific isolation and expansion in low serum, and the surprisingly low seeding density required to maintain their growth and phenotype (Zhao et al., 2002). How MAPCs relate to other mesenchymal stem cell preparations is not easily understood, considering the profound divergence of their culture conditions. If we compare the extensive phenotypic study (Zhao et al., 2002) with that reported for human MSCs (Pittenger et al., 1999), a few fundamental differences emerge in the expression pattern of MHC-I molecules, CD49d, and most importantly the expression of Oct4. Oct4 expression has never been reported in hMSC cultures, but this crucial gene associated with the maintenance of pluripotency of ES cells seems expressed, albeit at low levels, in MAPCs. Another remarkable feature of MAPCs is their robust telomerase activity, whereas telomerase expression appears less consistent in hMSCs. Although telomerase activity was detected in some cases (Pittenger et al., 1999), other reports argue the contrary (Okamoto et al., 2002; Banfi et al., 2002). This ambiguity justifies attempts to express this activity ectopically in hMSCs to increase their *in vitro* expansion capacity (Simonsen et al., 2002; Shi et al., 2002).

5.2 Differentiation towards mesenchymal lineages: lessons from hMSCs

5.2.1 In vitro *differentiation towards mesenchymal lineages*

As the title implies, this chapter is meant to focus on mesoderm lineages. Osteogenic, adipogenic, and chondrogenic differentiation in hMSC cultures represent standard *in vitro* assays (Pittenger et al., 1999; Muraglia et al., 2000), and well-established protocols can be briefly described. Adipocytes arise in hMSC cultures following 1–2 week exposure to a minimal cocktail containing a combination of glucocorticoid (dexamethasone), phosphodiesterase inhibitor (isobutylmethylxanthine, IBMX), and agonist of PPARgamma (rosiglitazone or indomethacin), supplemented with insulin (Pittenger et al., 1999; Muraglia et al., 2000; Halleux et al., 2001). This treatment induces a substantial level of adipogenic conversion, which varies depending on cell preparations, and is generally assessed by staining with Oil Red O or Nile Red. Stained cells can then be measured by flow cytometry, enzymatic activity, or by analysis of adipogenic marker expression (Sottile et al., 2002).

Osteogenic differentiation, sometimes considered a default pathway because of its spontaneous appearance in long-term hMSC culture (Muraglia et al., 2000), is induced by classic osteogenic treatments combining dexamethasone, extra-cellular matrix-inducer (ascorbic acid) and a source of inorganic phosphate (beta-glycerophosphate) (Pittenger et al., 1999). Staining for specific markers such as osteocalcin and alkaline phosphatase, and quantitative measurement of calcium deposition resulting from osteoblastic activity are widely used to monitor progression through this lineage. Molecular marker expression using a panel of genes such as collagen1, PTH-receptor, bone sialoprotein and the transcription factor cbfa-1 reflect the positive response of the culture to osteogenic conditions.

Chondrogenic pellets are prepared in micromass cultures maintained in serum-free conditions, in the presence of dexamethasone and TGFbeta (Mackay *et al.*, 1998; Pittenger *et al.*, 1999; Sekiya *et al.*, 2001). Chondrogenic markers such as collagen II, collagen X, aggrecan, syndecan-1 are looked for after 2–3 weeks of treatment, and Saffranin-O staining of pellet sections provides visual assessment of differentiation.

These three lineages have each been separately obtained from mouse ES cultures (Dani *et al.*, 1997; Kramer *et al.*, 2000; Buttery *et al.*, 2001; Phillips *et al.*, 2001), but have not yet been reported in the literature for human ES cells, although broad RT-PCR analysis indicates expression of mesenchymal lineage markers (Itskovitz-Eldor *et al.*, 2000). As mouse and human ES cells differ in many aspects (Odorico *et al.*, 2001), their requirements for *in vitro* differentiation may also differ. It is interesting to consider that the pharmacological basis for inducing cocktails described for mouse ES cell differentiation protocols is closely related to typical recipes used with MSCs. However, mouse ES cell experiments seem to reveal a progressive inducibility of the culture, presenting differentiation as a series of sequential steps to take rather than a single switch event. According to Buttery *et al.* (2001), a latent phase of 14 days before applying osteogenic treatment leads to an increased differentiation response. In a contrasting study, Phillipps *et al.* (2001) demonstrates that temporary exposure to RA between days 2 and 5 is required for osteogenic differentiation of mouse ES cells, and a similar window of susceptibility is revealed during adipogenesis experiments (Dani *et al.*, 1997). Reports on mouse ES cell-derived chondrogenesis also point towards a stage-dependent response to mesenchymal induction, leaving a critical need for the identification of these regulatory steps, which might differ between adult and embryonic material. Careful study of early adipogenesis has, for example, suggested ERK activation as a prerequisite to commitment (Bost *et al.*, 2002), contrary to what was observed for hMSCs (Jaiswal *et al.*, 2000) where ERK inhibition has a promoting role. An evaluation of the differential response of ES cells and MSCs to relevant signaling molecules and growth factors will undoubtedly facilitate the successful adaptation of known mesenchymal differentiation protocols to the particular case of hES cells.

Considering the broader potential assigned to ES cells, directed differentiation towards a particular lineage is likely to be much more complex because of the concomitant need to block unwanted lineages. One illustration of this phenomenon can be found by comparing the impact of the bone morphogenetic protein BMP2 in both cell systems. As suggested by its name, BMP2 strongly promotes osteogenic commitment in cultures of mesenchymal precursors. When applied to cultures of neuroectodermal origin however, BMP2 is known to have a neurogenic effect (Morrison *et al.*, 2000). In teratocarcinomas, BMP2 is additionally reported to promote endodermal maturation (Pera and Herszfeld, 1998). In the case of ES cell cultures, where the differentiation potential is wide and still largely untamed, it is likely that cellular responses to BMP2 will be multifaceted, presenting a more complex picture than that observed for MSCs. Specific induction of one particular lineage by BMP2 might therefore only become apparent for ES cells upon preliminary restriction of their potential to derivatives of this particular germ layer (Phillips *et al.*, 2001).

One particular mesoderm-derived lineage has, however, been thoroughly demonstrated from hES cultures: cardiomyogenic beating structures (see Chapter

12) are easily spotted in embryoid body preparations, where they appear spontaneously (Kehat et al., 2001; Xu et al., 2002). In hMSC cultures, commitment to this lineage is dependent on exposure to the demethylating agent 5-azacytidine (Makino et al., 1999; Hakuno et al., 2002; Toma et al., 2002), thus suggesting that it may be necessary to modulate the expression of important regulatory factors, prior to lineage induction. The contrast between an accessible (hES) and quite remote (hMSC) status for this differentiation program may illustrate the difference between the processes governing mesenchymal commitment in both systems.

5.2.2 Mesenchymal potential assessed through in vivo experiments

Diffusion chamber implantations, followed by subcutaneous or local *in vivo* introduction in a bone non-union gap has long been an important tool in differentiation experiments. This method has extensively exemplified the osteochondrogenic potential of MSCs, as well as their adipogenic capacity (Kuznetsov et al., 1997). The latest development of such functional assessment, which involves loading cells on a favorable scaffold and measuring their contribution to the bone healing process, has been tested in large animal models (Petite et al., 2000), and is even reported in a clinical application in patients (Quarto et al., 2001). Similarly, tissue engineering approaches are being investigated for the repair of cartilage defects (Diduch et al., 2000). Attempts to use MSCs systemically for skeletal applications are somewhat inconclusive, some reports claiming appropriate homing of cells to bones (Devine et al., 2002), others showing major invasion of the lung microvasculature (Gao et al., 2001), highlighting the importance of the site of injection. In cardiac applications, animal studies introducing MSCs at the infarcted location demonstrated successful contribution of the inoculum to myocardial recovery (Toma et al., 2002; Shake et al., 2002; Yau et al., 2003). In general, transplantation experiments show that MSCs seem able to engraft and contribute spontaneously to the functionality of the organ in which they reside while avoiding tumor formation. A distinctive approach involving the transplantation of hMSCs to fetal sheep *in utero* suggests that hMSCs are able to integrate into various tissues and display markers of appropriate lineage differentiation for cartilage, muscle, marrow stroma and cardiac muscle (Liechty et al., 2000).

5.2.3 Non-mesodermal lineages from hMSC

A new and largely unexpected set of observations from different laboratories has led the scientific community to re-examine the plasticity of MSCs and question developmental principles. Studies aimed at introducing marrow stromal cells into the brain have reported successful engraftment and *in situ* differentiation towards the neuroectodermal lineage (Kopen et al., 1999; Zhao et al., 2002). Such a germ layer transition raised considerable optimism that somatic cells may be more broadly therapeutically useful than initially envisaged (Prockop et al., 2000; Mezey and Chandross, 2000; Mahmood et al., 2002; Stewart and Przyborski, 2002). These plasticity data apparently challenge the concept of germ layers (Woodbury et al., 2000; Sanchez-Ramos, 2002). Although caution is imperative in addressing such fundamental issues, careful examination of these principles can only be encouraged after publication of the broad potential of adult MAPCs, and their multilineage contribution (Jiang et al., 2002).

5.3 Synergy of hMSC and hES research

In order to discuss hMSCs in the wider context of hES research, we would like to spend this last section considering how these two research areas can improve each other's knowledge, and synergistically contribute to our understanding of stem cell biology. A recent publication showing the successful culture of hES cells on human adult stromal cells illustrates the progressive convergence of these research fields (Cheng et al., 2003). On a methodological front, the intellectual approach to stem cell characterization in adult stem cell biology already seems informed by research on its ES cell counterpart. For example, clinically orientated hMSC research has historically focused on functional proof of concept rather than scrupulous systematic analysis. Most hMSC cultures have not been karyotyped, or clonally investigated. Increasingly, these questions are taken into account because of the new proximity between adult stem cell and embryo-derived stem cell research.

The present limits of hMSC biology can confidently expect to gain from ES cell-based research. Several laboratories have already taken advantage of telomerase, a hallmark of ES cells, to try to circumvent senescence in hMSC cultures where endogenous telomerase activity, despite initial reports (Pittenger et al., 1999), is commonly acknowledged as being low (Okamoto et al., 2002; Shi et al., 2002; Simonsen et al., 2002; Verfaillie et al., 2002). Introducing ectopic expression to increase their *in vitro* lifespan has also proven beneficial for the osteogenic commitment of theses cells, leading to intriguing speculations on alternative actions of telomerase in mesenchymal differentiation programs. Beyond telomerase itself, ongoing research in elucidating human ES cell self-renewal mechanisms will certainly provide working hypotheses to improve *in vitro* expansion and replenishment of hMSC populations *in vivo*. Technical solutions to overcome the problem of ES cell isolation from non-permissive mouse strains or other animal species (Kawase et al., 1994; McWhir et al., 1996) might have useful applications to improve hMSC expansion from variable sources, for which species and strain effects are also reported (Kuznetsov and Gehron Robey, 1996; Phinney et al., 1999).

In return, hMSC study might deliver some clues to prevent the tumorigenic activity of hES cells. The adult bone marrow stromal stem cell compartment is maintained throughout life, reflecting the physiological need for tissue homeostasis. Very little is known about the renewal or turnover rate of this tissue or the mechanisms that control this process *in vivo*, but it is remarkable that these stem cells remain under tight regulation from their environment to avoid malignancy. The identification of these crucial signals ensuring renewal and appropriate cycle control could inspire strategies to prevent tumor formation of hES cell-derived material. A pragmatic hypothesis is that upon appropriate treatment *in vitro*, hES cell cultures could be stimulated to give rise to a more mature cell population with hMSC characteristics. Such a protocol would contribute valuable fundamental scientific information, as well as an alternative solution to hMSC expansion variability by providing a robust supply of cells.

As previously mentioned, progressive establishment of hES directed differentiation towards mesodermal lineages undoubtedly finds inspiration in the existing data from adult stem cells. Although pharmacological requirements may be different between the two cell systems, effective inducer cocktails provide a useful base for prospective research. This is the approach developed in our laboratory, and we have established a protocol inducing osteogenic differentiation of human ES cells *in vitro* (Sottile et al., 2003). Alizarin Red staining of the mineral deposits induced in hES cell cultures, under osteogenic conditions, is shown in *Color Plate 2*, alongside hMSC cells differentiated under similar conditions. Once such appropriate culture conditions are determined, parallel comparisons of molecular events driving commitment of ES cells and MSCs towards a common lineage are expected to reveal key early events of differentiation and, potentially, to identify new primary regulators. This strategy could be applied to other lineages, and may be useful from the perspective of modulating and possibly broadening the potentiality assigned to hMSCs both for answering fundamental biological questions as well as generating therapeutics. The theoretical possibility has to be considered that what limits the isolation efficiency of hMSCs is only our own technical capacities, and what restricts their pluripotency is our primitive understanding of optimal culture conditions.

Similarly, current studies on somatic cell reprogramming convey both enthusiasm and hope. Techniques involving cell fusion (Tada et al., 2001) or incubation in cell extracts (Hakelien et al., 2002) will hopefully uncover the mechanism(s) revealed by the birth of Dolly (Wilmut et al., 1997), whereby an adult nucleus can undergo reprogramming to an embryonic state when introduced into an ooplast. Identifying the factor(s) capable of reverting adult cells to a more primitive undifferentiated state is among the most exciting prospects in cell biology, and will considerably enrich the application potential of hMSC cells.

5.4 Conclusion

Research on hMSC and hES cells appears complementary and will help us better understand the critical features of stem cell biology, which are contingent on a subtle balance. Self-renewal involves extensive cell division capacity as well as precise control of proliferation. On the other hand, multilineage differentiation involves retaining a broad differentiation potential and possessing the ability to regulate fate decisions. hES cells and hMSCs both hold key properties to achieve this equilibrium. Clinically, each of them may ultimately be favored for different applications. An important issue for the development of future therapeutic strategies will be the qualitative assessment of these two cell sources in parallel.

Acknowledgments

V. Sottile is deeply indebted to Dr K. Seuwen for his patient and generous guidance and discussions over the past few years. V. Sottile was supported by a grant from the Geron Corporation.

References

Abkowitz JL, Golinelli D, Harrison DE, Guttorp P (2000) In vivo kinetics of murine hemopoietic stem cells. *Blood* 96, 3399–3405.

Asakura A, Komaki M, Rudnicki M (2001) Muscle satellite cells are multipotential stem cells that exhibit myogenic, osteogenic, and adipogenic differentiation. *Differentiation* 68, 245–253.

Ashton BA, Allen TD, Howlett CR, Eaglesom CC, Hattori A, Owen M (1980) Formation of bone and cartilage by marrow stromal cells in diffusion chambers in vivo. *Clin. Orthop.* 151, 294–307.

Awad HA, Butler DL, Boivin GP, Smith FN, Malaviya P, Huibregtse B, Caplan AI (1999) Autologous mesenchymal stem cell-mediated repair of tendon. *Tissue Eng.* 5, 267–277.

Banfi A, Muraglia A, Dozin B, Mastrogiacomo M, Cancedda R, Quarto R (2000) Proliferation kinetics and differentiation potential of ex vivo expanded human bone marrow stromal cells: Implications for their use in cell therapy. *Exp. Hematol.* 28, 707–715.

Banfi A, Bianchi G, Notaro R, Luzzatto L, Cancedda R, Quarto R (2002) Replicative aging and gene expression in long-term cultures of human bone marrow stromal cells. *Tissue Eng.* 8, 901–910.

Bartholomew A, Patil S, Mackay A, Nelson M, Buyaner D, Hardy W *et al.* (2001) Baboon mesenchymal stem cells can be genetically modified to secrete human erythropoietin in vivo. *Hum. Gene Ther.* 12, 1527–1541.

Beresford JN, Bennett JH, Devlin C, Leboy PS, Owen ME (1992) Evidence for an inverse relationship between the differentiation of adipocytic and osteogenic cells in rat marrow stromal cell cultures. *J. Cell Sci.* 102, 341–351.

Bianchi G, Muraglia A, Daga A, Corte G, Cancedda R, Quarto R (2001) Microenvironment and stem properties of bone marrow-derived mesenchymal cells. *Wound Repair Regen.* 9, 460–466.

Bianco P, Gehron Robey P (2000) Marrow stromal stem cells. *J. Clin. Invest.* 105, 1663–1668.

Bianco P, Riminucci M, Kuznetsov S, Robey PG (1999) Multipotential cells in the bone marrow stroma: regulation in the context of organ physiology. *Crit. Rev. Eukaryot. Gene Expr.* 9, 159–173.

Bost F, Caron L, Marchetti I, Dani C, Le Marchand-Brustel Y, Binetruy B (2002) Retinoic acid activation of the ERK pathway is required for embryonic stem cell commitment into the adipocyte lineage. *Biochem. J.* 361, 621–627.

Bruder SP, Horowitz MC, Mosca JD, Haynesworth SE (1997) Monoclonal antibodies reactive with human osteogenic cell surface antigens. *Bone* 21, 225–235.

Bruder SP, Ricalton NS, Boynton RE, Connolly TJ, Jaiswal N, Zaia J, Barry FP (1998) Mesenchymal stem cell surface antigen SB-10 corresponds to activated leukocyte cell adhesion molecule and is involved in osteogenic differentiation. *J. Bone Miner. Res.* 13, 655–663.

Buttery LD, Bourne S, Xynos JD, Wood H, Hughes FJ, Hughes SP, Episkopou V, Polak JM (2001) Differentiation of osteoblasts and in vitro bone formation from murine embryonic stem cells. *Tissue Eng.* 7, 89–99.

Caplan AI (1991) Mesenchymal stem cells. *J. Orthop. Res.* **9**, 641–650.

Cheng L, Hammond H, Ye Z, Zhan X, Dravid G (2003) Human adult marrow cells support prolonged expansion of human embryonic stem cells in culture. *Stem Cells* **21**, 131–142.

Colter DC, Class R, DiGirolamo CM, Prockop DJ (2000) Rapid expansion of recycling stem cells in cultures of plastic-adherent cells from human bone marrow. *Proc. Natl Acad. Sci. USA* **97**, 3213–3218.

Colter DC, Sekiya I, Prockop DJ (2001) Identification of a subpopulation of rapidly self-renewing and multipotential adult stem cells in colonies of human marrow stromal cells. *Proc. Natl Acad. Sci. USA* **98**, 7841–7845.

Conget PA, Minguell JJ (1999) Phenotypical and functional properties of human bone marrow mesenchymal progenitor cells. *J. Cell Physiol.* **181**, 67–73.

Dani C (1999) Embryonic stem cell-derived adipogenesis. *Cells Tiss. Organs* **165**, 173–180.

Dani C, Smith AG, Dessolin S, Leroy P, Staccini L, Villageois P, Darimont C, Ailhaud G (1997) Differentiation of embryonic stem cells into adipocytes in vitro. *J. Cell Sci.* **110**, 1279–1285.

Devine MJ, Mierisch CM, Jang E, Anderson PC, Balian G (2002) Transplanted bone marrow cells localize to fracture callus in a mouse model. *J. Orthop. Res.* **20**, 1232–1239.

Diduch DR, Jordan LC, Mierisch CM, Balian G (2000) Marrow stromal cells embedded in alginate for repair of osteochondral defects. *Arthroscopy* **16**, 571–577.

Draper JS, Pigott C, Thomson JA, Andrews PW (2002) Surface antigens of human embryonic stem cells: changes upon differentiation in culture. *J. Anat.* **200**, 249–258.

Erices A, Conget P, Minguell JJ (2000) Mesenchymal progenitor cells in human umbilical cord blood. *Br. J. Haematol.* **109**, 235–242.

Friedenstein AJ (1976) Precursor cells of mechanocytes. *Int. Rev. Cytol.* **47**, 327–359.

Friedenstein AJ (1980) Stromal mechanisms of bone marrow: cloning in vitro and retransplantation in vivo. *Haematol. Blood Transfus.* **25**, 19–29.

Friedenstein AJ, Piatetzky-Shapiro II, Petrakova KV (1966) Osteogenesis in transplants of bone marrow cells. *J. Embryol. Exp. Morphol.* **16**, 381–390.

Friedenstein AJ, Petrakova KV, Kurolesova AI, Frolova GP (1968) Heterotopic of bone marrow. Analysis of precursor cells for osteogenic and hematopoietic tissues. *Transplantation* **6**, 230–247.

Friedenstein AJ, Chailakhyan RK, Gerasimov UV (1987) Bone marrow osteogenic stem cells: in vitro cultivation and transplantation in diffusion chambers. *Cell Tiss. Kinet.* **20**, 263–272.

Gao J, Dennis JE, Muzic RF, Lundberg M, Caplan AI (2001) The dynamic in vivo distribution of bone marrow-derived mesenchymal stem cells after infusion. *Cells Tiss. Organs* **169**, 12–20.

Gronthos S, Simmons PJ (1995) The growth factor requirements of STRO-1-positive human bone marrow stromal precursors under serum-deprived conditions in vitro. *Blood* **85**, 929–940.

Gronthos S, Graves SE, Ohta S, Simmons PJ (1994) The STRO-1+ fraction of adult human bone marrow contains the osteogenic precursors. *Blood* 84, 4164–4173.

Gundle R, Joyner CJ, Triffitt JT (1995) Human bone tissue formation in diffusion chamber culture in vivo by bone-derived cells and marrow stromal fibroblastic cells. *Bone* 16, 597–601.

Hakelien AM, Landsverk HB, Robl JM, Skalhegg BS, Collas P (2002) Reprogramming fibroblasts to express T-cell functions using cell extracts. *Nat. Biotechnol.* 20, 460–466.

Hakuno D, Fukuda K, Makino S, Konishi F, Tomita Y, Manabe T, Suzuki Y, Umezawa A, Ogawa S (2002) Bone marrow-derived regenerated cardiomyocytes (CMG Cells) express functional adrenergic and muscarinic receptors. *Circulation* 105, 380–386.

Halleux C, Sottile V, Gasser J, Seuwen K (2001) Multi-lineage potential of human mesenchymal stem cells following clonal expansion. *J. Musculoskel. Neuron Interact.* 2, 71–76.

Hanada K, Dennis JE, Caplan AI (1997) Stimulatory effects of basic fibroblast growth factor and bone morphogenetic protein-2 on osteogenic differentiation of rat bone marrow-derived mesenchymal stem cells. *J. Bone Miner. Res.* 12, 1606–1614.

Haynesworth SE, Baber MA, Caplan AI (1992a) Cell surface antigens on human marrow-derived mesenchymal cells are detected by monoclonal antibodies. *Bone* 13, 69–80.

Haynesworth SE, Goshima J, Goldberg VM, Caplan AI (1992b) Characterization of cells with osteogenic potential from human marrow. *Bone* 13, 81–88.

Hung SC, Chen NJ, Hsieh SL, Li H, Ma HL, Lo WH (2002a) Isolation and characterization of size-sieved stem cells from human bone marrow. *Stem Cells* 20, 249–258.

Hung SC, Cheng H, Pan CY, Tsai MJ, Kao LS, Ma HL (2002b) In vitro differentiation of size-sieved stem cells into electrically active neural cells. *Stem Cells* 20, 522–529.

Itskovitz-Eldor J, Schuldiner M, Karsenti D, Eden A, Yanuka O, Amit M, Soreq H, Benvenisty N (2000) Differentiation of human embryonic stem cells into embryoid bodies compromising the three embryonic germ layers. *Mol. Med.* 6, 88–95.

Jaiswal N, Haynesworth SE, Caplan AI, Bruder SP (1997) Osteogenic differentiation of purified, culture-expanded human mesenchymal stem cells in vitro. *J. Cell Biochem.* 64, 295–312.

Jaiswal RK, Jaiswal N, Bruder SP, Mbalaviele G, Marshak DR, Pittenger MF (2000) Adult human mesenchymal stem cell differentiation to the osteogenic or adipogenic lineage is regulated by mitogen-activated protein kinase. *J. Biol. Chem.* 275, 9645–9652.

Jessop HL, Noble BS, Cryer A (1994) The differentiation of a potential mesenchymal stem cell population within ovine bone marrow. *Biochem. Soc. Trans.* 22, 248S.

Jiang Y, Jahagirdar BN, Reinhardt RL, Schwartz RE, Keene CD, Ortiz-Gonzalez XR *et al.* (2002) Pluripotency of mesenchymal stem cells derived from adult marrow. *Nature* 418, 41–49.

Jones EA, Kinsey SE, English A, Jones RA, Straszynski L, Meredith DM, Markham AF, Jack A, Emery P, McGonagle D (2002) Isolation and characterization of bone marrow multipotential mesenchymal progenitor cells. *Arthritis Rheum.* 46, 3349–3360.

Kadiyala S, Young RG, Thiede MA, Bruder SP (1997) Culture expanded canine mesenchymal stem cells possess osteochondrogenic potential in vivo and in vitro. *Cell Transplant.* 6, 125–134.

Kawase E, Suemori H, Takahashi N, Okazaki K, Hashimoto K, Nakatsuji N (1994) Strain difference in establishment of mouse embryonic stem (ES) cell lines. *Int. J. Dev. Biol.* 38, 385–390.

Kehat I, Kenyagin-Karsenti D, Snir M, Segev H, Amit M, Gepstein A, Livne E, Binah O, Itskovitz-Eldor J, Gepstein L (2001) Human embryonic stem cells can differentiate into myocytes with structural and functional properties of cardiomyocytes. *J. Clin. Invest.* 108, 407–414.

Kopen GC, Prockop DJ, Phinney DG (1999) Marrow stromal cells migrate throughout forebrain and cerebellum, and they differentiate into astrocytes after injection into neonatal mouse brains. *Proc. Natl Acad. Sci. USA* 96, 10711–10716.

Kramer J, Hegert C, Guan K, Wobus AM, Muller PK, Rohwedel J (2000) Embryonic stem cell-derived chondrogenic differentiation in vitro: activation by BMP-2 and BMP-4. *Mech. Dev.* 92, 193–205.

Krause DS, Theise ND, Collector MI, Henegariu O, Hwang S, Gardner R, Neutzel S, Sharkis SJ (2001) Multi-organ, multi-lineage engraftment by a single bone marrow-derived stem cell. *Cell* 105, 369–377.

Krebsbach PH, Kuznetsov SA, Satomura K, Emmons RV, Rowe DW, Robey PG (1997) Bone formation in vivo: comparison of osteogenesis by transplanted mouse and human marrow stromal fibroblasts. *Transplantation* 63, 1059–1069.

Krebsbach PH, Kuznetsov SA, Bianco P, Robey PG (1999) Bone marrow stromal cells: characterization and clinical application. *Crit. Rev. Oral Biol. Med.* 10, 165–181.

Kuznetsov S, Gehron Robey P (1996) Species differences in growth requirements for bone marrow stromal fibroblast colony formation in vitro. *Calcif. Tissue Int.* 59, 265–270.

Kuznetsov SA, Krebsbach PH, Satomura K, Kerr J, Riminucci M, Benayahu D, Robey PG (1997) Single-colony derived strains of human marrow stromal fibroblasts form bone after transplantation in vivo. *J. Bone Miner. Res.* 12, 1335–1347.

Kuznetsov SA, Mankani MH, Gronthos S, Satomura K, Bianco P, Robey PG (2001) Circulating skeletal stem cells. *J. Cell Biol.* 153, 1133–1140.

Lawman MJ, Lawman PD, Bagwell CE (1992) Ex vivo expansion and differentiation of hematopoietic stem cells. *J. Hematother.* 1, 251–259.

Liechty KW, MacKenzie TC, Shaaban AF, Radu A, Moseley AM, Deans R, Marshak DR, Flake AW (2000) Human mesenchymal stem cells engraft and demonstrate site-specific differentiation after in utero transplantation in sheep. *Nat. Med.* 6, 1282–1286.

Locklin RM, Oreffo RO, Triffitt JT (1999) Effects of TGFbeta and bFGF on the differentiation of human bone marrow stromal fibroblasts. *Cell Biol. Int.* **23**, 185–194.

Lodie TA, Blickarz CE, Devarakonda TJ, He C, Dash AB, Clarke J, Gleneck K, Shihabuddin L, Tubo R (2002) Systematic analysis of reportedly distinct populations of multipotent bone marrow-derived stem cells reveals a lack of distinction. *Tissue Eng.* **8**, 739–751.

Mackay AM, Beck SC, Murphy JM, Barry FP, Chichester CO, Pittenger MF (1998) Chondrogenic differentiation of cultured human mesenchymal stem cells from marrow. *Tissue Eng.* **4**, 415–428.

Mahmood A, Lu D, Wang L, Chopp M (2002) Intracerebral transplantation of marrow stromal cells cultured with neurotrophic factors promotes functional recovery in adult rats subjected to traumatic brain injury. *J. Neurotrauma* **19**, 1609–1617.

Makino S, Fukuda K, Miyoshi S, Konishi F, Kodama H, Pan J *et al.* (1999) Cardiomyocytes can be generated from marrow stromal cells in vitro. *J. Clin. Invest.* **103**, 697–705.

Mankani MH, Kuznetsov SA, Fowler B, Kingman A, Robey PG (2001) In vivo bone formation by human bone marrow stromal cells: effect of carrier particle size and shape. *Biotechnol. Bioeng.* **72**, 96–107.

Mareschi K, Biasin E, Piacibello W, Aglietta M, Madon E, Fagioli F (2001) Isolation of human mesenchymal stem cells: bone marrow versus umbilical cord blood. *Haematologica* **86**, 1099–1100.

McWhir J, Schnieke AE, Ansell R, Wallace H, Colman A, Scott AR, Kind AJ (1996) Selective ablation of differentiated cells permits isolation of embryonic stem cell lines from murine embryos with a non-permissive genetic background. *Nat. Genet.* **14**, 223–226.

Merino A, Mazzara R, Fuste B, Diaz-Ricart M, Rozman M, Lozano M, Ordinas A (2003) Transfusion medicine illustrated. The mesenchymal stem cell revealed. *Transfusion* **43**, 1.

Mezey E, Chandross KJ (2000) Bone marrow: a possible alternative source of cells in the adult nervous system. *Eur. J. Pharmacol.* **405**, 297–302.

Micklem HS, Lennon JE, Ansell JD, Gray RA (1987) Numbers and dispersion of repopulating hematopoietic cell clones in radiation chimeras as functions of injected cell dose. *Exp. Hematol.* **15**, 251–257.

Morrison SJ, Shah NM, Anderson DJ (1997) Regulatory mechanisms in stem cell biology. *Cell* **88**, 287–298.

Morrison SJ, Perez SE, Qiao Z, Verdi JM, Hicks C, Weinmaster G, Anderson DJ (2000) Transient Notch activation initiates an irreversible switch from neurogenesis to gliogenesis by neural crest stem cells. *Cell* **101**, 499–510.

Muraglia A, Cancedda R, Quarto R (2000) Clonal mesenchymal progenitors from human bone marrow differentiate in vitro according to a hierarchical model. *J. Cell Sci.* **113**, 1161–1166.

Noth U, Osyczka AM, Tuli R, Hickok NJ, Danielson KG, Tuan RS (2002) Multilineage mesenchymal differentiation potential of human trabecular bone-derived cells. *J. Orthop. Res.* **20**, 1060–1069.

Odorico JS, Kaufman DS, Thomson JA (2001) Multilineage differentiation from human embryonic stem cell lines. *Stem Cells* 19, 193–204.

Okamoto T, Aoyama T, Nakayama T, Nakamata T, Hosaka T, Nishijo K, Nakamura T, Kiyono T, Toguchida J (2002) Clonal heterogeneity in differentiation potential of immortalized human mesenchymal stem cells. *Biochem. Biophys. Res. Commun.* 295, 354–361.

Orlic D, Kajstura J, Chimenti S, Limana F, Jakoniuk I, Quaini F, Nadal-Ginard B, Bodine DM, Leri A, Anversa P (2001) Mobilized bone marrow cells repair the infarcted heart, improving function and survival. *Proc. Natl Acad. Sci. USA* 98, 10344–10349.

Owen M (1988) Marrow stromal stem cells. *J. Cell Sci. Suppl.* 10, 63–76.

Owen ME, Cave J, Joyner CJ (1987) Clonal analysis in vitro of osteogenic differentiation of marrow CFU-F. *J. Cell Sci.* 87, 731–738.

Park SR, Oreffo RO, Triffitt JT (1999) Interconversion potential of cloned human marrow adipocytes in vitro. *Bone* 24, 549–554.

Pells S, Di Domenico AI, Gallagher EJ, McWhir J (2002) Multipotentiality of neuronal cells after spontaneous fusion with embryonic stem cells and nuclear reprogramming in vitro. *Cloning Stem Cells* 4, 331–338.

Pera MF, Herszfeld D (1998) Differentiation of human pluripotent teratocarcinoma stem cells induced by bone morphogenetic protein-2. *Reprod. Fertil. Dev.* 10, 551–555.

Petite H, Viateau V, Bensaid W, Meunier A, de Pollak C, Bourguignon M, Oudina K, Sedel L, Guillemin G (2000) Tissue-engineered bone regeneration. *Nat. Biotechnol.* 18, 959–963.

Phillips BW, Belmonte N, Vernochet C, Ailhaud G, Dani C (2001) Compactin enhances osteogenesis in murine embryonic stem cells. *Biochem. Biophys. Res. Commun.* 284, 478–484.

Phinney DG, Kopen G, Isaacson RL, Prockop DJ (1999) Plastic adherent stromal cells from the bone marrow of commonly used strains of inbred mice: variations in yield, growth, and differentiation. *J. Cell Biochem.* 72, 570–585.

Pittenger MF, Marshak DR (2001) Mesenchymal stem cells of human adult bone marrow. In: *Stem Cell Biology* (eds DR Marshak, RL Gardner, D Gottlieb). Cold Spring Harbor Laboratory Press, Cold Spring Harbor, NY, pp. 349–373.

Pittenger MF, Mackay AM, Beck SC, Jaiswal RK, Douglas R, Mosca JD, Moorman MA, Simonetti DW, Craig S, Marshak DR (1999) Multilineage potential of adult human mesenchymal stem cells. *Science* 284, 143–147.

Prockop DJ (1997) Marrow stromal cells as stem cells for nonhematopoietic tissues. *Science* 276, 71–74.

Prockop DJ, Azizi SA, Colter D, Digirolamo C, Kopen G, Phinney DG (2000) Potential use of stem cells from bone marrow to repair the extracellular matrix and the central nervous system. *Biochem. Soc. Trans.* 28, 341–345.

Quarto R, Mastrogiacomo M, Cancedda R, Kutepov SM, Mukhachev V, Lavroukov A, Kon E, Marcacci M (2001) Repair of large bone defects with the use of autologous bone marrow stromal cells. *N. Engl. J. Med.* 344, 385–386.

Quirici N, Soligo D, Bossolasco P, Servida F, Lumini C, Deliliers GL (2002) Isolation of bone marrow mesenchymal stem cells by anti-nerve growth factor receptor antibodies. *Exp. Hematol.* 30, 783–791.

Sanchez-Ramos JR (2002) Neural cells derived from adult bone marrow and umbilical cord blood. *J. Neurosci. Res.* 69, 880–893.

Sekiya I, Colter DC, Prockop DJ (2001) BMP-6 enhances chondrogenesis in a subpopulation of human marrow stromal cells. *Biochem. Biophys. Res. Commun.* 284, 411–418.

Sekiya I, Larson BL, Smith JR, Pochampally R, Cui JG, Prockop DJ (2002) Expansion of human adult stem cells from bone marrow stroma: conditions that maximize the yields of early progenitors and evaluate their quality. *Stem Cells* 20, 530–541.

Shake JG, Gruber PJ, Baumgartner WA, Senechal G, Meyers J, Redmond JM, Pittenger MF, Martin BJ (2002) Mesenchymal stem cell implantation in a swine myocardial infarct model: engraftment and functional effects. *Ann. Thorac. Surg.* 73, 1919–1925.

Shi S, Gronthos S, Chen S, Reddi A, Counter CM, Robey PG, Wang CY (2002) Bone formation by human postnatal bone marrow stromal stem cells is enhanced by telomerase expression. *Nat. Biotechnol.* 20, 587–591.

Simonsen JL, Rosada C, Serakinci N, Justesen J, Stenderup K, Rattan SI, Jensen TG, Kassem M (2002) Telomerase expression extends the proliferative life-span and maintains the osteogenic potential of human bone marrow stromal cells. *Nat. Biotechnol.* 20, 592–596.

Sottile V, Halleux C, Bassilana F, Keller H, Seuwen K (2002) Stem cell characteristics of human trabecular bone-derived cells. *Bone* 30, 699–704.

Sottile V, Thomson A, McWhir J (2003) *In vitro* osteogenic differentiation of human ES cells. *Cloning Stem Cells* 5, 149–155.

Stewart R, Przyborski S (2002) Non-neural adult stem cells: tools for brain repair? *Bioessays* 24, 708–713.

Tada M, Takahama Y, Abe K, Nakatsuji N, Tada T (2001) Nuclear reprogramming of somatic cells by in vitro hybridization with ES cells. *Curr. Biol.* 11, 1553–1558.

Terada N, Hamazaki T, Oka M, Hoki M, Mastalerz DM, Nakano Y, Meyer EM, Morel L, Petersen BE, Scott EW (2002) Bone marrow cells adopt the phenotype of other cells by spontaneous cell fusion. *Nature* 416, 54254–54255.

Toma C, Pittenger MF, Cahill KS, Byrne BJ, Kessler PD (2002) Human mesenchymal stem cells differentiate to a cardiomyocyte phenotype in the adult murine heart. *Circulation* 105, 93–98.

Tsutsumi S, Shimazu A, Miyazaki K, Pan H, Koike C, Yoshida E, Takagishi K, Kato Y (2001) Retention of multilineage differentiation potential of mesenchymal cells during proliferation in response to FGF. *Biochem. Biophys. Res. Commun.* 288, 413–419.

Verfaillie CM, Pera MF, Lansdorp PM (2002) Stem cells: hype and reality. *Hematology* (Am. Soc. Hematol. Educ. Program) 369–391.

Walsh S, Jefferiss C, Stewart K, Jordan GR, Screen J, Beresford JN (2000) Expression of the developmental markers STRO-1 and alkaline phosphatase in cultures of human marrow stromal cells: regulation by fibroblast growth factor

(FGF)-2 and relationship to the expression of FGF receptors 1-4. *Bone* **27**, 185–195.

Wilmut I, Schnieke AE, McWhir J, Kind AJ, Campbell KH (1997) Viable offspring derived from fetal and adult mammalian cells. *Nature* **385**, 810–813.

Woodbury D, Schwarz EJ, Prockop DJ, Black IB (2000) Adult rat and human bone marrow stromal cells differentiate into neurons. *J. Neurosci. Res.* **61**, 364–370.

Xu C, Police S, Rao N, Carpenter MK (2002) Characterization and enrichment of cardiomyocytes derived from human embryonic stem cells. *Circ. Res.* **91**, 501–508.

Yau TM, Tomita S, Weisel RD, Jia ZQ, Tumiati LC, Mickle DA, Li RK (2003) Beneficial effect of autologous cell transplantation on infarcted heart function: comparison between bone marrow stromal cells and heart cells. *Ann. Thorac. Surg.* **75**, 169–176.

Ying QL, Nichols J, Evans EP, Smith AG (2002) Changing potency by spontaneous fusion. *Nature* **416**, 545–548.

Zhao LR, Duan WM, Reyes M, Keene CD, Verfaillie CM, Low WC (2002) Human bone marrow stem cells exhibit neural phenotypes and ameliorate neurological deficits after grafting into the ischemic brain of rats. *Exp. Neurol.* **174**, 11–20.

Zohar R, Sodek J, McCulloch CA (1997) Characterization of stromal progenitor cells enriched by flow cytometry. *Blood* **90**, 3471–3481.

Zuk PA, Zhu M, Mizuno H, Huang J, Futrell JW, Katz AJ, Benhaim P, Lorenz HP, Hedrick MH (2001) Multilineage cells from human adipose tissue: implications for cell-based therapies. *Tissue Eng.* **7**, 211–228.

Zvaifler NJ, Marinova-Mutafchieva L, Adams G, Edwards CJ, Moss J, Burger JA, Maini RN (2000) Mesenchymal precursor cells in the blood of normal individuals. *Arthritis Res* **2**, 477–488.

6. Trophoblast differentiation from embryonic stem cells

Thaddeus G. Golos and Ren-He Xu

6.1 Origin and development of the placenta: introduction

During periimplantation embryonic development, a subset of blastomeres of the cleavage stage embryo will become restricted to the trophectoderm and subsequently give rise to the trophoblast lineage of the placenta. Upon implantation, these trophoblast cells diverge down several differentiation pathways. In humans some cells exhibit phenotypic characteristics essential for invasive trophoblast interaction with, and modification of, the endometrium, while the formation of chorionic villi during placental morphogenesis is defined by a characteristic organization of trophoblast and stromal cells. Current research indicates that the formation of the placenta requires both the initiation of intrinsic pathways and response to intercellular cues to coordinate lineage determination, trophoblast differentiation and formation of the chorionic villi. This can be summarized in three central events:

- 1.) **Lineage determination**: Data from the mouse indicates that the determination of the trophectoderm lineage requires both the withdrawal of Oct-4, and the initiation of expression of other transcriptional regulators, including caudal type homeobox transcription factor 2 (cdx2), eomesodermin (eomes) and estrogen related receptor beta (ERRβ).

- 2.) **Trophoblast differentiation**: Networks of transcription factors direct phenotypic differentiation of discrete trophoblast cell types. Trophoblast differentiation requires withdrawal of inhibitory transcription factors as well as expression of activating transcription factors.

- 3.) **Villous morphogenesis**: The formation of chorionic villi in the primate placenta involves the interaction of epithelial and stromal cells. The characteristic villous architecture is established as fetal mesenchyme and vascular endothelia invade into developing trophoblast outgrowths.

Given that the placenta is the first organ to achieve formation in embryonic development, and that the trophoblasts arise from totipotent blastomeres of the cleavage stage preimplantation embryo, these events should be ideally investigated by use of a 'blank palette' upon which putative intrinsic and extrinsic factors can act to set into motion the early embryonic events that ultimately direct placental morphogenesis.

While there has been substantial information gathered regarding the molecular decisions that contribute to trophoblast differentiation and placental development in the mouse, the significance of many of the relevant genes for human placental development remains elusive. This is in part due to the inability to investigate differentiation at the level of the embryo and the implantation site during the morula-blastocyst transition in the human embryo, and the first weeks of human implantation. Primate ES cells have now been shown to have the capacity to differentiate spontaneously to cells of the trophoblast lineage. Thus, human ES cells offer an exciting opportunity to address this critical developmental window at the cellular and molecular level.

The mechanisms that direct trophoblast differentiation and placenta formation are of central concern for human reproduction, fertility and development. An inability to initiate appropriate early placental function (embryo attachment, invasion, and hormone secretion) is likely a significant component of embryo loss in early pregnancy. Additionally, abnormal establishment of the maternal–fetal interface and inappropriate placental development are thought to contribute to the pathogenesis of diseases of later pregnancy (e.g., pre-eclampsia, fetal growth restriction). Finally, placental function is essential for support of fetal growth and development. Compromised or inappropriate placental development can impact not only on fetal well-being, but is also now thought to contribute to adult disease later in life. Thus, a more detailed understanding of human placental development is an important goal for investigators in maternal–fetal medicine.

6.1.1 *Primate trophoblast differentiation and placental morphogenesis*

One striking example of how early primate development differs from that of non-primates is the distinctly unique pattern of primate trophoblast differentiation and placenta formation. The trophectoderm of the pre-implantation primate blastocyst rapidly differentiates to multiple trophoblast phenotypes following attachment and initial invasion of the uterine endometrium. Trophoblast phenotypes include a multinucleated syncytiotrophoblast with critical endocrine activity in support of the ovarian progesterone to sustain the uterine endometrium in early pregnancy. In humans, proliferating trophoblasts contribute to the formation of the syncytiotrophoblasts shortly after attachment to the endometrium. The first week of development also marks the appearance of lacunae within the syncytial layer, which will fill with maternal blood as invasive trophoblasts tap the maternal blood vessels.

As the proliferating trophoblasts form columns separated by lacunae, these columns are 'invaded' first by extraembryonic fetal mesenchyme, and subsequently by endothelial cells forming the fetal villous capillaries (for a detailed description see Benirschke and Kaufmann, 1999). It is during this period of investment of the trophoblasts by the fetal mesenchyme that the characteristic cellular architecture of the chorionic villi is established (*Figure 6.1*). The primate chorionic villi are quite different from placental structures seen in non-primate laboratory or domestic species. The villous cytotrophoblasts (CTBs) represent a relatively undifferentiated mononuclear, replicating cell which continues throughout much of pregnancy to undergo fusion and terminal differentiation to form the syncytiotrophoblast (STB), a multinuclear cell layer which forms the primary fetal interface with the maternal blood space. The STB is also the major source of protein and steroid

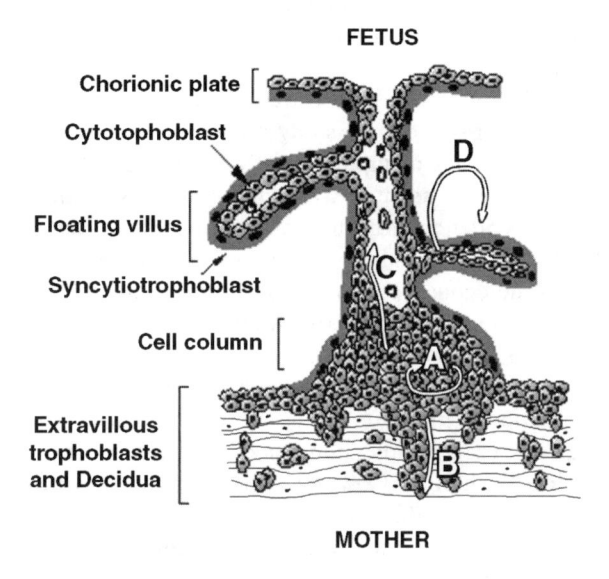

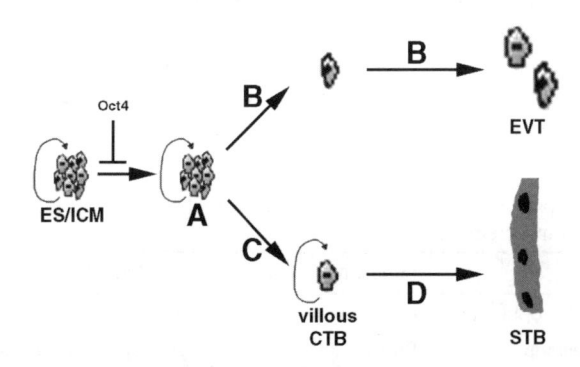

Figure 6.1: Summary of human placental histology (top panel), and cognate stages in human trophoblast differentiation (lower panel) from hES or inner cell mass (ICM) cells. (**A**) Self-renewing trophoblast stem cells. Human trophoblast stem cells remain a virtual cell type. (**B**) Commitment and differentiation to extravillous trophoblasts (EVT). (**C**) Commitment to the villous trophoblast lineage. (**D**) Syncytiotrophoblast (STB) terminal differentiation. Corresponding cells are indicated in both panels as points A–D. Circular arrows indicate self-renewing populations.

hormones produced by the placenta. The villous stroma containing fetal capillaries lie underneath the CTB, separated by a basement membrane.

In addition, the primate placenta includes extravillous trophoblastic elements (Benirschke and Kaufmann, 1999; Enders, 1993). A distinct cell population of extravillous cytotrophoblasts (EVT) develops from the cytotrophoblastic columns, which project from the distal tips of the anchoring villi (*Figure 6.1*). EVTs are highly invasive, penetrating to the level of the maternal uterine vasculature. It seems likely that EVTs play an important role in modifying the vasculature of the endometrium to accommodate the special needs of pregnancy (Cartwright *et al.*, 2002; Zhou *et al.*, 1997). Our understanding of the molecular control of the phenotype of each distinct population of human trophoblasts remains very incomplete, at best.

While early embryonic events are extraordinarily difficult to study in most species, the mouse is well-suited to the study of mammalian development, owing to the wide repertoire of genetic manipulation techniques available to developmental biologists. The capacity of selected transcription factors to direct trophoblast formation and differentiation from totipotent embryonic blastomeres (or their surrogates, ES cells) has been studied in the mouse, particularly through the use of gene deletion by homologous recombination (for excellent reviews see Hemberger and Cross, 2001; Rossant and Cross, 2001). Additional insight has been obtained to events that control subsequent interactions of trophoblasts and villous mesenchymal cells to promote placental morphogenesis in this laboratory species. Nonetheless, limitations remain in the use of the mouse for studying primate placental biology. For example, the mouse does not have an obvious villous 'tree' architecture, as does the human. However, although rodent and primate placentas have clear morphological differences, these mammals do share the general organizational scheme of a chorioallantoic, hemochorial monodiscoid placenta. Despite the obvious differences between human and rodent trophoblast cells, insightful homology has been suggested based on broad functional categories (*Table 6.1*; Rossant and Cross, 2001). Consideration of this perspective is useful since there is significant information that has become available on specific genes contributing to mouse placental development. There is an enormous diversity to mammalian placental structure, both in terms of the morphology of the fetal placenta as well as the nature of attachment and interactions with the maternal

Table 6.1: Comparison of human and mouse placental morphology, and proposed common functional elements.

Human	Function	Mouse
Extravillous cytotrophoblasts	Invasion	Trophoblast giant cells
Cell column; villous cytotrophoblast	Proliferation	Spongiotrophoblast
Syncytiotrophoblasts	Transport	Labryinth (syncytiotrophoblast, glycogen cells)
Proximal trophoblastic columns?	Source for trophoblast proliferation	Trophoblast stem cells: (extraembryonic ectoderm)

endometrium. Thus, we have necessarily limited this chapter to mouse and human species for which there is a sharing of the hemochorial mode of implantation, as well as for which there are well-characterized embryonic stem cells available.

6.1.2 Trophectoderm formation and trophoblast stem cells: initiating placental development

The formation of the trophectoderm is essential for initiating placental development. The POU domain protein Oct-4 (*Pou5f1*) plays a pivotal role in the maintenance of the pluripotent state of the ICM of the preimplantation embryo as well as sustaining undifferentiated proliferation of mouse ES cells (Pesce and Scholer, 2001). In the preimplantation embryo, Oct-4 expression is withdrawn coincident with the formation of the trophectoderm. Knockdown of Oct-4 by genetic and regulated transgene approaches with mouse ES cells has demonstrated that the withdrawal of Oct-4 gives rise to cells with a trophoblast giant cell phenotype (Niwa et al., 2000). This work is nicely complemented by the derivation of mouse trophoblast stem cells (TSC) from cultured mouse blastocysts by treatment with FGF4 and heparin (Tanaka et al., 1998). These cells sustain proliferation in the presence of FGF4, but upon withdrawal of FGF4, there is an increase in the expression of trophoblast giant cell markers, including placental lactogen-I and Hand1 (Tanaka et al., 1998). The role of FGF4 in the formation and maintenance of proliferation in TSC is relevant to the mouse blastocyst, where FGF4 is expressed in the inner cell mass (ICM), and the FGFR2 is expressed in the polar trophectoderm (Haffner-Krausz et al., 1999; Niswander and Martin, 1992; Rappolee et al., 1994). Additionally, deletion of FGF2 or the FGFR2 by homologous recombination results in similar preimplantation lethality (Haffner-Krausz et al., 1999; Niswander and Martin, 1992; Rappolee et al., 1994). Evidence for FGF-mediated MAP kinase signaling in the trophectoderm also supports this model (Rossant, 2001).

Upon TSC differentiation, there is also a decrease in the expression of the caudal-related gene *cdx2*, the T-box gene *eomes* and the *ERRβ*, transcription factors shown by knock-out studies to contribute to mouse placenta formation (Rossant, 2001; Russ et al., 2000; Tremblay et al., 2001). This expression pattern implies that they play a role in sustaining proliferation of a trophoblast stem cell-like phenotype. Examination of the mouse implantation site *in vivo* has provided evidence for a resident trophoblast stem cell population in the extraembryonic ectoderm, in the basal region of the ectoplacental cone (Tanaka et al., 1998). This region expresses *eomes*, *ERRβ*, and *cdx2*, factors, which have sustained expression in mouse trophoblast stem cells (Tanaka et al., 1998). Thus, *in vitro* and *in vivo* studies indicate that a subset of transcription factors is likely to be responsible for sustaining TSC populations upon exit from the totipotent stem cell cycle phenotype. This unique transcriptional milieu is appropriately modeled by TSCs, since TSCs can contribute to all stages of trophoblast differentiation when introduced into the mouse embryo, but not to nonplacental tissues (Tanaka et al., 1998).

It seems likely that the TSC *concept* should be applicable to primate embryos, however the elements by which proliferation is maintained are likely to be different. It is not clear that the FGF4/FGFR2 paradigm is operative in human or non-human primate embryos. Efforts with human embryos in the Rossant laboratory where the mouse TSC model was established have failed to develop TSCs using methods successful with mouse cells (Rossant, 2001). This is in keeping with the observation that both non-human primate as well as human ES cells do not require leukemia inhibitor factor (LIF) to sustain undifferentiated growth, whereas this factor is essential for maintenance of mouse ES cells. Thus, alternative models will be needed to provide insight into the factors that sustain and direct primate trophoblast differentiation.

6.1.3 Molecular control of trophoblast differentiation

Helix-loop-helix factors

So far, we have seen that the initial steps in the formation of the trophectoderm require at least the withdrawal of Oct-4. The trophectoderm, however, is a simple epithelium and the hemochorial placenta contains multiple cell types with responsibility for invasion, endometrial remodeling, and endocrine activity to direct changes in maternal physiology appropriate for sustaining pregnancy. Thus, placental biologists have turned to other differentiation and development systems for insight into the mechanisms that control placental development and trophoblast differentiation.

The decision between growth and development (or stated another way, proliferation and differentiation) is influenced by the transcriptional milieu present. In many cell types, basic helix-loop-helix (bHLH) transcription factors play an important role in differentiation. These factors contain a conserved HLH region that allows hetero- and homodimerization with other HLH factors, and a basic region which directs DNA binding and transactivation (Massari and Murre, 2000). The bHLH factors generally bind to a CANNTG element, or E box, which represents the core motif present in target genes. Networks of bHLH factors bind to and activate E boxes in target genes whose expression constitutes the phenotypic characteristics of the cell in question. In the paradigm as originally described, widely distributed positive regulatory factors (Class A factors, including the mammalian E2A gene products E12 and E47, ITF-2/E2-2, and daughterless (da) in Drosophila (Massari and Murre, 2000)) form dimers with cell and gene-specific Class B HLH factors, such as the myoD/myf family (Olson and Klein, 1994) critical in myogenesis, the insulin transcriptional regulatory factor BETA2/*neuroD* (Naya et al., 1995), the *Tal/Scl-1/lyl-1* factors which play a role in erythroid differentiation (Lister and Baron, 1998), and the neural determination factors *mash-1* (Johnson et al., 1990) and *neuroD* (Lee et al., 1995).

Basic HLH factors driving differentiation are subject to an important checkpoint. Id (inhibitor of differentiation) transcriptional regulators are HLH proteins that lack the basic region responsible for DNA binding in the basic bHLH family (Benezra et al., 1990). Id proteins can function as dominant-negative bHLH proteins by forming high affinity heterodimers with Class A bHLH proteins and preventing binding to DNA. Current evidence indicates that Id proteins not only block differentiation but also promote proliferation; mitogenic signals can upregulate Id (Deed et al., 1997; Persengiev and Kilpatrick, 1997), and cyclin-dependent kinases can phosphorylate and inactivate Id1-3 (Hara et al., 1997). Id proteins have overlapping but distinct patterns of expression. Interestingly, Id2 is typically expressed in epithelia, including the human placenta (Janatpour et al., 2000) while Id1 and 3 are typically expressed in mesenchyme of developing organs (Jen et al., 1996). Other negative regulators of bHLH factors, including I-mfa (Kraut et al., 1998) and E2-2 (Parrinello et al., 2001) have also been identified, thus within a given cell, there is likely to be complex interplay of factors to control cellular phenotype through selective gene expression.

There is information from the literature with mouse genetic models that bHLH factors play an important role in placental development and trophoblast differentiation. The bHLH gene *mash-2* was initially cloned by hybridization with a cDNA for *mash-1*, a bHLH factor gene expressed in the developing central nervous system (Johnson et al., 1990). Mash-2 mRNA was subsequently found to be expressed at high levels in the mouse placenta. Knock-out experiments (Guillemot et al., 1994) suggest that *mash-2* may primarily contribute to spongiotrophoblast proliferation, and that the overall phenotype of these placentas is the lack of a spongiotrophoblast zone, and a relative loss of the trophoblast giant cells (Guillemot et al., 1994). A novel bHLH transcription factor gene was cloned from mouse embryonic libraries and mice in which this factor, Hand1 (previously called Hxt, Thing-1, or eHand), was ablated demonstrate altered placental development, particularly failure of trophoblast giant cell differentiation (Riley et al., 1998), and early pregnancy failure. Consequently, overexpression of Hand1 in mouse cleavage stage embryos or mouse ES cells resulted in loss of embryonic cell division, suggesting that Hand1 may be involved in trophoblast terminal differentiation.

Homeobox factors and other factors in trophoblast differentiation

Homeodomain containing transcription factors play important roles in development, most widely recognized in the control of pattern formation, but also important in organogenesis and cellular differentiation. The role of the POU transcription factor Oct-4 as a gatekeeper for trophectoderm formation has already been discussed. Recently, there have been additional factors identified in the placenta that may play a role in trophoblast differentiation, based on their pattern of expression as well as results of mouse knockout studies. The mouse homolog of the Drosophila *caudal* gene, *Cdx-2*, is expressed in the extraembryonic ectoderm and in the spongiotrophoblast layer of the mouse placenta (Beck et al., 1995). Cdx-2 demonstrates sustained expression in mouse TSCs (Tanaka et al., 1998). Finally, disruption of both *cdx2* alleles is embryonic lethal; cdx2 -/- embryos develop to the blastocyst stage, but do not survive beyond day 3.5–4.5 post-coitum (Chawengsaksophak et al., 1997).

Other homeobox factors may be involved in the control of later stages of trophoblast differentiation. The human placenta expresses the *distal-less* class homeobox genes *dlx3, 4 and 5*, as well as *hlx-1, mox2*, and *msx-2* (Quinn et al., 1997, 1998). Deletion of *dlx-3* results in reduced labyrinth development (Morasso et al., 1999); however, knowledge of the role(s) of the other homeobox factors in the mouse placenta is incomplete.

Other transcription factors clearly play important roles in trophoblast differentiation in the mouse. The novel factor *gcm1* plays an important role in interactions of the chorion and the allantois in the process of labyrinth formation. *Gcm1* is expressed in trophoblasts of the chorionic plate, but in its absence, chorionic differentiation and fusion of the chorioallantois to initiate labyrinth formation does not proceed (Anson-Cartwright *et al.*, 2000). This indicates that trophoblasts play an active role in regulation of morphogenesis.

6.1.4 Morphogenesis in the mouse placenta

Studies with mouse knock-out models have revealed a handful of genes that play roles in the transcriptional control of trophoblast differentiation (summarized above). With the expanding functional and structural complexity of the placenta during its growth and development, the identification of specific roles for discrete genes in these processes becomes complex and daunting. This is logical, as the placenta becomes a mixture not only of multiple derivatives of the trophoblast lineage, but contains fetal mesoderm and endothelium as well. Transcription factors, cell adhesion molecules, soluble factors and their receptors, and signaling pathway molecules have been identified which contribute to appropriate placental morphogenesis, however it is beyond the scope of this review to summarize this information. The reader is referred to recent reviews (Hemberger and Cross, 2001; Rossant and Cross, 2001) for comprehensive summaries of mouse genetic information implicating specific genes that contribute to chorioallantoic fusion and branching, and branching and development in the mouse placental labyrinth.

6.2 Bridging the mouse–human gap in placental biology

While genetic data provide a compelling picture of placental development in the mouse, the relevance of these factors and their roles in human pregnancy and trophoblast differentiation remain unclear. For example, the human homolog of *Mash2*, *Hash2* (Westerman *et al.*, 2001) is expressed in the cytotrophoblastic columns of the placenta, which suggests that this proliferating population may be equivalent to the spongiotrophoblast zone in the mouse placenta, a primary site for *Mash2* expression. Conversely, while Hand1 is a critical component of trophoblast giant cell formation in the mouse, *Hand1* is not expressed in the definitive human placenta (Janatpour *et al.*, 1999; Knofler *et al.*, 2002), although it does appear to be expressed in the embryonic trophectoderm and the amnion (Knofler *et al.*, 2002). The expression of putative bHLH factors in the human placenta has been described (Janatpour *et al.*, 1999), but, in general, the binding partners and target genes for these factors in the human placenta are not known. For example, in the first trimester human placenta Id2 has been proposed to participate in directing proliferation and differentiation (Janatpour *et al.*, 2000). However, it is not clear which factors Id2 interacts with, and thus, precisely how differentiation may be modulated by a bHLH-directed mechanism remains to be established. Clearly new experimental models to address this gap between mouse and humans are needed.

6.2.1. Trophoblast differentiation from human ES cells

Mammalian ES cells can proliferate without a known limit and can form advanced derivatives of all three embryonic germ layers. What remains unclear is whether ES

Chapter 6 – Trophoblast differentiation from embryonic stem cells

cells can also form the extraembryonic tissues that differentiate from the embryo before gastrulation. When formed into chimeras with intact pre-implantation embryos, mouse ES cells rarely contribute to the trophoblast, and the manipulation of external culture conditions has, to date, failed to direct mouse ES cells to trophoblast. The failure to form trophoblast is consistent with the idea that mouse ES cells are developmentally similar to primitive ectoderm, which forms after delamination of the primitive endoderm from the inner cell mass and which no longer contributes to the trophoblast.

Human ES cell lines have been derived from blastocyst-stage preimplantation embryos produced by *in vitro* fertilization (Thomson *et al.*, 1998). Human and non-human primate ES cells express characteristic cell surface markers, including stage specific embryonic antigens, in a pattern quite distinct from rodent ES cell marker expression. LIF fails to prevent the differentiation of human or rhesus monkey ES cells in the absence of fibroblasts, however, conditioned medium from fibroblast feeder layers supplemented with FGF-2 sustains undifferentiated proliferation of human ES cells (Xu *et al.*, 2001). These features reflect fundamental embryological differences between primates and mice.

We have previously shown that ES cells derived from rhesus and common marmoset monkeys have the ability spontaneously to differentiate into trophoblast, as evidenced by secretion of chorionic gonadotrophin (CG) and expression of mRNAs for CG α and β subunits (Thomson *et al.*, 1995, 1996). Although CG secretion is detectable in spontaneously differentiating human ES cell cultures (Thomson *et al.*, 1998), flow cytometry and RT-PCR experiments indicate that spontaneous trophoblast differentiation, while reproducible, is not extensive (Xu *et al.*, 2002). Are there ways to modulate hES cell differentiation to trophoblasts? If the hES cell model is to be a productive system in which to investigate the control of trophoblast differentiation, it will be important to demonstrate that this differentiation can be experimentally manipulated. We have recently shown that this is indeed feasible. We have identified a growth factor, bone morphogenetic protein 4 (BMP4), that can drive human ES cells synchronously and uniformly to differentiate toward a trophoblast lineage (Xu *et al.*, 2002).

Bone morphogenetic proteins (BMPs) are members of the transforming growth factor beta (TGF β) superfamily and embryonic morphogens that have profound influences on patterning and polarity in embryonic development in vertebrates from *Xenopus* through mammals. Treatment of human ES cells with BMP4 resulted in the formation of a uniform population of cells with epithelial appearance within 7 days of treatment (*Figure 6.2A, B*). Other BMP family members, such as BMP2, BMP7, and growth and differentiation factor-5 (GDF5), induced similar morphological changes. However, other TGF-β superfamily members, such as TGF-β1 and activin A, did not induce any noticeable morphological changes. The addition of inhibitors of BMP signaling, such as the soluble BMP receptor (human BMPR-IB/Fc chimera) or the BMP-antagonizing protein noggin, blocked the morphological changes induced by the BMPs. These data suggest that the BMP-induced trophoblast differentiation is a BMP receptor-mediated action, which is probably transduced by the downstream effectors Smad1/Smad5, rather than Smad2/Smad3, to the nucleus.

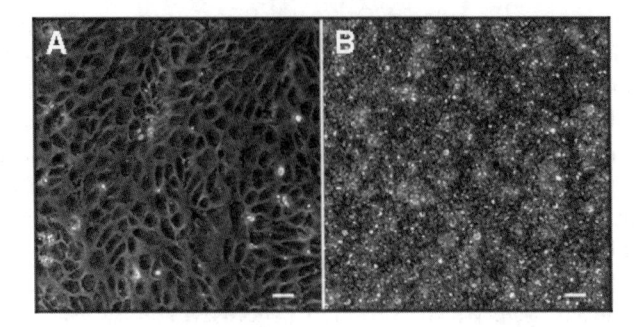

Figure 6.2: Morphological changes of BMP4-treated human ES cells. The ES cell line H1 (propagated in mouse embryonic fibroblast-conditioned medium supplemented with 4 ng/ml bFGF) was treated with (A) or without (B) 100 ng/ml BMP4 for seven days. Bars, 25 µm.

Microarray (Xu et al., 2002) and RT-PCR (*Figure 6.3*) analyses of RNA from undifferentiated and BMP-4-treated hES cells indicate that the expression of many trophoblast or placental marker genes, such as *CGα, CGβ, placental growth factor, Gcm1, HLA-G1,* and *CD9*, is dramatically enhanced by BMP4. Surprisingly, some genes whose homologs are known to be important for mouse trophoblast differentiation, such as *cytokeratin 7, Hash2, ERR-β,* and *Met*, are expressed both in the BMP4-treated ES cells and the untreated and undifferentiated ES cells at apparently similar levels (*Figure 6.3* and Xu et al., 2002). These results, particularly the unvarying expression of Hash2 and ERR β, for example, which would be predicted to be significantly upregulated in trophoblasts, further underscore the apparent differences between mouse ES cells and human ES cells in terms of trophoblast

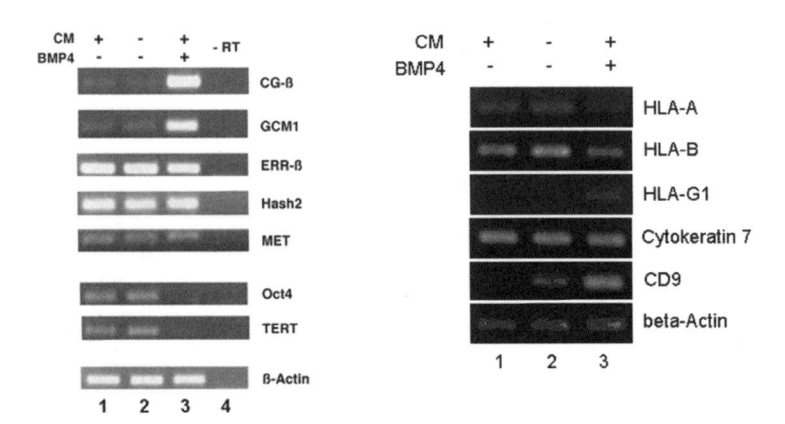

Figure 6.3: RT-PCR analysis of BMP4-treated human ES cells. The ES cell line H1 was cultured in conditioned medium, unconditioned medium, or conditioned medium + BMP4 (100 ng/ml) for 7 days, all in the continuous presence of 4 ng/ml bFGF. Genes known to be expressed in trophoblast (such as *CG-β, Gcm1, ERR-β, Hash2, Met, HLA-G, cytokeratin 7, and CD9*), pluripotent cell marker genes (such as Oct4 and telomerase (TERT)), and HLA class I genes (such as HLA-A and HLA-B) were examined. β-Actin expression was used as an internal control for equal RNA loading. Reactions processed without reverse transcriptase (-RT) serve as negative controls.

differentiation. For a full listing of genes up-regulated by BMP-4 as detected by cDNA microarray, please refer to online supplemental data (Xu et al., 2002). Trophoblast identity was confirmed by immunostaining and flow cytometry analyses for CG-β, and detection of placental hormones hCG, estradiol, and progesterone in media conditioned by the BMP-treated ES cells (Xu et al., 2002).

Moreover, the ES-derived trophoblast can further progress to terminal syncytiotrophoblast differentiation. The occurrence is as high as 7% among individual BMP4-treated ES cells plated at low density, and these syncytial cells contained different numbers of nuclei (from 2 to 100) and were positive for CG-β on immunostaining (Xu et al., 2002). Removal of trophoblast cells from the culture plate promoted vesicle formation while suspended in culture and continuous secretion of placental hormones like CG for many months (data not shown). Of potential physiological relevance is our recent observation that CG secretion from BMP4-differentiated ES cells was further enhanced by co-culture with a human uterine fibroblast line, HUF6 (*Figure 6.4*), suggesting the existence of interactions between the ES cell-derived trophoblasts and their physiological 'host' cells *in vitro*.

The effectors downstream of BMP4 that mediate its differentiation-inducing effects remain unidentified. However, the microarray analysis described above indicates that a subset of the genes upregulated by BMP4 include transcription factors previously implicated in trophoblast differentiation, such as *TFAP2*, *Msx2*, *Gcm1*, *GATA2*, *GATA3*, *Id2-4*, *Dlx4*, *Dlx5*, *SSI3* and *HEY1*. Expression of these

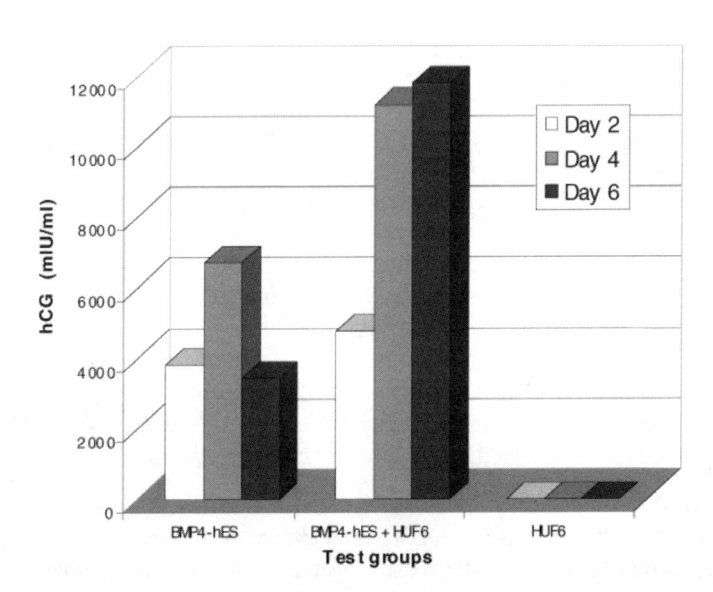

Figure 6.4: hCG levels in media conditioned by co-culture of BMP4-induced trophoblast and HUF6 cells. The hES cell line H1 was treated with 100 ng/ml BMP4 for 7 days to allow their differentiation to trophoblast. The cell colonies were removed from the culture dish by treatment with dispase and transferred onto a layer of a human uterine fibroblast (HUF6) cells. BMP4-treated hES cells transferred to culture dishes without HUF6 cells and HUF6 cells cultured without hES cells served as controls. Media were collected on days 2, 4, and 6 following initiation of co-culture.

genes is increased by BMP4 at various time points examined from 3 h to 7 days of the treatment, suggesting their sequential participation in the signal transduction of trophoblast formation. Further investigations of these transcription factors would provide great insight into the mechanisms by which ES cells exit their stem cell status and commit differentiation to trophoblast.

The question arises as to whether the BMP4-induced cell population represents human trophoblast stem cells. The human equivalent to mouse TS cells has not yet been derived, and it is likely that different growth factors will be required for their propagation. Although we show that BMP4 efficiently induced differentiation of human ES cells to trophoblast, these trophoblast cells propagated poorly, even in the continued presence of bFGF and fibroblast feeder layers (data not shown), suggesting that additional growth factors or other extrinsic modulators are required for their long-term proliferation.

6.2.2 Human ES cells as a model for placental morphogenesis

These studies indicate that trophoblast differentiation can proceed by the activation of specific pathways. However, trophoblast differentiation is only one component of the process of placental morphogenesis, which requires significant temporal and spatial coordination of multiple cell types. We have also investigated whether hES cells offer either *in vivo* or *in vitro* opportunities for studying placental morphogenesis.

Teratoma formation

When human ES cells are injected into immunocompromised mice, they form teratomas with the differentiation of multiple cell types. There is abundant evidence of coordinated interactions between cells, and even between cells originating from different embryonic germ layers. For example, the development of hair requires coordinated interactions between the overlying ectoderm and underlying mesenchyme (Thomson *et al.*, 1998). Thus, human ES cells exhibit excellent pluripotent characteristics. We hypothesized that trophoblasts may differentiate and be organized into villous structures upon teratoma formation from hES cells. We prepared teratomas by injecting human ES cells into the skeletal muscle of the hind limb of SCID mice, and collected solid teratomas 28–47 days post-injection. A wide range of differentiation was observed, including widespread differentiation of epithelia, as indicated by cytokeratin staining. However, immunostaining of adjacent sections for hCG β subunit (to detect syncytiotrophoblast) or HLA-G (to detect extravillous trophoblasts) has failed to detect at this time significant trophoblast differentiation, as judged by the expression of these two markers.

Embryoid body formation

We considered embryoid bodies (EBs) as an alternative to teratoma formation to evaluate trophoblast differentiation from human ES cells. For our studies, EBs were prepared generally as previously described (Itskovitz *et al.*, 2000). Briefly, human ES cells were cultured to 70% confluence and released as intact undifferentiated colonies by brief collagenase treatment. Colonies were then cultured on a continuous rocker plate. After 8 days of suspension culture, EBs will roll up and form spherical to cylindrical structures of up to several hundred μm in diameter.

Chapter 6 – Trophoblast differentiation from embryonic stem cells

We evaluated embryoid bodies for differentiation by immunostaining with cytokeratin (epithelium) and vimentin (mesoderm) antibodies. A substantial number of cells were positive for cytokeratin within embryoid bodies (*Color Plate 3*), whereas only a subset of cells stained for the mesoderm marker vimentin (data not shown).

We harvested culture medium at different times following the initiation of EB culture and assayed for the secretion of CG, progesterone, and estradiol. Within several days after the initiation of EB formation, the secretion of all three hormones was detectable in the culture medium, whereas levels were undetectable (CG) or very low (steroid hormones) in medium not exposed to embryoid bodies (*Figure 6.5*) or in medium from undifferentiated human ES cells (data not shown). The levels of secretion of these steroid hormones were generally correlated with the relative levels of CG secretion (*Figure 6.5*). This triad of reproductive hormones is

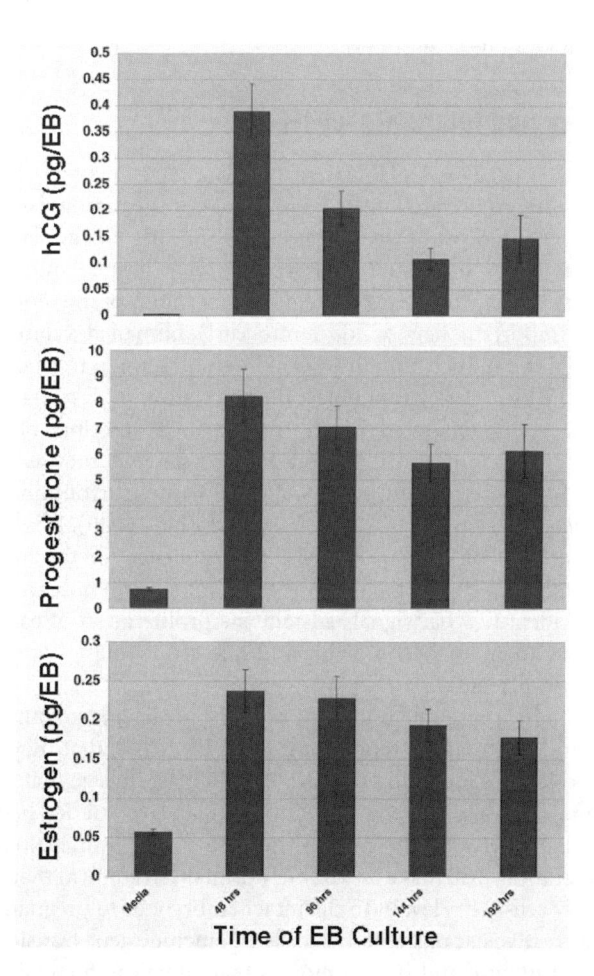

Figure 6.5: Hormone secretion by embryoid bodies. Immunoreactive CG, progesterone (P4) and estradiol 17-β (E2) during 192 h of culture are presented, in comparison with medium not exposed to EBs.

a hallmark of advanced trophoblast differentiation and has been noted in essentially all EB experiments evaluated (data not shown). These results suggest that EBs may be a useful model for the study of trophoblast differentiation. This is supported by the observation that cells within EBs also show positive immunostaining for CG (*Color Plate 3*).

If trophoblast differentiation is a reliable event in embryoid body formation from hES cells, this approach may provide an additional model for the study of placental morphogenesis. Epithelial-mesenchymal interactions have been explored in explant settings as well as *in vivo* in genetically modified mice (e.g., prostate and uterus) (Kurita *et al.*, 2001), and recent studies have shown effects of endothelia on organogenesis in the liver and the pancreas (Lammert *et al.*, 2001; Matsumoto *et al.*, 2001). Adaptation of these concepts to the interactions of ES and placental mesenchymal cells in a three-dimensional setting may provide a more physiological series of cues to generate cellular responses that may culminate in placental morphogenesis (Hagios *et al.*, 1998).

6.3 Summary and future prospects

Because it is not ethically acceptable to manipulate experimentally the postimplantation human embryo, we are largely ignorant about the mechanisms of very early human development. Most of what is known about early postimplantation human development is based on histological sections of a limited number of human embryos and by analogy to the experimental embryology of the mouse. Human ES cells offer an important new window into early human developmental events. Although BMP4 can induce human ES cell differentiation to trophoblast *in vitro*, a direct role of BMPs in early trophoblast differentiation *in vivo* has not yet, to our knowledge, been demonstrated in any mammal. Transcripts of various BMP receptors are present in morula- and blastocyst-stage mouse embryos, and transcripts of BMPs are present in the maternal tissues surrounding the embryos (Fujiwara *et al.*, 2001; Paria *et al.*, 2001; Ying and Zhao, 2000). It is clear that BMP receptors are present on human ES cells. The challenge for the future will be to determine whether BMP signals have a role in human trophoblast differentiation *in vivo* and to identify what signals sustain the proliferation of early trophoblast cells and direct them to become the multiple trophoblast populations of the definitive human placenta.

Thus, the studies presented underline both the power and an inherent weakness of this new model. A major strength of human ES cells is that they give access to early human cell types including the early trophoblast lineage that was heretofore essentially unobtainable, thereby allowing comparative studies between various species, especially, between the mouse and human. A distinct limitation is that ethical considerations will make it extremely difficult to confirm that *in vitro* results with these early cells are relevant to the intact embryo, or to pregnancy *in vivo*.

Thus, the creative use of human ES cells will include gene transfer into ES cells, development of *in vitro* and *in vivo* models that allow careful analysis of cell–cell interactions, and ultimately interfacing with non-human primate models which now allow transgene expression in embryos (Chan *et al.*, 2001; Wolfgang *et al.*,

2002) and in placentas (Wolfgang et al., 2001). Fetal growth restriction, gestational diabetes, preeclampsia and premature labor are conditions of human pregnancy with high societal and economic costs, and whose pathophysiology is intimately tied to placental development and trophoblast differentiation. The promise of human ES cells in maternal–fetal medicine is the opportunity to model placental development at its earliest stages with the long-term goal of applying experimental insights to therapeutic development.

Acknowledgments

We acknowledge the contributions of our colleagues, especially Dong Li (WiCell Research Institute), Gregory C. Addicks, Clay Glennon, Thomas P. Zwaka and James A. Thomson (Wisconsin Primate Research Center and Dept. of Anatomy, University of Wisconsin Medical School) for work on BMP4-induced human ES cell differentiation; and Behzad Gerami-Naini, Maureen Durning, Andy Ryan, Becky Norris and Oksana Dovzhenko (Wisconsin Primate Research Center) for work with embryoid bodies and teratomas. We thank our collaborators Xin Chen, Rui Li, and Patrick Brown (Stanford University School of Medicine, Stanford, CA) for microarray analysis, and Edgardo Carosella and Nathalie Rouas-Friess (Hopital Saint-Louis, Paris) for HLA-G antibodies. Richard Grendell and Robert Becker provided expert assistance in figure preparation.

References

Anson-Cartwright L, Dawson K, Holmyard D, Fisher SJ, Lazzarini RA, Cross JC (2000) The glial cells missing-1 protein is essential for branching morphogenesis in the chorioallantoic placenta. *Nature Genet.* 25, 311–314.

Beck F, Erler TA, Russell TA, James R (1995) Expression of Cdx-2 in the Mouse embryo and placenta: Possible role in patterning of the extra-embryonic membranes. *Develop. Dynamics* 204, 219–227.

Benirschke K, Kaufmann P (1999) *Pathology of the Human Placenta*, 4th edn. Springer-Verlag, New York.

Benezra R, Davis RL, Lockshon D, Turner DL, Weintraub H (1990) The protein Id: a negative regulator of helix-loop-helix DNA binding proteins. *Cell* 61, 49–59.

Cartwright JE, Kenny LC, Dash PR, Crocker IP, Aplin JD, Baker PN, Whitley GS (2002) Trophoblast invasion of spiral arteries: a novel in vitro model. *Placenta* 23, 232–235.

Chan AWS, Chong KY, Martinovich C, Simerly C, Schatten G (2001) Transgenic monkeys produced by retroviral gene transfer into mature oocytes. *Science* 291, 309–312.

Chawengsaksophak K, James R, Hammond VE, Kontgen F, Beck F (1997) Homeosis and intestinal tumours in Cdx2 mutant mice. *Nature* 386, 84–87.

Deed RW, Hara E, Atherton GT, Peters G, Norton JD (1997). Regulation of Id3 cell cycle function by Cdk-2-dependent phosphorylation. *Mol. Cell. Biol.* 17, 6815–6821.

Enders AC (1993) Overview of the morphology of implantation in primates. In: *Primates, in In vitro Fertilization and Embryo Transfer in Primates* (eds DP Wolf, RL Stouffer, RM Brenner). Springer-Verlag, New York, pp. 145–157.

Fujiwara T, Dunn NR, Hogan BLM (2001) Bone morphogenetic protein 4 in the extraembryonic mesoderm is required for allantois development and the localization and survival of primordial germ cells in the mouse. *PNAS* **98**, 13739–13744.

Guillemot F, Nagy A, Auerbach A, Rossant J, Joyner AL (1994) Essential role of Mash-2 in extraembryonic development. *Nature* **371**, 333–336.

Haffner-Krausz R, Gorivodsky M, Chen Y, Lonai P (1999) Expression of FGFR2 in the early mouse embryo indicates its involvement in preimplantation development. *Mech. Dev.* **85**, 167–172.

Hagios CA, Lochter A, Bissell MJ (1998) Tissue architecture: the ultimate regulator of epithelial function? *Philosophical Trans. Royal Soc. London Series B: Biol. Sci.* **353**, 857–870.

Hara E, Hall M, Peters G (1997) Cdk2-dependent phosphorylation of Id2 modulates activity of E2A-related transcription factors. *EMBO J.* **16**, 332–342.

Hemberger M, Cross JC (2001) Genes governing placental development. *Trends Endocrinol. Metab.* **12**, 162–168.

Itskovitz-Eldor J, Schuldiner M, Karsenti D, Eden A, Yanuka O, Amit M, Soreq H, Benvenisty N. (2000) Differentiation of human embryonic stem cells into embryoid bodies compromising the three embryonic germ layers. *Molec. Med.* **6**, 88–95.

Janatpour MJ, Utset MF, Cross JC, Rossant J, Dong J, Israel MA, Fisher SJ (1999) A repertoire of differentially expressed transcription factors that offers insight into mechanisms of human cytotrophoblast differentiation. *Develop. Gen.* **25**, 146–157.

Janatpour MJ, McMaster MT, Genbacev O, Zhou Y, Dong J, Cross JC, Israel MA, Fisher SJ (2000) Id-2 regulates critical aspects of human cytotrophoblast differentiation, invasion and migration. *Development* **127**, 549–558.

Jen Y, Manova K, Benezra R (1996) Expression patterns of Id1, Id2, and Id3 and highly related but distinct from that of Id4 during mouse embryogenesis. *Developmental Dynamics: an official publication of the Am. Assn. Anatomists* **207**, 235–252.

Johnson JE, Birren SJ, Anderson DJ (1990) Two rat homologues of *Drosophila achaete-scute* specifically expressed in neuronal precursors. *Nature* **346**, 858–861.

Knofler MG, Meinhardt G, Bauer S, Loregger T, Vasicek R, Bloor DJ, Kimber SJ, Husslein, P (2002) Human Hand1 basic helix-loop-helix (bHLH) protein: extra-embryonic expression pattern, interaction partners and identification of its transcriptional repressor domains. *Biochem. J.* **361** (Pt 3), 641–651.

Kraut N, Snider L, Amy-Chen CM, Tapscott SJ, Groudine M (1998) Requirement of the mouse I-mfa gene for placental development and skeletal patterning. *EMBO J.* **17**, 6276–6288.

Kurita T, Cooke PS, Cunha GR (2001) Epithelial-stromal tissue interaction in paramesonephric (Mullerian) epithelial differentiation. *Developmental Biol.* 240, 194–211.

Lammert E, Cleaver O, Melton D (2001) Induction of pancreatic differentiation by signals from blood vessels. *Science* 294, 564–567.

Lee JE, Hollenberg SM, Snider L, Turner DL, Lipnic N, Weintraub H (1995) Conversion of Xenopus ectoderm into neurons by NeuroD, a basic helix-loop-helix protein. *Science* 268, 836–844.

Lister JA, Baron MH (1998) Induction of basic helix-loop-helix protein-containing complexes during erythroid differentiation. *Gene Expression* 7, 25–38.

Massari ME, Murre C (2000) Helix-loop-helix proteins: regulators of transcription in eucaryotic organisms. *Molec. Cell. Biol.* 20, 429–440.

Matsumoto K, Yoshitomi H, Rossant J, Zaret KS (2001) Liver organogenesis promoted by endothelial cells prior to vascular function. *Science* 294, 559–563.

Morasso MI, Grinberg A, Robinson G, Sargent TD, Mahon KA (1999) Placental failure in mice lacking the homeobox gene Dlx3. *Proc. Natl Acad. Sci. USA* 96, 162–167.

Naya FJ, Stellrecht CCM, Tsai M-J (1995) Tissue-specific regulation of the insulin gene by a novel basic helix-loop-helix transcription factor. *Genes Dev.* 9, 1009–1019.

Niswander L, Martin GR (1992) FGF-4 expression during gastrulation, myogenesis, limb and tooth development in the mouse. *Development* 114, 755–768.

Niwa H, Miyazaki J, Smith AG (2000) Quantitative expression of Oct-3/4 defines differentiation, dedifferentiation or self-renewal of ES cells. *Nature Gen.* 24, 372–376.

Olson EN, Klein WH (1994) bHLH factors in muscle development: dead lines and commitments, what to leave in and what to leave out. *Genes Dev.* 8, 1–8.

Paria BC, Ma W, Tan J, Raja S, Das SK, Dey SK, Hogan BLM (2001) Cellular and molecular responses of the uterus to embryo implantation can be elicited by locally applied growth factors. *PNAS* 98, 1047–1052.

Parrinello S, Lin CQ, Murata K, Itahana Y, Singh J, Krtolica A, Campisi J, Desprez PY (2001) Id-1, ITF-2, and Id-2 comprise a network of helix-loop-helix proteins that regulate mammary epithelial cell proliferation, differentiation and apoptosis. *J. Biol. Chem.* 276, 39213–39219.

Persengiev SP, Kilpatrick DL (1997) The DNA methyltransferase inhibitor 5-azacytidine specifically alters the expression of helix-loop-helix proteins Id1, Id2, and Id3 during neuronal differentiation. *Neuroreport* 8, 2091–2095.

Pesce M, Scholer HR (2001): Oct-4, gatekeeper in the beginnings of mammalian development. *Stem Cells* 19, 271–278.

Quinn LM, Johnson BV, Nicholl J, Sutherland GR, Kalionis B (1997) Isolation and identification of homeobox genes from the human placenta including a novel member of the Distal-less family, DLX4. *Gene* 187, 55–61.

Quinn LM, Latham SE, Kalionis B (1998) A distal-less class homeobox gene, DLX4, is a candidate for regulating epithelial-mesenchymal cell interactions in the human placenta. *Placenta* 19, 87–93.

Rappolee DA, Basilico C, Patel Y, Werb Z (1994) Expression and function of FGF-4 in peri-implantation development in mouse embryos. *Development* **120**, 2259–2269.

Riley P, Anson-Cartwright L, Cross JC (1998) The Hand1 bHLH transcription factor is essential for placentation and cardiac morphogenesis. *Nature Gen.* **18**, 271–275.

Rossant J (2001) Stem cells from the Mammalian blastocyst. *Stem Cells* **19**, 477–482.

Rossant J, Cross JC (2001) Placental development: lessons from mouse mutants. *Nature Rev.: Genetics* **2**, 538–548.

Russ AP, Wattler S, Colledge WH, Aparicio SA, Carlton MB, Pearce JJ et al. (2000) Eomesodermin is required for mouse trophoblast development and mesoderm formation. *Nature* **404**, 95–99.

Tanaka S, Kunath T, Hadjantonakis AK, Nagy A, Rossant J (1998) Promotion of trophoblast stem cell proliferation by FGF4. *Science* **282**, 2072–2075.

Thomson JA, Kalishman J, Golos TG, Durning M, Harris CP, Becker RA, Hearn JP (1995) Isolation of a primate embryonic stem cell line. *Proc. Natl. Acad. Sci. USA* **92**, 7844–7848.

Thomson JA, Kalishman J, Golos TG, Durning M, Harris MC, Becker R, Hearn JP (1996) Pluripotent cell lines derived from common marmoset (*Callithrix jacchus*) blastocysts. *Biol. Reprod.* **55**, 254–259.

Thomson JA, Itskovitz-Eldor J, Shapiro SS, Waknitz MA, Swiergiel JJ, Marshall VS, Jones JM (1998) Embryonic stem cell lines derived from human blastocysts. *Science* **282**, 1145–1147.

Tremblay GB, Kunath T, Bergeron D, Lapointe L, Champigny C, Bader JA, Rossant J, Giguere V (2001) Diethylstilbestrol regulates trophoblast stem cell differentiation as a ligand of orphan nuclear receptor ERR*b*. *Genes Dev.* **15**, 833–838.

Westerman BA, Poutsma A, Looijenga LHJ, Wouters D, van Wijk IJ, Oudejans CBM (2001) The Human Achaete Scute Homolog 2 gene contains two promotors, generating overlapping transcripts and encoding two proteins with different nuclear localization. *Placenta* **22**, 511–518.

Wolfgang MJ, Eisele SG, Browne MA, Schotzko ML, Garthwaite MA, Durning M, Ramezani A, Hawley RG, Thomson JA, Golos TG (2001) Rhesus Monkey Placental Transgene Expression after Lentiviral Gene transfer into preimplantation embryos. *Proc. Natl Acad. Sci. USA* **98**, 10728–10732.

Wolfgang MJ, Marshall VS, Eisele SG, Schotzko ML, Thomson JA, Golos TG (2002) Efficient method for expressing transgenes in nonhuman primate embryos using a stable episomal vector. *Mol. Reprod. Dev.* **62**, 69–73.

Xu C, Inokuma MS, Denham J, Golds K, Kundu P, Gold JD, Carpenter MK (2001) Feeder-free growth of undifferentiated embryonic stem cells. *Nature Biotechnol.* **19**, 971–974.

Xu R-H, Chen X, Li DS, Li R, Addicks GC, Glennon C, Zwaka TP, Thomson JA (2002) BMP4 initiates human embryonic stem cell differentiation to trophoblast. *Nature Biotechnol.* Published online 11 November 2002.

Ying Y, Zhao G-Q (2000) Detection of multiple bone morphogenetic protein messenger ribonucleic acids and their signal transducer, Smad1, during mouse decidualization. *Biol. Reprod.* **63**, 1781–1786.

Zhou Y, Fisher SJ, Janatpour M, Genbacev O, Dejana E, Wheelock M, Damsky CH (1997) Human cytotrophoblasts adopt a vascular phenotype as they differentiate. A strategy for successful endovascular invasion? *J. Clin. Invest.* **99**, 2139–2151.

7. Current and future prospects for hematopoiesis studies using human embryonic stem cells

Dan S. Kaufman

7.1 Introduction

Cell-based medicine and other forms of 'regenerative medicine' have recently gained considerable interest as potential means to treat a variety of diseases. This attention results from the convergence of a variety of recent findings, including the derivation of human embryonic stem (ES) cells (Thomson et al., 1998) and experiments that suggest 'adult' stem cells may be less developmentally restricted than previously recognized (reviewed in refs. (Blau et al., 2001; Lagasse et al., 2001; Verfaillie, 2002). However, for a hematologist, this focus on cellular therapies is nothing new. Hematology is probably the most cell-based medical discipline. The body produces billions of individual blood cells every day. Blood, as a collection of cells and proteins, can be easily sampled and analyzed in minute detail. In cases where blood is lost (trauma), or blood production is diminished due to disease or iatrogenic events (chemotherapeutic drugs), transfusions of specific blood products such as red blood cells or platelets are routinely administered. Many diseases that affect blood cells such as leukemia, multiple myeloma, aplastic anemia, or immunodeficiencies can now be best treated by bone marrow transplantation, a form of 'stem cell therapy' that has been in clinical use for over 30 years (Thomas, 1999). This therapy is now more correctly called hematopoietic cell transplantation (HCT), since the hematopoietic stem cells are not always collected from bone marrow. Here, chemotherapy, at times combined with immunological methods, is used to both kill the diseased cells and allow engraftment of healthy hematopoietic stem cells (HSCs). The transplanted HSCs can either be from the affected patient (autologous transplantation), or another donor (allogeneic transplantation). Over 30 000 patients receive HCT each year (International Bone Marrow Transplant Registry, 2002). In large part, the clinical success of HCT has grown out of studies of basic blood development (hematopoiesis). In order to make further advances in HCT, an even more sophisticated knowledge of basic hematopoiesis is mandatory. Human embryonic stem (ES) cells now offer an invaluable starting point to make new strides towards understanding blood development, ultimately contributing to progress in clinical hematology and HCT.

Despite the clinical success of HCT in treating a host of malignant and non-malignant diseases, the procedure is still fraught with difficulties. Foremost among these is often the lack of a suitable donor. Whereas solid organ transplants do not require precise matching of human leukocyte antigens (HLA) between donor and host, HCT is most successful when the donor and host are perfect HLA matches. Depending on the degree of precision desired, 6, 8, or 10 HLA antigens may be typed at a molecular level to ensure the best histocompatibility. If an allogeneic donor is required, the first choice is always an HLA-matched sibling. Each sibling has only a one-in-four chance of being an appropriate match. Unfortunately, an HLA-matched sibling donor is often not available and alternative donors are sought. Typically, a search for an unrelated HLA-matched donor ensues. There are now over eight million donors who have volunteered to be potential HCT donors (Bone Marrow Donors Worldwide, 2003). However, despite this generous number, many patients, especially ethnic minorities or people of mixed racial heritage, do not have a suitable donor. Moreover, it can take several months to arrange collection of HSCs from a donor, and some patients suffer disease relapse in the interim. Other sources of HSCs for transplantation, such as umbilical cord blood and partially-HLA matched (haploidentical) marrow, are being studied in clinical trials as means to most effectively treat patients who would benefit form HCT. However, HCT using donor cells from any of these allogeneic sources continues to face complications such as graft failure, graft-versus-host disease, or disease relapse, often with a fatal result. The known clinical benefit of HCT can create a frustrating situation when patients do not have a suitable donor or suffer untoward side effects of the procedure. Therefore human ES cells are an extremely attractive area for hematologists to investigate in order to advance clinical prospects for future patients.

The benefits of hematopoiesis research on human ES cells extend well beyond clinical HCT. While any list that tries to enumerate reasons to pursue a particular area of research is bound to be incomplete, some of these reasons include the following. (1) Human ES cells allow study of the earliest stages of basic human hematopoietic development. While considerable knowledge of mammalian hematopoiesis has been gained from work on mouse embryos, mouse ES cells, and other model systems (such a zebrafish), there are important differences between mice and humans during embryonic development. This includes differences in yolk sac development, the site of primitive hematopoiesis (Palis and Yoder, 2001). Other studies of human hematopoiesis typically use phenotypically defined HSCs isolated from umbilical cord blood or bone marrow. However, these post-natal sources cannot be used to characterize earlier stages of hematopoiesis that leads to production of HSCs in the first place. (2) Understanding early stages of human hematopoiesis may permit better '*ex vivo* expansion' of human HSCs as an alternative source of cells for patients who require HCT. Considerable research has been carried out to define methods that allow the expansion of HSCs without further differentiation in culture, yet this remains an elusive goal. Use of human ES cells to characterize genes and proteins that regulate early hematopoiesis may translate into more successful expansion of HSCs for clinical applications. In this way, human ES cells will not directly be the source of cells for HCT or other therapies. However, these studies could lead to better utilization of other, more accessible cell types. (3) Human ES cells could be the starting material to produce blood cells (erythrocytes, platelets) for transfusion medicine. With advances in clinical medicine in areas such as surgery, transplantation, and chemotherapy for malignancies, the need for these blood products continues to increase. At the same time, the pool of volunteer blood donors continues to shrink due to increased concern about blood borne diseases that are increasingly difficult to detect. Most recently, anxiety about bovine spongiform encephalopathy ('Mad Cow Disease') has led to the elimination of individuals who have traveled for an extended time to Great Britain and some other areas of Western Europe from the blood donor pool. Also, West Nile Virus and other emerging infections may be transmitted via blood transfusions (Anonymous, 2002). The ability efficiently to induce human ES cells to differentiate into terminally differentiated red blood cells and platelets could provide a stable, defined source of cells known to be free of exogenous viruses or other pathogens. Indeed, production of these terminally differentiated blood cells may be technically easier then isolation of HSCs from human ES cells, and some progress in large-scale production of *in vitro* derived red cells and megakaryocytes (platelet precursors) has been described (Eto *et al.*, 2002; Neildez-Nguyen *et al.*, 2002; Pick *et al.*, 2002).

The ability to derive hematopoietic cells from human ES cells may also lead to an innovative means to prevent immunologic rejection of other human ES cell-derived cells. Speculation regarding the ability to co-transplant human ES cell-derived HSCs along with a second human ES cell-derived cell type of interest (such pancreatic beta-cells for diabetics) in order to induce tolerance has been described elsewhere (Kaufman and Thomson, 2002; Odorico *et al.*, 2001). Briefly, studies of HCT recipients have shown that HCT can induce immunological tolerance to other cells or tissues (typically a kidney) from the same donor that was the source of the blood cells (Down and White-Scharf, 2003; Millan *et al.*, 2002; Spitzer *et al.*, 1999). Similarly, it may be possible to derive HSCs and other cell types from the same parental ES cell line. These two cell types would be identical at all major and minor histocompatibility loci. Thus, the induction of hematopoietic chimerism from the ES cell-derived HSCs would theoretically prevent rejection of the other ES cell-derived cells without need for long-term immunosuppression.

7.2 Lessons from mouse ES cell-based hematopoiesis

Much of the interest and excitement regarding the prospects of human ES cells is a direct consequence of knowing the utility of mouse ES cells for defining cellular and molecular pathways of mammalian development. Not long after mouse ES cells were first characterized, Doetschman and colleagues demonstrated that when these cells were induced to differentiate *in vitro*, blood and other defined cell types could be characterized (Doetschman *et al.*, 1985). Subsequent studies have used several methods to promote differentiation of mouse ES cells in order to define the developmental pathways of a range of hematopoietic cell types (Keller, 1995; Smith, 2001). However, despite the ability to derive all types of blood cells from ES cells *in vitro*, it has been exceedingly difficult to demonstrate long-term, multi-lineage engraftment of mouse ES cell-derived blood cells *in vivo*. The reasons for this transplantation barrier are unclear. However, mouse ES cells that over-express genes that regulate hematopoiesis can be used to produce engraftable hematopoietic stem cells. Initial studies expressed bcr/abl, a fusion protein that is pathologic in chronic myelogenous leukemia (CML). Bcr/abl-positive mouse ES cells were induced to form embryoid bodies (EBs) that contained hemtaopoietic cells. Transplantation of these cells led to production of myeloid, erythroid and lymphoid cells *in vivo* (Perlingeiro et al., 2001). Unfortunately and not surprisingly, these cells were leukemogenic. However, subsequent work found productive hematopoietic progenitors could be produced by HoxB4 expression (Kyba et al., 2002). While these cells were not found to be leukemogenic, hematopoietic reconstitution was not entirely normal. Also, in an elegant *tour de force*, combined cell and gene therapy were used to 'cure' a mouse model of genetic disease. Using Rag 2-deficient mice, somatic cell nuclear transfer was done to produce embryos with this immunodeficiency. ES cell lines were then derived having the genetic make-up of the Rag-2$^{-/-}$ mouse. One mutant allele was genetically corrected via homologous recombination. These now corrected Rag-2 ES cells (Rag 2$^{+R/-}$) were induced to form HSCs that were then transfected with HoxB4. Finally, the Hox B4$^+$, Rag 2$^{+R/-}$ ES cell derived HSCs were transplanted into Rag 2$^{-/-}$ mice with successful engraftment and improved B cell and T cell function (Rideout et al., 2002). Together, these results show the utility and barriers to ES cell-based therapies.

7.3 Hematopoiesis from human ES cells: studies to date

Initial studies to derive hematopoietic cells from human ES cells sought to define a method or methods efficiently and reproducibly to induce blood cell development. Some particular cell lineages, such as pancreatic islet cells, cardiomyocytes or neurons can be characterized by specific cellular proteins that are central to their function (such as insulin, myosin, and neurotransmitters). However, hematopoietic lineages are commonly identified by surface antigens, such as CD34 and CD45, which may or may not be directly related to specific cell functions. Moreover, these surface antigens are not always specific or unique to a particular lineage, or even to blood cells. For example, CD34 is an excellent marker of HSCs and is used clinically to define engraftment potential for HSCT. However, CD34 is also expressed on other cell types such as endothelial cells, and some CD34 cells may also represent HSCs capable of long-term engraftment (Dao et al., 2003; Krause et al., 1996; Sato et al., 1999). The most definitive test of hematopoietic cell function is *in vivo* engraftment and function. However, as alluded to above, work in the mouse ES cell-hematopoiesis system has shown that long-term multilineage engraftment of ES cell-derived hematopoietic cells can be quite difficult without the addition of exogenous genes. Why there is such a barrier to engraftment of ES cell-derived blood cells remains unclear. NK cell-mediated killing of ES cell-derived cells that express low level of MHC class I molecules may play a role (Rideout et al., 2002). Others speculate that ES cell-derived blood cells may not home efficiently to the hematopoietic microenvironment, leading to poor engraftment and growth. With this daunting barrier to *in vivo* engraftment, initial studies focused on multiple complementary methods of *in vitro* characterization of hematopoietic lineages. These methods include gene expression, surface antigen expression and a functional *in vitro* assay for hematopoietic colony-forming cells that has been well validated to identify early hematopoietic precursors.

To induce hematopoietic development from these cells, our initial strategy was to co-culture human ES cells on stromal cells or accessory cells that were known to support the growth and expansion of human HSCs (Kaufman et al., 2001). Our initial studies focused on two mouse cell lines: S17 cells derived from mouse bone marrow (Collins and Dorshkind, 1987), and C166 cells derived from mouse yolk sac (Wang et al., 1996). Other cell lines including OP9 from the calvaria of an M-CSF-deficient mouse (Nakano et al., 1994) and primary human bone marrow stromal cells (Simmons and Torok-Storb, 1991) have also been successfully used to support the differentiation of hematopoietic cells from human ES cells (unpublished observations). S17 and C166 offer the advantage of being easy to maintain in culture and can be genetically modified if desired to characterize the role of stromal cell surface antigens or secreted factors. Also, since these are mouse-derived cell lines, proteins belonging to the human ES cell-derived cells can be easily distinguished with use of antibodies specific for human antigens. Human ES cells co-cultured with S17 or another of these stromal cells in medium containing fetal bovine serum (FBS) form colonies that become interdigitated with the stromal cells and can be seen to differentiate within about 3–5 days (*Figure 7.1*). After about 7–10 days, multiple types of differentiated cells can be seen, including some cells that resemble endothelial or hematopoietic cells (*Figure 7.1*). $CD34^+$ cells are seen first after about 7 days of differentiation under these conditions. Flow cytometric analysis for other hematopoietic surface antigens such as CD45, CD15, CD31 and glycophorin A can also be seen after 7–14 days of differentiation on S17 cells.

Hematopoietic colony-forming assays involve placing a single cell suspension in semi-solid methylcellulose-based media containing hematopoietic cytokines or growth factors. Typically blood or bone marrow cells are used, but in this case, differentiated human ES cell-derived cells can be analyzed. Under these conditions, a single hematopoietic colony-forming cell (CFC) will repeatedly divide to form a large cluster of cells that can be characterized based on the stereotypic appearance of the colony (Eaves and Lambie, 1995). Myeloid cells, granulocytes, erythrocytes, and megakaryocytes can all be readily identified in this manner. If desired, individual colonies can be picked, placed on glass slides and stained to confirm identity of the cells. In the case of ES cell-derived cells, this identification is important to be sure that the pluripotent cells are not just forming colonies that mimic the appearance of blood cells. Also, flow cytometric analysis and CFC assays can be combined to demonstrate that sorting for $CD34^+$ cells also enriches for CFCs, a finding expected from studies of blood and bone marrow-derived hematopoietic cells.

The CFC assay has probably been the most important means to characterize hematopoietic precursor cells derived from human ES cells. In the developmental system described above involving co-culture on S17 cells in FBS-containing media, CFC are seen after approximately 14 days of differentiation, peak around 17 days, and diminish after about 21 days of differentiation (Kaufman et al., 2001). Both

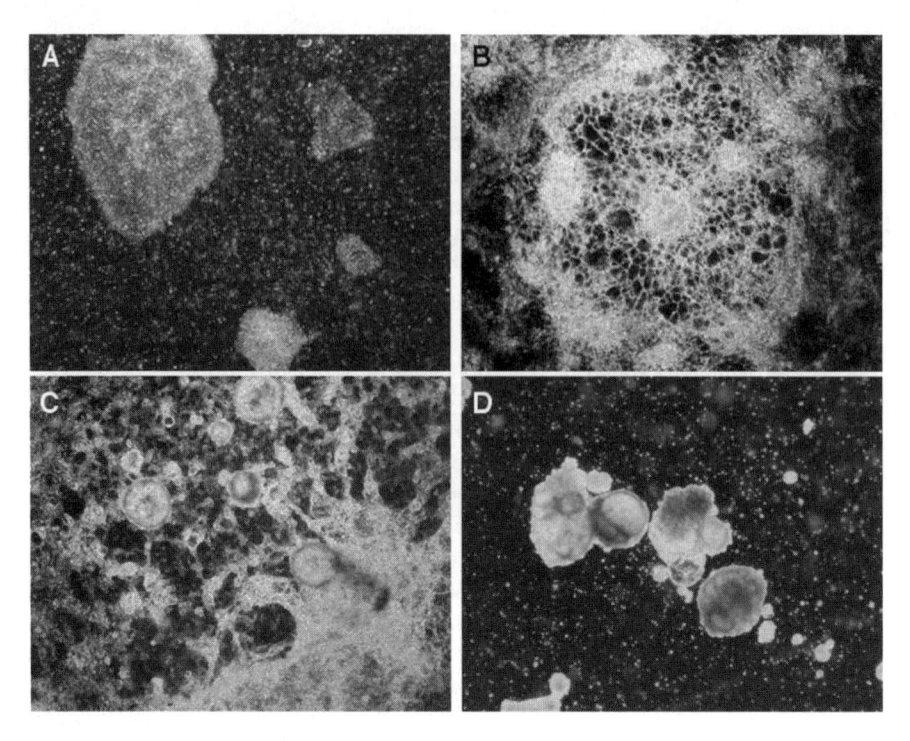

Figure 7.1: Hematopoietic differentiation of human ES cells. (A) Colonies of undifferentiated human ES cells demonstrating uniform morphology. (B) Human ES cells allowed to differentiate on S17 stromal cells for 8 days. The majority of cells in this image are derived from a single colony that has differentiated into multiple cell types including thin endothelial-type structures and more densely piled-up regions. (C) Human ES cells allowed to differentiate on S17 cells for 16 days. These cells are now seen to form spherical, cystic structures and a variety of other cell types. (D) Human ES cells induced to form embryoid bodies in suspension for 14 days. Cystic EBs are formed that resemble cystic structures in panel C. All images original magnification 20×.

myeloid and erythroid colonies can be identified. Importantly, if human ES cells are co-cultured on S17 cells in serum-free media, no CFCs are produced. Also, if human ES cells are induced to differentiate on fibroblast cells instead of S17 cells, few if any CFCs are produced. Therefore, a combination of serum factors and stromal cell factors are needed to produce hematopoietic CFCs in optimal numbers. Interestingly, $CD34^+$ cells are readily seen to develop from human ES cells under conditions that do not support the development of CFCs. Since CD34 is not a unique hematopoietic marker, other cells such as $CD34^+$ endothelial cells are likely produced under these less-specific conditions.

7.4 Hematopoiesis from human ES cells: the next stage

The experiments described above do not identify cellular and genetic mechanisms that promote hematopoietic development, but provide a potential *in vitro* model system to study normal and pathologic regulation of this process. At a very basic level, it is unclear whether the conditions (serum and stromal cells) that lead to differentiation into CD34$^+$ cells and CFCs actually direct the differentiation of human ES cells down a hematopoietic pathway, or whether these conditions simply support the survival and growth of hematopoietic precursor cells that arise randomly as the ES cells spontaneously differentiate. Studies of gene expression in other models of hematopoiesis suggest that this second scenario of stochastic or 'primed' development is more likely (Akashi et al., 2003; Delassus et al., 1999; Enver et al., 1998). The accessibility of human ES cells and the ability to define their environment will allow future studies to characterize gene expression in ES cells as they differentiate down specific developmental pathways.

Since a combination of serum factors and stromal factors are required for CFC development, the next stage is to define better the nature of these stimuli. Preliminary studies by our group have found that culture of human ES cells on fibroblasts, but fed with S17 conditioned media (containing FBS), does not sufficiently replicate growth of the ES cells cultured in direct contact with S17 cells. This suggests that cell surface protein interactions are needed to support the growth or survival of hematopoietic CFCs. However, when human ES cells were grown in transwell cultures with S17 cells somewhat different results were observed. The transwell culture physically separates the S17 cells and ES cells, but one micron sized pores allow diffusion of soluble factors. Under these conditions, CD34$^+$ cells and CFCs are identified at a rate similar to ES cells cultured in direct with S17 cells. Somewhat surprisingly, ES cells in transwell culture with MEFs (fibroblasts) also efficiently differentiate into CD34$^+$ cells and CFCs, unlike ES cells cultured in the same media, but in direct contact with MEFs. These results suggest that direct contact with MEFs may actually hinder development of CFCs. Alternatively, the nature of differentiation in transwell cultures, where the ES cells form more three-dimensional and cystic-type structures may be fundamentally different from ES cells in contact with a stromal layer. Indeed, the differentiated ES cells in transwell cultures seem to resemble embryoid bodies (EBs) more than cells that differentiate in the presence of a stromal layer. As described below, EBs derived from human ES cells grown in FBS do give rise to CD34$^+$ cells and CFCs without need for accessory cells.

The presence of serum obviously adds a considerable variable to hematopoietic differentiation of human ES cells. Several groups working with mouse ES cells or other HSC cultures have sought to use chemically-defined media that will support hematopoiesis. In these studies, serum-free media supplemented with defined hematopoietic cytokines or growth factors can support hematopoietic development (Adelman et al., 2002) and limited HSC expansion (Lebkowski et al., 1995). Similarly, our preliminary studies suggest that while culture of human ES cells on S17 cells in serum-free media does not lead to significant hematopoietic development, addition of a cytokine cocktail containing stem cell factor (SCF), Flt3-ligand (Flt3L) and thrombopoietin (TPO) will support development to CD34$^+$ cells and CFCs to levels comparable to culture in FBS (Kaufman et al., 2002).

EB formation can also be used as a means to promote hematopoietic differentiation of human ES cells. EBs are commonly used in studies of hematopoiesis with mouse ES cells. However, the nature of EB formation is different between mouse and human ES cells. Mouse ES cells kept in suspension in the absence of leukemia inhibitory factor (LIF) or other gp130 agonists, will readily form EBs within a few days (Keller, 1995). These mouse ES cell-derived EBs undergo an ordered process of differentiation that at least in part recapitulates events during early embryogenesis. In contrast human ES cell-derived EBs are best produced from intact human ES cell colonies, without individualization into single cells. These EBs will form complex structures with many cystic regions and other evidence of differentiation (*Figure 7.1*). Whether the developmental process is as ordered as mouse EB development has not yet been determined. However, addition of defined cytokines to developing human EBs can regulate their developmental potential (Schuldiner *et al.*, 2000). In our studies, human EBs grown for approximately 2 weeks form many $CD34^+$ cells and CFCs. Not surprisingly, EBs allowed to develop in serum-free media do not show these markers of hematopoietic development. However, in contrast to human ES cell differentiation on S17 cells, the addition of SCF, Flt3L, and TPO to serum-free media does not support hematopoietic development. These results suggest that part of the mechanism by which FBS promotes or supports hematopoiesis is by indirect action on the S17 or other stromal cells. Alternatively, FBS may function to promote an equivalent stromal-type cell within the EB, and the substituted cytokines do not have this same effect.

Other recent studies have examined the effects of cytokines plus FBS on hematopoietic differentiation of human ES cell-derived EBs (Chadwick *et al.*, 2003). Here, EBs were cultured in a cytokine cocktail in media containing 20% FBS. By using a greater number and higher concentration of hematopoietic supporting cytokines in combination with FBS, a culture system enriched for $CD45^+$ cells and CFCs could be produced. Addition of BMP4 seemed particularly effective at increasing the efficiency of hematopoietic differentiation in this EB-based system. A similar role for BMP4 was demonstrated for hematopoietic differentiation of rhesus monkey ES cells in the S17 stromal co-culture model(Li *et al.*, 2001).

The results described from the aforementioned studies of hematopoiesis clearly demonstrate that $CD34^+$ cells and hematopoietic CFCs can be derived from human ES cells. While these studies identify cells having important characteristics of early hematopoietic precursor cells, they do not yet define a hematopoietic stem cell (HSC) population originating from human ES cells. CFCs remain an intermediary stage of development between HSC and more terminally differentiated lineages and cells other than HSCs can express CD34. While an HSC population must exist at some time between the ES cell stage and development of downstream lineages such as erythrocytes and monocytes, these cells may only transiently survive within the culture conditions used. HSCs are best demonstrated by *in vivo* engraftment studies. As mentioned earlier in this chapter, these transplantation experiments with mouse ES cell derivatives have been notoriously difficult for reasons that are somewhat unclear and controversial. *In vitro* surrogate assays for HSCs such as long-term culture initiating cells (LTC-ICs), myeloid-lymphoid initiating cells (ML-IC), and cobblestone area-forming cells (CAFCs) are useful in characterizing the phenotype of cells with HSC-like properties. However, the best characterization of human HSCs is achieved by transplantation into immunodeficient animals (NOD/SCID mice) or pre-immune animals (fetal sheep) (Dao and Nolta, 1999; Lapidot *et al.*, 1997; Zanjani, 2000).

Reports using human bone marrow or umbilical cord blood demonstrate that phenotypically homogeneous cells can sustain long-term engraftment as xenografts into these animals. Moreover, bone marrow or cord blood HSCs, or similar populations such as SCID reconstituting cells (SRCs) can be serially transplanted between animals. These results fulfill the definition of HSCs as a population of cells that can maintain long-term self-renewal and multi-lineage differentiation. Whether a similar population can be purified from human ES cells remains to be seen. Preliminary results by Zanjani and colleagues suggest that human ES cells induced to form hematopoietic cells by co-culture with S17 stromal cells can engraft and give rise to human blood cells after transplantation into fetal sheep (Narayan et al., 2002). Similar studies using the NOD/SCID mouse model are underway. Successful engraftment of bulk populations of human ES cell-derived hematopoietic cells will then permit more careful dissection of the HSC population. For example, while HSCs are commonly defined by CD34 expression, some studies suggest $CD34^-$ HSCs may also exist. Alternatively, CD34 expression on HSCs may vary by the activation status of the cell. A human ES cell-based hematopoietic differentiation system, such as we have described above, should facilitate the identification of HSCs from a population of cells whose growth conditions and stimuli can be carefully manipulated and monitored during development. Eventually, these results may be applied to promote *ex vivo* expansion of bone marrow or cord blood-derived HSCs for clinical therapies.

7.5 Human embryonic stem cells, preimplantation genetic diagnosis, and hematopoiesis

Perhaps even more important than the potential to provide an unlimited source of tissue for generating cellular blood elements for transfusion or hematopoietic reconstitution, studies of hematopoiesis from human ES cells will be uniquely valuable to define better the basic mechanisms of human blood development. While the potential to use human ES cell-derived blood cells (or other human ES cell-derived cells or tissues) to treat a litany of diseases is an inherently enticing prospect, there are considerable hurdles to overcome before this type therapy becomes a reality. However, many diseases could be significantly better treated or cured via the capacity to regulate gene expression and cell fate determination. This knowledge could be used to correct defects in hematopoietic stem cells or other hematopoietic lineages that are readily accessible. For example, sickle cell anemia arises from a homozygous genetic defect in the β-globin gene. This abnormal protein leads to red blood cell sickling in hypoxic conditions resulting in numerous physiologic problems that cause significant morbidity and even mortality. HCT can be used as a definitive treatment for sickle cell anemia, though the potential complications make this therapy suitable for only a subset of patients whose disease is among the most difficult to control (Walters et al., 2000). Other patients can be treated by oral hydroxyurea to stimulate production of red blood cells with increased γ-globin gene expression. γ-globin replaces the abnormal β-globin protein, resulting in production of hemoglobin F, which does not lead to red cell sickling. While hydroxyurea treatment can reduce the incidence of painful sickle crises, this treatment does not benefit all patients equally (Steinberg, 1999). Other drugs to promote normal globin gene expression could potentially benefit a great number of patients.

A potential means to define the molecular mechanisms that regulate gene expression is through the creation of ES cells lines with defined genetic mutations and then to study how these genetic mutations affect development of ES cells *in vitro*. This is quite likely to be feasible using embryos produced for patients undergoing preimplantation genetic diagnosis (PGD) (*Figure 7.2*). Since PGD was first described in 1990 (Handyside *et al.*, 1990), this process is becoming an increasingly common method to prevent heritable genetic diseases such as sickle cell anemia, Fanconi anemia, cystic fibrosis, muscular dystrophies, immunodeficiencies, and many others (*Table 7.1*). To date, over 500 children have been born through this process (Kuliev and Verlinsky, 2002). PGD involves removing one or two blastomeres at the approximately 8-cell stage of embryo development. The genetic status of these blastomeres can be evaluated by a variety of means to test for a defined genetic allele, HLA type, or Y chromosome (for sex-linked disorders) (Braude *et al.*, 2002). The genetic testing is usually accomplished in one day, and the genetically unaffected embryos are transferred to the mother on day 4 or 5. In some cases, PGD has been done to evaluate embryos for both disease status (such as Fanconi anemia) and HLA haplotype. In these cases, a couple can have a child

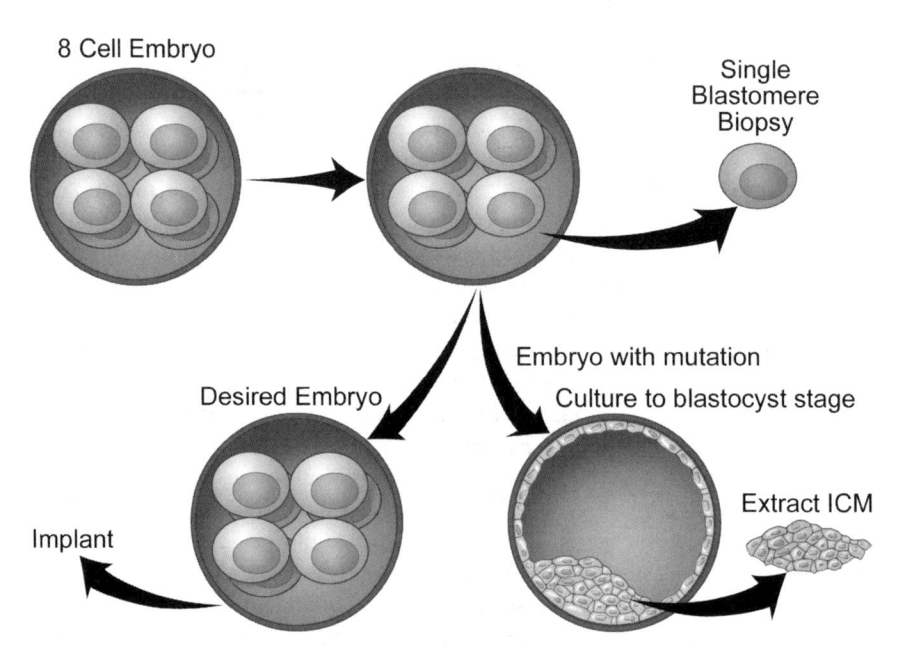

Figure 7.2: Combined pre-implantation genetic diagnosis (PGD) and derivation of ES cell lines with defined genetic mutations. *In vitro* fertilized oocytes are cultured to the 8-cell stage. One blastomere is removed by micropipet and single cell genetic analysis is done via PCR, FISH or other methods. If genetically unaffected, the embryo can be implanted for normal development. If genetically abnormal, the embryo can be cultured to blastocyst stage, the ICM isolated and used to derive new ES cell lines with the defined genetic mutation.

Table 7.1: X-linked or single gene disorders potentially screened by preimplantation genetic diagnosis (partial list).

Cystic Fibrosis
Sickle Cell Anemia
Thalassemia
Fanconi Anemia
Lesch Nyhan Syndrome
Osteogenesis Imperfecta
Tay Sachs Disease
Huntington Disease
Myotonic Dystrophy
Adrenoleukodystrophy
Spinal Cerebellar Ataxia
Spinal Muscular Atrophy
Duchenne Muscular Dystrophy
Fragile X Syndrome
Hemophilia

that is unaffected by a potentially fatal illness, and whose umbilical cord blood cells become a potentially curative therapy for an affected sibling (Robertson et al., 2002; Verlinsky et al., 2001).

As a result of PGD, embryos with defined genetic abnormalities are identified but remain unused, though often stored. These embryos could be further cultured to blastocyst stage and the inner cell mass isolated to derive ES cell lines with specific genetic mutations. These unique ES cell lines would provide a powerful tool to study novel aspects of gene regulation and potential new therapies (*Figure 7.2*). Human ES cell lines derived from 'leftover' PGD embryos are an especially attractive means to study diseases such as Fanconi anemia, Sickle cell anemia, or thalassemias that primarily affect the hematopoietic system. Although these well-characterized genetic diseases can be cured by HCT, this option is most effective for those patients with an HLA-matched sibling. Unfortunately, an appropriate sibling donor is not available for most patients. Attempts to cure these diseases by genetic repair of autologous HSCs have not yet been successful (Sorrentino and Nienhuis, 2001). The ability to derive HSCs from PGD embryos with these defined mutations would allow a unique means to evaluate how early stages of hematopoiesis are disrupted by these mutations. For example, numerous mutations that prevent normal expression of β-globin genes have been identified in patients with β-thalassemia. These abnormalities have been difficult to replicate in animal (mouse) models and studies of humans affected by these disorders are limited by the availability of HSCs from these patients. Human ES cell lines with defined mutations would permit an almost unlimited supply of cells with uniform genetic characteristics. These cell lines would offer an ideal starting point to evaluate novel methods to induce expression of globin genes (typically γ-globin) that can functionally replace the defective β-globin. Methods of gene repair can also be assessed in this model. Unfortunately, these experiments require derivation of new human ES cell lines. At this time, these studies are thus not eligible for US federal funding.

However, with other funding or work done in other countries, derivation of cell lines from PGD embryos could represent the future of human ES cell research.

7.6 Summary

Hematopoiesis is one of the most thoroughly studies areas of developmental biology. Important aspects of our understanding of this field have come from studying organisms as diverse as frogs (*Xenopus*), zebrafish, mice, and humans. Each model has its own particular strengths and weaknesses. However, only by research on human ES cells can we obtain a more thorough understanding of how human HSCs arise during the earliest stages of development. No other approach combines the availability, uniformity, and relevance to human biology and disease that human ES cells afford. More research groups gaining access to human ES cell lines will lead to exciting advances in the next few years, which will ultimately lead to greater insight into human blood formation and new methods to treat hematologic diseases.

Acknowledgments

I thank Robert and Wanda Auerbach for editorial assistance. Julie Morgan provided the photograph of human ES cell-derived embryoid bodies.

References

Anonymous (2002) Investigation of blood transfusion recipients with West Nile virus infections. *MMWR Morb. Mortal. Wkly Rep.* **51**, 823.

Adelman CA, Chattopadhyay S, Bieker JJ (2002) The BMP/BMPR/Smad pathway directs expression of the erythroid-specific EKLF and GATA1 transcription factors during embryoid body differentiation in serum-free media. *Development* **129**, 539–549.

Akashi K, He X, Chen J, Iwasaki H, Niu C, Steenhard B, Zhang J, Haug J, Li L (2003) Transcriptional accessibility for genes of multiple tissues and hematopoietic lineages is hierarchically controlled during early hematopoiesis. *Blood* **101**, 383–389.

Blau HM, Brazelton TR, Weimann JM (2001) The evolving concept of a stem cell: entity or function? *Cell* **105**, 829–841.

Bone Marrow Donors Worldwide, www.bmdw.org, 2003.

Braude P, Pickering S, Flinter F, Ogilvie CM (2002) Preimplantation genetic diagnosis. *Nat. Rev. Genet.* **3**, 941–953.

Chadwick K, Wang L, Li L, Menendez P, Murdoch B, Rouleau A, Bhatia M (2003). Cytokines and BMP-4 promote hematopoietic differentiation of human embryonic stem cells. *Blood* **102**, 906–915.

Collins LS, Dorshkind K (1987) A stromal cell line from myeloid long-term bone marrow cultures can support myelopoiesis and B lymphopoiesis. *J. Immunol.* **138**, 1082–1087.

Dao MA, Nolta JA (1999) Immunodeficient mice as models of human hematopoietic stem cell engraftment. *Curr. Opin. Immunol.* **11**, 532–537.

Dao MA, Arevalo J, Nolta JA (2003) Reversibility of CD34 expression on human hematopoietic stem cells that retain the capacity for secondary reconstitution. *Blood* **101**, 112–118.

Delassus S, Titley I, Enver T (1999) Functional and molecular analysis of hematopoietic progenitors derived from the aorta-gonad-mesonephros region of the mouse embryo. *Blood* **94**, 1495–1503.

Doetschman TC, Eistetter H, Katz M, Schmidt W, Kemler R (1985) The *in vitro* development of blastocyst-derived embryonic stem cell lines: formation of visceral yolk sac, blood islands and myocardium. *J. Embryol. Exp. Morphol.* **87**, 27–45.

Down JD, White-Scharf ME (2003) Reprogramming immune responses: enabling cellular therapies and regenerative medicine. *Stem Cells* **21**, 21–32.

Eaves C, Lambie K (1995) *Atlas of Human Hematopoietic Colonies*. StemCell Technologies, Inc., Vancouver, BC.

Enver T, Heyworth CM, Dexter TM (1998) Do stem cells play dice? *Blood* **92**, 348–351.

Eto K, Murphy R, Kerrigan SW, Bertoni A, Stuhlmann H, Nakano T, Leavitt AD, Shattil SJ (2002) Megakaryocytes derived from embryonic stem cells implicate CalDAG-GEFI in integrin signaling. *Proc. Natl Acad. Sci. USA* **99**, 12819–12824.

Handyside AH, Kontogianni EH, Hardy K, Winston RM (1990) Pregnancies from biopsied human preimplantation embryos sexed by Y-specific DNA amplification. *Nature* **344**, 768–770.

International Bone Marrow Transplant Registry, www.ibmtr.org, 2002.

Kaufman DS, Thomson JA (2002) Human ES cells–haematopoiesis and transplantation strategies. *J. Anat.* **200**, 243–248.

Kaufman DS, Hanson ET, Lewis RL, Auerbach R, Thomson JA (2001). Hematopoietic colony-forming cells derived from human embryonic stem cells. *Proc. Natl Acad. Sci. USA* **98**, 10716–10721.

Kaufman DS, Lewis RL, Thomson JA (2002) Cell contact and cytokine requirements for hematopoietic differentiation of human ES cells. *Blood* **100**(Suppl. 2), 151b.

Keller GM (1995) *In vitro* differentiation of embryonic stem cells. *Curr. Opin. Cell Biol.* **7**, 862–869.

Krause DS, Fackler MJ, Civin CI, May WS (1996) CD34: structure, biology, and clinical utility. *Blood* **87**, 1–13.

Kuliev A, Verlinsky Y (2002) Current features of preimplantation genetic diagnosis. *Reprod. Biomed. Online* **5**, 294–299.

Kyba M, Perlingeiro RC, Daley GQ (2002) HoxB4 confers definitive lymphoid-myeloid engraftment potential on embryonic stem cell and yolk sac hematopoietic progenitors. *Cell* **109**, 29–37.

Lagasse E, Shizuru JA, Uchida N, Tsukamoto A, Weissman IL (2001) Toward regenerative medicine. *Immunity* **14**, 425–436.

Lapidot T, Fajerman Y, Kollet O (1997) Immune-deficient SCID and NOD/SCID mice models as functional assays for studying normal and malignant human hematopoiesis. *J. Mol. Med.* **75**, 664–673.

Lebkowski JS, Schain LR, Okarma TB (1995) Serum-free culture of hematopoietic stem cells: a review. *Stem Cells* **13**, 607–612.

Li F, Lu S, Vida L, Thomson JA, Honig GR (2001) Bone morphogenetic protein 4 induces efficient hematopoietic differentiation of rhesus monkey embryonic stem cells in vitro. *Blood* **98**, 335–342.

Millan MT, Shizuru JA, Hoffmann P, Dejbakhsh-Jones S, Scandling JD, Carl Grumet F, Tan JC, Salvatierra O, Hoppe RT, Strober AS (2002) Mixed chimerism and immunosuppressive drug withdrawal after HLA-mismatched kidney and hematopoietic progenitor transplantation. *Transplantation* **73**, 1386–1391.

Nakano T, Kodama H, Honjo T (1994) Generation of lymphohematopoietic cells from embryonic stem cells in culture. *Science* **265**, 1098–1101.

Narayan AD, Thomson JA, Lewis RL, Kaufman DS, Almeida-Porada MG, Zanjani ED (2002) In vitro and *in vivo* potential of human embryonic stem cells. *Blood* **100** (Suppl. 2), 154b.

Neildez-Nguyen TM, Wajcman H, Marden MC, Bensidhoum M, Moncollin V, Giarratana MC, Kobari L, Thierry D, Douay L (2002) Human erythroid cells produced *ex vivo* at large scale differentiate into red blood cells *in vivo*. *Nat. Biotechnol.* **20**, 467–472.

Odorico JS, Kaufman DS, Thomson JA (2001) Multilineage differentiation from human embryonic stem cell lines. *Stem Cells* **19**, 193–204.

Palis J, Yoder MC (2001) Yolk-sac hematopoiesis: the first blood cells of mouse and man. *Exp. Hematol.* **29**, 927–936.

Perlingeiro RC, Kyba M, Daley GQ (2001) Clonal analysis of differentiating embryonic stem cells reveals a hematopoietic progenitor with primitive erythroid and adult lymphoid-myeloid potential. *Development* **128**, 4597–4604.

Pick M, Eldor A, Grisaru D, Zander AR, Shenhav M, Deutsch VR (2002) *Ex vivo* expansion of megakaryocyte progenitors from cryopreserved umbilical cord blood. A potential source of megakaryocytes for transplantation. *Exp. Hematol.* **30**, 1079–1087.

Rideout WM 3rd, Hochedlinger K, Kyba M, Daley GQ, Jaenisch R (2002) Correction of a genetic defect by nuclear transplantation and combined cell and gene therapy. *Cell* **109**, 17–27.

Robertson JA, Kahn JP, Wagner JE (2002) Conception to obtain hematopoietic stem cells. *Hastings Cent. Rep.* **32**, 34–40.

Sato T, Laver JH, Ogawa M (1999) Reversible expression of CD34 by murine hematopoietic stem cells. *Blood* **94**, 2548–2554.

Schuldiner M, Yanuka O, Itskovitz-Eldor J, Melton DA, Benvenisty N (2000) Effects of eight growth factors on the differentiation of cells derived from human embryonic stem cells. *Proc. Natl Acad. Sci. USA* **97**, 11307–11312.

Simmons PJ, Torok-Storb B(1991) Identification of stromal cell precursors in human bone marrow by a novel monoclonal antibody, STRO-1. *Blood* **78**, 55–62.

Smith AG (2001) Embryo-derived stem cells: of mice and men. *Ann. Rev. Cell Dev. Biol.* **17**, 435–403.

Sorrentino BP, Nienhuis AW (2001) Gene therapy for hematopoietic diseases. In: *The Molecular Basis of Blood Diseases* (ed. H Varmus). W.B. Saunders Company, Philadelphia, pp. 969–1003.

Spitzer TR, Delmonico F, Tolkoff-Rubin N, McAfee S, Sackstein R, Saidman S, Colby C, Sykes M, Sachs DH, Cosimi AB (1999) Combined histocompatibility leukocyte antigen-matched donor bone marrow and renal transplantation for multiple myeloma with end stage renal disease: the induction of allograft tolerance through mixed lymphohematopoietic chimerism. *Transplantation* **68**, 480–484.

Steinberg MH (1999) Management of sickle cell disease. *N. Engl. J. Med.* **340**, 1021–1030.

Thomas ED (1999) Bone marrow transplantation: a review. *Semin. Hematol.* **36**, 95–103.

Thomson JA, Itskovitz-Eldor J, Shapiro SS, Waknitz MA, Swiergiel JJ, Marshall VS, Jones JM (1998) Embryonic stem cell lines derived from human blastocysts. *Science* **282**, 1145–1147.

Verfaillie CM (2002) Adult stem cells: assessing the case for pluripotency. *Trends Cell Biol.* **12**, 502–508.

Verlinsky Y, Rechitsky S, Schoolcraft W, Strom C, Kuliev A (2001) Preimplantation diagnosis for Fanconi anemia combined with HLA matching. *JAMA* **285**, 3130–3133.

Walters MC, Storb R, Patience M, Leisenring W, Taylor T, Sanders JE *et al.* (2000) Impact of bone marrow transplantation for symptomatic sickle cell disease: an interim report. Multicenter investigation of bone marrow transplantation for sickle cell disease. *Blood* **95**, 1918–1924.

Wang SJ, Greer P, Auerbach R (1996) Isolation and propagation of yolk-sac-derived endothelial cells from a hypervascular transgenic mouse expressing a gain-of-function fps/fes proto-oncogene. *In Vitro Cell Develop Biol. – Animal* **32**, 292–299.

Zanjani ED (2000) The human sheep xenograft model for the study of the *in vivo* potential of human HSC and in utero gene transfer. *Stem Cells* **18**, 151.

Plate 1. General outline of MAPC isolation and differentiation into three major lineages; ectoderm, endoderm, and mesoderm. Human MAPC isolation involves depletion of CD45 and Glycophorin A positive cells using micromagnetic beads. The remaining CD45-/Glycophorin A cells are plated and after a period of time are subcultured at 10 cells/well. Similarly but with some modifications, MAPC can be isolated from rodent bone marrow. Rodent bone marrow mononuclear fraction is initially plated. After 2–4 weeks, CD45-/Ter119 mouse cells or CD45-/RBC rat cells are selected and subcultured at 10 cells/well. In both cases, resulting clones have pluripotent capability and can differentiate into ectoderm as demonstrated by neurofilament 200 staining, endoderm as demonstrated by albumin staining, and mesoderm as demonstrated by von Willebrand factor staining. These differentiations have been shown at the clonal level and have been coupled with more extensive functional and phenotypic characterization. (See Chapter 3.)

Plate 2. Osteogenic differentiation of human embryonic stem (hES) cells and human mesenchymal stem cells (hMSC) *in vitro* using similar differentiation protocols. Stained cells exhibit characteristic bone mineralization. (See Chapter 5.)

Plate 3. Embryoid body (EB) formation and immunostaining. IgG = nonspecific antibody controls, CG = chorionic gonadotropin, ctk = cytokeratin. (See Chapter 6)

Plate 4. *In vitro* and *in vivo* differentiation of hES-derived neuroepithelial cells. (**A**) ES cell-derived neural rosettes (H1.1, p68), when plated onto laminin substrate in differentiation medium in the absence of FGF2 (Zhang *et al.*, 2001), generated cells of epithelial morphology, some of which began to extend out neurites at 3 days. (**B**) After 2 weeks in differentiation culture, neurons with multiple long neurites developed and the neurites connected to each other, forming networks. (**C**) Immunostaining after 45 days of differentiation showed many GFAP+ astrocytes (green) appearing along NF200+ neurites (red). (**D–F**) Neurons expressing glutamate (**D**), GABA (**E**), and tyrosine hydroxylase (**F**) were also observed. (**G**) Oligodendrocytes expressing O4 antigen were observed only after treatment with sonic hedgehog in long-term culture (7–8 weeks). (**H–K**) After transplantation of the neuroepithelial cells into cerebral ventricles of neonatal mice, grafted cells, identified by labeling with human nuclear protein (in green), were localized to ventricles and subventricular areas (**H**). Grafted cells (green in **I–K**) in the brain parenchyma expressed neuronal markers βIII-tubulin (red in **I**), or MAP2 (red in **J**), and astrocytic marker GFAP (red in **K**). Bar = 100 μm (**A–G**), 10 μm (**I–K**). D–G and I–K are replicates of parts of Figs. 2 and 4 in Zhang *et al.* (2001) with permission. (See Chapter 9.)

Plate 5. ES cells differentiated by the protocol of Hori *et al.* (2002). Cells in the final stages are cultured in medium supplemented with exogenous insulin. Fixation and immunostaining was performed on cells continuously cultured in high insulin containing medium (**A and C**) or on cells similarly treated, except for a short (~18 h) chase period of culture in medium lacking exogenous insulin (**B and D**). Cells were stained for the co-expression of insulin and C-peptide 1 (**A and B**) or C-peptide 2 (**C and D**). Insulin immunostaining in the non-chased cultures demonstrates a non-cytoplasmic staining pattern and cells are pyknotic. Insulin immunostaining is eliminated by a washout period and no C-peptide immunostaining is detected. Scale bar, 50 μm. (See Chapter 10.)

Plate 6. Murine ES cell-derived cells have characteristics of pancreatic progenitors and cells at early stages of islet differentiation. (**A**) Graph showing the increasing numbers of PDX1+ cells appearing during the differentiation of mouse ES cells *in vitro*, following plating of 7-day EBs. (**B**) 7-day post-plating culture showing a focus of PDX1+ cells (red) surrounded by YY+ (blue) and glucagon+ (green) cells. (**C**) Another focus on day 21 showing a central area of PDX1+ cells (red) and cells stained with a cocktail of antibodies to insulin, glucagon, and somatostatin (green) and antibody to PP (blue). (**D**) Cluster of cells expressing IAPP (red, top panel) and YY (green, middle panel). Lower panel shows merged image. Many cells co-express YY and IAPP in cytoplasmic granules. (**E**) Insulin+ cells (red panel) co-express both YY (green panel) and IAPP (blue panel), as seen in merged image (lower right panel). (**F**) Early cells (before day 21 post-plating) co-express insulin with glucagon in cytoplasmic granules. Scale bars, 20 μm. (See Chapter 10.)

Plate 7. Insulin-positive cells from late murine ES cell-derived cultures have characteristics of mature β cells. **(A)** Cells are characterized by nuclear PDX1 staining. **(B)** Insulin and glucagon are no longer co-expressed after day 17 post-plating, but now occur in separate cells. Co-expression of insulin with C-peptide 1 **(C)** and C-peptide 2 **(D)** in cytoplasmic granules, is observed. Scale bars, 50 μm. (See Chapter 10.)

Plate 8. Human ES cells show stages of differentiation similar to mouse ES cells. **(A)** PDX1+ cells at two weeks post-plating of EBs derived from the H9 (WA09) cell line. **(B)** Somatostatin+ and glucagon+ cells at 8–10 weeks post-plating. Insulin+ cells **(C)** co-express both YY **(D)** and IAPP **(E)**, as seen in merged image **(F)** at 12 weeks post-plating. Scale bars, 50 μm. (See Chapter 10.)

Plate 9. Phenotypic characterization of the human ES cell-derived cardiomyocytes. A) Positive immunostaining of dispersed cells from the contracting EB with anti-cardiac α/β myosin heavy chain (MHC), anti-sarcomeric α-actinin, anti-cardiac troponin I (cTnI), anti-desmin, and anti-ANP antibodies; B) Transmission electron micrograph showing the ultrastructural properties of the human ES cell-derived cardiomyocytes; C) Calcium transients, recorded from the ES cell-derived cardiomyocytes by means of Fura 2 fluorescence; D) Extracellular electrogram recording from the beating EB; E,F) Plating of the contracting EB on top of microelectrode array (MEA) plates (E) allowed the generation of a detailed activation maps (F) showing the presence of spontaneous pacemaking activity (red area in the map) and an excitable syncytium with action potential propagation; and G) Immunostaining for connexin 43 (red) and 45 (green) showing the co-localization of these proteins to the same gap junctions. (Part of the figure was adapted from *J. Clin. Invest.* **108**, 407–414, 2001.) (See Chapter 11.)

Plate 10. Expression of enhanced green fluorescent protein (eGFP) following *a*) Transient transfection of an eGFP expressing plasmid, *b*) Infection of a lentiviral vector expressing eGFP and *c*) Stable transfection of an eGFP expressing plasmid (Itskovitz-Eldor *et al.*, 2000). (See Chapter 12.)

Plate 11. Therapeutic cloning. Schema to produce differentiated cell types from human ES cells derived from somatic cell nuclear transfer (SCNT) generated blastocysts. (See Chapter 14.)

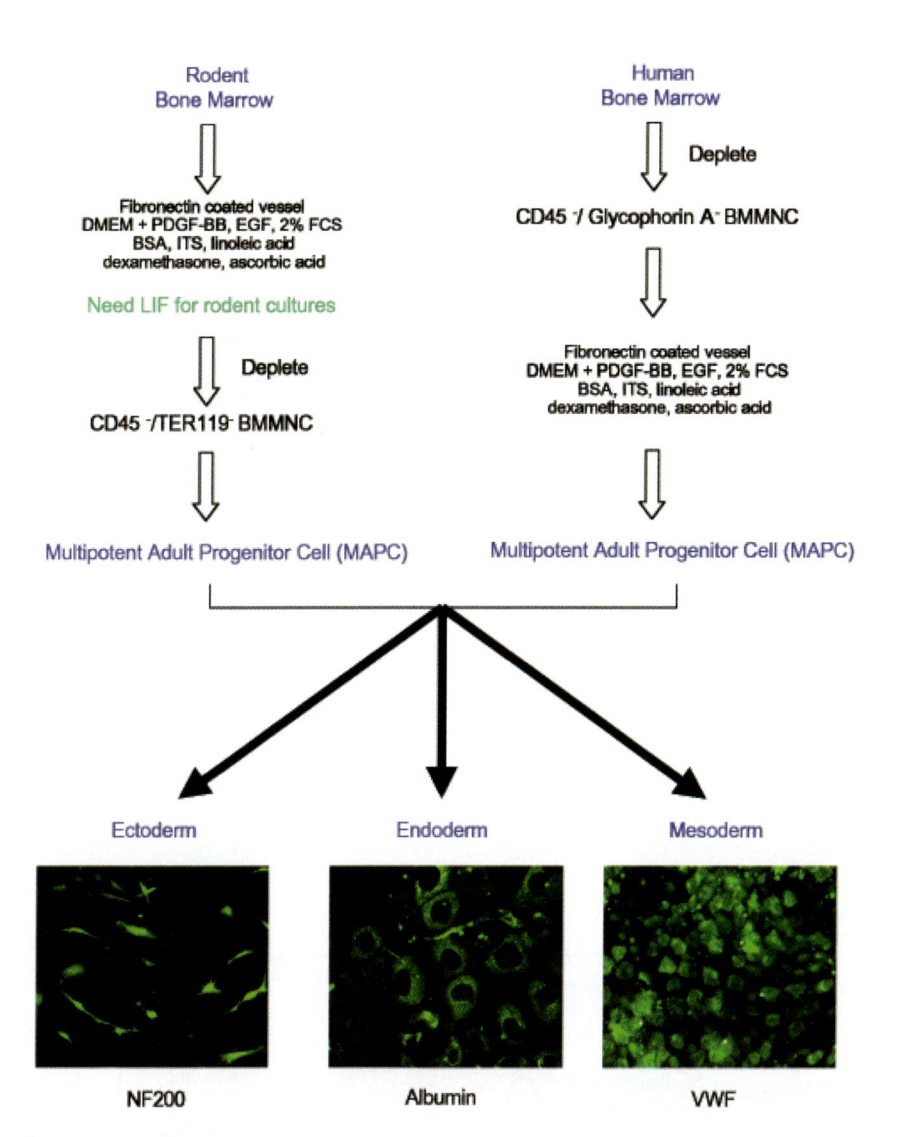

```
                    Rodent                              Human
                 Bone Marrow                         Bone Marrow
                      ⇓                                   ⇓  Deplete

         Fibronectin coated vessel            CD45 -/ Glycophorin A- BMMNC
       DMEM + PDGF-BB, EGF, 2% FCS
            BSA, ITS, linoleic acid                       ⇓
         dexamethasone, ascorbic acid
                                                Fibronectin coated vessel
        Need LIF for rodent cultures          DMEM + PDGF-BB, EGF, 2% FCS
                                                   BSA, ITS, linoleic acid
                      ⇓  Deplete                dexamethasone, ascorbic acid

           CD45 -/TER119- BMMNC                           ⇓

                      ⇓
   Multipotent Adult Progenitor Cell (MAPC)   Multipotent Adult Progenitor Cell (MAPC)
```

Ectoderm   Endoderm   Mesoderm

NF200   Albumin   VWF

Plate 1.

Osteogenic differentiation in vitro
Alzarin Red S stain at day 15

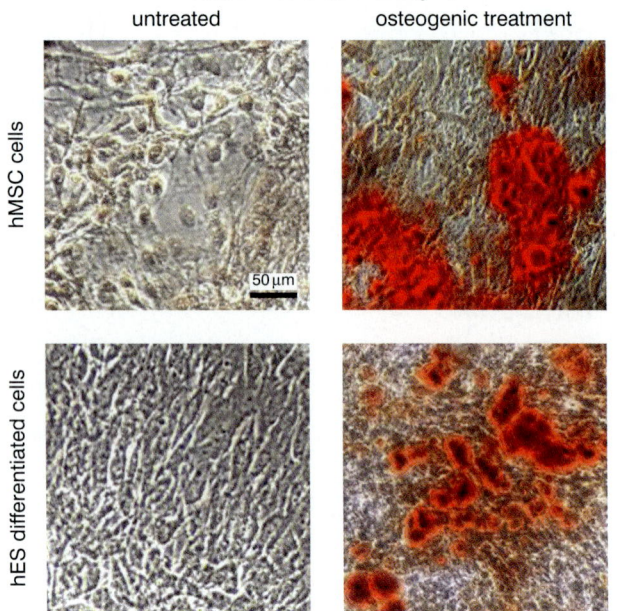

Plate 2.

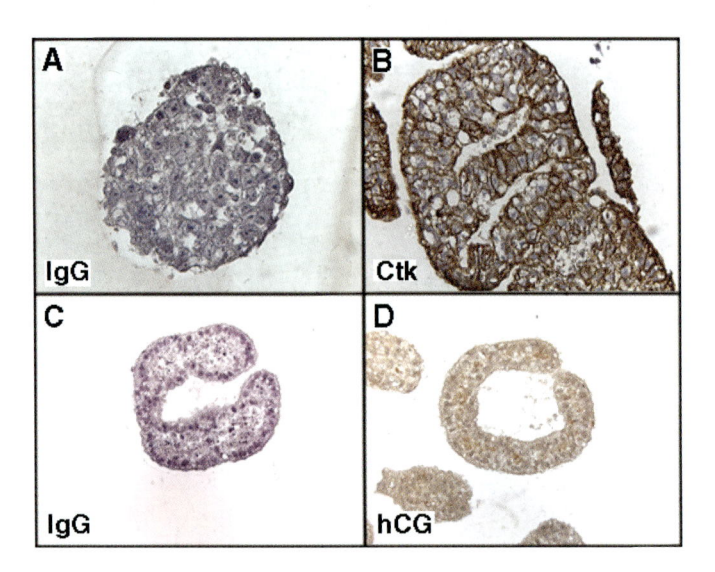

Plate 3.

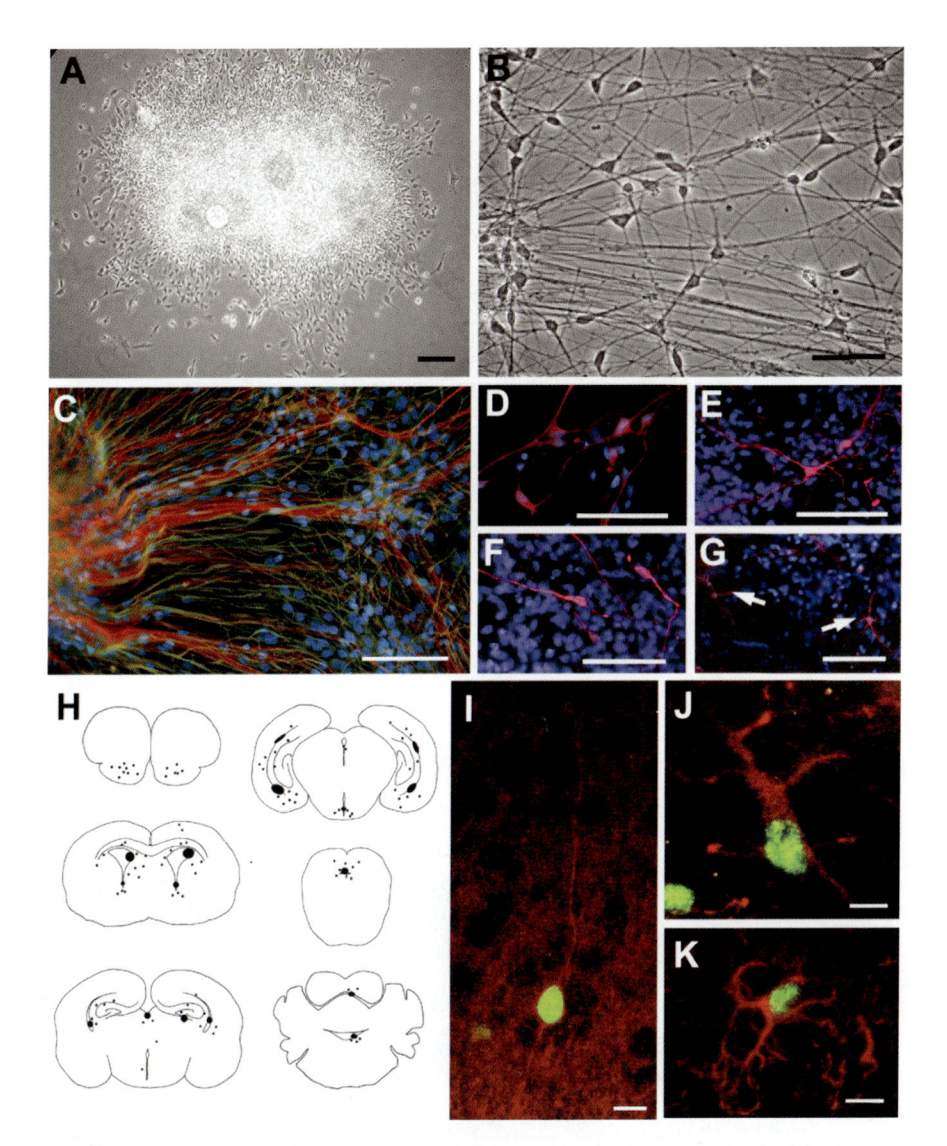

Plate 4.

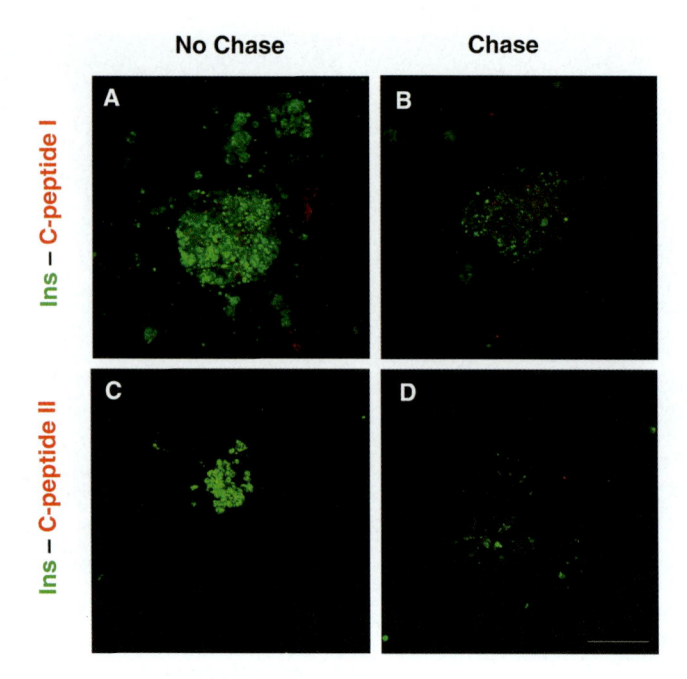

Plate 5.

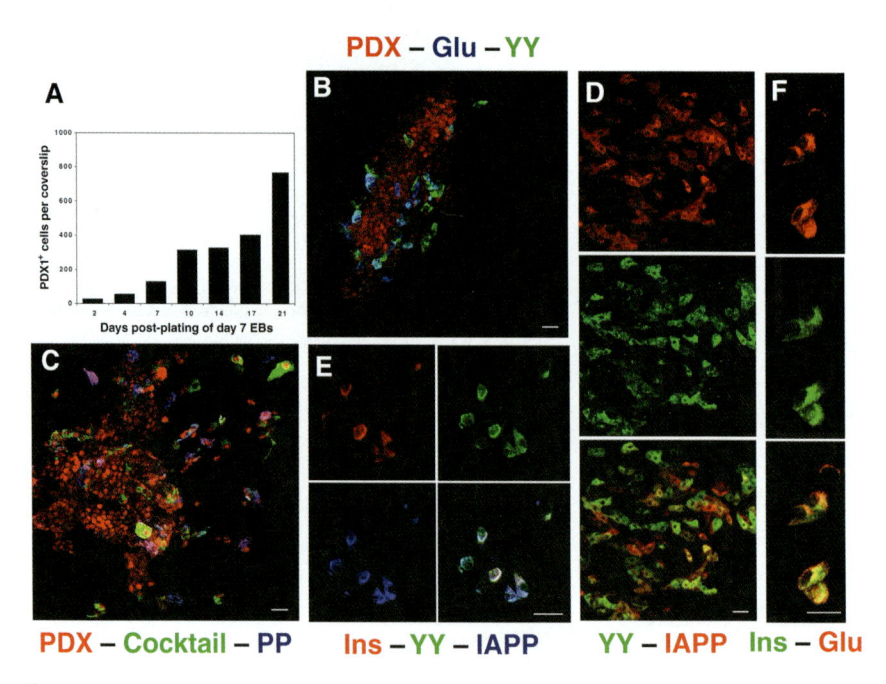

Plate 6.

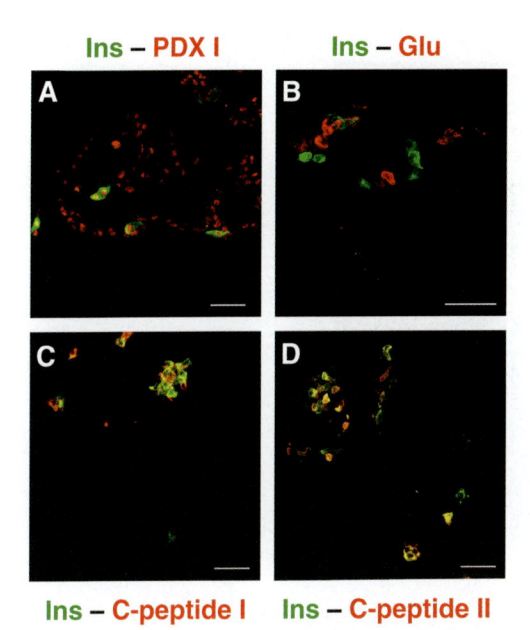

Plate 7.

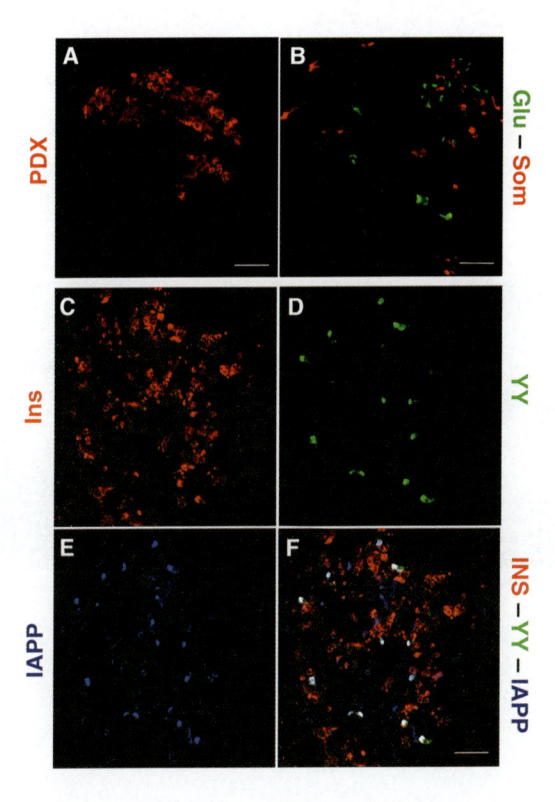

Plate 8.

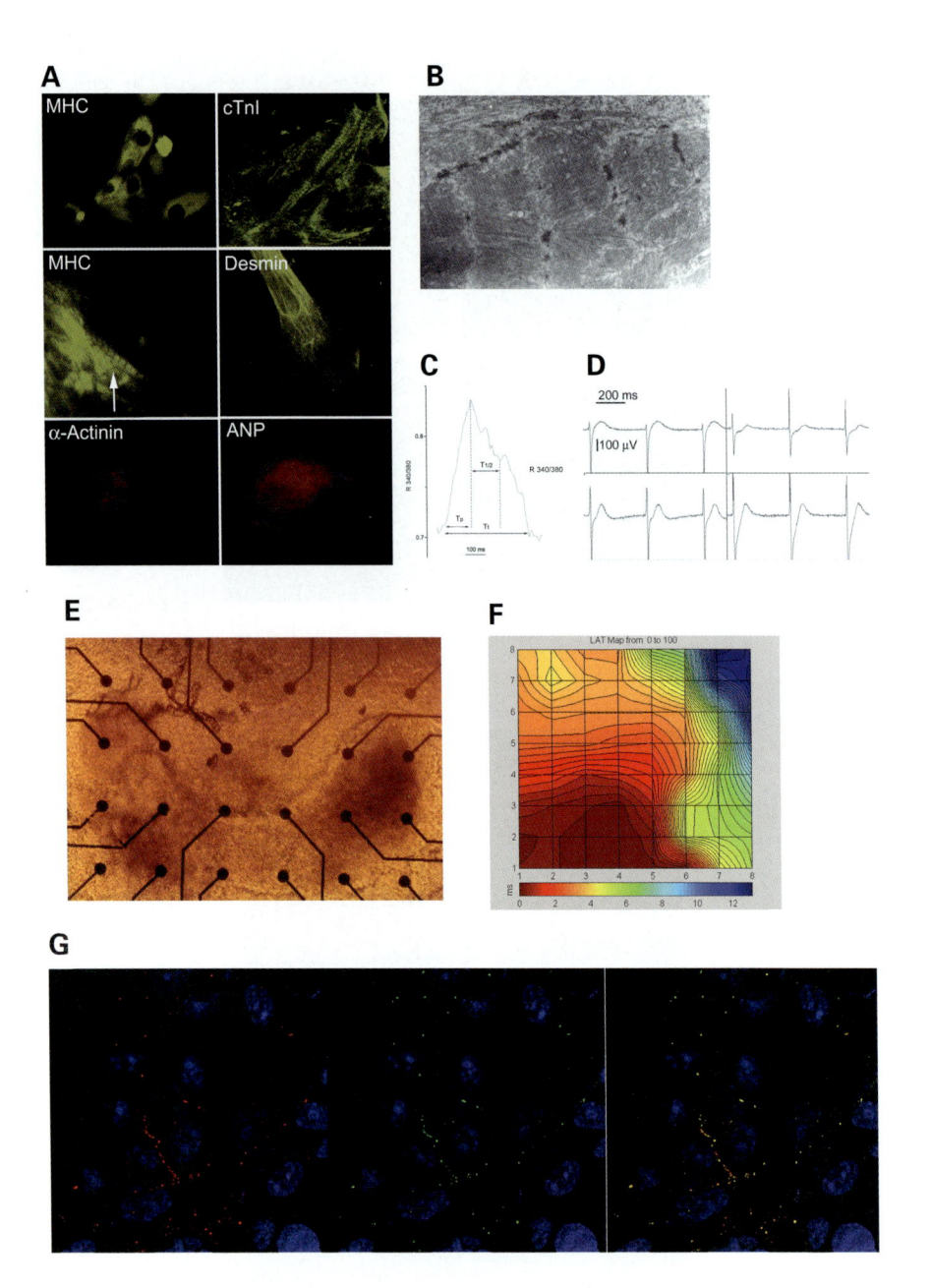

Plate 9.

A B C

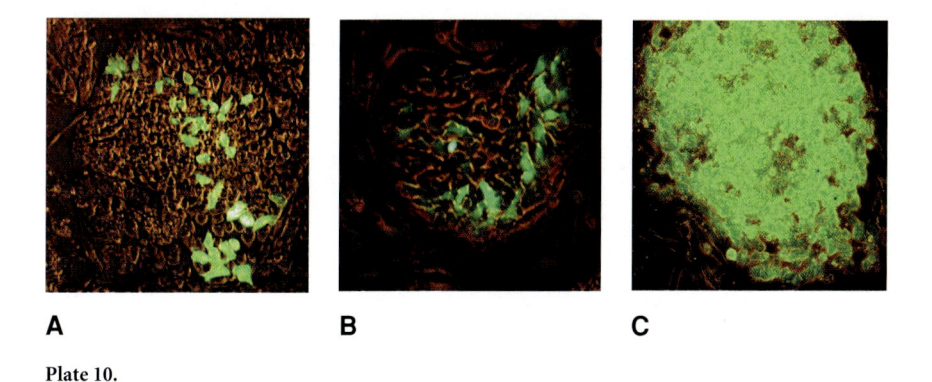

Plate 10.

Plate 11.

8. Derivation of endothelial cells from human embryonic stem cells

Shulamit Levenberg, Ngan F. Huang and Robert Langer

8.1 Introduction

Endothelial cells are involved in regulating physiological processes such as angiogenesis, inflammation, and thrombosis. They are of great research interest because of their potential to treat vascular diseases and stimulate growth of ischemic tissue. One cell source for such applications is embryonic stem (ES) cells, which have been shown to differentiate into endothelial cells through a process known as vasculogenesis. While the isolation and differentiation of endothelial cells from ES cells have been extensively studied in animal models, the lack of a reliable experimental system has hindered such examination in humans. In 1998, the isolation of human ES (hES) cell lines from the inner cell mass of blastocysts established a system for studying differentiation towards all embryonic lineages (Thomson et al., 1998). Recently, we have described the differentiation of hES cells into endothelial cells. We have isolated the hES-derived endothelial cells and characterized their differentiation *in vitro* and *in vivo* (Levenberg et al., 2002). These cells are candidates for a wide range of therapeutic applications involving tissue engineering and cell transplantation. Here, we will describe the development and differentiation of endothelial progenitors in ES cell models, highlighting various methods for isolation and characterization of endothelial cells and their progenitors. We also discuss therapeutic applications of endothelial cells in novel tissue engineering constructs, therapeutic neovascularization, and gene therapy.

8.2 Development of endothelial cell progenitors

8.2.1 Embryonic vasculogenesis

In the developing vertebrate embryo, the cardiovascular system is the first organ system to form. In response to induction of the mesoderm, the yolk sac (Haar and Ackerman, 1971) is formed. Progenitor endothelial cells (angioblasts) and hematopoietic cells differentiate from the extraembryonic mesoderm of the yolk sac, forming blood islands that later fuse to form vascular networks (Risau, 1995). Due to the close association between nascent endothelial cells and blood cells, it has been suggested that both populations may be derived from a common hemangioblast precursor (Murray, 1932; Nishikawa et al., 2001).

The formation of blood vessels involves signaling and regulatory pathways that are not fully elucidated. Recent reports have begun to provide insight into this complicated process and have identified hedgehog signaling as an important mediator for pattern formation during vasculogenesis in several species (McMahon, 2000). In the mouse embryo, posterior epiblast cells differentiate into endothelial and red blood cells through Indian hedgehog (IHH) signaling from the visceral endoderm (Baron, 2001). Decreased vascularization is observed in *sonic hedgehog* (*Shh*) null mutant mice. *In vitro* analysis of *Ihh*-deficient murine embryoid bodies (EBs) and embryo yolk sacs from *Ihh*-1 mice suggests similar findings of decreased vascularization (Byrd et al., 2002).

Besides hedgehog signaling, endothelial cell differentiation is also regulated by interactions with the extracellular matrix (ECM). It has been reported that integrin–fibronectin interactions are necessary for normal angiogenesis and vasculogenesis (Francis et al., 2002). Fibronectin deficiency leads to a reduction in capillary plexus formation during *in vitro* vasculogenesis in EBs, and integrin deficiency *in vivo* results in abnormal vascular network patterning. Additionally, maturation of blood vessels requires the recruitment of mesenchymal stem cells and their differentiation into vascular smooth muscle cells. Targeted disruption in animal embryos has identified SMAD5, myocyte enhancer binding factor 2C (MEF2C), and lung Kruppel-like factor (LKLF) to be important transcription factors for smooth muscle cell differentiation (Kuo et al., 1997; Lin et al., 1998; Yang et al., 1999).

Growth factors are also known to be involved in embryonic vasculogenesis. Vascular endothelial growth factor (VEGF) is crucial for vasculogenesis as VEGF-deficient mutants develop defects in early embryonic vessel formation and die in early embryonic life (Carmeliet et al., 1996; Ferrara et al., 1996). Basic fibroblast growth factor (bFGF) is involved in generating angioblasts from the mesoderm (Beck and D'Amore, 1997), and both transforming growth factor beta (TGFβ) and platelet-derived growth factor (PDGF) appear to be important inducers of smooth muscle cell differentiation (Hellstrom et al., 1999; Hirschi et al., 1998).

8.2.2 Endothelial cell differentiation and vascularization in ES cells

In mouse ES cell systems, it has been shown that following formation of EBs and induction of cell differentiation, cells will spontaneously start to differentiate into hemangioblasts, forming blood islands that contain endothelial and hematopoietic progenitors (Choi et al., 1998; Risau et al., 1988). Further differentiation results in vascularization of the EBs and formation of vessel-like networks (Vittet et al., 1996). Many studies have analyzed the timing of endothelial cell differentiation in EBs mainly by immunohistochemistry, flow cytometry, and gene expression analysis (Faloon et al., 2000; Kabrun et al., 1997; Robertson et al., 2000; Vittet et al., 1996).

A variety of markers that are associated with endothelial cells are used in these studies including: CD31/PECAM1, CD34, VE-cadherin (CD144), Tie-1, Tie-2, GATA-2, GATA-3, Flk-1 (VEGFR-2/KDR), Flt-1 (VEGFR-1), and von Willebrand factor (vWF). CD31, also known as platelet/endothelial cell adhesion molecule 1(PECAM1), is expressed in mammals primarily on cells of the vasculature, including endothelial cells, monocytes, neutrophils, platelets, and some T-cell subgroups (Albelda et al., 1990). CD34 is a transmembrane surface glycoprotein that is expressed in endothelial cells and hematopoietic stem cells (Andrews et al., 1986). VE-cadherin, a member of the cadherin family of adhesion receptors, is a specific and constitutive marker for endothelial cells (Dejana et al., 2001). Targeted mutagenesis studies have revealed that VE-cadherin plays an important role in early vascular assembly (Vittet et al., 1997). Tie-1 and Tie-2 are receptor tyrosine kinases that are expressed in the endothelium of blood vessels during embryonic development (Korhonen et al., 1995). GATA-2 regulates transcriptional checkpoints for the determination and/or survival of pluripotent hematopoietic stem cells (Tsai et al., 1994), and also plays a role in endothelial cell differentiation (Lee et al., 1991). VEGF receptors Flk-1 (VEGFR-2/KDR) and Flt-1 (VEGFR-1) play a significant role in embryonic vascular and hematopoietic development (Fong et al., 1999; Schuh et al., 1999). Flk-1 is considered an early differentiation marker for endothelial cells and blood cells (Yamaguchi et al., 1993). Recently, it has been shown that Flk-1 positive cells derived from mouse embryonic stem cells can differentiate into both endothelial and mural cells and can reproduce the vascular organization process (Yamashita et al., 2000). In late stages of vasculogenesis, characteristics of mature endothelial cells emerge such as the synthesis and secretion of vWF (Jaffe et al., 1974).

Mouse ES cell differentiation studies demonstrated that endothelial cell-associated markers are expressed in sequential steps, which closely recapitulate endothelial cell development *in vivo*. Endothelial cell-restricted genes are not generally expressed in mouse ES cells (or they are only expressed at very low levels that disappear immediately during early EB cell differentiation, as is observed for CD31 and Tie-2). *In vitro*, endothelial cell differentiation begins as the expression of endothelial cell-restricted genes commences on days 3–5 of EB formation (Flk-1 at day 2–3; CD31 and Tie-2 at day 4; VE-cad and Tie-1 at day 5) (Robertson et al., 2000; Vittet et al., 1996). Subsequently, mRNA levels of these endothelial genes gradually increases, reaching a maximum around day 8–11. Similarly in human ES cell cultures, the expression of endothelial markers (CD31, VE-cad and CD34 and GATA-2) gradually increase during EB differentiation, reaching a maximum by day 13–15 (GATA-2 around day 18) (Levenberg et al., 2002). However, in hES cells, some genes were found to be expressed in the undifferentiated cultures at either high levels (Flk-1, AC133, Tie-2) or low levels (GATA-3, CD34), and others became notable following EB formation and differentiation (CD31, VE-cad, GATA-2) (Kaufman et al., 2001; Levenberg et al., 2002).

The functional significance of this pattern of endothelial gene expression in undifferentiated human ES cell cultures and whether it represents any important difference between mouse and human ES cells is not known. However, given that humans and mice have significantly different behavior and development, the pattern of human endothelial gene expression described here might indicate critical differences in the underlying mechanisms of embryonic endothelial development. It is also possible, however, that a more simple explanation accounts for these distinctions, namely, that the expression of endothelial marker genes in hES cells could be related to the 'escape' of some cells from the undifferentiated state. RT-PCR analysis is not able to distinguish these possibilities, nor is it able to determine whether the entire population of undifferentiated cells express these marker genes or only a subpopulation. Studies utilizing *in situ* hybridization could further elucidate these issues.

Nevertheless, it was shown that similar to mouse ES cells, hES cells could also spontaneously differentiate and organize within EBs into three-dimensional vessel-like structures that were positive for CD31 and that resembled embryonic vascularization. The capillary area in the hEBs increases during subsequent maturation steps starting as small cell clusters that later sprout into capillary-like structures and eventually become organized in a network-like arrangement (Levenberg *et al.*, 2002).

Differentiation into hematopoietic and endothelial lineages can also be induced in two-dimensional systems (without EB formation) by seeding ES cells onto feeder cells or an ECM component. Murine endothelial progenitors (Flk-1+ cells) were isolated following differentiation of ES cells on collagen IV coated plates (Nishikawa *et al.*, 1998). Flk-1+ cells have been shown in mouse systems to be precursors for hematopoietic, endothelial, and smooth muscle cells (Yamashita *et al.*, 2000). It is not clear at present whether human Flk-1+ cells will manifest these multipotent features as well. Using a bone marrow yolk sac-derived stromal feeder layer, Kaufman *et al.* showed that hES cells are able to differentiate into CD34+ cells (1–2%), which are known as precursors both for endothelial and hematopoietic cells (Kaufman *et al.*, 2001; Asahara *et al.*, 1997). Interestingly, in their study, about 50% of the CD34+ cells also expressed CD31. The CD34+ cells were isolated and differentiated toward mature hematopoietic lineages by the addition of cytokines (Kaufman *et al.*, 2001). Further studies will be required to determine whether isolated human embryonic CD34+ cells can be directed down an endothelial cell lineage pathway.

8.3 Isolation of endothelial cells and their progenitors

8.3.1 Separation using surface receptors

Endothelial cells and their progenitors have been successfully isolated by taking advantage of characteristic surface receptors. The isolation techniques are well established for animal systems, particularly in mouse, and have recently been applied to human models.

Isolation of murine endothelial cells and progenitors

The isolation of murine endothelial cells and their progenitors varies slightly between cell lines and growth conditions, but are based on the following general procedure. Murine ES cells were maintained in gelatin-coated tissue dishes over a feeder layer of mitomycin-treated mouse embryonic fibroblasts (MEF). To induce differentiation, LIF and MEF are first removed from the cells. In some studies, the ES cells were suspended in media to allow the formation of EB aggregates (Vittet *et al.*, 1996), while in others they were seeded onto feeder cells or ECM components (Nishikawa *et al.*, 1998). In order to induce differentiation toward the endothelial lineage, the medium is supplemented with growth factors including VEGF, bFGF, TGFβ, interleukin-6 (IL-6), and/or erythropoietin (EPO), applied alone or in combination to induce differentiation toward an endothelial lineage (Choi *et al.*, 1998; Kabrun *et al.*, 1997; Vittet *et al.*, 1996). A similar approach can also be applied to isolated ES cell-derived endothelial progenitors rather than bulk cultures. While these growth factors have been extensively studied in murine ES cell systems, the effects on hES cells remain to be explored.

Endothelial cells and their progenitors can be isolated from mES cells by flow cytometry and cell sorting by labeling lineage-restricted surface receptors such as Flk1 (Hirashima *et al.*, 1999; Yamashita *et al.*, 2000) and CD31 (Balconi *et al.*, 2000). To accomplish this, cells are generally dissociated by trypsin or collagenase, incubated with fluorescently-labeled antibodies, and then sorted by a flow cytometer. Magnetic bead separation techniques have also been used effectively for endothelial cell isolation(Balconi *et al.*, 2000).

Isolation of human endothelial cells and progenitors

Recently, we have established the successful isolation of endothelial cells from human ES cells (Levenberg *et al.*, 2002). The isolation procedure is briefly described here. Human ES cells (H9 clone) were grown on gelatin-coated dishes containing mitomycin-treated MEF feeder layers. The growth medium consisted of 80% KnockOut DMEM and 20% KnockOut SR serum-free formulation, supplemented with glutamine, 2-mercaptoethanol, bFGF, LIF, and non-essential amino acids (Schuldiner *et al.*, 2000). To form hEB aggregates, the hES colonies were dissociated with 1 mg/ml collagenase type IV and grown in Petri dishes. hEBs at 13–15 days were dissociated with trypsin and incubated with fluorescently labeled CD31 antibody before cell sorting. The CD31+ cells were replated and grown in endothelial cell growth medium.

Alternatively, dissociated cells can also be isolated by magnetic column separation (MACS). For isolation of CD34+ cells, cells were labeled with anti-CD34 antibody QBEND/10 and then with a magnetically labeled secondary antibody and isolated by a MACS column (Kaufman *et al.*, 2001).

8.3.2 Separation using a selectable marker

Enriched populations of endothelial cells could potentially be isolated from heterogeneous bulk cultures by using a selectable marker whose expression is controlled by an endothelial gene promoter or by providing endothelial cells with a selective growth advantage. Using this latter approach, Balconi *et al.* (2000) obtained a number of mouse ES cell-derived endothelial cell lines. They transfected EB-dissociated cells with PmT, an oncogene that specifically immortalizes endothelial cells but not other cell types and, taking advantage of the presence of a transgene conferring neomycin resistance, were able to purify immortalized endothelial cell lines. Limiting the application of this strategy is the oncogenic potential of the derivative cells. To derive non-transformed cells, procedures involving the expression of a selectable marker gene (antibiotic resistance or fluorescent protein) under the control of an endothelial gene promoter, such as vWF or PECAM1, could be used.

8.4 Characterization techniques for isolated endothelial cells

8.4.1 Expression of endothelial markers

Through studies in animal models, and more recently in humans, a number of related markers, transcriptional factors, adhesion molecules and growth factor receptors for endothelial cells have been identified (see section 8.2.2). These molecules have been used to characterize endothelial cells by RNA/gene expression assays (RT-PCR, Northern blot, *in situ* hybridization) or by immunostaining for protein expression and localization in cell structures. Analysis of endothelial marker expression in hES-derived endothelial cells indicated characteristics similar to vessel endothelium. Flow cytometric analysis revealed a similar CD34/Flk-1 expression profile in isolated hES-derived CD31+ cells and HUVEC cells. Using immunofluorescence microscopy to analyze adhesion molecules distribution, CD31+ cells appeared to present a correct organization of endothelial cell junctions. N-cadherin and the endothelium-specific VE-cadherin were distributed at adherent type junctions. Actin stress fibers, which were found throughout the cells, terminated in both cell–cell adherence junctions and focal contacts as seen by double staining with vinculin. The tight junction component, CD31, was distributed at the intercellular clefts, and the endothelial marker vWF was highly expressed in the cytoplasm (Levenberg *et al.*, 2002) (*Figure 8.1*).

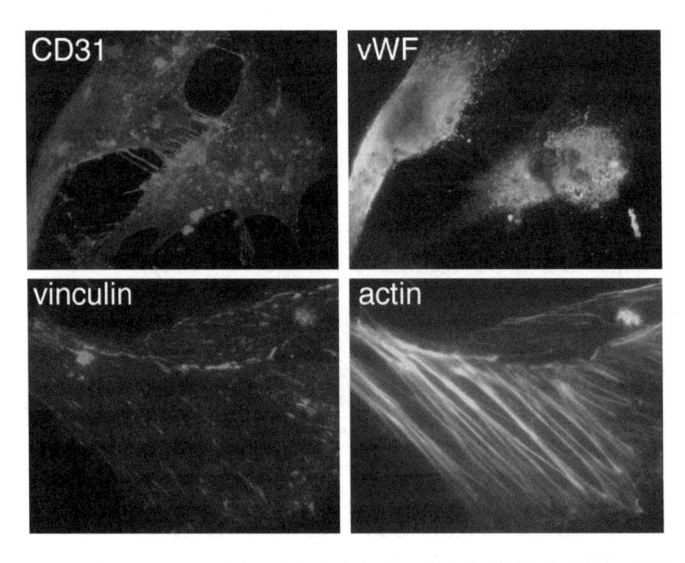

Figure 8.1: Immunofluorescence staining of hES-derived endothelial cells. Staining of CD31 at cell–cell junctions and vWF in the cytoplasm. Vinculin is found at both focal contacts and cell–cell adherent junctions where it associates with actin stress fiber ends. (Orig. mag. ×1000).

8.4.2 LDL incorporation

To characterize endothelial cells derived from ES cells, a functional method is generally used that involves measuring the uptake of acetylated-low density lipoprotein (ac-LDL) using a fluorescent probe 1,1′-dioctadecyl-3,3,3′,3′-tetramethylindocarbocyanine perchlorate (Dil-Ac-LDL). This compound does not have a deleterious effect on the endothelial cell growth rate at incubation conditions of 10 µg/ml for 4 h at 37°C (Voyta et al., 1984), but provides a rapid, simple assay for LDL incorporation by cells. Using this method, we have shown that hES-derived CD31+ cells incubated with Dil-Ac-LDL stained brightly, and therefore readily took up LDL from the media (Levenberg et al., 2002).

8.4.3 Tube formation on matrigel/collagen

Three-dimensional matrices such as collagen or matrigel are often used to analyze endothelial cell differentiation, vascularization potential, and organization into tube-like structures *in vitro*. In this method, cells are seeded either on top of or within the gel itself, either by mixing the cells with the gel or seeding cells between two gel layers (Levenberg et al., 2002; Yamashita et al., 2000). Capillary tube formation can be evaluated by static or real-time phase-contrast microscopy after seeding the cells for hours or days, and the effect of growth factors on these processes (such as network density, elongation rate, lumen site) can be studied. Electron microscopy of the tube cross-sections can be used to characterize the luminal diameter and ultrastructural morphology of individual capillary tubes (Grant et al., 1991; Vernon et al., 1995) (*Figure 8.2*).

8.4.4 In vivo *vessel formation*

While establishing that hES cell derived-CD31+ endothelial cells have phenotypic and biochemical characteristics of mature endothelial cells and are capable of forming a capillary network *in vitro* is important, a more clinically relevant question is whether hES cell-derived endothelial cells are able to participate in vasculogenesis and angiogenesis *in vivo*. A variety of strategies have been used to determine whether implanted endothelial cells integrate into newly-forming host blood vessels. One method involves injecting endothelial cells into embryos to analyze their ability to incorporate into vascular structures in the developing embryo (Hatzopoulos et al., 1998). Endothelial precursors have also been injected into infarcted myocardium or ischemic hind limb to analyze the effect of the cells on neovascularization and angiogenesis (Kalka et al., 2000; Kocher et al., 2001). Another method involves seeding endothelial cells into polymer scaffolds and then implanting the cell-scaffold construct subcutaneously in mice to assess the ability of donor cells to form capillaries in an *in vivo* environment and to determine whether these blood vessels establish new connections with host-derived blood vessels involved in the neovascularization of wounding (Nor et al., 2001). This latter technique has been used to characterize the endothelial cells derived from hES cells. Sponges seeded with hES cell-derived CD31+ cells were implanted in the subcutaneous tissue of SCID mice and analyzed by immunostaining with human specific endothelial markers following one week and two weeks of implantation.

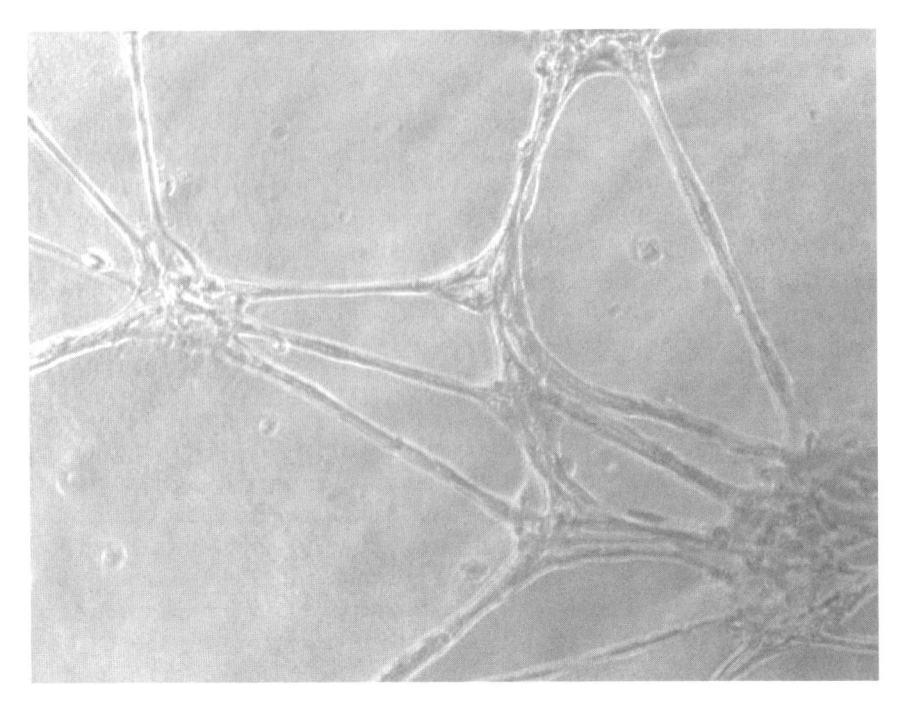

Figure 8.2: Tube-like structure formation by hES-derived endothelial cells. CD31+ cells formed vascular-like structures in Matrigel. Cord formation was evaluated by phase-contrast microscopy 3 days after seeding the cells on Matrigel.

We have shown that the implanted cells formed blood vessels *in vivo* that appeared to anastomose with the host murine vasculature (Levenberg et al., 2002) (*Figure 8.3*).

8.5 Therapeutic applications of endothelial cell progenitors

The goal of tissue engineering and therapeutic cell transplantation is to replace lost or damaged tissue with biologically compatible substitutes that restore or improve tissue function (Soker et al., 2000). Endothelial cells and their precursors are important players in tissue repair and regeneration. Endothelial cell progenitors have been identified as an integral component of many tissue engineering applications such as the formation of blood vessels, cardiac valves, and liver (Harimoto et al., 2002; Niklason et al., 1999; Shinoka, 2002) and as a means of improving blood flow to ischemic tissues (Asahara et al., 1997; Kocher et al., 2001). Recent advances have created the possibility of using hES cells as a cell source for these therapeutic strategies. Although most preclinical therapeutic studies to date have been carried out using adult-derived endothelial and progenitor cells, it would be important to examine the efficacy of hES cell-derived endothelial cells in these models because of their unlimited expansion capabilities. Similarly, disease states which could be treated by adult-derived endothelial cells could also potentially be

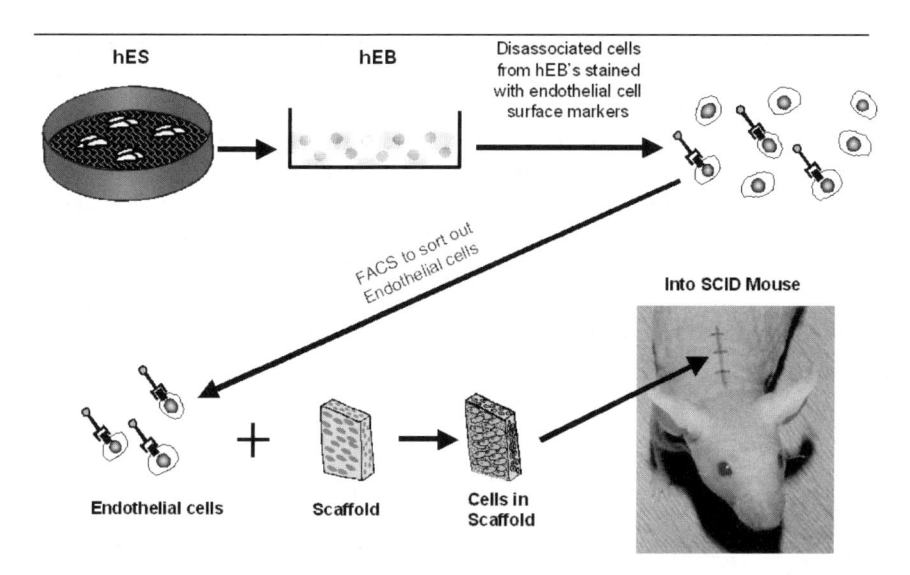

Figure 8.3: Embryonic endothelial cells *in vivo*. Embryonic endothelial cells were isolated from hEBs by staining EB cells with endothelial surface marker (CD31) and sorting out positive cells using flow cytometric cell sorting. Isolated endothelial cells were seeded on polymer scaffolds and implanted into SCID mice to analyze vessel formation *in vivo*.

treated by hES cell-derived endothelial cell therapy. Here, we will describe three burgeoning therapeutic applications of endothelial cells and their progenitors: tissue engineering novel constructs (with the use of polymer matrices), therapeutic neovascularization, and gene therapy.

8.5.1 Tissue engineered blood vessels and other constructs

Endothelial cells in engineered blood vessels

Numerous artificial vessels have been tried and tested. One technique is to isolate and grow endothelial cells on polymer scaffolds *in vitro*, followed by *in vivo* implantation. This method has been tested in an ovine model, in which expanded autologous pulmonary arterial cells were grown on biodegradable polyglactin/polyglycolic acid tubular scaffolds for one week *in vitro* and then were used to replace a 2 cm segment of the pulmonary artery (Shinoka *et al.*, 1998). During a period of 24 weeks *in vivo*, the vascular grafts showed growth and development of endothelial lining, as well as production of extracellular matrix components like collagen and elastic fibers. Functionally, grafts remained patent and non-aneurysmal. Another approach is to subject the vascular construct to pulsatile flow during *ex vivo* vessel formation while sequentially applying different cellular components that normally comprise the vessel wall (Niklason and Langer, 1997). This biomimetic system involves seeding bovine aortic smooth muscle cells into hollow tubular polyglycolic acid (PGA) scaffolds, later followed by injection of bovine aortic endothelial cells into the lumen. In comparison to native arteries, engineered arteries constructed in this way have relatively normal wall thickness and collagen content after 8 weeks of culture in a bioreactor ((Niklason et al., 1999). Apart from using differentiated cell types, endothelial progenitor cells (EPCs) represent another cell source. They normally circulate in the blood and migrate to sites of injury, such as trauma or ischemia (Takahashi et al., 1999). When EPCs from the peripheral blood of sheep were grown on decellularized porcine iliac vessels, the grafts exhibited endothelial cell morphology, contractile activity, and patency for 150 days (Kaushal et al., 2001). These studies show that endothelial cells grown on matrices have some of the structural and functional capabilities of normal blood vessels. HES cell-derived endothelial cells would be a relevant material to test in these bioartificial vascular constructs.

Endothelial cells in other engineered tissue constructs

Recently, endothelial cells have also been incorporated into cardiac valve leaflets, liver tissue, and engineered skin. In cardiac valves, engineered leaflets were created using PGA scaffolds seeded with endothelial cells and myofibroblasts from ovine arteries (Shinoka, 2002; Shinoka et al., 1996). After surgical implantation into the right posterior leaflet of the pulmonary valve for 8 weeks, the transplanted autologous cells generated a proper matrix on the polymer. In liver engineering, combining hepatocytes with endothelial cells may ultimately improve the function of bioartificial or cellular liver grafts. A recent study showed that when human aortic endothelial cells and hepatocytes were co-cultured in double layered sheets using thermo-responsive culture dishes grafted with poly (N-isopropylacrylamide) hepatocyte function (i.e., albumin expression) was better maintained than in hepatocytes grown in culture alone (Harimoto et al., 2002). Another type of vascularized tissue construct was fabricated by co-culturing human keratinocytes, dermal fibroblasts, and umbilical vein endothelial cells together in a collagen matrix (Black et al., 1998). This vascularized skin equivalent demonstrated capillary formation and extracellular matrix production. These studies indicate that engineered tissues can be fabricated with endothelial cells closely to mimic native tissue function.

8.5.2 Therapeutic neovascularization

Besides incorporation into tissue-engineered constructs, endothelial cells and their progenitors can also be directly injected *in vivo* to promote neovascularization of diseased and ischemic organs. Using this approach, therapies for conditions such as ischemia and restenosis are beginning to be explored.

Ischemia

Angiogenesis, or the growth of new capillaries, is a natural response by the body designed to maintain tissue perfusion during periods of tissue ischemia, but this capacity is often impaired or inadequate (Kalka et al., 2000). In attempts to augment or accelerate the body's angiogenic mechanisms, investigators have studied the ability of injected cells to contribute to neovascularization and functionally improve blood flow in ischemic models. Kocher et al. (2001) studied the neovascularization of ischemic myocardium in athymic nude mice following inoculation with human bone marrow-derived EPCs. They found that after myocardial infarction, the precursor endothelial cells induced both vasculogenesis in the infarct-bed as well as angiogenesis in the existing vasculature. The formation of new vessels also reduced apoptosis in hypertrophied myocytes and prevented heart failure. Hematopoietic stem cells in adult mice were also found clonally to differentiate into endothelial cells that contributed to new vessel formation in ischemic retina (Grant et al., 2002). These studies illustrate that transplantation of stem cells may be a promising therapy for ischemia, but the use of hES cells as the cell source for such therapeutic strategies is still under consideration.

Restenosis

Injury of the arterial endothelium due to intravascular stenting, balloon angioplasty or surgery often results in restenosis, which is characterized by intimal hyperplasia and may ultimately cause thrombosis. It has been shown that injury resulting in the loss of endothelial integrity and endothelial cell damage initiates a cascade of events, beginning with the egress of leukocytes into the vessel wall and adherence of platelets that elaborate PDGF, causing migration and proliferation of vascular smooth muscle cells in the vessel wall and collagen and proteoglycan deposition. It has been hypothesized that rapid re-endothelialization after vessel injury may delay or inhibit intimal hyperplasia and re-stenosis. Thus, a therapeutic approach to accelerate re-endothelialization and prevent intimal hyperplasia involves delivering endothelial cells to the luminal surface of injured blood vessels or grafts. Modified endothelial cells were seeded onto stent surfaces and introduced to denuded arteries using catheters (Nabel et al., 1989). It was reported that approximately 70% of seeded endothelial cells adhered to the stent surface when exposed to pulsatile flow *in vitro* (Flugelman et al., 1992), although other studies suggest lower figures (Shayani et al., 1994). Another approach uses endothelial cells seeded on scaffolds and implanted adjacent to injured arteries. It was shown that the implanted endothelial cells significantly reduced experimental restenosis and provided long-term control of vascular repair (Nugent and Edelman, 2001). These studies highlight the potential of stem cells for treating restenosis, but also identify important challenges.

8.5.3 Applications for gene therapy

Although angiogenesis plays an important role in physiological repair, it is also involved in the progression of diseases like cancer, in which blood vessels are necessary for tumor expansion. For this reason, vascular cells such as endothelial cells and their progenitors can be useful vehicles for anti-angiogenic gene therapy. The goal of this type of treatment is to transfer suicide genes to target cells and hinder the formation of vessels (Scappaticci, 2002). While many forms of gene therapy use viruses or naked DNA as vectors, an alternative vector is a patient's own cells. This approach involves removing cells from the body, inserting therapeutic genes extracorporally, and then reimplanting them back into the patient (Arafat et al., 2000). In one study, primate CD34+ EPCs were modified by a non-replicative thymidine kinase transduced herpes virus vector and then administered intravenously to primates undergoing skin grafting. After administration of ganciclovir, the skin grafts underwent necrosis. This study shows the feasibility of using CD34+ cell vehicles for systemic gene therapy (Gomez-Navarro *et al.*, 2000). As research in this area continues, it is possible that hES cell-derived endothelial cells will prove to be a suitable vector for gene therapy.

8.6 Challenges today and hopes for tomorrow

Before hES cell-derived endothelial cells are used in human diseases, several challenges need to be met. First, it will be important to elucidate and characterize the mechanisms underlying endothelial cell differentiation. Secondly, it will be essential to understand how vascularization is regulated in diseased states. Knowledge of the cues governing vascularization will undoubtedly be useful for developing novel therapeutic applications. Finally, efficient isolation methods are needed so that more cells can be made available.

Endothelial cells are key players in the normal and diseased vascularization process. As a result, they have tremendous potential to treat a wide range of diseases. Research in hES cells has only begun in the past half decade, but it has already been shown that hES cells are a promising source of endothelial cells. As research using hES cell-derived endothelial cells expands, it is hoped that these cells will prove themselves to be important for therapeutic applications.

References

Albelda SM, Oliver PD, Romer LH, Buck CA (1990) EndoCAM: a novel endothelial cell-cell adhesion molecule. *J. Cell Biol.* **110**, 1227–1237.

Andrews RG, Singer JW, Bernstein ID (1986) Monoclonal antibody 12-8 recognizes a 115-kd molecule present on both unipotent and multipotent hematopoietic colony-forming cells and their precursors. *Blood* **67**, 842–845.

Arafat WO, Casado E, Wang M, Alvarez RD, Siegal GP, Glorioso JC, Curiel DT, Gomez-Navarro J (2000) Genetically modified CD34+ cells exert a cytotoxic bystander effect on human endothelial and cancer cells. *Clin. Cancer Res.* **6**, 4442–4448.

Asahara T, Murohara T, Sullivan A, Silver M, van der Zee R, Li T, Witzenbichler B, Schatteman G, Isner JM (1997) Isolation of putative progenitor endothelial cells for angiogenesis. *Science* **275**, 964–967.

Balconi G, Spagnuolo R, Dejana E (2000) Development of endothelial cell lines from embryonic stem cells: A tool for studying genetically manipulated endothelial cells *in vitro*. *Arterioscler. Thromb. Vasc. Biol.* **20**, 1443–1451.

Baron MH (2001) Molecular regulation of embryonic hematopoiesis and vascular development: a novel pathway. *J. Hematother. Stem Cell Res.* **10**, 587–594.

Beck L Jr, D'Amore PA (1997) Vascular development: cellular and molecular regulation. *Faseb. J.* **11**, 365–373.

Black AF, Berthod F, L'Heureux N, Germain L, Auger FA (1998) *In vitro* reconstruction of a human capillary-like network in a tissue-engineered skin equivalent. *Faseb J.* **12**, 1331–1340.

Byrd N, Becker S, Maye P, Narasimhaiah R, St-Jacques B, Zhang X, McMahon J, McMahon A, Grabel L (2002) Hedgehog is required for murine yolk sac angiogenesis. *Development* **129**, 361–372.

Carmeliet P, Ferreira V, Breier G, Pollefeyt S, Kieckens L, Gertsenstein M et al. (1996) Abnormal blood vessel development and lethality in embryos lacking a single VEGF allele. *Nature* 380, 435–439.

Choi K, Kennedy M, Kazarov A, Papadimitriou JC, Keller G (1998) A common precursor for hematopoietic and endothelial cells. *Development* 125, 725–732.

Dejana E, Spagnuolo R, Bazzoni G (2001) Interendothelial junctions and their role in the control of angiogenesis, vascular permeability and leukocyte transmigration. *Thromb. Haemost.* 86, 308–315.

Faloon P, Arentson E, Kazarov A, Deng CX, Porcher C, Orkin S, Choi K (2000) Basic fibroblast growth factor positively regulates hematopoietic development. *Development* 127, 1931–1941.

Ferrara N, Carver-Moore K, Chen H, Dowd M, Lu L, O'Shea KS, Powell-Braxton L, Hillan KJ, Moore MW (1996) Heterozygous embryonic lethality induced by targeted inactivation of the VEGF gene. *Nature* 380, 439–442.

Flugelman MY, Virmani R, Leon MB, Bowman RL, Dichek DA (1992) Genetically engineered endothelial cells remain adherent and viable after stent deployment and exposure to flow *in vitro*. *Circ. Res.* 70, 348–354.

Fong GH, Zhang L, Bryce DM, Peng J (1999) Increased hemangioblast commitment, not -vascular disorganization, is the primary defect in flt-1 knock-out mice. *Development* 126, 3015–3025.

Francis SE, Goh KL, Hodivala-Dilke K, Bader BL, Stark M, Davidson D, Hynes RO (2002) Central roles of alpha5beta1 integrin and fibronectin in vascular development in mouse embryos and embryoid bodies. *Arterioscler. Thromb. Vasc. Biol.* 22, 927–933.

Gomez-Navarro J, Contreras JL, Arafat W, Jiang XL, Krisky D, Oligino T et al. (2000) Genetically modified CD34+ cells as cellular vehicles for gene delivery into areas of angiogenesis in a rhesus model. *Gene Ther.* 7, 43–52.

Grant DS, Lelkes PI, Fukuda K, Kleinman HK (1991) Intracellular mechanisms involved in basement membrane induced blood vessel differentiation *in vitro*. *In Vitro Cell Dev. Biol.* 27A, 327–336.

Grant MB, May WS, Caballero S, Brown GA, Guthrie SM, Mames RN et al. (2002) Adult hematopoietic stem cells provide functional hemangioblast activity during retinal neovascularization. *Nat. Med.* 8, 607–612.

Haar JL, Ackerman GA (1971) A phase and electron microscopic study of vasculogenesis and erythropoiesis in the yolk sac of the mouse. *Anat. Rec.* 170, 199–223.

Harimoto M, Yamato M, Hirose M, Takahashi C, Isoi Y, Kikuchi A, Okano T (2002) Novel approach for achieving double-layered cell sheets co-culture: overlaying endothelial cell sheets onto monolayer hepatocytes utilizing temperature-responsive culture dishes. *J. Biomed. Mater. Res.* 62, 464–470.

Hatzopoulos AK, Folkman J, Vasile E, Eiselen GK, Rosenberg RD (1998) Isolation and characterization of endothelial progenitor cells from mouse embryos. *Development* 125, 1457–1468.

Hellstrom M, Kalen, M, Lindahl P, Abramsson A, Betsholtz C (1999) Role of PDGF-B and PDGFR-beta in recruitment of vascular smooth muscle cells and pericytes during embryonic blood vessel formation in the mouse. *Development* 126, 3047–3055.

Hirashima M, Kataoka H, Nishikawa S, Matsuyoshi N (1999) Maturation of embryonic stem cells into endothelial cells in an *in vitro* model of vasculogenesis. *Blood* **93**, 1253–1263.

Hirschi KK, Rohovsky SA, D'Amore PA (1998) PDGF, TGF-beta, and heterotypic cell-cell interactions mediate endothelial cell-induced recruitment of 10T1/2 cells and their differentiation to a smooth muscle fate. *J. Cell Biol.* **141**, 805–814.

Jaffe EA, Hoyer LW, Nachman RL (1974) Synthesis of von Willebrand factor by cultured human endothelial cells. *Proc. Natl Acad. Sci. USA* **71**, 1906–1909.

Kabrun N, Buhring HJ, Choi K, Ullrich A, Risau W, Keller G (1997) Flk-1 expression defines a population of early embryonic hematopoietic precursors. *Development* **124**, 2039–2048.

Kalka C, Masuda H, Takahashi T, Kalka-Moll WM, Silver M, Kearney M, Li T, Isner JM, Asahara T (2000) Transplantation of *ex vivo* expanded endothelial progenitor cells for therapeutic neovascularization. *Proc. Natl Acad. Sci. USA* **97**, 3422–3427.

Kaufman DS, Hanson ET, Lewis RL, Auerbach R, Thomson JA (2001) Hematopoietic colony-forming cells derived from human embryonic stem cells. *Proc. Natl Acad. Sci. USA* **98**, 10716–10721.

Kaushal S, Amiel GE, Guleserian KJ, Shapira OM, Perry T, Sutherland FW *et al.* (2001) Functional small-diameter neovessels created using endothelial progenitor cells expanded *ex vivo*. *Nat. Med.* **7**, 1035–1040.

Kocher AA, Schuster MD, Szabolcs MJ, Takuma S, Burkhoff D, Wang J, Homma S, Edwards NM, Itescu S (2001) Neovascularization of ischemic myocardium by human bone-marrow-derived angioblasts prevents cardiomyocyte apoptosis, reduces remodeling and improves cardiac function. *Nat. Med.* **7**, 430–436.

Korhonen J, Lahtinen I, Halmekyto M, Alhonen L, Janne J, Dumont D, Alitalo K (1995) Endothelial-specific gene expression directed by the tie gene promoter *in vivo*. *Blood* **86**, 1828–1835.

Kuo CT, Veselits ML, Barton KP, Lu MM, Clendenin C, Leiden JM (1997) The LKLF transcription factor is required for normal tunica media formation and blood vessel stabilization during murine embryogenesis. *Genes Dev.* **11**, 2996–3006.

Lee ME, Temizer DH, Clifford JA, Quertermous T (1991) Cloning of the GATA-binding protein that regulates endothelin-1 gene expression in endothelial cells. *J. Biol. Chem.* **266**, 16188–16192.

Levenberg S, Golub JS, Amit M, Itskovitz-Eldor J, Langer R (2002) Endothelial cells derived from human embryonic stem cells. *Proc. Natl Acad. Sci. USA* **99**, 4391–4396.

Lin Q, Lu J, Yanagisawa H, Webb R, Lyons GE, Richardson JA, Olson EN (1998) Requirement of the MADS-box transcription factor MEF2C for vascular development. *Development* **125**, 4565–4574.

McMahon AP (2000) More surprises in the Hedgehog signaling pathway. *Cell* **100**, 185–188.

Murray P (1932) Development *in vitro* of the blood of the early chick embryo. *Proc. R. Soc. Lond.* **111**, 497–521.

Nabel EG, Plautz G, Boyce FM, Stanley JC, Nabel GJ (1989) Recombinant gene expression *in vivo* within endothelial cells of the arterial wall. *Science* **244**, 1342–1344.

Niklason LE, Langer RS (1997) Advances in tissue engineering of blood vessels and other tissues. *Transpl. Immunol.* **5**, 303–306.

Niklason LE, Gao J, Abbott WM, Hirschi KK, Houser S, Marini R, Langer R (1999) Functional arteries grown *in vitro*. *Science* **284**, 489–493.

Nishikawa S, Nishikawa S, Fraser S, Fujimoto T, Yoshida H, Hirashima M, Ogawa M (2001) Developmental relationship of hematopoietic stem cells and endothelial cells. In: *Hematopoiesis: A Developmental Approach* (ed. LI Zon). Oxford University Press, New York, pp. 171–179.

Nishikawa SI, Nishikawa S, Hirashima M, Matsuyoshi N, Kodama H (1998) Progressive lineage analysis by cell sorting and culture identifies FLK1+VE-cadherin+ cells at a diverging point of endothelial and hemopoietic lineages. *Development* **125**, 1747–1757.

Nor JE, Peters MC, Christensen JB, Sutorik MM, Linn S, Khan MK, Addison CL, Mooney DJ, Polverini PJ (2001) Engineering and characterization of functional human microvessels in immunodeficient mice. *Lab. Invest.* **81**, 453–463.

Nugent HM, Edelman ER (2001) Endothelial implants provide long-term control of vascular repair in a porcine model of arterial injury. *J. Surg. Res.* **99**, 228–234.

Risau W (1995) Differentiation of endothelium. *Faseb J.* **9**, 926–933.

Risau W, Sariola H, Zerwes HG, Sasse J, Ekblom P, Kemler R, Doetschman T (1988) Vasculogenesis and angiogenesis in embryonic-stem-cell-derived embryoid bodies. *Development* **102**, 471–478.

Robertson SM, Kennedy M, Shannon JM, Keller G (2000) A transitional stage in the commitment of mesoderm to hematopoiesis requiring the transcription factor SCL/tal-1. *Development* **127**, 2447–2459.

Scappaticci FA (2002) Mechanisms and future directions for angiogenesis-based cancer therapies. *J. Clin. Oncol.* **20**, 3906–3927.

Schuh AC, Faloon P, Hu QL, Bhimani M, Choi K (1999) *In vitro* hematopoietic and endothelial potential of flk-1(-/-) embryonic stem cells and embryos. *Proc. Natl. Acad. Sci. USA* **96**, 2159–2164.

Schuldiner M, Yanuka O, Itskovitz-Eldor J, Melton DA, Benvenisty N (2000) From the cover: effects of eight growth factors on the differentiation of cells derived from human embryonic stem cells. *Proc. Natl Acad. Sci. USA* **97**, 11307–11312.

Shayani V, Newman KD, Dichek DA (1994) Optimization of recombinant t-PA secretion from seeded vascular grafts. *J. Surg. Res.* **57**, 495–504.

Shinoka T (2002) Tissue engineered heart valves: autologous cell seeding on biodegradable polymer scaffold. *Artif. Organs* **26**, 402–406.

Shinoka T, Ma PX, Shum-Tim D, Breuer CK, Cusick RA, Zund G, Langer R, Vacanti JP, Mayer JE Jr (1996) Tissue-engineered heart valves: autologous valve leaflet replacement study in a lamb model. *Circulation* **94**, II164–II168.

Shinoka T, Shum-Tim D, Ma PX, Tanel RE, Isogai N, Langer R, Vacanti JP, Mayer JE Jr (1998) Creation of viable pulmonary artery autografts through tissue engineering. *J. Thorac. Cardiovasc. Surg.* **115**, 536–545.

Soker S, Machado M, Atala A (2000) Systems for therapeutic angiogenesis in tissue engineering. *World J. Urol.* **18**, 10–18.

Takahashi T, Kalka C, Masuda H, Chen D, Silver M, Kearney M, Magner M, Isner JM, Asahara T (1999) Ischemia- and cytokine-induced mobilization of bone marrow-derived endothelial progenitor cells for neovascularization. *Nat. Med.* **5**, 434–438.

Thomson JA, Itskovitz-Eldor J, Shapiro SS, Waknitz MA, Swiergiel JJ, Marshall VS, Jones JM (1998) Embryonic stem cell lines derived from human blastocysts. *Science* **282**, 1145–1147.

Tsai FY, Keller G, Kuo FC, Weiss M, Chen J, Rosenblatt M, Alt FW, Orkin SH (1994) An early haematopoietic defect in mice lacking the transcription factor GATA-2. *Nature* **371**, 221–226.

Vernon RB, Lara SL, Drake CJ, Iruela-Arispe ML, Angello JC, Little CD, Wight TN, Sage EH (1995) Organized type I collagen influences endothelial patterns during 'spontaneous angiogenesis *in vitro*': planar cultures as models of vascular development. *In Vitro Cell Dev. Biol. Anim.* **31**, 120–131.

Vittet D, Prandini MH, Berthier R, Schweitzer A, Martin-Sisteron H, Uzan G, Dejana E (1996) Embryonic stem cells differentiate *in vitro* to endothelial cells through successive maturation steps. *Blood* **88**, 3424–3431.

Vittet D, Buchou T, Schweitzer A, Dejana E, Huber P (1997) Targeted null-mutation in the vascular endothelial-cadherin gene impairs the organization of vascular-like structures in embryoid bodies. *Proc. Natl Acad. Sci. USA* **94**, 6273–6278.

Voyta JC, Via DP, Butterfield CE, Zetter BR (1984) Identification and isolation of endothelial cells based on their increased uptake of acetylated-low density lipoprotein. *J. Cell Biol.* **99**, 2034–2040.

Yamaguchi TP, Dumont DJ, Conlon RA, Breitman ML, Rossant J (1993) flk-1, an flt-related receptor tyrosine kinase is an early marker for endothelial cell precursors. *Development* **118**, 489–498.

Yamashita J, Itoh H, Hirashima M, Ogawa M, Nishikawa S, Yurugi T, Naito M, Nakao K (2000) Flk1-positive cells derived from embryonic stem cells serve as vascular progenitors. *Nature* **408**, 92–96.

Yang X, Castilla LH, Xu X, Li C, Gotay J, Weinstein M, Liu PP, Deng CX (1999) Angiogenesis defects and mesenchymal apoptosis in mice lacking SMAD5. *Development* **126**, 1571–1580.

9. Neural specification from human embryonic stem cells

Su-Chun Zhang

9.1 Introduction

Development of the vertebrate central nervous system (CNS) involves multiple steps, beginning with the induction of neuroepithelia from the embryonic ectoderm and the patterning of the neural plate into complex regional compartments along rostro-caudal and dorso-ventral axes. How neuroepithelia are specified from the embryonic ectoderm and how the neural ectoderm is compartmentalized are primarily investigated using stage-specific *Xenopus* and chick embryos (Wilson and Edlund, 2001). These embryos are easily accessible and are amenable to genetic and surgical manipulations. In addition, the formation of the neural ectoderm occurs within 24 h and the whole process of neural induction can be visualized continuously. In mammals, these studies are compromised by experimental inaccessibility to early embryos. Further, the process of neural induction takes place in the womb over a substantially longer period. In mice, the neural plate forms at embryonic day 7, whereas in humans it develops around embryonic day 18 (O'Rahilly and Muller, 1994). Hence, the cellular and molecular mechanisms of neural induction in mammals remain elusive and whether a similar principle applies to human has not been explored.

Embryonic stem (ES) cells are the *in vitro* counterparts of the inner cell mass of a preimplantation embryo at the blastocyst stage (Evans and Kaufman, 1981; Martin, 1981). They are, at least theoretically, capable of giving rise to all cell types that constitute an animal. Hence, ES cells provide a simple *in vitro* alternative for molecular and cellular analyses of neural induction and cell lineage specification in mammals. Recent studies in mouse ES cells indicate that the process of neural induction and cell specification may be recapitulated in a petri dish based on the principles of neural induction and patterning learned from other vertebrates (Tropepe *et al.*, 2001; Wichterle *et al.*, 2002; Mizuseki *et al.*, 2003). This underscores the usefulness of stem cells in understanding early neural development and reinforces the need to integrate the principles of developmental biology and stem cell biology (Anderson, 2001). Similarly, the establishment of continuous human ES (hES) cell lines (Thomson et al., 1998; Reubinoff et al., 2000) provides an otherwise inaccessible system for modeling aspects of early human development. Understanding how neural specification occurs in humans will facilitate the generation of selective neural cells for regenerative medicine.

9.2 Neural induction in vertebrates

The initial step in the generation of the vertebrate nervous system is the specification of neuroepithelia from ectodermal cells, a process known as neural induction. In amphibians, neuroepithelia form on the dorsal side of the ectoderm, whereas epidermis comes from the ventral ectoderm. Seminal studies conducted by Spemann and Mangold demonstrated that the dorsal lip of the amphibian blastopore, later known as the organizer, could induce a secondary body axis with a fully developed nervous system when transplanted to the ventral side of a host embryo at the gastrula stage (reviewed in Weinstein and Hemmati-Brivanlou, 1999). Explant cultures of dorsal ectoderm alone taken at the gastrula stage form epidermis, but generate neural tissues when co-cultured with organizer grafts. These findings led to the idea that the organizer region is a local source of neural inductive signals. However, disaggregation of the dorsal ectoderm leads to generation of neural cells, raising the possibility that a neural fate could be derived from competent ectoderm in the absence of signals from the organizer and that signaling between early ectodermal cells normally suppresses neural specification (Grunz and Tacke, 1989). A mechanistic explanation for this phenomenon comes from a study employing a truncated activin receptor which blocks the signaling by activin and other related ligands of the transforming growth factor-β (TGFβ) family, including bone morphogenetic proteins (BMPs). Injection of this mutant receptor into the dorsal ectoderm resulted in the generation of neural tissues (Hemmati-Brivanlou and Melton, 1992). This triggered the idea that signaling by BMP inhibitors, secreted from the organizer, act as neural inducers. Consistent with this idea, BMP antagonists, including Noggin, Follistatin, and Chordin, are expressed in the organizer region and can induce neural markers in dorsal ectoderm explant cultures (Lamb et al., 1993). Thus, in amphibians, the neural fate of ectoderm is regarded as a default state, i.e., dorsal ectoderm will form neural plate as long as the inhibitory effect of BMPs in the ectoderm is suppressed by factors secreted from the organizer.

In chick and mouse embryos, similar transplantation studies have demonstrated that the organizer (Hensen's node) can induce ectopic neural tissues. However, mutants that fail to develop the organizer or its derivatives still generate a neural plate, suggesting that the organizer is not required for neural induction (Episkopou et al., 2001). Explant cultures of primitive ectoderm (epiblast) at the blastula stage result in the generation of neural tissue, indicating that neural induction in chickens may be initiated before gastrulation (Wilson et al., 2000). Using a differential expression screen, Streit et al. (2000) identified an *early response to neural induction (ERNI)* gene which is expressed in the prospective neural plate at pre-primitive streak stages. Using this marker, they demonstrated that neural induction takes place before the gastrula stage and that fibroblast growth factor (FGF) signaling is required for initiating neural induction. The effect of FGF signaling in chick neural induction appears to be achieved at least in part by repressing BMP expression (Wilson *et al.*, 2000). Hence, neural induction in amphibians and chick may differ in terms of the initiation of neural induction and the nature of the neural inducing signals. However, in both species, anti-BMP signaling appears to be involved. In addition to FGF and anti-BMP signaling events, Wnt signaling is also likely to be involved in the regulation of neural induction (Patapoutian and Reichardt, 2000; Wilson and Edlund, 2001).

9.3 Embryonic stem cells as a window to mammalian neural development

Because experimental manipulation of the post-implantation human embryo is ethically unacceptable, our knowledge about human development is limited to observations on static histological sections of human embryos. It is almost impossible to obtain human embryos within the first several weeks of development, a period critical for embryonic induction and patterning. Consequently, how developmental events such as neural specification and patterning take place in human is based upon analogy to experimental embryological studies in animal models, primarily the mouse model. It is generally accepted that the basic principles underlying early embryonic development are conserved across species. However, this needs to be verified independently. As discussed above, differences may exist between one species and another in terms of when, where, and how these fundamental principles are implemented in neural induction. From an evolutionary standpoint, it is not unreasonable to expect that neural specification in humans may involve unique molecules at particular windows of development. Hence, the principles of neural induction gained from animal studies can form the foundation for exploring novel processes that occur in neural specification in humans.

How can the dynamic neural specification process in humans be studied without the use of stage-specific embryos? The answer appears to lie in human embryonic stem (ES) cells (Thomson *et al.*, 1998; Reubinoff *et al.*, 2000). ES cells are equivalent to the inner cell mass of an embryo at the blastocyst stage. Hence, ES cells are precursors to all embryonic lineages. These cells should allow tracing the history from the root to individual branches of the cell lineage tree in a simplified and controllable culture environment. One concern is that cell culture does not have the complex cell and tissue interactions that are critical to embryonic induction at distinct developmental stages. These cellular interactions, however, can be largely recreated in culture to reflect the *in vivo* environment, allowing studies on the earliest stages of development such as neural specification.

Like studies in mouse ES cells over the past two decades (Evans and Kaufman, 1981; Martin, 1981), much attention has been paid to the therapeutic potential of human ES cells in regenerative medicine. This, to some degree, has overshadowed the potential of human ES cells to aid in our understanding of normal and abnormal human development, and the need for basic research upon which future therapeutic applications of human ES cells can be built. A glance over the past decade's studies on neural (or other) lineage differentiation from mouse ES cells reveals that the majority of protocols for deriving neural cells from mouse ES cells are empirically formulated. Only recently have we seen a trend of applying the principles of developmental biology to directing neural differentiation from mouse ES cells (Tropepe et al., 2001; Rathjen et al., 2002; Wichterle et al., 2002). In order to harness the therapeutic potential of hES cells, it is essential to gain a clear understanding of how human ES cells are restricted and patterned to a specific fate. This will require the integration of developmental neurobiology and stem cell biology (Anderson, 2001) as well as lessons learned from mouse ES cell studies.

9.4 Neural differentiation from mouse ES cells

9.4.1 Neural differentiation: common methodology

ES cells tend to differentiate spontaneously under conditions that do not favor self-renewal, e.g., upon removal from feeder cells or withdrawal of the growth factor leukemia inhibitory factor (LIF). To bias the differentiation process toward a neural fate, culture conditions are modified to promote the differentiation, survival, and/or proliferation of neural cells from ES cells. The most commonly used approach for neural differentiation from mouse ES cells is the spontaneous aggregation of ES cells into so-called embryoid bodies and treatment of these ES cell aggregates with retinoic acid (RA). This method is a modified version of neuronal differentiation from teratocarcinoma cells (Jones-Villeneuve et al., 1982). The procedure involves culturing suspended ES cell aggregates in regular ES cell growth medium without the growth factor LIF for 4 days, followed by the addition of RA (0.1–1 µM) for another 4 days. Hence, this procedure is often regarded as a 4-/4+ protocol (Bain et al., 1995; McDonald et al., 1999). Other RA-induced neural differentiation protocols are variations of the 4-/4+ protocol (Wobus et al., 1988; Fraichard et al., 1995; Strubing et al., 1995; Dinsmore et al., 1996; Renoncourt et al., 1998). Embryoid bodies can be easily obtained by culturing the dissociated ES cells in non-adherent culture dishes or in the form of 'hanging drops' (Wobus et al., 1991). The latter method generates embryoid bodies of the same size with a similar number of cells. It thus allows quantitative analyses. Embryoid bodies treated with RA yield a good proportion (38%) of neuronal cells upon differentiation (Bain et al., 1995). The predominant population of neuronal cells comprises glutaminergic and GABAergic neurons (Bain et al., 1995; Strubing et al., 1995; Fraichard et al., 1995).

Signaling through RA and its cognate receptors is important during development, particularly in rostral/caudal patterning of the neural tube (Muhr et al., 1999; Maden, 2002). Whether neurons generated from RA-treated ES cells are committed to a caudal fate is not clearly defined. Nevertheless, there is little evidence to suggest that RA in these protocols acts to induce neural specification from ES cells (see section 9.2). RA is a strong morphogen that appears to push ES cells toward post-mitotic neurons although RA-treated cultures contain neural cells at various developmental stages including neural progenitors (Gottlieb and

Huettner, 1999). Hence, maintenance and expansion of neural progenitors derived from RA-treated mouse ES cells is difficult. One exception appears to be the culture of oligodendroglial progenitors as 'oligospheres' isolated using this protocol (Liu *et al.*, 2000). Again, this 'oligosphere' culture contains both progenitors and a substantial proportion of immature oligodendrocytes (O4+) as compared with the uniform progenitor or pre-progenitor nature of oligospheres generated using other approaches (Avellana-Adalid *et al.*, 1996; Zhang *et al.*, 1998, 1999).

In addition to the RA-stimulated neural differentiation approaches, conditioned media from mesoderm-derived cell lines have been used effectively to promote the generation of neural cells from ES cells. The rationale behind this is that signals from mesodermal cells are required to induce neural specification from the ectoderm during early development. While the identity of the neural-promoting molecules in the conditioned media remains unknown, these approaches allow for the production of large numbers of specialized neural cells. In particular, Sasai and colleagues are able to induce mouse ES cells to differentiate into a large proportion of midbrain dopamine neurons using a bone marrow stroma cell line, PA6 cells (Kawasaki *et al.*, 2000). The same conditioned medium can also induce differentiation of dopamine neurons from non-human primate ES cells (Kawasaki *et al.*, 2002), suggesting that the neural differentiation process is similar between rodents and primates. It would be interesting to see whether human ES cells can be similarly induced to generate dopamine neurons and whether other mesodermal cell lines have a similar effect on neural differentiation. In addition, Rathjen *et al.* (2002) use a conditioned medium from a hepatic cell line to induce mouse ES cells to differentiate into neuroepithelial cells through an intermediate stage called primitive ectoderm-like cells.

9.4.2 ES cells as a tool to model neural lineage development

ES cells sit on top of the cell lineage tree. Hence, ES cells provide an ideal system to analyze early embryonic induction, especially cell lineage specification in mammals. Understanding how an individual cell lineage is specified will be instrumental for effective generation of a particular cell type for therapy. To understand how pluripotent ES cells are sequentially restricted to neural cells, McKay and colleagues have developed a protocol for the generation of an enriched population of neuroepithelial cells from mouse ES cells. This protocol first entails the formation of embryoid bodies for 4 days, followed by culturing dissociated cells in a chemically defined culture system with FGF2 selectively to support the survival and proliferation of neuroepithelial cells (Okabe *et al.*, 1996). In contrast to the RA-induced neural differentiation process, treatment of ES cells with FGF2 produces a high proportion of neuroepithelial cells (80%) at a synchronous developmental stage (Okabe *et al.*, 1996). These neural precursor cells can be expanded for a period of time. More importantly, they do not appear to commit to a regional fate and can differentiate into neurons and glia in response to environmental cues (Okabe *et al.*, 1996; Brustle *et al.*, 1999; Lee *et al.*, 2000; Kim *et al.*, 2002). While FGF2 is a survival and proliferation factor for early neuroepithelial cells generated from ES cells, it has also been shown to play an important role in neural induction (Streit *et al.*, 2000; Wilson *et al.*, 2000). It is possible that in the

FGF2-treated ES cell differentiation model, FGF2 may also play a role in inducing neuroepithelial differentiation from ES cells. Hence, this model, to some degree, reflects early neural development.

The 'default' neural induction model deduced from experimental studies in *Xenopus* implies that an embryonic ectodermal cell would become a neural cell as long as neural-inhibiting factors are removed. To determine whether neural induction in the mouse is similar to that in *Xenopus*, Tropepe et al. (2001) have developed a clonal culture system to follow neural differentiation. In such cultures, mouse ES cells transform into primitive neural stem cells within hours in the absence of any exogenous inductive factors. This suggests that mouse ES cells have an innate tendency to become neural cells without the presence of inhibitory signals from neighboring cells. This is further supported by the observation that neural differentiation is enhanced in ES cells lacking Smad4, a key component in the bone morphogenetic protein (BMP) signaling pathway. Studies using non-clonal cultures of mouse ES cells also suggest the involvement of anti-BMP signaling in early neural specification in mice (Finley et al., 1999; Kawasaki et al., 2000). Hence, the default mechanism of neural induction may be a conserved process. The low efficiency (0.2%) of neural induction in the clonal culture condition, however, may suggest that an active induction, in addition to anti-BMP signaling, is necessary. Indeed, a simple withdrawal of LIF from a monolayer culture of mouse ES cells in a neural medium leads to a high proportion of neuroepithelia. This neural induction appears to involve FGF signaling as blocking FGFs produced in culture inhibits neural differentiation (Ying et al., 2003). From a therapeutic perspective, the clonal induction protocol has its advantage for generating clonally derived, pure populations of neural stem cells that can be expanded and differentiated into neurons and glial cells. Nevertheless, it is not clear whether neural stem cells generated in this way are committed to neural cells of a particular regional identity.

One of the characteristic features of neural cells is their positional identity acquired during neural induction and patterning. This positional information is imparted upon neuroepithelial cells via morphogen gradients secreted from surrounding tissues. Partially to mimic the positional information in a culture petri dish, morphogens that affect rostro-caudal and dorso-ventral fate choices are applied together or in sequence. Thus, by applying FGF8, which influences the mid-hindbrain fate, and sonic hedgehog (SHH), a ventralizing molecule, Lee et al. (2000) further induce ES cell-derived neuroepithelial cells into midbrain dopamine neurons. Similarly, Jessell and colleagues have developed a system to guide mouse ES cells to differentiate into spinal motor neurons in a stepwise fashion based on the current understanding of motoneuron development. This procedure entails formation of embryoid bodies and induction of neuroectodermal cells by RA, followed by caudalization of the ES-derived neuroectodermal cells by contact with bone marrow stromal cells and exposure to RA, and finally induction of motoneurons by the ventralizing molecule SHH (Wichterle et al., 2002). In this protocol, it is questionable whether RA induces neuroectodermal differentiation (see above). Since RA strongly pushes post-mitotic and caudalizing neural differentiation, it is critical to catch the narrow window of neuroepithelial cell stage in order to achieve a motoneuron differentiation. This also explains why it is difficult to direct neural differentiation of forebrain and midbrain phenotypes using RA-treated ES cells. These studies reinforce the importance of applying developmental insights toward advancing the potential of ES cells.

9.4.3 Isolation of ES-derived neural cells

The neural differentiation protocols summarized above, with the exception of the clonal differentiation culture, yield a mixed population of cells including neural cells. It is therefore necessary to isolate the cells of interest from the mixture. A simple and efficient way of separating cells of choice is immunoseparation based on known cell surface epitopes. For example, Rao and colleagues sort neuronal progenitor cells from mouse ES cell-differentiated progenies based on the fact that neuronal restricted progenitors express an embryonic form of neural cell adhesion molecule (NCAM) (Mujtaba *et al.*, 1999). However, cell surface markers are not always available for neural cells at various stages. Also, while cell surface markers are useful in isolating a certain population of neural cells from a mixture of brain-derived cells, the effectiveness of this approach may be compromised by the fact the cell surface molecule may also be expressed by non-neural lineage cells in the ES cell-generated mixture. An alternative approach is to isolate these cells based on their physio-chemical properties. McKay and colleagues use a medium that preferentially promotes the survival and proliferation of neural precursor cells (Okabe *et al.*, 1996; Lee *et al.*, 2000) although how non-neural cells are eliminated is unclear. A more complicated approach incorporates genetic selection. By coupling identifiable markers, such as a green fluorescent protein gene under the control of a cell type specific promoter, cells of choice can be isolated through immunosorting (Li *et al.*, 1998; Kim *et al.*, 2002; Wichterle *et al.*, 2002; Ying *et al.*, 2003).

9.5 Neural differentiation from human ES cells

9.5.1 Neural differentiation from hES cells: modification of mouse protocols

Similar to mouse ES cells, hES can spontaneously differentiate into a variety of cells, including neural cells, as demonstrated by teratoma formation (Thomson *et al.*, 1998). However, hES cells behave differently from their mouse counterparts in many respects (Thomson and Odorico, 2000). Human ES colonies grow more slowly than mouse ES cell colonies, and hES cells do not rely on LIF for self-renewal. Technically, the cloning efficiency of hES is low due to the poor survival of disaggregated individual hES cells (Amit *et al.*, 2000). Hence, direct neural differentiation from disaggregated individual hES cells, as employed by Tropepe *et al.* (2001) in mouse ES cell differentiation, is technically challenging. Our initial attempt to use this approach resulted in no survival of hES cells. Modification of the culture system, including the use of substrates and addition of B27 and growth factors such as FGF2 yielded only occasional formation of neurospheres (Zhang, unpublished). The neural differentiation protocol used by Reubinoff *et al.* (2001) does not involve EB formation due to difficulty in making healthy EBs in their initial studies. Instead, they grew hES cells at high density for a prolonged period so that the ES cells spontaneously differentiated into various cell types including neural cells. Neural precursor cells differentiated in this manner formed clusters characteristic of neurospheres. These neurosphere-like clusters could be mechanically dissected with a micropipette under a microscope and expanded as neurospheres in suspension cultures. While hES cells, like their mouse counterparts, tend to differentiate spontaneously into various cell types including neural cells, high-density culture is generally not favorable to neural induction or differentiation. Thus, the efficiency of neural differentiation in this system may be low. Nevertheless, it is possible that a particular hES cell line tends to differentiate spontaneously into a preferential lineage.

To date, the protocols used for neural differentiation from hES cells generally involve the initial aggregation of hES cells as embryoid bodies (EB). Carpenter *et al.* (2001) used the traditional mouse ES cell differentiation protocol to derive neural precursor cells. Human ES cells are first aggregated to form EBs, which are subsequently treated with RA to induce neural differentiation in suspension cultures. As discussed above with mouse ES cells, RA treatment leads to the differentiation of neural cells from progenitor to mature stages. This is displayed by the mixture of neural precursors expressing poly-sialylated neural cell adhesion molecule (PSA-NCAM) and cells with apparent neuronal morphology and expression of β_{III}-tubulin within 4 days of RA treatment and 3 days of differentiation (Carpenter *et al.*, 2001). In order to select neural progenitor cells, the differentiated EBs are dissociated and neural progenitor cells isolated by immunoseparation (magnetic sorting) based on the fact that a certain population of neural progenitor cells expresses PSA-NCAM (Carpenter *et al.*, 2001). In this study, treatment of hES cells with RA does not appear to increase the neural cell fraction to the same degree as seen in mouse ES cells. Even though 10–20 times more RA (10 µM) was used, only a modest enrichment of the neural cell population is achieved (Carpenter *et al.*, 2001). Similarly, in a study by Schuldiner *et al.* (2001), 10 µM of RA was applied in order to increase the proportion of neurons. In our hands, treatment of adherent hES cell colonies with RA at a dosage higher than 1 µM causes cell detachment and subsequent cell death, indicating possible toxicity of RA to hES cells at this concentration (Zhang, unpublished). These observations suggest a potential species difference in neural differentiation in response to RA. Thus, novel approaches are necessary to induce hES cells to a neural fate.

9.5.2 Neural specification from hES cells: application of neural induction principles

Embryonic induction such as neural induction is likely a conserved process across vertebrates. Hence, neural induction in humans may be similar to that in *Xenopus* and chickens. Based on this hypothesis, we have designed a chemically defined culture system to induce hES cells toward a neural fate using FGF2. Cell fate determination is the outcome of the interplays between cell intrinsic programs and environmental influences (Edlund and Jessell, 1999). Considerations were given to the timing of neural development in humans, an obvious factor that differs between humans and other species. The neural differentiation process is initiated through the aggregation of ES cells in suspension for 4 days. In long-term culture, these aggregates, or embryoid bodies (EBs) develop into a cystic cavity surrounded by three germ layers, resembling an embryo at gastrulation. In practice, mouse ES cells are usually trypsinized and placed in a bacteriological grade petri dish or in hanging drops so that individual ES cells aggregate together. As hES cells survive better in aggregates, we have designed an approach simply to 'lift' the ES cell colonies from the feeder layer or matrigel by treatment with a low concentration of dispase or collagenase. The detached ES cell colonies form round aggregates when placed in a suspension culture (Zhang et al., 2001).

In the mouse system, simple aggregation of ES cells appears to induce a transition of ES cells to epiblasts or primitive ectoderm cells (Lake et al., 2000; Rathjen et al., 2002). The primitive ectodermal stage is critical for embryonic induction and patterning. Since embryonic induction, including neural specification, generally takes place between blastula and gastrula stages, it would be counterproductive to utilize late, cyst-forming EBs to 'guide' specific lineage differentiation. We thus apply neural inducing agents to ES cell aggregates that have been cultured for 3–4 days. These cell aggregates lack a cystic cavity or three germ layer-like structures, thus the name EB may be somewhat misleading.

Most of the lineage induction protocols employ the addition of morphogens or growth factors to the ES cell aggregates in suspension cultures. This is technically straightforward as it is a simple extension of the suspension culture. However, it has some drawbacks. As mentioned above, extended culture of ES cell aggregates in suspension often leads to cyst formation, resulting in a differentiation culture that is difficult to control. An unusually high concentration of morphogens or growth factors is required in order for the factors to reach cells inside the aggregates (Bain et al., 1995; Carpenter et al., 2001; Schuldiner et al., 2001; Wichterle et al., 2002). Depending upon the size of the aggregates, cells on the surface and those inside the aggregates will have a varied degree of exposure to morphogens, thus creating a wide range of cell lineages or cells at various developmental stages. Because of the cluster nature, it is impossible to visualize the continual change of cell morphology in response to treatments. To overcome these drawbacks and to preserve the cellular interactions within the aggregate (or colony), we plated the ES cell aggregates onto a culture dish in a chemically defined medium. In this way, cells in the aggregate formed a colony of monolayer cells in a low-density culture. Morphologically, these cells resemble ES cells grown on feeder layers or matrigel (*Figure 9.1*). The monolayer nature of these cultures permits continual assessment of changes in cell morphology. The chemically defined culture system also allows testing the effects of signaling molecules on neural specification. In the presence of FGF2, cells in the colony center transformed to small elongated cells whereas those in the periphery gradually became flattened (*Figure 9.1D*). This small columnar cell population expanded in the presence of FGF2 and organized into rosette formations by 7–10 days after plating the aggregates (*Figure 9.1E*). These rosette formations are reminiscent of the early neural tube viewed from coronal sections. Hence, the small, columnar cells in the rosettes are likely neuroepithelial cells. This is confirmed by their expression of the early neuroepithelial markers, nestin and Musashi-1 (Zhang et al., 2001).

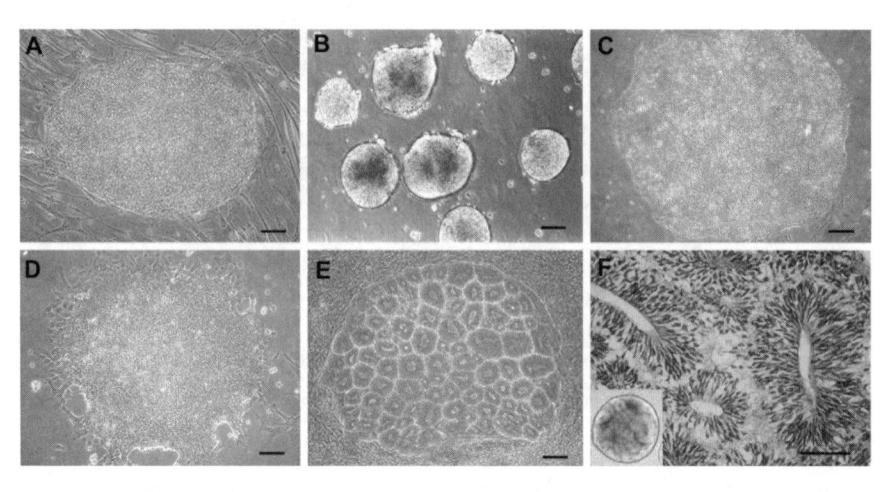

Figure 9.1: Induction of neuroepithelia from hES cells. (A) Phase contrast photograph of hES cell (H9, p82) colonies growing on fetal mouse embryonic fibroblasts. (B) ES cell colonies were detached and grown in suspension as aggregates called embryoid bodies for 4 days. (C) ES cell aggregates grew as a colony of cells (monolayer) 2 days after plating onto a cell culture dish in a chemically defined neural induction medium (Zhang et al., 2001). (D) 5 days after adherent culture, cells in the colony center became small columnar morphology whereas those in the periphery appeared flat, epithelial looking. (E) After 7–10 days of adherent culture, the columnar cells in the colony center developed into multiple contiguous rosette formations. (F) Isolated clusters of rosette cells were expanded in suspension culture as cell clusters resembling neurospheres (inset in F). Cryostat sections of the expanded clusters (after eight passages), after staining with H & E, displayed neural tube-like structures with columnar cells surrounding a lumen. Bar = 100 μm.

Considering human ES cells are equivalent to the inner cell mass of a 5–6-day-old embryo, rosette formation *in vitro* translates into day 18–20 in a human embryo, the time when neural plate (neuroectoderm) begins to develop (O'Rahilly and Muller, 1994). Thus, the *in vitro* neural specification recapitulates *in vivo* neural ectoderm formation with respect to temporal development, suggesting that the intrinsic program of neural specification is preserved in this culture system. This notion is substantiated by our observation that the temporal course of neural rosette formation from Rhesus monkey ES cells is consistent with that of neuroectoderm specification in monkey embryos (Piscitelli and Zhang, 2002).

The *in vitro* generated neuroepithelial cells, identified by columnar morphology and expression of neuroepithelial markers nestin and Musashi-1, invariably organize into neural tube-like rosettes in the center of the colony. These neural rosettes segregate themselves from the surrounding non-neural cells in extended culture (Zhang *et al.*, 2001). Hence, the spatial arrangement of neuroepithelial cells in the adherent colony culture mirrors positional organization in an embryo. Neuroepithelial cells in the form of neural rosettes can also be induced from ES cell colonies grown on matrigel without the involvement of cell aggregation (Zhang *et al.*, unpublished). However, the location of rosettes in a colony is random, i.e., not always in the colony center. This suggests that cell–cell interactions in the aggregate, even for a short period of time, may position the cells into an inside–outside pattern. This has been shown in mouse EBs, in which ectoderm resides interiorly, endoderm sits exteriorly, and the mesoderm is in the middle (Wiles, 1995; O'Shea, 1999).

The process of neuroepithelial differentiation from hES cells mimics *in vivo* neural development in terms of timing, spatial organization, and potentially the mechanism of neural specification. In addition, hES cell-derived neuroepithelial cells comprise over 70% of the total ES cell derivative population (Zhang *et al.*, 2001), suggesting the high efficiency of the differentiation protocol. Therefore, the ES cell neural differentiation culture system offers an ideal model to dissect the effect of signaling molecules on neural specification in humans.

9.5.3 Isolation of neural cells from differentiated human cell population

The principles for isolating neural precursors from hES cell differentiated progenies are essentially the same as those applied to the mouse ES cell differentiation system. In the case of spontaneously differentiated neural precursors in the form of neurospheres (Reubinoff *et al.*, 2001), the neurosphere can be mechanically picked up by a micropipette under a microscope. Similarly, neural rosettes can be readily isolated with the help of a micropipette since neuroepithelial cells aggregate together as rosette formations and the aggregate delineates from surrounding non-neural cells in extended culture. This approach yields an almost pure population of neural precursors that is very useful for further cellular and molecular characterization. However, the efficiency of cell production using this method is low. Thus, we have designed an approach to isolate neuroepithelial cells based on differential enzymatic response and adhesion. Neuroepithelial cells in the form of rosettes detach faster than surrounding cells in response to a low concentration of dispase. Contaminating cells along the outside edges of the neural rosettes can be removed by a differential adhesion step since the flat cells more readily attach to plastic surfaces. Neuroepithelial cell populations isolated in this way are enriched by at least 95% for cells that are positive for nestin. The isolated neuroepithelial cells can be expanded in the form of 'neurospheres', although cells reorganize into rosette formations within the cluster (*Figure 9.1F*). Theoretically, it would be easier and more efficient to sort targeted cells by flow cytometric cell sorting based on specific expression of molecules on the target cell surface. Carpenter *et al.* (2001) isolate a PSA-NCAM-expressing population and an A2B5 positive fraction from hES cell differentiated progenies. The PSA-NCAM+ cells appear to be neuronal restricted progenitors and the A2B5 positive fraction seems to generate both neurons and glia. This is similar to what has been shown in mouse ES cell derivative cultures (Mujtaba *et al.*, 1999). However, cell surface markers are not readily available for multipotent neuroepithelial cells, although some selected markers have been used to sort progenitors from a pool of brain cells (Uchida *et al.*, 2000; Capela and Temple, 2002). It is worth noting that while cell surface markers such as NCAM and A2B5 are relatively specific for selecting progenitor populations from a mixture of brain-derived cells, they might prove less specific for isolating the same cells from ES cell-derivatives because these molecules are also expressed by non-neural cells.

An alternative approach is to use genetic means to isolate the desired population of differentiated progenies as mentioned with mouse ES cells (see section 9.3.3). An example of this is homologous recombination. This technology has been widely applied to mouse ES cells to generate transgenic mice for the past two decades. However, homologous recombination in human ES cells was technically difficult because of the low transfection and cloning efficiency of hES cells. This difficulty has been overcome recently by Zwaka and Thomson (2003) through the successful targeting of the genes encoding hypoxanthine phosphoribosyltransferase-1 (HPRT1) and the octmer-binding transcription factor 4 (Oct4). In principle, this technology will allow the generation of 'knock-in' cell lines with a selectable marker introduced into a locus with a cell type-specific expression pattern. This should allow the purification of a targeted cell population from the mixture of hES cell-differentiated progenies. It will thus revolutionize molecular analysis and potential clinical application of hES cells, as it did in the mouse ES cell field.

9.5.4 Identity of hES cell-generated neural cells

Neural progenies differentiated from hES cells differ from each other depending on the induction protocols used. Similar to mouse ES cells, RA treatment of hES cells results in the generation of neural cells at various developmental stages (Carpenter *et al.*, 2001; Schuldiner *et al.*, 2001), as does the spontaneous differentiation from hES cells using high-density cultures (Reubinoff *et al.*, 2001). In both cases, however, neural progenitor cells can be isolated based either on their expression of neural cell adhesion molecules on the cell surface (Carpenter *et al.*, 2001), or on their characteristic morphological features (Reubinoff *et al.*, 2001). In contrast, FGF treatment induces a synchronized differentiation of hES to neuroepithelial cells (Zhang *et al.*, 2001). While the nature of those neural precursors induced with different protocols remains to be clarified, they appear to be at different developmental stages. Neural precursor cells generated by RA treatment and isolated through immunoseparation with PSA-NCAM and A2B5 are neuronal and/or glial restricted progenitors (Carpenter *et al.*, 2001). Neurosphere cells selected from spontaneous hES cell differentiation cultures express PSA-NCAM and behave similarly to brain-derived cells. In both cases, the neural precursor cells proliferate in response to both FGF2 and EGF (Carpenter *et al.*, 2001; Reubinoff *et al.*, 2001). In contrast, those induced by FGF2 do not express PSA-NCAM and require FGF2 instead of EGF for proliferation (Zhang *et al.*, 2001). However, the PSA-NCAM negative cells will express PSA-NCAM over time in culture (Zhang *et al.*, 2001), suggesting that the PSA-NCAM negative cells are at an earlier developmental stage than the PSA-NCAM positive cells. Given the fact that the FGF2-induced neural precursors appear at a time roughly equivalent to the birth of neuroectoderm *in vivo*, and that these cells invariably organize into neural tube-like rosette formations, it is likely that the NCAM negative precursors are neuroectodermal cells.

Interestingly, the neural precursors/progenitors differentiated and isolated from hES cells using the above three approaches appear to lie on consecutive developmental stages along the neural differentiation pathway. Neural progenitors isolated based on expression of PSA-NCAM appear to be restricted to a neuronal fate. Those isolated based on A2B5 expression appear to be either glial or neuronal restricted progenitors (Carpenter *et al.*, 2001). Neural precursor cells grown as neurospheres in the spontaneous hES cell cultures resemble brain-derived neurospheres in terms of morphology, growth factor response, and differentiated progenies. Hence, these neural precursors are likely a mixture of neuronal and glial progenitors with few self-renewing neural stem cells. This comparison of neural precursor cells differentiated from the same source, ES cells, by different methodologies in different laboratories is rather interesting. It suggests that the neural differentiation process is a rather stereotypic process. From an application perspective, the neural progenitors are well committed to a neuronal or glial lineage, thus providing a predictable cell lineage outcome. However, they may not offer a substantial advantage over brain-derived neurosphere cultures if they lack the potential to generate projection neurons such as motoneurons and dopaminergic neurons.

The hES-derived neuroepithelial cells induced by FGF2 treatment appear to be at the earliest stage of the neural lineage based on marker expression, growth requirements, and differentiation potential (Zhang et al., 2001). They are equivalent to neuroepithelial cells at the neural plate stage. The primitive state of the neural rosette cells derived from hES cells suggests a broader potency in lineage differentiation. Indeed, we have found that the neural rosette cells can be directed to glia and neurons including the projection neurons that are normally born during early development, such as midbrain dopaminergic neurons and spinal motor neurons (*Color Plate 4A–G*, Zhang et al., 2001; and unpublished). Hence, ES cell-derived neuroepithelial cells could provide an intermediate source for generating specialized neuronal and glial types.

9.5.5 Functional properties of hES cell-derived neurons

Neurons generated from hES cells have never been in the brain. Under most culture conditions, ES cell-derived neurons do not 'live' in a three-dimensional environment and do not normally receive 'peripheral' inputs (or targets) that are critical for functional maturation. Hence, whether and how the *in vitro* generated cells mature into electrophysiologically active neurons as those 'grown up' in the brain is an important question from both a fundamental biological standpoint and an application perspective. Relevant data is available for mouse ES cell-derived neurons. Using an RA treatment protocol, mouse ES cell-derived neurons have been shown to express sodium and potassium channels and to excite spontaneous and induced action potentials similar to neurons in primary cultures (Bain et al., 1995). Similarly, neurons generated from hES cells using a comparable differentiation protocol (Carpenter et al., 2001) express voltage-gated potassium and sodium currents when depolarized. These neurons also appear to fire action potentials in response to depolarizing stimuli. On the other hand, neurons differentiated from hES cells using the FGF2 induction protocol (Zhang et al., 2001) remain physiologically inactive for 3 weeks in differentiation cultures. These ES cell-generated neurons express potassium and sodium currents, fire spontaneous action potential, as well as synaptic currents when the cells mature in a long-term differentiation culture (Zhang and Pearce, unpublished). Therefore, hES cell-derived neurons can be electrophysiologically active. The delayed functional maturation of neurons generated through the FGF protocol rather than using the RA protocol again may reflect the rapid differentiation of neurons from ES cells in the presence of RA. In addition, under the RA protocol conditions, neurons mature in a three-dimensional EB structure together with other potential target cells such as muscle cells. In the FGF2 induction protocol, neuroepithelia are induced in a two-dimensional adherent colony culture, and the isolated neuroepithelial cells differentiate into neurons in the absence of non-neural cells. Thus, the cultures of neurons derived through FGF induction protocol offer an opportunity to follow the intrinsic neuronal maturation program, and to study environmental effects on functional neuronal maturation by reconstituting the cellular compartments that are present during development.

9.5.6 Engraftability of hES cell-derived neural precursors

Public and scientific interest in hES cells lies mainly in the therapeutic potential of ES cell derivatives. Whether hES cell-generated neural precursors are able to survive, differentiate into mature neurons and glia, and functionally integrate into the brain environment will determine the applicability of hES cells for the treatment of neurodegenerative diseases. As a proof-of-concept, hES-derived neural precursors have been transplanted into the ventricles of neonatal mouse brains. The grafted cells migrate into the brain parenchyma and incorporate into both neurogenic regions, such as the hippocampus and the rostral migratory pathway, and non-neurogenic areas such as the cerebral cortex (*Color Plate 4H*). More importantly, they differentiate into mature neurons and glia, which are indistinguishable from endogenous cells unless otherwise marked (Reubinoff et al., 2001; Zhang et al., 2001, Figure 2I–K). These results suggest that the *in vitro* generated neural precursors, similar to their mouse counterparts, are able to mature in response to normal developmental cues. Since the neural precursors are generated from several hES cell lines using a wide variety of approaches in different laboratories, the engraftability and responsiveness to local cues may be inherent to neural cells at a particular developmental stage.

Whether the hES cell-derived neural cells incorporate into the adult brain and contribute to neural function is an important indicator for potential future application of hES cells in restoring neurological deficits. We have transplanted hES cell-derived neuroepithelial cells into the striatum of adult rats that have been subjected to 6-hydroxy dopamine treatment to create a Parkinsonian state. In the adult brain environment, hES cell-derived neuroepithelial cells largely remain as immature, nestin-expressing progenitors for at least 3 months. A small subpopulation of the grafted cells are able to differentiate into neurons that express β_{III}-tubulin and/or MAP2. An even smaller number of cells mature and express tyrosine hydroxylase, the rate-limiting enzyme that is required for dopamine synthesis (Yan et al., 2002). Thus, hES cell-generated neuroepithelial cells are capable of generating mature neurons in adult brains but are largely dependent upon their intrinsic maturation program in the absence of developmental cues.

9.6 Outstanding questions

9.6.1 How is the neural fate specified from hES cells?

Neural induction in vertebrates, particularly in *Xenopus* and chicks, is achieved through FGF and/or anti-BMP signalings. Whether the same signaling mechanisms are used in mammalian and human neuroectoderm induction needs to be confirmed. In particular, how these and other signaling pathways interact in time and space to result in neural specification needs further dissection. Studies using mouse ES cells as a model system suggest that anti-BMP signaling is, at least in part, involved. In humans, ES cells offer the only approach to the inaccessible experimental paradigm at the moment. Development of neuroectoderm in humans takes about 3 weeks whereas it occurs in 1 week in mouse and within a day in chick. Understanding how individual signaling molecules at a particular developmental stage of this protracted period orchestrate neural fate specification from hES cells will still be a challenge to developmental biologists. Yet these studies are essential in order to devise optimal strategies that will ultimately lead to the application of hES cells in repairing neurological disorders.

9.6.2 How to direct hES cells to glia and neurons with regional identities?

The generation of neurons and glial cells is a stereotypic process of gradual fate restriction. Once the neuroectoderm is specified, neuroepithelial cells are further conferred with regional identities through a process called neural patterning. Hence, neural epithelial cells or neural stem cells derived from embryonic mouse and human brain tissues are generally regionally specified (Hiroshi *et al.*, 2002; Ostenfeld *et al.*, 2002). That explains, at least in part, why it is difficult to direct the differentiation of fetal tissue-derived neural stem/progenitor cells into neurons of other regions. This emphasizes the importance of generating neuroepithelial cells from hES cells that are 'plastic' enough to be directed to neurons having varied regional and functional characteristics. As discussed above, this requires a unique neural induction and/or selection protocols. The question is how to guide these intermediates further down a particular neuronal pathway to obtain a defined regional and neurotransmitter identity. Developmental studies in vertebrates such as *Xenopus*, zebrafish, and chicks have laid down a beautiful framework of neural patterning. The principles learned from these studies have proven useful in differentiating mouse ES cells to neurons with regional identities such as midbrain dopaminergic neurons and spinal cord motor neurons (Lee *et al.*, 2000; Wichterle *et al.*, 2002). It is expected that similar principles may be applied to generate specialized neurons and glial cells from hES cells. In humans, the translation of these principles may turn out to be difficult due to the length of time required for certain cell lineage development. In particular, the differentiation of oligodendroglia in humans does not begin until the third month of embryonic development. Identification of the pathway(s) that leads to the development of oligodendroglia from hES cells could be a challenging endeavor.

9.6.3 Are in vitro *generated neurons functional?*

Enthusiasm about hES cells is partly attributed to the expectation that their derivatives, such as neurons, may be useful in regenerative medicine. The key to the application of hES cells in replacement therapies for neurological conditions is the demonstration that these mature neurons and glial cells can integrate into the neural circuitry in a functional manner and consequently contribute to the correction of functional deficits. Limited information indicates that hES cell-derived neurons can be electrophysiologically active and that *in vitro* generated human neural precursors are able to differentiate into neurons and glial cells following transplantation into the CNS of rodents. A full battery of electrophysiological analyses and a wide range of animal models will need to be employed to test the functionality of these *in vitro* generated human neurons before these cells can be applied to patients. While it seems to be a simple replication of what has been done in rodents, there are important technical difficulties that need to be overcome. The maturation time of human neural precursor cells is substantially longer than that of mouse neural precursors. The 'time' factor again could pose serious challenges in a xenotransplant pre-clinical model. In addition, differentiation of hES cell-derived neural precursors may become an independent process in the xeno-environment due to the inability of molecular cues to function across species. Therefore, novel approaches or experimental paradigms are necessary to deal with these problems.

9.6.4 Will ES cell-derived neural cells be safe for cell therapy?

The pluripotency of hES cells is a concerning aspect of hES cells in therapy due to the potential generation of undesirable cells or tissues or even the formation of teratomas. Hence, hES cells need to be instructed to become a particular cell type. For example, hES cells need to be restricted to at least a neural fate in order for an application in neurological conditions. Since most current approaches for directed neural differentiation yield a mixture of cells, sorting out the desirable cell population appears necessary to avoid unpredictable outcomes. Knock-in of a selectable marker into a cell type-specific gene, as described by Zwaka and Thomson (2003), should allow the positive selection of differentiated, post-mitotic cells of choice and/or removal of remaining undifferentiated stem cells, thereby minimizing the risk of teratoma formation. Only if hES cell derivatives are safe will therapies based on hES cells be brought to the clinic.

Acknowledgment

Studies in my laboratory have been supported by the NIH-NCRR (RR16588-01), NIH-NINDS (NS045926-01), the Michael J. Fox Foundation, the National ALS Association, and the Myelin Project.

References

Amit M, Carpenter MK, Inokuma MS, Chiu CP, Harris CP, Waknitz MA, Itskovitz-Eldor J, Thomson JA (2000) Clonally derived human embryonic stem cell lines maintain pluripotency and proliferative potential for prolonged periods of culture. *Dev. Biol.* **227**, 271–278.

Anderson DJ (2001) Stem cells and pattern formation in the nervous system: the possible versus the actual. *Neuron* **30**, 19–35.

Avellana-Adalid V, Nait-Oumesmar B, Lachapelle F, Evercooren AB (1996) Expansion of rat oligodendrocyte progenitors into proliferative 'oligospheres' that retain differentiation potential. *J. Neurosci. Res.* **45**, 558–570.

Bain G, Kitchens D, Yao M, Huettner JE, Gottlieb DI (1995) Embryonic stem cells express neuronal properties *in vitro*. *Dev. Biol.* **168**, 342–357.

Brustle O, Jones KN, Learish RD, Karram K, Choudhary K, Wiestler OD, Duncan ID, McKay RD (1999) Embryonic stem cell-derived glial precursors: a source of myelinating transplants. *Science* **285**, 754–756.

Capela A, Temple S (2002) LeX/ssea-1 is expressed by adult mouse CNS stem cells, identifying them as nonependymal. *Neuron* **35**, 865–875.

Carpenter MK, Inokuma MS, Denham J, Mujtaba T, Chiu CP, Rao MS (2001) Enrichment of neurons and neural precursors from human embryonic stem cells. *Exp. Neurol.* **172**, 383–397.

Dinsmore J, Ratliff J, Deacon T, Pakzaban P, Jacoby D, Galpern W, Isacson O (1996) Embryonic stem cells differentiated in vitro as a novel source of cells for transplantation. *Cell Transpl.* **5**, 131–143.

Edlund T, Jessell TM (1999) Progression from extrinsic to intrinsic signaling in cell fate specification: A view from the nervous system. *Cell* **96**, 211–224.

Episkopou V, Arkell R, Timmons PM, Walsh JJ, Andrew RL, Swan D (2001) Induction of the mammalian node requires Arkadia function in the extra-embryonic lineages. *Nature* **410**, 825–830.

Evans MJ, Kaufman MH (1981) Establishment in culture of pluripotential cells from mouse embryos. *Nature* **292**, 154–156.

Finley MF, Devata S, Huettner JE (1999) BMP-4 inhibits neural differentiation of murine embryonic stem cells. *J. Neurobiol.* **40**, 271–287.

Fraichard A, Chassande O, Bilbaut G, Dehay C, Savatier P, Samarut J (1995) In vitro differentiation of embryonic stem cells into glial cells and functional neurons. *J. Cell Sci.* **108**, 3181–3188.

Gottlieb DI, Heuttner JE (1999) An in vitro pathway from embryonic stem cells to neurons and glia. *Cells Tiss. Org.* **165**, 165–172.

Grunz H, Tacke I (1989) Neural differentiation of *Xenopus laevis* ectoderm takes place after disaggregation and delayed reaggregation without inducer. *Cell Diff. Dev.* **28**, 211–217.

Hemmati-Brivanlou A, Melton DA (1992) A truncated activin receptor inhibits mesoderm induction and formation of axial structures in Xenopus embryos. *Nature* **359**, 609–614.

Hitoshi S, Tropepe V, Ekker M, van der Kooy D (2002) Neural stem cell lineages are regionally specified, but not committed, within distinct compartments of the developing brain. *Development* **129**, 233–244.

Jones-Villeneuve EMV, McBurney MW, Rogers KA, Kalnins VI (1982) Retinoic acid induces embryonic carcinoma cells to differentiate into neurons and glial cells. *J. Cell Biol.* **94**, 253–262.

Kawasaki H, Mizuseki K, Nishikawa S, Kaneko S, Kuwana Y, Nakanishi S, Nishikawa S-I, Sasai Y (2000) Induction of midbrain dopaminergic neurons from ES cells by stromal cell-derived inducing activity. *Neuron* **28**, 31–40.

Kawasaki H, Suemori H, Mizuseki K, Watanabe K, Urano F, Ichinose H et al. (2002) Generation of dopaminergic neurons and pigmented epithelia from primate ES cells by stromal cell-derived inducing activity. *Proc. Natl Acad. Sci. USA* **99**, 1580–1585.

Kim J-H, Auerbach JM, Rodriguez-Gomez JA, Velasco I, Gavin D, Lumelsky N et al. (2002) Dopamine neurons derived from embryonic stem cells function in an animal model of Parkinson's disease. *Nature* **418**, 50–56.

Lake J-A, Rathjen J, Remiszewski J, Rathjen PD (2000) Reversible programming of pluripotent cell differentiation. *J. Cell Sci.* **113**, 555–566.

Lamb T, Knecht AK, Smith WC, Stachel SE, Ecoonomides AN, Stahl N, Yancopolous GD, Harland RM (1993) Neural induction by the secreted polypeptide noggin. *Science* **262**, 713–718.

Lee S-H, Lumelsky N, Studer L, Auerbach JM, McKay RD (2000) Efficient generation of midbrain and hindbrain neurons from mouse embryonic stem cells. *Nature Biotechnol.* **18**, 675–679.

Li M, Pevny L, Lovell-Badge R, Smith A (1998) Generation of purified neural precursors from embryonic stem cells by lineage selection. *Curr. Biol.* **8**, 971–977.

Liu S, Qu Y, Stewart T, Howard M, Chakrabortty S, Holekamp T, McDonald JW (2000) Embryonic stem cells differentiate into oligodendrocytes and myelinate in culture and after spinal cord transplantation. *Proc. Natl Acad. Sci. USA* **97**, 6126–6131.

Maden M (2002) Retinoid signaling in the development of the central nervous system. *Nature Rev. Neurosci.* **3**, 843–853.

Martin GR (1981) Isolation of a pluripotent cell line from early mouse embryos cultured in medium conditioned by teratocarcinoma stem cells. *Proc. Natl. Acad. Sci. USA* **78**, 7634–7638.

McDonald JW, Liu XZ, Qu Y, Liu S, Mickey SK, Turetsky D, Gottlieb DI, Choi DW (1999) Transplanted embryonic stem cells survive, differentiate and promote recovery in injured rat spinal cord. *Nature Med.* **5**, 1410–1412.

Mizuseki K, Sakamoto T, Watanabe K, Muguruma K, Ikeya M, Nishiyama A et al. (2003) Generation of neural crest-derived peripheral neurons and floor plate cells from mouse and primate embryonic stem cells. *Proc. Natl Acad. Sci. USA* **100**, 5828–5833

Muhr J, Graziano E, Wilson S, Jessell TM, Edlund T (1999) Convergent inductive signals specify midbrain, hindbrain, and spinal cord identity in gastrula stage chick embryos. *Neuron* **23**, 689–702.

Mujtaba T, Piper DR, Kalyani A, Groves AK, Lucero MT, Rao MS (1999) Lineage-restricted neural precursors can be isolated from both the mouse neural tube and cultured ES cells. *Dev. Biol.* **214**, 113–127.

Okabe S, Forsberg-Nilsson K, Spiro AC, Segal M, McKay RDG (1996) Development of neuronal precursor cells and functional postmitotic neurons from embryonic stem cells *in vitro*. *Mech. Dev.* **59**, 89–102.

O'Rahilly R, Muller F (ed.) (1994) *The Embryonic Human Brain*. Wiley-Liss, New York.

O'Shea KS (1999) Embryonic stem cell models of development. *Anat. Rec.* **257**, 32–41.

Ostenfeld T, Joly E, Tai YT, Peters A, Caldwell M, Jauniaux E, Svendsen CN (2002) Regional specification of rodent and human neurospheres. *Dev. Brain Res.* 134, 43–55.

Patapoutian A, Reichardt LF (2000) Roles of Wnt proteins in neural development and maintenance. *Curr. Opin. Neurobiol.* 10, 392–399.

Piscitelli GM, Zhang S-C (2002) Differentiation of neural precursors from Rhesus monkey embryonic stem cells. *Soc. Neurosci. Abs.* 7.5.

Rathjen J, Haines BP, Hudson KM, Nesci A, Dunn S, Rathjen PD (2002) Directed differentiation of pluripotent cells to neural lineages: homogeneous formation and differentiation of a neuroectoderm population. *Development* 129, 2649–2661.

Reubinoff BE, Pera MF, Fong CY, Trounson A, Bongso A (2000) Embryonic stem cell lines from human blastocysts: somatic differentiation in vitro. *Nature Biotechnol.* 18, 399–404.

Reubinoff BE, Itsykson P, Turetsky T, Pera MF, Reinhartz E, Itzik A, Ben-Hur T (2001) Neural progenitors from human embryonic stem cells. *Nature Biotechnol.* 19, 1134–1140.

Renoncourt Y, Carroll P, Filippi P, Arce V, Alonso S (1998) Neurons derived in vitro from ES cells express homeoptroteins characteristic of motoneurons and interneurons. *Mech. Dev.* 79, 185–197.

Schuldiner M, Eiges R, Eden A, Yanuka O, Itskovitz-Eldor J, Goldstein RS, Benvenisty N (2001) Induced neuronal differentiation of human embryonic stem cells. *Brain Res.* 913, 201–205.

Streit A, Berliner AJ, Papanayotou C, Sirulnik A, Stern CD (2000) Initiation of neural induction by FGF signaling before gastrulation. *Nature* 406, 74–78.

Strubing C, Ahnert-Hlger G, Shan J, Wiedenmann B, Hescheler J, Wobus AM (1995) Differentiation of pluripotent embryonic stem cells into the neuronal lineage in vitro gives rise to mature inhibitory and excitatory neurons. *Mech. Dev.* 53, 275–287.

Thomson JA, Odorico JS (2000) Human embryonic stem cell and embryonic germ cell lines. *Trends Biotechnol.* 18, 53–57.

Thomson JA, Itskovitz-Eldor J, Shapiro SS, Waknitz MA, Swiergiel JJ, Marshall VS, Jones JM (1998) Embryonic stem cell lines derived from human blastocysts. *Science* 282, 1145–1147.

Tropepe V, Hitoshi S, Sirard C, Mak TW, Rossant J, van der Kooy D (2001) Direct neural fate specification from embryonic stem cells: a primitive mammalian neural stem cell stage acquired through a default mechanism. *Neuron* 30, 65–78.

Uchida N, Buck DW, He D, Reitsma MJ, Masek M, Phan TV, Tsukamoto AS, Gage FH, Weissman IL (2000) Direct isolation of human central nervous system stem cells. *Proc. Natl Acad. Sci. USA* 97, 14720–14725.

Weinstein DC, Hemmati-Brivanlou A (1999) Neural induction. *Ann. Rev. Cell Dev. Biol.* 15, 411–433.

Wichterle H, Lieberam I, Porter JA, Jessell TM (2002) Directed differentiation of embryonic stem cells into motor neurons. *Cell* 110, 385–397.

Wiles MV (1995) Embryonic stem cell differentiation in vitro. *Meth. Enzymol.* 225, 900–918.

Wilson SI, Edlund T (2001) Neural induction: toward a unifying mechanism. *Nature Neurosci.* 4, 1161–1168.

Wilson SI, Graziano E, Harland R, Jessell TM, Edlund T (2000) An early requirement for FGF signaling in the acquisition of neural cell fate in the chick embryo. *Curr. Biol.* 10, 421–429.

Wobus AM, Grosse R, Schoneich J (1988) Specific effects of nerve growth factor on the differentiation pattern of mouse embryonic stem cells in vitro. *Biomed. Biochim. Acta* 47, 965–973.

Wobus AM, Wallukat G, Hescheler J (1991) Pluripotent mouse embryonic stem cells are able to differentiate into cardiomyocytes expressing chronotropic responses to adrenergic and cholinergic agents and Ca2+ channel blockers. *Differentiation* 48,173–182.

Yan YP, Lyons E, Moreno P, Zhang S-C (2002) Survival and differentiation of human embryonic stem cell-derived neural precursors in a rat model of Parkinson's disease. *Soc. Neurosci. Abst.* 429.8.

Ying Q-L, Stavridis M, Griffiths D, Li M, Smith A (2003) Conversion of embryonic stem cells into neuroectodermal precursors in adherent monoculture. *Nature Biotechnol.* 21, 183–186.

Zhang S-C, Ge B, Duncan ID (1999) Adult brain retains the potential to generate oligodendroglial progenitors with extensive myelination capacity. *Proc. Natl Acad. Sci. USA* 96, 4089–4094.

Zhang S-C, Lundberg C, Lipsitz D, O'Connor LT, Duncan ID (1998) Generation of oligodendroglial progenitors from neural stem cells. *J. Neurocytol.* 27, 475–489.

Zhang S-C, Wernig M, Duncan ID, Brüstle O, Thomson JA (2001) *In vitro* differentiation of transplantable neural precursors from human embryonic stem cells. *Nature Biotechn.* 19, 1129–1133.

Zwaka TP, Thomson JA (2003) Homologous recombination in human embryonic stem cells. *Nature Biotechnol.* 21, 319–321.

10. Modeling islet development through embryonic stem cell differentiation

Jon S. Odorico, Brenda Kahan, Debra A. Hullett, Lynn M. Jacobson and Victoria L. Browning

10.1 Introduction

Perhaps the most common childhood pancreatic disease is type I diabetes, an autoimmune disorder in which β cells are selectively destroyed by an errant immune system, rendering millions of children and young adults dependent on exogenous insulin injections. In developed countries, diabetes is the leading cause of blindness, kidney failure, neuropathy, foot ulcers and amputations among young people. As a result of accelerated heart disease, patients can expect a severely compromised quality of life and a drastically shortened life span. With the incidence nearly doubling in successive generations, diabetes is reaching epidemic rates throughout the world (Bingley and Gale, 1989; Gale, 2002; Stovring et al., 2003; Yudkin and Beran, 2003) thereby putting severe economic strains on national health care systems (Currie et al., 1997; Rubin et al., 1994).

In insulin-deficient diabetic patients, an ideal treatment goal is to replace the lost β cells with physiologically normal β cells or islet tissue from other sources with the hope of restoring normal glucose control. Currently, this can be readily accomplished through transplantation of the pancreas or isolated islets of Langerhans derived from human cadaver donors (Shapiro et al., 2000; Sollinger et al., 1998; Sutherland et al., 2001). These treatments can frequently restore normoglycemia, and have been shown to improve quality of life; in many cases, they can also forestall end-organ complications, such as retinopathy and neuropathy, and prolong life (Odorico and Sollinger, 2002). However, because of the limited number of available cadaver donors this source is unlikely to provide enough organs or tissue for all patients with diabetes. Consequently, there is an urgent need for a renewable and readily-available source of functional insulin-producing cells, such as those that could potentially be generated from embryonic stem (ES) cells (Kahan et al., 2003; Soria et al., 2000). A comprehensive understanding of the mechanisms regulating pancreaticogenesis and islet differentiation from endoderm will be critical to manipulate cells effectively to differentiate into this lineage while excluding differentiation into unrelated lineages. Large-scale isolation of human pancreatic or islet progenitor cells from human ES cells could potentially alleviate the shortage of material available for transplantation into diabetic patients.

Our ability to direct differentiation of human ES cells into pancreatic stem cells and ultimately functional islet endocrine cells is hindered by an incomplete understanding of both pancreas development and the specific mechanisms by which this process might differ among humans and other vertebrates. A model system would improve our understanding of human pancreatic development, which is currently limited by access to material and ethical considerations of performing studies on human embryos.

This chapter will review current concepts of pancreas development, describe the current status of islet differentiation from mouse and human ES cells, and outline how one might begin to recapitulate developmental signals in culture to direct ES cell specification to islet lineages. Remaining challenges toward generating transplantable functional islet tissue for treating diabetes will also be discussed.

10.2 Development of the pancreas and islets of Langerhans in vertebrates

10.2.1 Endoderm formation and pancreatic morphogenesis

Pancreatic development is a multi-step process that requires correct endoderm patterning and pancreas specification, bud formation, branching morphogenesis and islet formation (*Figure 10.1*). During gastrulation, totipotent cells of the epiblast divide, differentiate, and rearrange into three distinct germ layers: ectoderm, mesoderm, and endoderm. This process, which starts at embryonic day (e) 6.5 in mouse, involves the formation of the primitive streak, a structure located at the posterior end of the epiblast. In normal embryogenesis, epiblast cells migrate through the primitive streak to become endoderm and mesoderm precursors. The endoderm at this point is a one cell-layer thick sheet that will ultimately form the epithelium of the entire gastrointestinal tract, including lungs, liver, bile ducts, and pancreas. *Nodal* signaling is necessary for formation of the node and for the proper formation of the primitive streak (Beddington, 1994; Conlon *et al.*, 1994). Consequently, *Nodal* mutant embryos fail to form most mesoderm tissue and form no definitive endoderm (Stemple, 2001).

Do epiblast cells acquire their new germ layer cell fate before, during or after migration? Studies by Carey *et al.* (1995) demonstrate that individual cells of the mid-streak epiblast (e7.5) may be already committed to a single germ layer, whereas progeny of epiblast cells from early streak (e6.5) embryos were not necessarily confined to a single germ layer. This suggests that the lineages are probably not separated at the beginning of gastrulation but adopt a fate as they make a transition to a new germ layer by the late streak stage (Lawson *et al.*, 1991). Following formation of the endoderm sheet on the outside of the mouse embryo, the endoderm layer undergoes complex movements and foldings to form a closed gut tube, a process aided by the rotation of the embryo to invert the developing endoderm to an interior position.

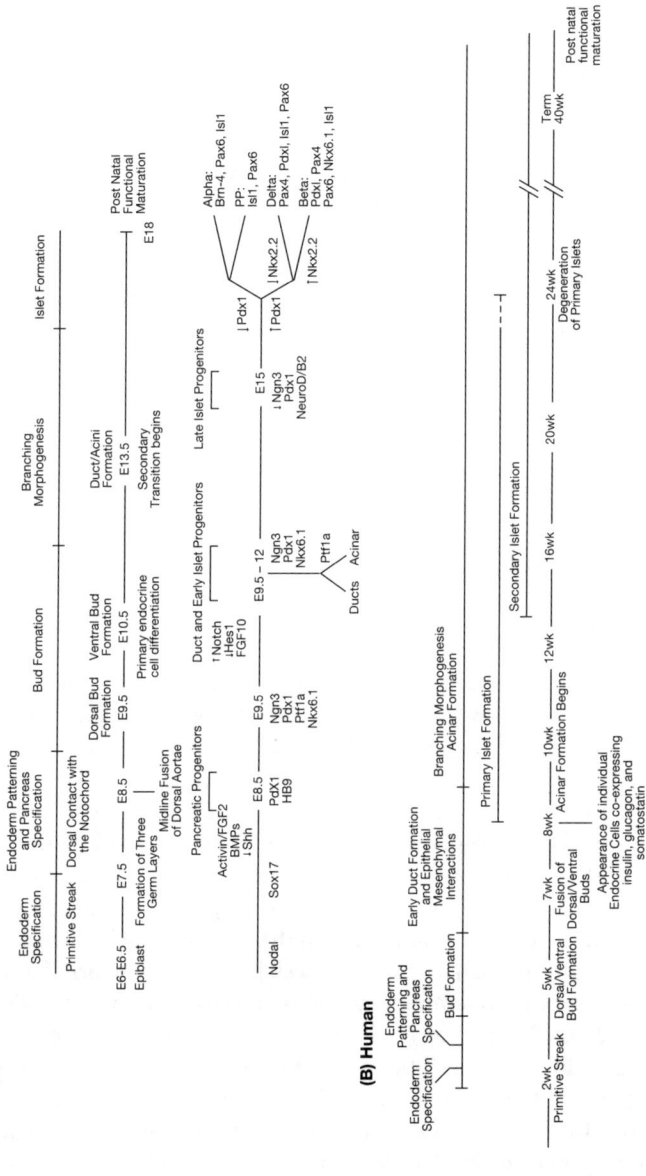

Figure 10.1: Sequence of major developmental events of the pancreas and islets of Langerhans in the mouse (A) and human (B). In both species, islet endocrine cells appear in two waves. The first is characterized by the appearance of isolated glucagon and insulin co-expressing endocrine cells within the early pancreatic bud epithelium. In rodents, these cells also generally express peptide YY and are found intermixed with PDX1-expressing cells. Endocrine cells at this stage are post-mitotic and existing evidence suggests they do not give rise to the islet cells in the adult. The second wave, termed the secondary transition, begins ~e13.5 in mice and in the 4th month in man, and accounts for most of the adult islet tissue. Cells expressing a single hormone begin to emerge, which results in a dramatic increase in the β cell mass. Based on lineage tracing studies, these cells are believed to arise from NGN3+ cells and have exited the cell cycle by the time they express the hormone proteins. Acini and ducts differentiate, proliferate, and branch in response to mesenchymal to epithelial signaling. Known key transcription factors that are involved in important cell type transitions from pancreatic progenitors to islet progenitors and adult endocrine cells are indicated. The data summarized here on human pancreas development is compiled from (Conklin, 1962; Liu and Potter, 1962; Polak *et al.*, 2000; Skandalakis *et al.*, 1994). Murine pancreas development was recently reviewed (Edlund, 2002).

The pancreas forms in the embryo as a dorsal and ventral outgrowth of the foregut endoderm (Slack, 1995). At the 20–25 somite stage, around e9.5–10 in the mouse, mesenchyme coalesces around the dorsal side of the gut tube and forms a bulge that is the dorsal bud. Slightly later, the ventral bud, whose development parallels that of the dorsal bud, emerges. The buds undergo rapid growth to develop into branched structures that fuse at ~e14 after gut rotation. Ductal differentiation occurs concomitantly with branching around e12.5; exocrine cells differentiate and acini form a short time later. Late in murine development intact islets form by the migration of individual endocrine cells from the epithelium and aggregation of the four endocrine cell types (α, β, δ, and PP) into a vascularized micro-organ with a defined relationship among endocrine cells.

The first hormone-expressing cells in the early pancreatic epithelium co-express endocrine hormones. Do these cells represent progenitors of the endocrine cells present in adult islets? Lineage tracing studies (Herrera, 2000) suggest that α and β endocrine cells do not arise from a cell that co-expresses both hormones simultaneously. A serial immunohistochemical study (Jensen et al., 2000a) also suggests that mature single hormone positive cells do not develop through a co-expressing progenitor, but rather arise independently. Additional lineage tracing studies demonstrate that hormone-negative cells that emerge later from the pancreatic epithelium during mid-development are the progenitors of adult islet endocrine cells (Gu et al., 2003). At present, the precise identity of the cells that give rise exclusively to mature endocrine cells remains unclear.

10.2.2 Endodermal origin of islets

For many years, a neuroectodermal origin of pancreatic endocrine cells had been suggested based largely on expression of neurally-restricted proteins, such as neuron-specific enolase and synaptophysin. Early studies involving the heterospecific transplantation of quail neural primordia and neural crest tissue into chick embryo host, however, demonstrated that neural crest cells do not give rise to islet endocrine cells in the chicken (Andrew, 1976; Fontaine et al., 1977; LeDouarin, 1988). More recent studies in mice have used lineage tracing with *Cre-LoxP* methodology to identify and follow progenitor cells that give rise to specific mature pancreatic cells. These studies demonstrate that mature islet cells derive from embryonic day 8 pancreatic endoderm (Gu et al., 2003; Kawaguchi et al., 2002).

Recently, lineage tracing studies have re-addressed the question of whether islet cells can be generated from or through a neural stem cell lineage, particularly from embryonic nestin-expressing cells resembling neural stem cells (Humphrey et al., 2003; Treutelaar et al., 2003). Treutelaar et al. (2003) demonstrated that nestin-positive cells were found in the pancreatic mesenchyme in embryos and were restricted to the vascular endothelium in post-natal mice; no insulin-positive cells were found to be derived from this lineage. In addition, it has been shown that nestin is expressed in the pancreatic mesenchyme in early mouse embryos (e10 and e13), but not in the epithelium where the first hormone-positive and presumed precursor cells appear (Edlund, 2002).

Humphrey et al. (2003) isolated nestin+ cells from 12–24 week human fetal pancreatic tissue using a nestin promoter lineage selection transgene. In these experiments, nestin was not expressed in pan cytokeratin+ ductal epithelium or insulin+ cells, but did co-localize with expression of PECAM, α smooth muscle actin, and vimentin, suggesting that nestin is not a specific marker of β cell precursors in the developing human pancreas. Transplantation of isolated nestin+ cells with fetal pancreatic fibroblasts into mice did not develop into β cells, whereas transplantation of nestin-negative fetal pancreatic epithelial cells gave rise to insulin+ β cells. These data support the idea that functional islet and ductal lineages of the pancreas do not derive from nestin-positive precursors.

The issue of nestin in the pancreatic lineage remains controversial however. Two additional recent studies (Delacour *et al.*, 2004; Esni *et al.*, 2004) used similar *Cre-loxP* lineage tracing experiments to follow the progeny of nestin-expressing cells. These groups found nestin expressed not only in mesenchymal cells in the pancreas but also in early epithelial cells on day e10.5 that were precursors of the exocrine lineage, not of the endocrine lineage. A number of studies have suggested that islet development in differentiating ES cell cultures occurs from nestin-positive precursors (Blyszczuk *et al.*, 2003; Hori *et al.*, 2002; Lumelsky *et al.*, 2001). As none of these studies gave direct evidence of lineage, one cannot conclude that nestin-positive cells found within the early stages of culture give rise to the endocrine cells seen in later stages. Indeed, the lineage tracing studies described above suggest that this is not the case.

10.2.3 Transcription factors involved in pancreas development

Differentiation from an uncommitted endoderm cell to a pancreatic β cell is characterized by a sequential and tissue-specific expression of a series of transcription factors (*Figure 10.1A*). The study of 'knock-out' mice has revealed valuable information on the roles of a variety of transcription factors involved in pancreatic development, which has recently been reviewed (Edlund, 2002). Several genes are known to be involved in endoderm specification, including *FoxA3* (Horne-Badovinac *et al.*, 2003) and *Sox17*, whose importance is highlighted by the observation that *Sox 17 -/-* mice have profound deficits in gut formation, particularly the mid- and hind-gut, and abnormal development of the foregut epithelium (Kanai-Azuma *et al.*, 2002).

One of the first transcription factors to be expressed in the foregut region is *pancreatic duodenal homeobox 1* (*pdx1*). Although this homeodomain protein is absolutely required for pancreas development (Jonsson *et al.*, 1994; Offield *et al.*, 1996), pancreatic bud formation is initiated in *Pdx1*-deficient mice, and the appearance of early glucagon-positive cells proceeds normally, suggesting that there are other critical factors acting in concert with *Pdx1* (Ahlgren *et al.*, 1996). Formation of the main pancreatic ducts is not dependent on *Pdx1* (Gu *et al.*, 2002; Holland *et al.*, 2002). *P48/Ptf1a* is also expressed in the early bud epithelium with PDX1, and may be an important marker of a putative 'switch' between intestinal enteroendocrine fate or pancreatic fate (Kawaguchi *et al.*, 2002). In mid-gestation (e9.5–15.5) in the mouse, a subset of pancreatic epithelial cells begins to express *neurogenin 3* (*Ngn3*), a bHLH transcription factor that is required for endocrine lineage specification of the pancreas (Gradwohl *et al.*, 2000). It is generally accepted that PDX1+ pancreatic epithelial cells that also express NGN3 are the endocrine precursors (Gu et al., 2002; McKinnon and Docherty, 2001), although the precise factors that promote *Ngn3* expression and repress HES1 are not known. Notch signaling, HES1 activity, and *Ngn3* expression are believed to be involved in the epithelial fate choices between differentiating into endocrine cells versus acinar cells and may be linked to the cellular choice of proliferation versus differentiation (Apelqvist et al., 1999; Hitoshi et al., 2002).

The dual primordia of the pancreas, which form the ventral and dorsal anlagen, have initial contact with different embryonic tissues. This has generally been considered an indication that some of the early inductive signaling events may be different between the two lobes. Dorsal endoderm is in contact with notochord until the dorsal aortae fuse, at which time the endoderm separates from notochord and shortly thereafter forms the dorsal pancreatic bud. The ventral bud forms later and is not in direct contact with these structures. Several studies have suggested that distinct molecular mechanisms may underlie these morphological differences. Whereas dorsal bud formation is dependent on fibroblast growth factor 2 (FGF2), homeobox transcription factor HB9 expression in the epithelium, and N-cadherin expression in the dorsolateral pancreatic mesenchyme, formation of ventral lobe is not (Deutsch et al., 2001; Esni et al., 2001; Harrison et al., 1999; Hebrok et al., 1998; Li et al., 1999).

A variety of transcription factors, including *Pax4*, *Nkx2.2* and *Nkx6.1*, are known to be involved in commitment to endocrine cell subtypes. Mice lacking *Pax4* are deficient in both β and δ cells (Sosa-Pineda et al., 1997), whereas mice lacking *Nkx2.2* have no insulin-positive cells and reduced numbers of α and PP cells; it is thought that *Nkx2.2* plays a role in the terminal maturation of β cell function including insulin production (Sussel et al., 1998). In contrast, knockout mice lacking *Nkx6.1* expression contain δ cells but not β cells (Sander et al., 2000).

Another set of transcription factors including *Pax6*, *Brn4*, *HB9*, *Pdx1* and *Isl1* are found in more mature hormone expressing cells (Wilson et al., 2003). *Isl1* and *Pax6* are expressed in all four subtypes, whereas PDX1 is restricted to β cells. Currently, it is thought that some combination of *Pax4*, *Pax6*, *Nkx6.1*, *Brn4* expression determines α cell versus β cell fate, as *Brn4* is expressed predominately in α cells (Hussain et al., 1997) and *Nkx6.1* is restricted to β cells in the mature islet (Sander et al., 2000).

10.2.4 Patterning of the endoderm and inductive tissue interactions in the early embryo

Extra-embryonic tissues play an inductive role in endoderm germ layer genesis. In mice, the anterior visceral endoderm (AVE) is required for proper formation of definitive endoderm (Beddington and Robertson, 1999). There is a continuously changing spatial relationship between the two populations, as first the AVE actively migrates away from the future site of node formation to establish the AP axis (Srinivas et al., 2004), and later, as the primitive endoderm of the late gastrula is displaced to the extra-embryonic region by the newly-formed definitive endoderm, which has expanded from the anterior end of the early primitive streak (Lawson and Pedersen, 1987). Critical reciprocal embryonic inductive signals between the epiblast or mesoendodermal precursors and extra-embryonic tissues involving

TGFβ signaling are required for proper endoderm formation (Robertson *et al.*, 2003).

Following endodermal sheet and tube formation, the pancreatic epithelial compartment is specified, and then undergoes proliferation, differentiation, and branching morphogenesis. As in other endodermally-derived organs, it is believed that mesenchymal-to-epithelial signaling participates in some or all of these processes in the pancreas (Gittes *et al.*, 1996; Golosow and Grobstein, 1962; Scharfmann, 2000). The mechanisms and growth factors that mediate these events are not completely resolved, although recent studies have suggested that a number of factors play critical roles in early pancreas specification. Important mesoderm-to-endoderm signaling may originate from neighboring tissues such as the notochord, dorsal aorta, and lateral plate mesoderm (*Figure 10.2*). Early dorsal endoderm mechanically separated from mesoderm and ectoderm does not express *pdx1*, whereas the same endoderm co-cultured with notochord does express *pdx1*, suggesting that soluble factors from the notochord signal dorsal pancreatic specification (Wells and Melton, 2000). It has been proposed that downregulation of *sonic hedgehog* (*shh*) in the region of the endoderm in contact with the notochord allows expression of *pdx1* within the foregut epithelium (Hebrok *et al.*, 2000; Hebrok, 2003). Factors expressed by the notochord, including FGF2 and activin βB, are capable of repressing *shh* in this region (Hebrok *et al.*, 1998). Lateral plate mesoderm (LPM) may also provide key instructive signals directing non-pancreatic endoderm to initiate pancreatic differentiation. Factors such as BMP4, BMP7, and activin A are capable of inducing *pdx1* gene expression in anterior chick endoderm not fated to become pancreas, suggesting that these factors may mediate the inductive signals of the LPM (Kumar *et al.*, 2003).

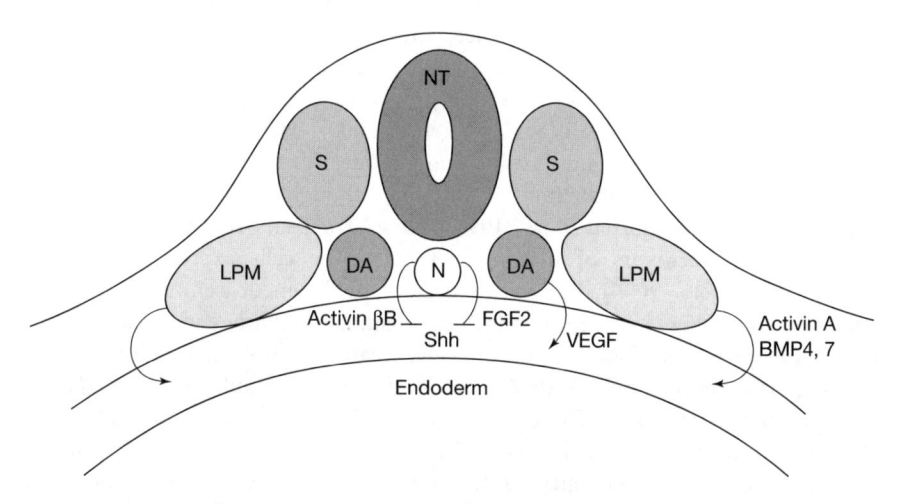

Figure 10.2: Schematic representation of tissue interactions in the early embryo (Stage 12, 8.5 dpc) involved in pancreas induction. For differentiation of dorsal endoderm towards a pancreatic fate, signals from the notochord (N) repress endodermal expression of Shh, and the dorsal aortae produce VEGF and possibly other non-blood element mediated signals. Lateral plate mesoderm (LPM) induces differentiation of the ventral endoderm, perhaps through Activin A and BMP4/7. NT, neural tube; S, somite (Kumar and Melton, 2003).

Other studies have found that completion of the pancreatic program, including bud development, expression of *Ptf1a*, maintenance of *Pdx1* expression, and induction of insulin and glucagon expression is dependent on the presence of aortic endothelial cells (Yoshitomi and Zaret, 2004). In addition, inhibitory effects of adjacent mesenchyme on early foregut endoderm have been demonstrated. Cardiac mesenchyme in contact with the ventral foregut endoderm appears to alter a 'default' program from pancreatic gene expression to liver specification, acting through inhibitory FGF signaling (Deutsch et al., 2001). Whereas BMP4 and BMP7 appear to be important for patterning the early endoderm to a pancreatic fate (Kumar et al., 2003), BMP6 expression is not required in the pancreatic epithelium and, in fact, may be inhibitory. Dichmann et al. (2003) demonstrated that aberrant expression of BMP6 under control of the pdx1 promoter resulted in complete agenesis of the pancreas and abnormalities of the stomach, liver, spleen, and duodenum.

FGF10, which is normally present in the pancreatic mesenchyme directly adjacent to the dorsal and ventral pancreatic epithelial buds, is required for normal pancreatic development and may act by inducing proliferation of PDX1-positive epithelial progenitor cells (Bhushan et al., 2001). Ectopic expression of FGF10 in the foregut epithelium results in increased proliferation of a pool of undifferentiated pancreatic stem cells and a hyperplastic pancreas (Norgaard et al., 2003). FGF10 may also mediate inductive mesenchymal-to-epithelial interactions during pancreas development.

In the embryo, differentiation of the four major islet endocrine cell types precedes formation of the islet organ and is thought to be promoted by the Notch and TGFβ signaling pathways (Apelqvist et al., 1999; Jensen et al., 2000b; Kim and Hebrok, 2001; reviewed in Wells and Melton, 1999). Notch inactivation may promote an endocrine fate choice, and enhances endocrine development (Apelqvist et al., 1999). In contrast, constitutive Notch activation may lead to preservation and/or expansion of certain progenitor cell populations (Hitoshi et al., 2002; Ohishi et al., 2001; Wakamatsu et al., 2000). Early endocrine cells (cells immunostaining for glucagon, insulin, and peptide YY) first appear in the e9–10 PDX1-expressing bud epithelium. It was once thought these represented islet progenitor cells (Upchurch et al., 1994). Now, it appears that early endocrine cells are not precursors of mature islet endocrine cells expressing a single hormone (Herrera, 2000), but may represent phylogenetic remnants of more primitive endocrine tissue (Kim and MacDonald, 2002). Early endocrine cells are devoid of expression of *Pdx1* and *Nkx6.1*, whereas β cell precursors express these genes. Furthermore, small clusters of endocrine cells are still present in *Pdx1*-deficient and *Ptf1a*-deficient mice in the under-developed pancreatic rudiments that persist in these animals, indicating that formation of early 'primitive' endocrine cells is independent of *Pdx1* and *Ptf1a* expression (Kawaguchi et al., 2002). Beginning at ~e13.5, there is a secondary wave of endocrine differentiation that results in a dramatic increase in beta cell mass toward the end of gestation. Virtually nothing is known about the growth factors and tissue interactions involved in this secondary transition leading to the development of mature β cells.

10.2.5 Human pancreas development compared with mice

Numerous mechanistic studies in the mouse have revealed significant insights into the regulation of murine pancreas development, however relatively little information is available regarding human pancreas development. What is known has been obtained from serial static immunohistochemical studies on human embryos (Clark and Grant, 1983; Conklin, 1962; Fukayama *et al.*, 1986; Githens, 1989; Hahn *et al.*, 1989; Like and Orci, 1972; Liu and Potter, 1962; Moore and Persaud, 1998; Polak *et al.*, 2000; Skandalakis *et al.*, 1994; Stefan *et al.*, 1983). Even these data have been difficult to obtain owing primarily to the scarcity of material and imprecision in determining gestational age (Polak *et al.*, 2000). In humans, endoderm specification begins by about 2 weeks with formation of the primitive streak and the onset of gastrulation (*Figure 10.1B*). Dorsal and ventral bud formation occurs in the 5th week of gestation; rotation and migration occur in the 6th week, and fusion of the buds in the 7th week of gestation. At these early stages, the pancreas is composed of an epithelium surrounded by loose and dense mesenchyme.

Liu and Potter first confirmed the existence of two populations of islet endocrine cells that were suggested by Laguesse in 1896, the first appearing at the end of the second month and the latter arising possibly from centroacinar cells during the 4th month (Skandalakis *et al.*, 1994). The first group of hormone positive cells can be identified in the epithelium between 8 and 10 weeks, the majority of which co-express some or all of the classical endocrine hormones (Conklin, 1962; Polak *et al.*, 2000). As in mice, the first endocrine cells to emerge in humans are glucagon-expressing cells (Conklin, 1962). In mice, most of the initial endocrine cells to appear in the first wave are Glu+ Ins+ PYY+, whereas in humans, Glu+ Ins+ Som+ cells appear first (Polak *et al.*, 2000; Slack, 1995; Upchurch *et al.*, 1994). It is clear that human endocrine cells, whether associated with the first or second wave, are rarely proliferative at any stage of development (Bouwens *et al.*, 1997; Polak *et al.*, 2000). In contrast to the low proliferative rate of endocrine cells, the non-hormone-positive cells within the early pancreatic epithelium are highly proliferative (Polak *et al.*, 2000). The early human pancreatic epithelial cells probably express PDX1 as in the mouse; the importance of PDX1 in humans mirrors its importance in murine development as a patient born with pancreatic agenesis was found to harbor a deleterious mutation in the *pdx1* gene (Stoffers *et al.*, 1997). No studies have been performed to evaluate the role of other pancreatic transcription factors in human development.

The formation of acini from primitive duct cells in humans begins by ~12 weeks of gestation; however, zymogen granules are only present from 20 weeks onward and trypsin is seen as early as 22 weeks (Skandalakis *et al.*, 1994). Little data exist on the ontogeny of ducts (Githens, 1989). Interestingly, when fragments of human pancreatic rudiments are implanted into nude mice, endocrine and duct-like cells accumulate, but acinar cells do not (Tuch *et al.*, 1984). Most growth factor studies have been performed with late gestation human fetal pancreatic tissue and have revealed beneficial effects of nicotinamide (Otonkoski *et al.*, 1993), hepatocyte growth factor (Otonkoski *et al.*, 1994, 1996), insulin-like growth factor-1 (Eckhoff *et al.*, 1991), and exendin-4 (Movassat *et al.*, 2002) on maturation into glucose-responsive tissue.

Few studies directly compare mouse and human pancreas development primarily because of the lack of adequate access to human embryonic tissue from all developmental stages. Human and mouse embryos differ in the timing and/or nature of embryonic gene expression (e.g., insulin-like growth factor family genes, pancreatic polypeptide, insulin, tissue factor, and LIF among others; Gregor et al., 1996; Liu et al., 1997; Luther et al., 1996). β cells of the human fetus are relatively resistant to the β cell toxin, streptozotocin, compared with adult human β cells and to fetal or adult mouse β cells (Tuch and Chen, 1993). In adults, there are some morphological differences between the organs of the two species: the mouse pancreas exists as a thin film of acinar tissue between two leaves of the peritoneal membrane whereas the human organ is a solid organ composed of a fibrous stroma. Furthermore, the relative arrangement of α and β cells within the mouse and primate islet differs, exhibiting a more random distribution in primates, including humans (Orci, 1982; Wieczorek et al., 1998).

The essential dissimilarities between human and mouse development establish that it cannot be assumed that all aspects of development, and in particular pancreatic organogenesis, are identical in the two species. In fact, a comprehensive comparison of pancreatic differentiation between mice and humans has not been undertaken. Human ES cells could provide an opportunity to study human pancreas and islet development. For example, the isolation of pancreatic progenitor cells or islet progenitor cells from both human and murine ES cells will facilitate a direct comparative analysis of these precursor cell populations. Also, genetic and epigenetic regulation of human pancreatic differentiation and growth could be studied in a controlled culture environment. Studies of this type may ultimately expand our limited knowledge of how pancreatic islets develop in humans.

10.3 Islet differentiation from embryonic stem cells

The differentiation of ES cells to specific lineages can provide both a tool for the study of developmental pathways and a source of transplantable tissue, as multipotent precursor cells can be identified and isolated from ES cells in large numbers *ex vivo*. During cellular differentiation *in vitro*, individual ES cell descendents coordinately express specific lineage-restricted genes in their proper temporal sequences (Bain et al., 1995; Baker and Lyons, 1996; Fraichard et al., 1995; Robertson et al., 2000). Ultimately, committed multipotent precursor cells develop that can complete lineage-specific differentiation pathways *in vivo* or *in vitro* (Keller, 1995; Liu et al., 2000; McDonald et al., 1999). For example, proliferative neural precursor cells can give rise to functional post-mitotic neurons, astrocytes, and oligodendrocytes *in vitro* (Brustle et al., 1997; Brustle and McKay, 1996; Mujtaba et al., 1999; Okabe et al., 1996; Zhang et al., 2001), and hematopoietic stem cells cultured from ES cells have been shown to reconstitute lymphoid, myeloid, and erythroid lineages after transfer into irradiated mice (Keller, 1995; Kyba et al., 2002; Robertson et al., 2000). Taken together these studies indicate that the differentiation of ES cells in culture re-establishes many aspects of normal embryogenesis, including the specification of lineage progenitor cells. If pancreatic islet progenitor cells could be generated from murine and human ES cells respectively, an ES cell system could be used for the first time to study and directly compare the mechanisms regulating pancreatic islet development and pancreaticogenesis in the two species.

The first goal towards a stem cell-based transplantation therapy is to demonstrate either the directed differentiation or selection of a specific cell type. Previous successes in deriving specific neural sub-populations that were subsequently shown to be capable of integrating into the brain or spinal cord (McDonald et al., 1999) has inspired research to derive enriched or purified populations of pancreatic islet cells, in particular β cells, that would be of obvious value for treating diabetic patients. Indeed, reports to this effect have emerged. However, these claims of success in generating β cells *en masse* by certain protocols may have been premature (Rajagopal et al., 2003).

10.3.1 Identification of insulin-producing cells

A number of reports have indicated that insulin-producing cells could be produced with remarkable efficiency from differentiating mouse ES cells, ranging from 30–95% of the total population (Blyszczuk et al., 2003; Hori et al., 2002; Lumelsky et al., 2001). This was achieved using a multiple-step protocol, or variant thereof, first devised for the selection of nestin-positive neural stem cells from differentiating ES cells (Okabe et al., 1996). Initial selection required survival for 6 days in a supplemented serum-free medium designed to enhance the appearance of neural stem cells. The resulting population was then expanded using bFGF and subsequently encouraged to differentiate by removing the mitogen. As part of the neuronal-enhancing culture medium developed in the original protocol (Johe et al., 1996), high levels of insulin (~25–29 μg/ml total concentration) were supplied to cells throughout selection and during the final stages of differentiation.

It now appears that cells maintained in the presence of high levels of exogenous insulin can take up insulin from the medium and retain it over considerable periods of time (Rajagopal et al., 2003 and our own unpublished observations). Evidence for this includes the failure to detect reliably significant levels of *insulin I* or *pdx1* mRNA in cell cultures derived by this protocol, and the observation that cells (including fibroblasts) can concentrate exogenous FITC-conjugated insulin from an otherwise insulin-free medium, resulting in insulin-stained cells similar to those obtained by antibody staining of cells maintained in a high-insulin containing medium (Rajagopal et al., 2003). In addition, insulin-positive cells generated through this protocol demonstrated condensed nuclei that were TUNEL+, suggesting that the cells were undergoing apoptosis (Rajagopal et al., 2003). We and others have also found that in cells selected and grown by the protocols of Lumelsky et al. (2001) or Hori et al. (2002), anti-insulin immunostaining of cells readily disappears after an 18–24 h incubation in insulin-free medium (*Color Plate 5*). Additionally, insulin+ cells produced by either protocol failed to stain for either C-peptide I or II (*Color Plate 5*), both of which are normally produced by cleavage of the corresponding pro-hormone and are present in authentic β cells.

How then can the apparent insulin secretion observed in the aforementioned studies be explained? It is possible that the discharge of absorbed insulin or insulin fragments from dead or dying cells could account for some of the immunoreactive material detected in the medium. The methodologies used in these studies did not eliminate this possibility, and the level of insulin released into the medium was very low and did not approach that of normal beta cells even if cell numbers are taken into account. Consistent with this hypothesis is the observation that transfer of the derived cells to diabetic mice failed to normalize hyperglycemia (Lumelsky et al., 2001), and histology of the transplants revealed cells with pyknotic nuclei (Hori et al., 2002).

In order to avoid potential misinterpretations, it is important to understand the possible sources of exogenous insulin. Supplements from commercial sources for serum-free culture, including ITS (insulin, transferrin and selenium), N2, and patented formulations such as B27 and Knockout Serum Replacement (KSR)©, often contain insulin. Generally, B27, ITS and N2 supplements diluted 100-fold contain 4 or 5 µg/ml insulin. The insulin content of KSR© is higher (~100 mg/l), resulting in a concentration of ~20 µg/ml in the 20% solutions that are frequently employed in serum-free culture recipes (Patent # WO 98/30679 and PCT/US98/00467), as compared with approximately 0.1 ng/ml in 10% fetal calf serum. This last source must be kept in mind when evaluating the only report to date describing the derivation of insulin-producing cells from human ES cells (Assady et al., 2001). Because KSR©, rather than serum, was used as the base for the differentiation medium in this report, the results of the immunohistochemical and insulin release studies should be interpreted with caution. The level of exogenous insulin in culture media that can be taken up by cells and result in a false positive in an immunohistochemical assay is not known.

Such findings make it imperative to substantiate by additional means whether the insulin detected under these conditions is synthesized endogenously. Methodologies such as RT-PCR, specific C-peptide staining, or metabolic labeling could be used to verify an intracellular origin and endogenous synthesis of insulin. Immunostaining for other β cell-related transcription factor proteins and other phenotypic markers of β cells, including PDX1, glucose transporter 2, β-glucokinase, pro-convertase 1/3 and 2, and characteristic potassium and calcium channels can provide corroborating evidence.

An additional way to eliminate the question of insulin uptake from the media is to identify insulin-producing cells based on activation of the insulin gene. Several studies have identified and/or isolated particular cell lineages by tagging or selecting cells that activate unique lineage-restricted genes (Klug et al., 1996; Li et al., 1998; Soria et al., 2000). One method to achieve this is through the insertion of a selectable marker, either a drug resistance gene or fluorescent tag, into the endogenous gene locus.

Using such a cell-trapping technology, Soria et al. (2000), transfected murine ES cells with a knock-in construct containing the human insulin promoter driving a fused lacZ and neomycin gene (βgeo). Following differentiation of embryoid bodies (EBs) into a variety of cell types, those cells in which the human insulin promoter became activated could be selected on the basis of their resistance to the antibiotic G418. Under conditions of low glucose (5 mM) culture for 5 days in the presence of nicotinamide, cell clusters produced from a cloned high insulin-producer were apparently capable of normalizing glycemia when implanted into diabetic mice and demonstrated insulin secretion *in vitro*. Although the strategy successfully generated insulin-producing cells and reversed diabetes in a few animals, questions remain as to the nature of selected clones regarding their authenticity and phenotypic stability. A similar strategy employing a fluorescent marker could be used to detect endogenous insulin gene activation without relying on immunostaining or other extrinsic methodologies.

10.3.2 Islet lineage differentiation from mouse and human ES cells

Based on the ability of ES cells to differentiate readily *in vivo* and *in vitro* into cellular derivatives of all three germ layers, and faithfully to reproduce normal development when implanted into blastocysts, it might be expected that they would have the ability to recapitulate normal development under proper conditions *in vitro*. Indeed, hematopoietic, cardiac, and neural lineage commitment pathways, including the formation of precursor cells, can be reproduced faithfully as murine ES cells differentiate in culture (Baker and Lyons, 1996; Kennedy and Keller, 2003; Green *et al.*, 2003; Wichterle *et al.*, 2002). For the treatment of diabetic patients, the ultimate goal is to obtain purified, functionally normal β cells differentiated from ES cells in large enough numbers to be used as a transplantation therapy. Towards this goal, we recently demonstrated that islet differentiation in murine ES cell cultures proceeds through stereotypical stages recapitulating many salient features of normal embryonic development, and results in cells resembling all four islet cell types (Kahan *et al.*, 2003).

We used a culture protocol that includes a period of embryoid body (EB) formation, which promotes early embryonic tissue interactions, and medium containing serum as the only source of exogenous insulin. In murine cells, the initial morphogenesis during EB formation mimics that of morula and blastula stage embryos *in vivo*. Generally, within 5–7 days of suspension culture, a layer of columnar ectoderm forms beneath a layer of primitive endoderm, and often surrounds a fluid-filled cavity, thereby creating a structure resembling the egg cylinder of the mouse. At this stage, EBs begin to express genes of primitive extra-embryonic endoderm (Abe *et al.*, 1996; Levinson-Dushnik and Benvenisty, 1997), ectoderm, and mesoderm lineages. The tissue juxtaposition in EBs may promote correct cell–cell interactions and inductive events necessary to enhance differentiation towards definitive endoderm and ultimately pancreatic lineages. Indeed, recent studies in our laboratory have indicated that a period of EB formation is required for the generation of PDX1-positive cells (unpublished results). Although these conditions allow the generation of PDX1+ cells and islet lineages, they do not necessarily promote directed differentiation of these cell types. Not surprisingly then, ES cells differentiating under these conditions generate heterogeneous cultures containing numerous cell types derived from ectoderm, endoderm, and mesoderm.

How other cell types might promote or inhibit pancreatic differentiation from ES cells is not clear. It is known, however, that critical inductive events mediated by mesoderm-derived embryonic tissues *in vivo* (i.e., notochord, dorsal aortae, lateral plate mesoderm) play an important role in patterning of the foregut endoderm and specification of a pancreatic fate (Hebrok *et al.*, 1998; Kim *et al.*, 1997;

Kumar et al., 2003; Lammert et al., 2001; Wells and Melton, 2000). It is possible that the presence of these other cell types may promote or induce pancreatic lineage differentiation from ES cells. On the other hand, it is also possible that the presence of other cell types in ES cell cultures, such as cardiac mesoderm, may elaborate inhibitory signals or induce non-pancreatic endoderm cell fates (Deutsch et al., 2001; Rossi et al., 2001). Just as the development of a particular cell type *in vivo* is likely to result from signals that are both inductive and inhibitory, the differentiation of ES cells *in vitro* is determined by a variety of positive and negative signals. Despite their rarity, it is notable that pancreatic lineage cells, including PDX1+ cells, early endocrine cells marked by YY immunostaining, and endocrine hormone expressing cells, all appear concomitantly within discrete foci, rather than being randomly distributed in ES cell cultures (*Color Plate 5*). The observation that foci appear to increase in size over the culture period, even though proliferating precursor cells could not be detected within the foci, suggests that locally inductive microenvironments probably exist within these heterogeneous cultures (Kahan et al., 2003).

We investigated whether features of normal pancreatic cytodifferentiation are recreated in these cultures. The earliest detectable areas representing pancreatic fate specification in ES cell cultures consist solely of PDX1-positive cells, which first appear approximately 4 days after plating EBs previously developed in suspension for 7 days (*Color Plate 6A*). PDX1+ cells appear in discrete foci that also contain cells identified by other markers of early progenitor cell types (*Color Plate 6B and C*). While these foci are rare, constituting less than 0.1% of the total population in dense cultures, in later stages they can contain hundreds of cells expressing one or more pancreatic markers. In most foci, the majority of cells express both YY and IAPP (*Color Plate 6D and E*), which are markers of early endocrine cells and appear in the pancreatic bud epithelium during normal development (Wilson et al., 2002; Upchurch et al., 1994). Interestingly, we observed a subset of PDX1+ cells that coexpressed YY, a subpopulation that may provide some clues to endocrine lineage development. We did not detect expression of *Ptf1a*, a bHLH transcription factor gene that appears to be required for complete acinar development and functional islet development (Kawaguchi et al., 2002; Krapp et al., 1998). The absence of *Ptf1a* may have contributed to the low frequency of islet hormone-positive cells and/or the absence of exocrine pancreas gene transcripts that we observed.

The first definitive islet hormone to appear is glucagon, consistent with normal development within the pancreatic epithelium. We observe that most hormone-positive cells appear within PDX1+ clusters that also contain PDX1+/hormone-negative cells (*Color Plate 6B and C*). Although definitive lineage relationships of hormone-positive cells in ES cell cultures await tracking studies, the co-staining patterns and the sequence of appearance of stained cell types we observed are compatible with the derivation of islet cell types from common pancreatic progenitor cells. In late stage cultures, PDX1 is expressed in the nucleus of nearly all ES cell-derived insulin+ cells (*Color Plate 7A*), and many somatostatin+ cells, a feature of mature β and δ cells, respectively. Most islet-specific cell types also co-express YY and/or IAPP, which may reflect a relative level of immaturity. As is seen in the normal development of β cells, the primary to secondary transition was recapitulated in these cultures: over a period of a few days, we observed an abrupt transition from double-positive cells expressing both glucagon and insulin to single-hormone positive cells (*Color Plate 7B*). Ultimately, cells emerge that, in addition to PDX1 and insulin expression, are characterized by features expected of normal β cells including co-expression of insulin with IAPP and both C-peptide I and II (*Color Plate 7C and D*). The cells transcribe both *insulin* genes, insulin protein is co-localized with C-peptide in cytoplasmic granules within the cells, and electron microscopy shows the presence of electron-dense secretory granules (Kahan *et al.*, 2003). Thus, it appears that many stereotypical features of normal islet cytodifferentiation are reproduced as ES cells differentiate under these culture conditions.

Considering the slower growth rate of human ES cells and longer gestation time of human embryos compared with their mouse counterparts, it might be predicted that differentiation of human ES cells would take longer than for mouse ES cells. Thus, we have allowed human EBs to develop for 14–21 days versus 7 days for mouse EBs, and observed the first hormone-positive cells after an additional 6–8 weeks of differentiation. Similar to the sequential appearance of typical pancreatic marker proteins in differentiating mouse ES cell cultures, we observed that in human ES cell-derived cultures, PDX1+ cells appear first at one to two weeks after plating, followed by the clustered emergence of hormone-positive cells within distinct regions (*Color Plate 8*). Cells expressing early endocrine cell markers YY and IAPP are also found generally clustered together. As in mouse ES cell cultures, it appears that the derivation of PDX1+ cells from human ES cells requires a period of EB formation. The frequency of insulin+ cells derived from human ES cells under non-selective culture conditions is very low, similar to what was observed in mouse ES cell cultures. Given the many as yet unknown cellular interactions regulating pancreatic and islet development, it is not surprising that differentiation of ES cells *in vitro* under non-selective conditions without growth factor supplementation generally results in relatively few cells assuming pancreatic lineage fates. However, the culture conditions can now be modified systematically in ways consistent with known developmental signals to achieve an enriched population of pancreatic lineage cells.

10.4 Recapitulating developmental pathways of islet differentiation in ES cells

Despite their multilineage differentiative potential, it has proven challenging to direct the differentiation of ES cells into specific lineages. Hematopoietic, cardiomyocyte, and neural lineages were the first embryonic cell types to be described from mouse and now more recently human ES cells (Doetschman *et al.*, 1985; He *et al.*, 2003; Kaufman *et al.*, 2001; Zhang *et al.*, 2001). This is likely due to the fact that these lineages are more easily formed in a non-manipulated environment; additionally, these cell types have distinct morphologies detectable by simple microscopy, unlike the relatively non-descript islet endocrine cell. Thus, it is not surprising that nearly 20 years elapsed between the derivation of mouse ES cells and reports of differentiation reports to an islet lineage.

The low frequencies of pancreatic cell types and the heterogeneity of the ES cell cultures has hampered in-depth study and functional characterization. The presence of varied cell types may lead to unpredictable cell–cell interactions that could inhibit endoderm formation and/or pancreatic specification. If inductive signals could be precisely re-established in a temporally and spatially-restricted fashion to provide the correct microenvironment to differentiating ES cells, a directed differentiation into PDX1-positive pancreatic progenitor cells might be achieved. For the enrichment of islet endocrine cells, sequential manipulations at several levels could be involved. First, to enrich for endodermal precursors one might choose to over-express genes encoding signaling molecules required for endoderm development, such as *Nodal* or *Sox17*, or endodermally-restricted transcription factor genes, such as *FoxA2*, *FoxA3*, and/or *Ptf1a*. Then, in order to pattern the ES cell-derived endoderm and induce expression of *Pdx1* in a wider expression domain, it might be important to inhibit the function of key developmental signaling molecules, such as *Shh*, which is normally down-regulated in the region of foregut endoderm that is specified to become pancreas. Mesenchymal-to-epithelial signaling, also critically important to proper specification of the pancreatic epithelium, could be recapitulated by adding specific growth factors, such as BMP4 or FGF10, or by co-culturing ES cells with cell lines (Kaufman *et al.*, 2001; Buttery *et al.*, 2001; Kitajima *et al.*, 2003) or embryonic tissues. By providing the proper developmental signals, it may be possible to promote the differentiation of a proliferative precursor population that could be isolated and expanded.

Correct cues for differentiating progenitors into post-mitotic, single hormone-expressing differentiated islet cells or β cells remain to be determined. Less is known about the regulation of the later stages of islet development, including terminal differentiation, migration of islet endocrine cells from the epithelium into the acinar lobules, or the mechanisms regulating their aggregation into micro-organs. Growth factors that have been shown to accelerate the functional maturation of human fetal pancreas tissue may be worthwhile to test in the ES cell differentiation system.

This basic knowledge of developmental mechanisms can be directly applied to ES cells as they differentiate in culture to promote lineage specific differentiation. Using such an approach and building on prior *in vivo* developmental studies, Wichterle *et al.* (2002) demonstrated robust, directed differentiation of motor neurons from mouse ES cells. This and other studies (Kyba *et al.*, 2002) demonstrate how applying known developmental cues to ES cells can direct differentiation to a desired phenotype and recapitulate normal ontogeny.

10.5 Remaining questions

There are many remaining challenges, including: (1) How can islet differentiation be optimized? (2) How do we determine if an ES cell-derived insulin-staining cell is a β cell? (3) What functional measure should be used? (4) Are β cells alone sufficient for normal glucose-responsive insulin secretion, or are other islet cell types required? These questions will be addressed below.

Although the system we describe demonstrates that ES cell cultures can differentiate into pancreatic progenitors and cells synthesizing insulin I in the absence of embryonic implantation, it is limited in important respects, particularly in the number of cells of this kind that are produced. We estimate the percentage of β-like cells to be from less than 0.01% to several orders of magnitude higher, depending on the criterion used. Furthermore, under these conditions, maintaining the growth potential of pancreatic precursor cells or islet progenitor cells is probably not optimal. Clearly, the development of robust methods for enrichment of progenitor cell types from ES cells is an important goal.

What should the benchmarks be for a 'real' β cell? This includes not only the issue of correctly identifying cells, but also the fact that cells from a variety of tissues induced in ES cultures can potentially synthesize insulin, including extraplacental membranes (Giddings and Carnaghi, 1989) and neuronal cells (Devaskar et al., 1994; Rulifson et al., 2002), among others (Giddings and Carnaghi, 1990; Goldfine et al., 1997). Additionally, β cells in various stages of functional maturity might be expected to exist in ES cell cultures, because the differentiation process may be temporally expanded or interrupted due to environmental deficiencies.

What criteria should be used for determining functionality, when, in fact, the response of islet tissue in standard assays may differ depending on the developmental stage (Hullett et al., 1995) or whether other islet endocrine cell types are present (Schuit and Pipeleers, 1985)? It has been suggested that all islet subtypes are required to achieve regulated functionality in normal *in situ* conditions (Schuit and Pipeleers, 1985; Tourrel et al., 2001), although there is some evidence that β cells alone can rescue STZ-induced diabetes in mice (Pericin et al., 2002). Normal β cell function has not yet been convincingly demonstrated for ES-derived cells, and this should be carefully evaluated with dynamic, single cell studies once enriched populations of insulin-producing tissue are reliably generated. Ideally, increased numbers of functional β cells will be obtained through new selective techniques and better culture conditions.

10.6 Summary

An ES cell-based strategy could permit the generation of an unlimited supply of islet stem cells or β cells from an abundant, renewable, and readily-accessible source for transplantation. Recent progress has been made in identifying pancreatic precursor cells and differentiated islet cells from mouse and human ES cells. In order to generate a robust ES cell-based replacement therapy for diabetes, a better understanding of the sequential genetic and epigenetic signals occurring during normal mouse and human development is necessary. Particularly relevant is the need to understand the nature and identity of true embryonic pancreatic precursor cells and islet progenitor cells, and to identify conditions that allow their efficient and large-scale isolation. An ES cell-based *in vitro* differentiation system can facilitate these goals by providing a straightforward means to select and purify progenitor cells and to investigate conditions that promote their expansion and differentiation *ex vivo*. Specifically, a human ES cell-based *in vitro* model system would be invaluable for studying human islet development and for providing cells for transplantation.

Acknowledgments

The authors wish to thank Karen Heim for assistance with manuscript preparation; Nick Weber for production of the figures; and Janet Fox for coordinating the editorial review process. Research on this topic in the author's laboratory is funded by the Juvenile Diabetes Research Foundation, Roche Organ Transplant Research Foundation, the National Institutes of Health-NIDDK, and Geron, Inc.

References

Abe K, Niwa H, Iwase K, Takiguchi M, Mori M, Abe SI, Yamamura KI (1996) Endoderm-specific gene expression in embryonic stem cells differentiated to embryoid bodies. *Exp. Cell Res.* **229**, 27–34.

Ahlgren U, Jonsson J, Edlund H (1996) The morphogenesis of the pancreatic mesenchyme is uncoupled from that of the pancreatic epithelium in IPF1/PDX1-deficient mice. *Development* **122**, 1409–1416.

Andrew A (1976) An experimental investigation into the possible neural crest origin of pancreatic APUD (islet) cells. *J. Embryol. Exp. Morphol.* **35**, 577–593.

Apelqvist A, Li H, Sommer L, Beatus P, Anderson DJ, Honjo T, Hrabe DA, Lendahl U, Edlund H (1999) Notch signalling controls pancreatic cell differentiation. *Nature* **400**, 877–881.

Assady S, Maor G, Amit M, Itskovitz-Eldor J, Skorecki KL, Tzukerman M (2001) Insulin production by human embryonic stem cells. *Diabetes* **50**, 1691–1697.

Bain G, Kitchens D, Yao M, Huettner JE, Gottlieb DI (1995) Embryonic stem cells express neuronal properties in vitro. *Dev. Biol.* **168**, 342–357.

Baker RK, Lyons GE (1996) Embryonic stem cells and in vitro muscle development. *Curr. Top. Dev. Biol.* **33**, 263–279.

Beddington RS (1994) Induction of a second neural axis by the mouse node. *Development* **120**, 613–620.

Beddington RS, Robertson EJ (1999) Axis development and early asymmetry in mammals. *Cell* **96**, 195–209.

Bhushan A, Itoh N, Kato S, Thiery JP, Czernichow P, Bellusci S, Scharfmann R (2001) Fgf10 is essential for maintaining the proliferative capacity of epithelial progenitor cells during early pancreatic organogenesis. *Development* **128**, 5109–5117.

Bingley PJ, Gale EA (1989) Rising incidence of IDDM in Europe. *Diabetes Care* **12**, 289–295.

Blyszczuk P, Czyz J, Kania G, Wagner M, Roll U, St Onge L, Wobus AM (2003) Expression of Pax4 in embryonic stem cells promotes differentiation of nestin-positive progenitor and insulin-producing cells. *Proc. Natl Acad. Sci. USA* **100**, 998–1003.

Bouwens L, Lu WG, De Krijger R (1997) Proliferation and differentiation in the human fetal endocrine pancreas. *Diabetologia* **40**, 398–404.

Brustle O, McKay RD (1996) Neuronal progenitors as tools for cell replacement in the nervous system. *Curr. Opin. Neurobiol.* **6**, 688–695.

Brustle O, Spiro AC, Karram K, Choudhary K, Okabe S, McKay RD (1997) In vitro-generated neural precursors participate in mammalian brain development. *Proc. Natl Acad. Sci. USA* **94**, 14809–14814.

Buttery LD, Bourne S, Xynos JD, Wood H, Hughes FJ, Hughes SP, Episkopou V, Polak JM (2001) Differentiation of osteoblasts and in vitro bone formation from murine embryonic stem cells. *Tissue Eng.* **7**, 89–99.

Carey FJ, Linney EA, Pedersen RA (1995) Allocation of epiblast cells to germ layer derivatives during mouse gastrulation as studied with a retroviral vector. *Dev. Genet.* **17**, 29–37.

Clark A, Grant AM (1983) Quantitative morphology of endocrine cells in human fetal pancreas. *Diabetologia* **25**, 31–35.

Conklin JL (1962) Cytogenesis of the human fetal pancreas. *Am. J. Anat.* **111**, 181–193.

Conlon FL, Lyons KM, Takaesu N, Barth KS, Kispert A, Herrmann B, Robertson EJ (1994) A primary requirement for nodal in the formation and maintenance of the primitive streak in the mouse. *Development* **120**, 1919–1928.

Currie CJ, Kraus D, Morgan CL, Gill L, Stott NC, Peters JR (1997) NHS acute sector expenditure for diabetes: the present, future, and excess in-patient cost of care. *Diabet. Med.* **14**, 686–692.

Delacour A, Nepote V, Trumpp A, Herrera PL (2004) Nestin expression in pancreatic exocrine cell lineages. *Mech. Dev.* **121**, 3–14.

Deutsch G, Jung J, Zheng M, Lora J, Zaret KS (2001) A bipotential precursor population for pancreas and liver within the embryonic endoderm. *Development* **128**, 871–881.

Devaskar SU, Giddings SJ, Rajakumar PA, Carnaghi LR, Menon RK, Zahm DS (1994) Insulin gene expression and insulin synthesis in mammalian neuronal cells. *J. Biol. Chem.* **269**, 8445–8454.

Dichmann DS, Miller CP, Jensen J, Scott HR, Serup P (2003) Expression and misexpression of members of the FGF and TGFbeta families of growth factors in the developing mouse pancreas. *Dev. Dyn.* **226**, 663–674.

Doetschman TC, Eistetter H, Katz M, Schmidt W, Kemler R (1985) The in vitro development of blastocyst-derived embryonic stem cell lines: formation of visceral yolk sac, blood islands and myocardium. *J. Embryol. Exp. Morphol.* **87**, 27–45.

Eckhoff DE, Sollinger HW, Hullett DA (1991) Selective enhancement of beta cell activity by preparation of fetal pancreatic proislets and culture with insulin growth factor 1. *Transplantation* **51**, 1161–1165.

Edlund H (2002) Pancreatic organogenesis–developmental mechanisms and implications for therapy. *Nat. Rev. Genet.* **3**, 524–532.

Esni F, Johansson BR, Radice GL, Semb H (2001) Dorsal pancreas agenesis in N-cadherin- deficient mice. *Dev. Biol.* **238**, 202–212.

Esni F, Stoffers DA, Takeuchi T, Leach SD (2004) Origin of exocrine pancreatic cells from nestin-positive precursors in developing mouse pancreas. *Mech. Dev.* **121**, 15–25.

Fontaine J, Le Lievre C, Le Douarin NM (1977) What is the developmental fate of the neural crest cells which migrate into the pancreas in the avian embryo? *Gen. Comp. Endocrinol.* 33, 394–404.

Fraichard A, Chassande O, Bilbaut G, Dehat C, Savatier P, Samarut J (1995) In vitro differentiation of embryonic stem cells into glial cells and functional neurons. *J. Cell Sci.* 108, 3181–3188.

Fukayama M, Ogawa M, Hayashi Y, Koike M (1986) Development of human pancreas. Immunohistochemical study of fetal pancreatic secretory proteins. *Differentiation* 31, 127–133.

Gale EA (2002) The rise of childhood type 1 diabetes in the 20th century. *Diabetes* 51, 3353–3361.

Giddings SJ, Carnaghi L (1989) Rat insulin II gene expression by extraplacental membranes. A non-pancreatic source for fetal insulin. *J. Biol. Chem.* 264, 9462–9469.

Giddings SJ, Carnaghi LR (1990) Selective expression and developmental regulation of the ancestral rat insulin II gene in fetal liver. *Mol. Endocrinol.* 4, 1363–1369.

Githens S (1989) Development of duct cells. In: *Human Gastrointestinal Development* (ed. E Lebenthal). Raven Press, New York, pp. 669–683.

Gittes GK, Galante PE, Hanahan D, Rutter WJ, Debase HT (1996) Lineage-specific morphogenesis in the developing pancreas: role of mesenchymal factors. *Development* 122, 439–447.

Goldfine ID, German MS, Tseng H-C, Wang J, Bolaffi JL, Chen JW, Olson DC, Rothman SS (1997) The endocrine secretion of human insulin and growth hormone by exocrine glands of the gastrointestinal tract. *Nat. Biotechnol.* 15, 1378–1382.

Golosow N, Grobstein C (1962) Epitheliomesenchymal interaction in pancreatic morphogenesis. *Dev. Biol.* 4, 242–255.

Gradwohl G, Dierich A, LeMaur M, Guillemot F (2000) Neurogenin 3 is required for the development of the four endocrine cell lineages of the pancreas. *Proc. Natl Acad. Sci. USA* 97, 1607–1611.

Green H, Easley K, Iuchi S (2003) Marker succession during the development of keratinocytes from cultured human embryonic stem cells. *Proc. Natl Acad. Sci. USA* 100, 15625–15630.

Gregor P, Feng Y, DeCarr LB, Cornfield LJ, McCaleb ML (1996) Molecular characterization of a second mouse pancreatic polypeptide receptor and its inactivated human homologue. *J. Biol. Chem.* 271, 27776–27781.

Gu G, Brown JR, Melton DA (2003) Direct lineage tracing reveals the ontogeny of pancreatic cell fates during mouse embryogenesis. *Mech. Dev.* 120, 35–43.

Gu G, Dubauskaite J, Melton DA (2002) Direct evidence for the pancreatic lineage: NGN3+ cells are islet progenitors and are distinct from duct progenitors. *Development* 129, 2447–2457.

Hahn von Dorsche H, Falt K, Titlbach M, Reiher H, Hahn HJ, Falkmer S (1989) Immunohistochemical, morphometric, and ultrastructural investigations of the early development of insulin, somatostatin, glucagon, and PP cells in foetal human pancreas. *Diabetes Res.* 12, 51–56.

Harrison KA, Thaler J, Pfaff SL, Gu H, Kehrl JH (1999) Pancreas dorsal lobe agenesis and abnormal islets of Langerhans in Hlxb9-deficient mice. *Nat. Genet.* **23**, 71–75.

He JQ, Ma Y, Lee Y, Thomson JA, Kamp TJ (2003) Human embryonic stem cells develop into multiple types of cardiac myocytes: action potential characterization. *Circ. Res.* **93**, 32–39.

Hebrok M (2003) Hedgehog signaling in pancreas development. *Mech. Dev.* **120**, 45–57.

Hebrok M, Kim SK, Melton DA (1998) Notochord repression of endodermal Sonic hedgehog permits pancreas development. *Genes Dev.* **12**, 1705–1713.

Hebrok M, Kim SK, St Jacques B, McMahon AP, Melton DA (2000) Regulation of pancreas development by hedgehog signaling. *Development* **127**, 4905–4913.

Herrera PL (2000). Adult insulin- and glucagon-producing cells differentiate from two independent cell lineages. *Development* **127**, 2317–2322.

Hitoshi S, Alexson T, Tropepe V, Donoviel D, Elia AJ, Nye JS, Conlon RA, Mak TW, Bernstein A, van der KD (2002) Notch pathway molecules are essential for the maintenance, but not the generation, of mammalian neural stem cells. *Genes Dev.* **16**, 846–858.

Holland AM, Hale MA, Kagami H, Hammer RE, MacDonald RJ (2002) Experimental control of pancreatic development and maintenance. *Proc. Natl Acad. Sci. USA* **99**, 12236–12241.

Hori Y, Rulifson IC, Tsai BC, Heit JJ, Cahoy JD, Kim SK (2002) Growth inhibitors promote differentiation of insulin-producing tissue from embryonic stem cells. *Proc. Natl Acad. Sci. USA* **99**, 16105–16110.

Horne-Badovinac S, Rebagliati M, Stainier DY (2003) A cellular framework for gut-looping morphogenesis in zebrafish. *Science* **302**, 662–665.

Hullett DA, MacKenzie DA, Alam T, Sollinger HW (1995) Preparation of fetal islets for transplantation: importance of growth factors. In: *Fetal Islet Transplantation* (eds CM Peterson, L Jovanovic-Peterson, B Formby). Plenum Press, New York, pp. 27–36.

Humphrey RK, Bucay N, Beattie GM, Lopez A, Messam CA, Cirulli V, Hayek A (2003) Characterization and isolation of promoter-defined nestin-positive cells from the human fetal pancreas. *Diabetes* **52**, 2519–2525.

Hussain MA, Lee J, Miller CP, Habener JF (1997) POU domain transcription factor brain 4 confers pancreatic alpha-cell-specific expression of the proglucagon gene through interaction with a novel proximal promoter G1 element. *Mol. Cell Biol.* **17**, 7186–7194.

Jensen J, Heller RS, Funder-Nielsen T, Pedersen EE, Lindsell C, Weinmaster G, Madsen OD, Serup P (2000a) Independent development of pancreatic alpha- and beta-cells from nurogenin3-expressing precursors. A role for the notch pathway in repression of premature differentiation. *Diabetes* **49**, 163–176.

Jensen J, Pedersen EE, Galante P, Hald J, Heller RS, Ishibashi M, Kageyama R, Guillemot F, Serup P, Madsen OD (2000b) Control of endodermal endocrine development by Hes-1. *Nat. Genet.* **24**, 36–44.

Johe KK, Hazel TG, Muller T, Dugich-Djordjevic MM, McKay RD (1996) Single factors direct the differentiation of stem cells from the fetal and adult central nervous system. *Genes Dev.* **10**, 3129–3140.

Jonsson J, Carlsson L, Edlund T, Edlund H (1994) Insulin-promoter-factor 1 is required for pancreas development in mice. *Nature* **371**, 606–609.

Kahan BW, Jacobson LM, Hullett DA, Oberley TD, Odorico JS (2003) Pancreatic precursors and differentiated islet cell types from murine embryonic stem cells: an in vitro model to study islet differentiation. *Diabetes* **52**, 2016–2024.

Kanai-Azuma M, Kanai Y, Gad JM, Tajima Y, Taya C, Kurohmaru M *et al.* (2002) Depletion of definitive gut endoderm in Sox17-null mutant mice. *Development* **129**, 2367–2379.

Kaufman DS, Hanson ET, Lewis RL, Auerbach R, Thomson JA (2001) Hematopoietic colony-forming cells derived from human embryonic stem cells. *Proc. Natl Acad. Sci. USA* **98**, 10716–10721.

Kawaguchi Y, Cooper B, Gannon M, Ray M, MacDonald RJ, Wright CVE (2002) The role of the transcriptional regulator Ptf1a in converting intestinal to pancreatic progenitors. *Nat. Genet.* **32**, 128–134.

Keller GM (1995). In vitro differentiation of embryonic stem cells. *Curr. Opin. Cell Biol.* **7**, 862–869.

Kennedy M, Keller GM (2003) Hematopoietic commitment of ES cells in culture. *Methods Enzymol.* **365**, 39–59.

Kim SK, Hebrok M (2001) Intercellular signals regulating pancreas development and function. *Genes Dev.* **15**, 111–127.

Kim SK, MacDonald RJ (2002) Signaling and transcriptional control of pancreatic organogenesis. *Curr. Opin. Genet. Dev.* **12**, 540–547.

Kim SK, Hebrok M, Melton DA (1997) Notochord to endoderm signaling is required for pancreas development. *Development* **124**, 4243–4252.

Kitajima K, Tanaka M, Zheng J, Sakai-Ogawa E, Nakano T (2003) In vitro differentiation of mouse embryonic stem cells to hematopoietic cells on an OP9 stromal cell monolayer. *Methods Enzymol.* **365**, 72–83.

Klug MG, Soonpaa MH, Koh GY, Field LJ (1996) Genetically selected cardiomyocytes from differentiating embryonic stem cells form stable intracardiac grafts. *J. Clin. Invest.* **98**, 216–224.

Krapp A, Knofler M, Ledermann B, Burki K, Berney C, Zoerkler N, Hagenbuchle O, Wellauer PK (1998) The bHLH protein PTF1-p48 is essential for the formation of the exocrine and the correct spatial organization of the endocrine pancreas. *Genes Dev.* **12**, 3752–3763.

Kumar M, Melton D (2003) Pancreas specification: a budding question. *Curr. Opin. Genet. Dev.* **13**, 401–407.

Kumar M, Jordan N, Melton D, Grapin-Botton A (2003) Signals from lateral plate mesoderm instruct endoderm toward a pancreatic fate. *Dev. Biol.* **259**, 109–122.

Kyba M, Perlingeiro RC, Daley GQ (2002) HoxB4 confers definitive lymphoid-myeloid engraftment potential on embryonic stem cell and yolk sac hematopoietic progenitors. *Cell* **109**, 29–37.

Lammert E, Cleaver O, Melton D (2001) Induction of pancreatic differentiation by signals from blood vessels. *Science* **294**, 564–567.

Lawson KA, Pedersen RA (1987) Cell fate, morphogenetic movement and population kinetics of embryonic endoderm at the time of germ layer formation in the mouse. *Development* 101, 627–652.

Lawson KA, Meneses JJ, Pedersen RA (1991) Clonal analysis of epiblast fate during germ layer formation in the mouse embryo. *Development* 113, 891–911.

LeDouarin NM (1988) On the origin of pancreatic endocrine cells. *Cell* 53, 169–171.

Levinson-Dushnik M, Benvenisty N (1997) Involvement of hepatocyte nuclear factor 3 in endoderm differentiation of embryonic stem cells. *Mol. Cell Biol.* 17, 3817–3822.

Li H, Arber S, Jessell TM, Edlund H (1999) Selective agenesis of the dorsal pancreas in mice lacking homeobox gene Hlxb9. *Nat. Genet.* 23, 67–70.

Li M, Pevny L, Lovell-Badge R, Smith A (1998) Generation of purified neural precursors from embryonic stem cells by lineage selection. *Curr. Biol.* 8, 971–974.

Like AA, Orci L (1972) Embryogenesis of the human pancreatic islets: a light and electron microscopic study. *Diabetes* 21, 511–534.

Liu HC, He ZY, Tang YX, Mele CA, Veeck LL, Davis O, Rosenwaks Z (1997) Simultaneous detection of multiple gene expression in mouse and human individual preimplantation embryos. *Fertil. Steril.* 67, 733–741.

Liu HM, Potter EL (1962) Development of the human pancreas. *Arch. Pathol.* 74, 439–452.

Liu S, Qu Y, Stewert TJ, Howard MJ, Chakrabortty S, Holekamp TF, McDonald JW (2000) Embryonic stem cells differentiate into oligodendrocytes and myelinate in culture and after spinal cord transplantation. *Proc. Natl Acad. Sci. USA* 97, 6126–6131.

Lumelsky N, Blondel O, Laeng P, Velasco I, Ravin R, McKay R (2001) Differentiation of embryonic stem cells to insulin-secreting structures similar to pancreatic islets. *Science* 292, 1389–1394.

Luther T, Flossel C, Mackman N, Bierhaus A, Kasper M, Albrecht S *et al.* (1996) Tissue factor expression during human and mouse development. *Am. J. Pathol.* 149, 101–113.

McDonald JW, Liu XZ, Qu Y, Liu S, Mickey SK, Turetsky D, Gottlieb DI, Choi DW (1999) Transplanted embryonic stem cells survive, differentiate and promote recovery in injured rat spinal cord. *Nat. Med.* 5, 1410–1412.

McKinnon CM, Docherty K (2001) Pancreatic duodenal homeobox-1, PDX-1, a major regulator of beta cell identity and function. *Diabetologia* 44, 1203–1214.

Moore KL, Persaud TVN (1998) *The Developing Human: Clinically Oriented Embryology*. W. B. Saunders, Philadelphia, PA.

Movassat J, Beattie GM, Lopez AD, Hayek A (2002) Exendin 4 up-regulates expression of PDX 1 and hastens differentiation and maturation of human fetal pancreatic cells. *J. Clin. Endocrinol. Metab.* 87, 4775–4781.

Mujtaba T, Piper DR, Kalyani A, Groves AK, Lucero MT, Rao MS (1999) Lineage-restricted neural precursors can be isolated from both the mouse neural tube and cultured ES cells. *Dev. Biol.* 214, 113–127.

Norgaard GA, Jensen JN, Jensen J (2003) FGF10 signaling maintains the pancreatic progenitor cell state revealing a novel role of Notch in organ development. *Dev. Biol.* **264**, 323–338.

Odorico JS, Sollinger HW (2002) Technical and immunosuppressive advances in transplantation for insulin-dependent diabetes mellitus. *World J. Surg.* **26**, 194–211.

Offield MF, Jetton TL, Labosky PA, Ray M, Stein RW, Magnuson MA, Hogan BL, Wright CV (1996) PDX-1 is required for pancreatic outgrowth and differentiation of the rostral duodenum. *Development* **122**, 983–995.

Ohishi K, Varnum-Finney B, Serda RE, Anasetti C, Bernstein ID (2001) The Notch ligand, Delta-1, inhibits the differentiation of monocytes into macrophages but permits their differentiation into dendritic cells. *Blood* **98**, 1402–1407.

Okabe S, Forsberg-Nilsson K, Spiro AC, Segal M, McKay RD (1996) Development of neuronal precursor cells and functional postmitotic neurons from embryonic stem cells in vitro. *Mech. Dev.* **59**, 89–102.

Orci L (1982) Macro- and micro-domains in the endocrine pancreas. *Diabetes* **31**, 538–565.

Otonkoski T, Beattie GM, Mally MI, Ricordi C, Hayek A (1993) Nicotinamide is a potent inducer of endocrine differentiation in cultured human fetal pancreatic cells. *J. Clin. Invest.* **92**, 1459–1466.

Otonkoski T, Beattie GM, Rubin JS, Lopez AD, Baird A, Hayek A (1994) Hepatocyte growth factor/scatter factor has insulinotropic activity in human fetal pancreatic cells. *Diabetes* **43**, 947–953.

Otonkoski T, Cirulli V, Beattie M, Mally MI, Soto G, Rubin JS, Hayek A (1996) A role for hepatocyte growth factor/scatter factor in fetal mesenchyme-induced pancreatic beta-cell growth. *Endocrinology* **137**, 3131–3139.

Pericin M, Althage A, Freigang S, Hengartner H, Rolland E, Dupraz P, Thorens B, Aebischer P, Zinkernagel RM (2002) Allogeneic beta-islet cells correct diabetes and resist immune rejection. *Proc. Natl Acad. Sci. USA* **99**, 8203–8206.

Polak M, Bouchareb-Banaei L, Scharfmann R, Czernichow P (2000) Early pattern of differentiation in the human pancreas. *Diabetes* **49**, 225–232.

Rajagopal J, Anderson WJ, Kume S, Martinez OI, Melton DA (2003) Insulin staining of ES cell progeny from insulin uptake. *Science* **299**, 363.

Robertson EJ, Norris DP, Brennan J, Bikoff EK (2003) Control of early anterior-posterior patterning in the mouse embryo by TGF-beta signalling. *Philos. Trans. R. Soc. Lond. B Biol. Sci.* **358**, 1351–1357.

Robertson SM, Kennedy M, Shannon JM, Keller G (2000) A transitional stage in the commitment of mesoderm to hematopoiesis requiring the transcription factor SCL/tal-1. *Development* **127**, 2447–2459.

Rossi JM, Dunn NR, Hogan BL, Zaret KS (2001) Distinct mesodermal signals, including BMPs from the septum transversum mesenchyme, are required in combination for hepatogenesis from the endoderm. *Genes Dev.* **15**, 1998–2009.

Rubin RJ, Altman WM, Mendelson DN (1994) Health care expenditures for people with diabetes mellitus, 1992. *J. Clin. Endocrinol. Metab.* **78**, 809A–F.

Rulifson EJ, Kim SK, Nusse R (2002) Ablation of insulin-producing neurons in flies: growth and diabetic phenotypes. *Science* 296, 1118–1120.

Sander M, Sussel L, Conners J, Scheel D, Kalamaras J, Dela CF, Schwitzgebel V, Hayes-Jordan A, German M (2000) Homeobox gene Nkx6.1 lies downstream of Nkx2.2 in the major pathway of beta-cell formation in the pancreas. *Development* 127, 5533–5540.

Scharfmann R (2000) Control of early development of the pancreas in rodents and humans: implications of signals from the mesenchyme. *Diabetologia* 43, 1083–1092.

Schuit FC, Pipeleers DG (1985) Regulation of adenosine 3′,5′-monophosphate levels in the pancreatic B cell. *Endocrinology* 117, 834–840.

Shapiro AM, Lakey JR, Ryan EA, Korbutt GS, Toth E, Warnock GL, Kneteman NM, Rajotte RV (2000) Islet transplantation in seven patients with type 1 diabetes mellitus using a glucocorticoid-free immunosuppressive regimen. *N. Engl. J. Med.* 343, 230–238.

Skandalakis JE, Gray SW, Ricketts R, Skandalakis LJ (1994) The pancreas. In: *Embryology for Surgeons: The Embryological Basis for the Treatment of Congenital Anomalies* (eds JE Skandalakis, SW Gray). Williams & Wilkins, Baltimore, pp. 366–404.

Slack JM (1995) Developmental biology of the pancreas. *Development* 121, 1569–1580.

Sollinger HW, Odorico JS, Knechtle SJ, D'Alessandro AM, Kalayoglu M, Pirsch JD (1998) Experience with 500 simultaneous pancreas-kidney transplants. *Ann. Surg.* 228, 284–296.

Soria B, Roche E, Berna G, Leon-Quinto T, Reig JA, Martin F (2000) Insulin-secreting cells derived from embryonic stem cells normalize glycemia in streptozotocin-induced diabetic mice. *Diabetes* 49, 157–162.

Sosa-Pineda B, Chowdhury K, Torres M, Oliver G, Gruss P (1997) The Pax4 gene is essential for differentiation of insulin- producing beta cells in the mammalian pancreas. *Nature* 386, 399–402.

Srinivas S, Rodriguez T, Clements M, Smith JC, Beddington RS (2004) Active cell migration drives the unilateral movements of the anterior visceral endoderm. *Development* 131, 1157–1164.

Stefan Y, Grasso S, Perrelet A, Orci L (1983) A quantitative immunofluorescent study of the endocrine cell populations in the developing human pancreas. *Diabetes* 32, 293–301.

Stemple DL (2001) Vertebrate development: the subtle art of germ-layer specification. *Curr. Biol.* 11, R878-81.

Stoffers DA, Zinkin NT, Stanojevic V, Clarke WL, Habener JF (1997) Pancreatic agenesis attributable to a single nucleotide deletion in the human IPF1 gene coding sequence. *Nat. Genet.* 15, 106–110.

Stovring H, Andersen M, Beck-Nielsen H, Green A, Vach W (2003) Rising prevalence of diabetes: evidence from a Danish pharmaco-epidemiological database. *Lancet* 362, 537–538.

Sussel L, Kalamaras J, Hartigan-O'Connor DJ, Meneses JJ, Pedersen RA, Rubenstein JLR, German MS (1998) Mice lacking the homeodomain transcription factor Nkx2.2 have diabetes due to arrested differentiation of pancreatic β cells. *Development* 125, 2213–2221.

Sutherland DE, Gruessner RW, Dunn DL, Matas AJ, Humar A, Kandaswamy R *et al.* (2001) Lessons learned from more than 1,000 pancreas transplants at a single institution. *Ann. Surg.* 233, 463–501.

Tourrel C, Bailbe D, Meile MJ, Kergoat M, Portha B (2001) Glucagon-like peptide-1 and exendin-4 stimulate beta-cell neogenesis in streptozotocin-treated newborn rats resulting in persistently improved glucose homeostasis at adult age. *Diabetes* 50, 1562–1570.

Treutelaar MK, Skidmore JM, Dias-Leme CL, Hara M, Zhang L, Simeone D, Martin DM, Burant CF (2003) Nestin-lineage cells contribute to the microvasculature but not endocrine cells of the islet. *Diabetes* 52, 2503–2512.

Tuch BE, Chen J (1993) Resistance of the human fetal beta-cell to the toxic effect of multiple low-dose streptozotocin. *Pancreas* 8, 305–311.

Tuch BE, Ng AB, Jones A, Turtle JR (1984) Histologic differentiation of human fetal pancreatic explants transplanted into nude mice. *Diabetes* 33, 1180–1187.

Upchurch BH, Aponte GW, Leiter AB (1994) Expression of peptide YY in all four islet cell types in the developing mouse pancreas suggests a common peptide YY-producing progenitor. *Development* 120, 245–252.

Wakamatsu Y, Maynard TM, Weston JA (2000) Fate determination of neural crest cells by NOTCH-mediated lateral inhibition and asymmetrical cell division during gangliogenesis. *Development* 127, 2811–2821.

Wells JM, Melton DA (1999) Vertebrate endoderm development. *Ann. Rev. Cell Develop. Biol.* 15, 393–410.

Wells JM, Melton DA (2000) Early mouse endoderm is patterned by soluble factors from adjacent germ layers. *Development* 127, 1563–1572.

Wichterle H, Lieberam I, Porter JA, Jessell TM (2002) Directed differentiation of embryonic stem cells into motor neurons. *Cell* 110, 385–397.

Wieczorek G, Pospischil A, Perentes E (1998) A comparative immunohistochemical study of pancreatic islets in laboratory animals (rats, dogs, minipigs, nonhuman primates). *Exp. Toxicol. Pathol.* 50, 151–172.

Wilson ME, Kalamaras JA, German MS (2002) Expression pattern of IAPP and prohormone convertase 1/3 reveals a distinctive set of endocrine cells in the embryonic pancreas. *Mech. Dev.* 115, 171–176.

Wilson ME, Scheel D, German MS (2003) Gene expression cascades in pancreatic development. *Mech. Dev.* 120, 65–80.

Yoshitomi H, Zaret KS (2004) Endothelial cell interactions initiate dorsal pancreas development by selectively inducing the transcription factor Ptf1a. *Development* 131, 807–817.

Yudkin JS, Beran D (2003) Prognosis of diabetes in the developing world. *Lancet* 362, 1420–1421.

Zhang SC, Wernig M, Duncan ID, Brustle O, Thomson JA (2001) In vitro differentiation of transplantable neural precursors from human embryonic stem cells. *Nat. Biotechnol.* 19, 1129–1133.

11. Cardiomyocyte differentiation in human embryonic stem cell progeny

Izhak Kehat, Joseph Itskovitz-Eldor and Lior Gepstein

11.1 Introduction

Adult cardiomyocytes are, to a large extent, terminally differentiated and therefore have limited regenerative capacity. Consequently, the loss of viable myocardium that is associated with myocardial infarction triggers a sequence of cellular and physiological processes leading to left ventricular dilatation and development of progressive heart failure. Replacement of the dysfunctional myocardium by implantation of exogenous myogenic cells is emerging as a novel paradigm for myocardial repair (Reinlib and Field, 2000) but clinical application has been hampered by the absence of a renewable source of cells for tissue grafting. The absence of an adequate *in vitro* source of human cardiac tissue also imposes significant limitations for several other basic cardiovascular research areas.

Recent advances in the field of stem cell research suggest a possible solution for the aforementioned cell-sourcing problem. Stem cells can be broadly divided into two categories: adult- (somatic) and embryo-derived stem cells, and are defined by their capacity for self-renewal and the potential to differentiate into one or more mature cell types. In adults, a highly regulated process of stem cell self-renewal and differentiation sustains tissues with high cell turnover. In recent years, adult-derived stem cells were also described in tissues believed to have relatively limited regenerative capacity such as the brain, pancreas, and possibly also in the heart (Anversa and Nadal-Ginard, 2002). Although the differentiation potential of adult stem cells may be more versatile than originally believed (Krause *et al.*, 2001), they are still thought to be relatively limited in the inability to differentiate into a plurality of cell types. Furthermore, the mechanism(s) underlying their change in phenotype is still controversial, and they can not be readily propagated outside the body.

In contrast, cells in the early mammalian embryo have the potential to contribute to all adult tissues. At the blastocyst stage, a group of cells on the interior of the embryo begins to segregate from the outer cells, forming what is called the inner cell mass (ICM). Whereas the outer cells become the trophoectoderm, the ICM cells will ultimately give rise, through specialized progenitor cells, to all the tissues in the body and are therefore truly pluripotent. In 1981, ICM cells isolated from mouse blastocysts were used to generate pluripotent stem cell lines that were termed embryonic stem (ES) cells (Evans and Kaufman, 1981; Martin, 1981). Mouse ES cell lines are characteristically capable of prolonged *in vitro* proliferation and self-renewal, but also retain the ability to differentiate into derivatives of all three germ layers both *in vitro* and *in vivo*.

The process used to generate human ES lines (Reubinoff *et al.*, 2000; Thomson *et al.*, 1998) was similar to that used in deriving mouse (Evans and Kaufman, 1981; Martin, 1981) and rhesus (Thomson *et al.*, 1995) ES cells and is described in detail in Chapter 4. The human ES cell lines were shown to fulfil criteria defining ES cell lines (Itskovitz-Eldor *et al.*, 2000; Thomson *et al.*, 1998): (1) they were derived from pre-implantation human blastocysts; (2) they were capable of prolonged undifferentiated proliferation when grown on mitotically inactivated mouse embryonic fibroblast (MEF) feeder layers; and (3) they demonstrated a stable developmental potential to form derivatives of all three germ layers including cardiomyocytes (Kehat *et al.*, 2001). The cell lines and their clonal derivatives (Amit *et al.*, 2000) were also shown to express high levels of telomerase and to retain a normal karyotype for prolonged culture periods.

The current chapter will focus on describing the processes involved in the derivation and characterization of cardiomyocytes using the unique human ES cell differentiation system. Particular emphasis will be placed on describing the potential research and clinical applications of this new technology in cardiovascular medicine. These applications will be discussed in the context of our current understanding of the inductive signals and transcriptional regulation involved in early embryonic cardiomyogenesis. Through growing knowledge of cardiac development in the embryo and the extensive experience in cardiomyocyte differentiation from murine ES cells, we can hope to harness the full research and clinical potential of these cells.

11.2 Early signals in cardiac development

The heart is the first major organ to form during embryogenesis and its circulatory function is essential for the viability of the developing mammalian embryo. Heart formation comprises multiple developmental steps that include determination of the cardiac field in the mesoderm, differentiation of cardiac precursor cells into cardiomyocytes, and morphogenesis of the chambered heart (Harvey, 2002). Our current understanding of early development of the vertebrate heart has been gained to a large extent through studies in a number of model organisms including the chick, amphibians, zebrafish and the mouse (Cripps and Olson, 2002; Harvey, 2002; Olson, 2001; Zaffran and Frasch, 2002).

In contrast to the fairly well characterized process of the morphologic transformation of the primitive heart tube into a four-chambered contractile structure, the inductive cues that lead to specification and terminal differentiation of cardiomyocytes are less well understood. The heart in vertebrates is derived from a subpopulation of mesodermal precursor cells that become committed to a cardiogenic fate in response to inductive signals from adjacent cell types. The heart arises from cells in the anterior lateral plate mesoderm of the early embryo, where they are arranged in bilateral fields on either side of the prechordal plate, in close proximity to the anterior endoderm (Cripps and Olson, 2002; Harvey, 2002; Olson, 2001; Zaffran and Frasch, 2002).

Specification of the vertebrate cardiac mesoderm occurs concurrently with gastrulation. Data derived from amphibians, chicks, and more recently also in mice suggest that signals emanating from the primitive endoderm may play a key role in the processes of cardiomyocyte induction of the precursor cells in the adjacent anterior mesoderm. The heart does not form if anterior endoderm is removed from embryos pointing to an instructive role for the anterior endoderm in this process (Nascone and Mercola, 1995). Moreover, when cells from the posterior non-cardiogenic mesoderm are transplanted to the cardiogenic region they differentiate to form heart instead of blood cells.

Experimental results from a variety of *in vitro* and *in vivo* models suggest that bone morphogenic proteins (BMPs), expressed in the endoderm adjacent to the heart-forming region, may play an important instructive role in cardiogenic induction as well as in maintaining the cardiac lineage once it is specified (Monzen *et al.*, 2002). Interestingly, the boundaries of the heart-forming region (the cardiac crescent) are also delineated by repressive signals mediated by members of the Wnt signaling molecule family (Olson, 2001). These proteins are secreted from the underlying neural tube and notochord and inhibit cardiomyogenesis in the posterior mesoderm. Coordination of these signaling gradients is further accomplished by secretion of the Wnt-binding proteins (Crescent and Dkk-1) in the anterior endoderm, inhibiting Wnt activity and thereby defining the heart field as an area of low Wnt activity and high BMP strength signals.

The extracellular inductive signals, discussed above, are interpreted in the nucleus by transcription factors that activate the myocardial gene program (Zaffran and Frasch, 2002). Two important transcriptional regulators of cardiac gene expression are the homeodomain protein Nkx2.5 and the zinc-finger transcription factor GATA4. Other important transcription factors include the MEF2 family, the homeodomain protein Tbx5 and the MADS box protein serum response factor (SRF), which may act in concert with Nkx2.5 and GATA4 to activate cardiac restricted genes (Sepulveda *et al.*, 2002). Recently, a new transcriptional activator expressed in cardiac and smooth muscle cells has been identified and termed myocardin (Wang *et al.*, 2001). Myocardin, a member of the SAP domain family of nuclear proteins, was shown to co-activate transcription of several cardiac specific gene promoters in conjunction with SRF. Interestingly, while several cardiac-restricted transcription factors have been identified, none has been found to possess the ability to induce cardiac identity on their own.

11.3 *In vitro* differentiation of mouse and human ES cells to cardiomyocytes

ES cells can be propagated continuously in the undifferentiated state when grown on MEF feeder layers. In the mouse model this can also be achieved in the feeder-free setting by supplementing the medium with leukemia inhibitory factor (LIF). The most common method used for inducing differentiation of the ES cells requires an initial aggregation step to form three-dimensional structures termed embryoid bodies (EBs, *Figure 11.1*). Differentiation is initiated by first removing cells from MEF feeder layers, and then by cultivating them in suspension (Doetschman *et al.*, 1985). Among other differentiating cell types within the EBs, cardiomyocyte tissue can be identified by the appearance of spontaneously contracting areas. Several factors may influence the ability of the ES cells to differentiate into cardiomyocytes including the specific type of ES line used, the starting number of cells within the EBs, the duration of the suspension phase until plating of the EBs, and the types of culture media, serum and growth factors used (Boheler *et al.*, 2002).

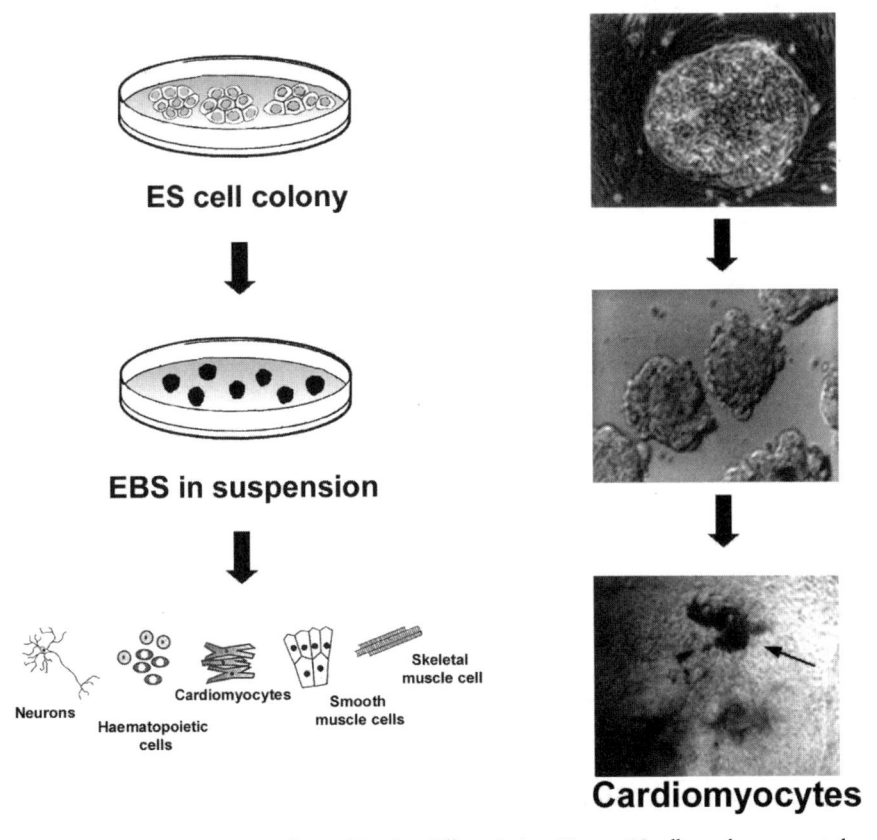

Figure 11.1: Human ES propagation and *in vitro* differentiation. Human ES cells can be propagated continuously in the undifferentiated state when grown on MEF feeder layers. To induce differentiation, ES cells are removed from the MEF feeder layer and grown in suspension where they form three-dimensional differentiating cell aggregates (embryoid bodies, EBs). This *in vitro* differentiating system can be used to generate a plurality of tissue types, including cardiomyocytes. The left panel shows a schematic representation of this process while the right panel shows micrographs of an ES cell colony (top), EBs in suspension (middle), and a spontaneously contracting area (arrow) within an EB following plating.

The generation of cardiomyocyte tissue within differentiating mouse EBs provides a unique *in vitro* tool for investigating early cardiomyogenesis. During *in vitro* differentiation, cardiomyocytes within the EB were shown to express cardiac specific genes, proteins, ion channels, receptors, and signal transduction machinery in a developmental pattern that closely recapitulates the developmental pattern of early *in vivo* murine cardiomyogenesis (Boheler *et al.*, 2002; Hescheler *et al.*, 1997). The mouse ES cell-derived cardiomyocytes were shown to express cardiac-restricted genes in a temporally regulated fashion with the mesodermal genes such as BMP-4 expressed initially and followed by the expression of cardiac-restricted transcription factors, Nkx2.5 and GATA4. This was followed by the expression of the cardiac-specific structural proteins such as atrial naturetic factor (ANF), myosin heavy chains (α-MHC and β-MHC), and phospholamban, with the chamber-specific genes such as myosin light chain-2V (MLC-2V) expressed last (Boheler *et al.*, 2002). Similarly, a developmental pattern was also noted for the expression of the different sarcomeric proteins (Boheler *et al.*, 2002).

The advent of the murine ES cell model has also provided important insights into the physiological processes involved in the development of excitability and electromechanical coupling in early cardiac tissue including patterns of gene expression, myofibrillogenesis, ion channel development and function, calcium handling, receptor expression, and signaling mechanisms involved in these processes (Boheler *et al.*, 2002; Hescheler *et al.*, 1997). Detailed electrophysiological studies of cardiomyocytes within developing murine EBs revealed a developmental cascade of ion channel expression and modulation. In some of these studies, genetically modified ES cell clones expressing a reporter gene under the control of early cardiac-specific promoters permitted the identification and analysis of cardiac precursor cells that were devoid of spontaneous contractions (Kolossov *et al.*, 1998). The non-contracting precursor cells already displayed the presence of the voltage-dependent L-type Ca^{2+} channels at very low densities. Cardiomyocytes at a very early differentiation stage possessed action potentials typical of primary myocardium that were generated by only two ionic currents, the L-type Ca^{2+} channel and the transient outward potassium channel. In contrast, cardiomyocytes derived from late stage EBs exhibited electrophysiological characteristics typical of postnatal cardiomyocytes. In these cells, the entire repertoire of ionic channels was present resulting in a diversification of cardiac phenotypes and the generation of cardiomyocytes with ventricular-like, atrial-like, and Purikinje-like properties.

We have recently used slightly different methodologies than those reported in the mouse model to generate a reproducible spontaneous cardiomyocyte differentiating system from human ES cells (Kehat *et al.*, 2001). Human ES cells were dissociated into small clumps of 3–20 cells and grown in suspension for 7–10 days where they formed EBs. The EBs were then plated on gelatin coated culture dishes, and observed microscopically for the appearance of spontaneous contractions. Rhythmically contracting areas appeared at 4–22 days after plating with the majority appearing between days 7–15 post-plating. The presence of spontaneous contraction within the EBs persisted from several days to several weeks.

Several lines of evidence confirmed the cardiomyocyte phenotype of these contracting areas (Kehat *et al.*, 2001) (*Color Plate 9*). Cells isolated from the beating areas were shown by RT-PCR to express cardiac-specific gene products such as transcription factors and structural proteins. Immunostaining studies demonstrated the presence of the cardiac-specific sarcomeric proteins (myosin heavy chain, α-actinin, desmin, and cardiac troponin I) as well as atrial natriuretic peptide (ANP). Electron microscopy revealed varying degrees of myofibrillar organization, consistent with the typical ultrastructural properties of early-stage cardiomyocytes.

Interestingly, during EB development we could observe a process of ultrastructural maturation. Early-stage cardiomyocytes were overwhelmingly mononucleated and consisted of relatively small and round cells situated in round accumulations within the EBs. In later-stage EBs, the morphology of these cells gradually changed to cells that were larger and more elongated and tended to accumulate in strands. The morphological changes were coupled with a similar ultrastructural maturation process characterized by a progressive increase in the amount and organization of the contractile material within the cells. Hence, the initial irregular myofibrillar distribution observed in early-stage cardiomyocytes gradually changed into parallel myofibril arrays that ultimately aligned into well-defined sarcomeres in late stage EBs.

The functional properties of human ES cell-derived cardiomyocytes also correlated with the functional phenotype of embryo-derived early-stage cardiomyocytes including typical electrical activity, calcium transients, and chronotropic response to adrenergic agents (*Color Plate 9*). More recently, we have demonstrated that this system is not limited to the differentiation of isolated cells having these properties. Using a high-resolution microelectrode array (MEA) mapping technique, we demonstrated the presence of a functional cardiomyocyte syncytium with stable focal activation (pacemaker activity) and synchronous action potential propagation (Kehat *et al.*, 2002) (*Color Plate 9*). Consistent with the development of a cardiac syncytium, morphological analysis revealed the presence of isotropic distribution of gap junctions homogeneously along the cell borders. Gap junctions, identified by immunostaining, were mainly composed of connexin 43 and 45 that were occasionally co-localized to the same junctional complex. This finding suggests a potential role for connexin 45 (usually absent in adult cardiac tissue) in early cardiac development.

11.4 Prospects for myocardial regeneration

The most attractive application of human ES cells, and the one that receives the most attention, is in cell replacement therapy: to replace diseased, missing or degenerative tissue. The adult heart has only limited regenerative capacity and therefore any significant myocardial cell death resulting from ischemic heart disease, viral infection or immunopathological conditions may lead to permanent impairment of myocardial performance, and ultimately, heart failure. Congestive heart failure is a growing epidemic in the Western world that afflicts many millions in the population. It results in significant morbidity and mortality, while placing a significant economic burden on health care systems (Cohn *et al.*, 1997). Despite advances in medical, interventional, and surgical therapeutic measures, the prognosis for patients with chronic heart failure remains poor, with more than half of the patients dying within 5 years of the initial diagnosis.

Cellular cardiomyoplasty is a promising new experimental strategy that offers the creation of new functional tissue for the replacement of lost or failing myocardium (Reinlib and Field, 2000). The rationale behind this approach is based on the assumption that an increase in the number of functional myocytes within the depressed areas may potentially improve the mechanical properties of these compromised regions. Based on this concept a number of cell sources have been suggested for tissue grafting. These include: (1) non-cardiomyocytes such as skeletal myoblasts (Menasche et al., 2001; Taylor et al., 1998), fibroblasts and smooth muscle cells (Yoo et al., 2000), (2) fetal and neonatal cardiomyocytes (Muller-Ehmsen et al., 2002a; Reinecke et al., 1999; Soonpaa et al., 1994), (3) mouse embryonic stem cells (Klug et al., 1996), and (4) bone marrow derived hematopoeitic and mesenchymal stem cells (Orlic et al., 2001; Wang et al., 2000). Recent animal studies have shown that cells derived from all sources may survive, differentiate, and may even improve myocardial performance following cell grafting. More recently cardiac cell transplantation studies using autologous skeletal myoblasts (Menasche et al., 2001) and bone marrow derived hematopoeitic stem cells (Assmus et al., 2002) have already entered the clinical arena.

Although varied cell sources have been used in the aforementioned studies, the inherent structural and functional properties of cardiomyocytes strongly suggest that they may be the ideal donor cell type. In early mouse studies, fetal cardiomyocytes transplanted into healthy hearts were demonstrated to survive and to form cell-to-cell contacts with host myocardium (Soonpaa et al., 1994). Later, studies examined the feasibility of cardiomyocyte cell grafting into infarcted or cryoinjured hearts. In these settings, cardiomyocyte transplantation could be shown to reduce infarct size (Li et al., 1997), prevent ventricular dilatation and pathological remodeling (Etzion et al., 2001), or improve ventricular systolic function (Scorsin et al., 1997). Despite the encouraging results in these animal studies, a major challenge for the clinical application of this strategy is the inability to obtain sufficiently large quantities of suitable human cardiomyocytes.

The development of the human ES cell lines and establishment of a cardiomyocyte differentiation system offers a number of advantages for cell therapy procedures. First, as human ES cells possess unlimited proliferative capacity as well as diverse differentiating capabilities, they are currently the only cell source that can provide, *ex vivo*, large numbers of human cardiac cells for transplantation. Secondly, the ability of human ES cells to differentiate into a plurality of cell lineages may be utilized for transplantation of different cell types such as endothelial progenitor cells for induction of angiogenesis, and even specialized cardiomyocytes subtypes (pacemaking cells, atrial, ventricular, etc.) tailored for specific applications. Thirdly, due to their clonal origin, human ES cell-derived cardiomyocytes could lend themselves to extensive characterization and genetic manipulation to promote desirable characteristics such as resistance to ischemia and apoptosis, improved contractile function, and specific electrophysiological properties. Fourthly, ES cell-derived cells could also serve as a platform or cellular vehicle for different gene therapy strategies aiming to alter the myocardial environment by the local secretion of growth promoting factors, various drugs, or angiogenic factors.

Finally, the ability to generate potentially unlimited numbers of cardiomyocytes *ex vivo* from human ES cells may also bring a unique value to the engineering of tissue substitutes which combine functional cells with 3D polymeric scaffolds to create bioartificial tissue replacements (Leor *et al.*, 2000). The possible advantages of this approach versus direct cell transplantation may lie in the ability to generate significantly thicker myocardial tissue grafts, to control graft shape and size, to provide adequate biomechanical support for the cell graft, to manipulate the cell composition and alignment, and to promote vascularization of the graft.

Although the development of human ES cell technology holds great potential for the field of myocardial regeneration, several hurdles must be overcome before any clinical applications can be expected. Some of the milestones that need to be achieved include: (1) development of strategies for directing differentiation of the human ES cells into the cardiac lineage; (2) derivation of selection protocols to allow generation of pure populations of cardiomyocytes for transplantation; (3) scale-up of the differentiation process to yield clinically relevant numbers of cells for transplantation, and (4) development of methods for circumventing the expected immune rejection of the grafted cells and resolving several technical and conceptual issues regarding their implantation.

11.4.1 Directing cardiomyocyte differentiation

Both undifferentiated ES cells and their more differentiated derivatives express receptors for various growth factors (Schuldiner *et al.*, 2000), and hence supplementation of the culture medium with appropriate growth factors may affect their differentiation pattern. A number of approaches are currently being actively investigated for their effects on promoting cardiomyocyte differentiation. These approaches include testing different culturing conditions and employing different growth factors, over-expressing cardiac-restricted transcription factors, and exploring the effects of various feeder layers and physical perturbations. A recent study demonstrated that the type of human ES line or clone used in the experiment and the culture conditions used during both the stem cell and EB cultivation stages may have a significant effect on the cardiomyocyte yield (Xu *et al.*, 2002).

In contrast to the effect of the muscle-specific *MyoD* family of transcription factors that are able by themselves to promote skeletal myogenesis in a variety of cells, there is no single transcription factor that can activate the entire cardiac gene program and convert a precursor cell into a cardiac phenotype. Indeed, as discussed above, cardiomyocyte differentiation is governed by a complex process of extracellular signaling, by multiple intracellular signal transduction pathways, and by the activation of multiple transcription factors that then act to activate a repertoire of cardiac-specific genes.

There is evidence to suggest, however, that knowledge gained from early cardiac differentiation in several model systems may also be applicable to human ES cells. The cardiogenic inductive role of the primitive endoderm, described above, was also evident in a number of *in vitro* models. Co-culture experiments of undifferentiated P19 EC cells, mouse ES cells, and human ES cells with END-2, an endoderm-like cell line promoted their differentiation into immature cardiomyocytes (Mummery et al., 1991, 2002). Similarly, mouse EBs depleted of primitive endoderm or parietal endoderm did not develop beating cardiomyocytes. Other studies have indicated an important role for extracellular matrix and cytoskeletal proteins in the regulation of ES cell-derived cardiac differentiation (Boheler et al., 2002). For example, the important role of β1 integrin for cardiac differentiation was determined by demonstrating impaired and delayed cardiac differentiation, delayed expression of cardiac-specific genes, and impaired sarcomeric organization in β1 integrin-deficient mouse ES lines (Boheler et al., 2002).

Some evidence also suggests a possible role for a number of soluble factors in promoting cardiomyocyte differentiation (Parker and Schneider, 1991). These include BMP and members of the TGF-β family, retinoic acid, LIF and other factors when provided at the appropriate timing and concentration. The extent to which these different factors, however, may actually promote cardiogenesis in human ES cells is still unknown.

11.4.2 Cardiomyocyte purification

Although the use of different growth conditions may enhance differentiation of human ES cells toward a specific lineage, the degree of purity that is likely to be achieved would still be insufficient for clinical purposes. Hence, one of the major challenges in using human ES cells for cell replacement strategies is to devise methods in which homogeneous cell populations can be selected from the heterogeneous cell mixture within the EB.

An elegant and relatively simple strategy for selecting cardiomyocytes was demonstrated in murine ES cells by Field's group (Klug et al., 1996). This strategy is based on using a tissue-specific promoter to drive a selectable marker such as an antibiotic resistance gene. In this approach two transcriptional units are incorporated in one vector. The first unit includes a promoter that is active in undifferentiated ES cells that controls expression of a selectable marker (such as an antibiotic resistant gene) and is used to identify cells in which the vector has stably integrated in undifferentiated ES cells. The second transcriptional unit includes a cardiac-specific promoter (for example, the alpha-myosin heavy chain promoter) that controls expression of a second selectable marker, such as a different antibiotic resistance gene.

Once a clone that stably expresses the vector is isolated, undifferentiated genetically modified ES cells could be propagated and expanded. The ES cells are then allowed to differentiate *in vitro* and are subjected to selection with the appropriate antibiotic. Using this approach, >99% pure cardiomyocyte cultures could be generated in a murine model (Klug et al., 1996). The selected cardiomyocytes were further demonstrated to form stable grafts following transplantation into adult dystrophic mice hearts.

An alternative approach could involve the introduction of a gene construct that allows expression of a fluorophore or other reporter gene under the control of a tissue-specific promoter/enhancer. This allows identification and subsequent sorting (for example by fluorescence-activated cell sorting, FACS) of the cells expressing the tissue-specific promoter. Using this approach, Muller et al. transfected murine ES cells with a construct comprising the ventricular-restricted MLC-2V promoter controlling the expression of EGFP (Muller et al., 2000). The MLC-2V constructs were expressed exclusively in the ventricular myocytes making it possible to identify, sort, and study a relatively pure population of ventricular cardiomyocytes.

11.4.3 Scale-up

It is estimated that a typical myocardial infarction that induces heart failure results in the death of up to 1 billion cardiomyocytes. Furthermore, evidence from animal cell transplantation studies suggests that the vast majority of the transplanted cells do not survive following cell grafting. Consequently, a major barrier for the possible use of human ES cells in cell transplantation strategies is the generation of sufficient numbers of cardiomyocytes. Large numbers of cells to meet this goal could be theoretically achieved by increasing the initial number of ES cells used for differentiation, by increasing the percentage of cells differentiating to the cardiac lineage, by increasing the ability of the cells to proliferate following cardiomyocyte differentiation, and/or by scaling up the entire process using bioreactors and related technologies.

Until recently, the undifferentiated propagation of human ES cells required the presence of a mouse MEF feeder layer. Aside from the obvious disadvantage of contamination with non-human tissues, this culture technique is not amendable to scale up. In a recent study, human ES cells were propagated on Matrigel in the presence of media conditioned by MEF feeder layers (Xu et al., 2001). This technique appears to yield equivalent results to cells cultured directly on feeder layers, suggesting a soluble factor produced by the MEFs supports undifferentiated growth. Alternatively, human feeder layers are also capable of supporting the undifferentiated propagation of human ES cells (Richards et al., 2002). A better understanding of the molecular mechanisms of self-renewal will lead to more efficient control of this process, which will facilitate development and expansion of a manufacturing process.

11.4.4 In vivo *transplantation and development of anti-rejection strategies*

Although changes in the remodeling process following myocardial infarction were demonstrated following the transplantation of myocardial cells in animal models, optimal systolic augmentation would require the functional integration between grafted and host cardiomyocytes. Thus, any expected functional improvement would require the long-term survival of the grafted cells, the presence of a critical tissue mass, and the structural and functional integration of host and donor tissue.

Several questions and issues remain to be addressed in this area (Reinlib and Field, 2000). First, the size of the cell graft may have important implications for its ultimate success in improving the ventricular mechanical function. Cell death occurring after engraftment is believed to have a major negative impact on graft size (Muller-Ehmsen et al., 2002b; Zhang et al., 2001). Cell survival in the ischemic myocardial area may depend on the adequate vascularization of the graft (through additional revascularization procedures or induction of angiogenesis) and by the properties of the grafted cells themselves such as their proliferative capacity or resistance to ischemia and apoptosis.

Additional factors that remain to be determined in future studies include the ideal nature of the graft (individual cells, small cell clumps, or combined with scaffolding biomaterials); the degree of maturity of the transplanted ES cell-derived cardiomyocytes and whether the cells will eventually develop into the adult phenotype; the appropriate delivery method (epicardial, endocardial, or via the coronary circulation); and the timing of cell delivery relative to the timing of the infarct.

An important aspect related to the possible utilization of ES cell-derived cardiomyocytes for future cell transplantation strategies relates to the safety of these procedures. A major safety concern is the possible development of ES cell-derived tumors such as teratomas. This potential problem could theoretically be prevented by assuring the absence of remaining pluripotent cells in the cardiomyocyte graft. A second major concern relates to the development of cardiac arrhythmias following cell transplantation. It is well known that patients with reduced ventricular function or scar tissue are at increased risk for developing malignant ventricular arrhythmias. Cell transplantation may modify the electrophysiological properties of the scar and could potentially increase the propensity for development of life-threatening arrhythmias.

Another barrier, which has to be overcome, is the prevention of immune rejection of the human ES cell-derived cardiomyocytic graft. This important issue, discussed in detail elsewhere in this book and in a number of review papers, is not limited to the cardiac lineage, but rather impacts all strategies aiming to use human ES cell derivatives to replace dysfunctional tissues. An anti-rejection treatment will probably be needed for this type of allogeneic cell transplantation. However, the immunosupressive regimens required may be mild because of the expected small pool size of alloreactive T cells to pure stem cell-derived cardiomyocytes that are believed to express MHC class I, but not class II, at the time of transplantation (Drukker *et al.*, 2002). Strategies aimed at reducing the number of alloreactive T cells are being developed and these and other novel therapies with particular relevance to the anticipated immune response mounted against ES cell-derived cell transplants will probably be employed. These strategies may include establishing 'banks' of major histocompatibility complex antigen-typed human ES cells, genetically altering ES cells to suppress the immune response, induction of tolerance to the graft, and possibly also by using somatic nuclear transfer techniques (Bradley *et al.*, 2002).

11.5 Summary

The development of human ES lines and their ability to differentiate to cardiomyocyte tissue holds great promise for several research and clinical areas in the cardiovascular field including developmental biology, functional genomics, drug discovery and testing, cell therapy and tissue engineering. Nevertheless, several key questions remain and much experimental work will be necessary in pre-clinical animal models. In addition, several methodologic aspects need to be resolved, and several milestones have to be achieved in order to fully harness the enormous research and clinical potential of this unique technology.

References

Amit M, Carpenter MK, Inokuma MS, Chiu CP, Harris CP, Waknitz MA, Itskovitz-Eldor J, Thomson JA (2000) Clonally derived human embryonic stem cell lines maintain pluripotency and proliferative potential for prolonged periods of culture. *Dev. Biol.* **227**, 271–278.

Anversa P, Nadal-Ginard B (2002) Myocyte renewal and ventricular remodelling. *Nature* **415**, 240–243.

Assmus B, Schachinger V, Teupe C, Britten M, Lehmann R, Dobert N, Grunwald F, Aicher A, Urbich C, Martin H. *et al* (2002) Transplantation of Progenitor Cells and Regeneration Enhancement in Acute Myocardial Infarction (TOPCARE-AMI). *Circulation* **106**, 3009–3017.

Boheler KR, Czyz J, Tweedie D, Yang HT, Anisimov SV, Wobus AM (2002) Differentiation of pluripotent embryonic stem cells into cardiomyocytes. *Circ. Res.* **91**, 189–201.

Bradley JA, Bolton EM, Pedersen RA (2002) Stem cell medicine encounters the immune system. *Nat. Rev. Immunol.* **2**, 859–871.

Cohn JN, Bristow MR, Chien KR, Colucci WS, Frazier OH, Leinwand LA, Lorell BH, Moss AJ, Sonnenblick EH, Walsh RA *et al.* (1997) Report of the National Heart, Lung, and Blood Institute Special Emphasis Panel on Heart Failure Research. *Circulation* **95**, 766–770.

Cripps RM, Olson EN (2002) Control of cardiac development by an evolutionarily conserved transcriptional network. *Dev. Biol.* **246**, 14–28.

Doetschman TC, Eistetter H, Katz M, Schmidt W, Kemler R (1985) The in vitro development of blastocyst-derived embryonic stem cell lines: formation of visceral yolk sac, blood islands and myocardium. *J. Embryol. Exp. Morphol.* **87**, 27–45.

Drukker M, Katz G, Urbach A, Schuldiner M, Markel G, Itskovitz-Eldor J, Reubinoff B, Mandelboim O, Benvenisty N (2002) Characterization of the expression of MHC proteins in human embryonic stem cells. *Proc. Natl Acad. Sci. USA* **99**, 9864–9869.

Etzion S, Battler A, Barbash IM, Cagnano E, Zarin P, Granot Y, Kedes LH, Kloner RA, Leor J (2001) Influence of embryonic cardiomyocyte transplantation on the progression of heart failure in a rat model of extensive myocardial infarction. *J. Mol. Cell Cardiol.* **33**, 1321–1330.

Evans MJ, Kaufman MH (1981) Establishment in culture of pluripotential cells from mouse embryos. *Nature* **292**, 154–156.

Harvey RP (2002). Patterning the vertebrate heart. *Nat. Rev. Genet.* **3**, 544–556.

Hescheler J, Fleischmann BK, Lentini S, Maltsev VA, Rohwedel J, Wobus AM, Addicks K (1997) Embryonic stem cells: a model to study structural and functional properties in cardiomyogenesis. *Cardiovasc. Res.* **36**, 149–162.

Itskovitz-Eldor J, Schuldiner M, Karsenti D, Eden A, Yanuka O, Amit M, Soreq H, Benvenisty N (2000) Differentiation of human embryonic stem cells into embryoid bodies compromising the three embryonic germ layers. *Mol. Med.* 6, 88–95.

Kehat I, Kenyagin-Karsenti D, Snir M, Segev H, Amit M, Gepstein A, Livne E, Binah O, Itskovitz-Eldor J, Gepstein L (2001) Human embryonic stem cells can differentiate into myocytes with structural and functional properties of cardiomyocytes. *J. Clin. Invest.* 108, 407–414.

Kehat I, Gepstein A, Spira A, Itskovitz-Eldor J, Gepstein L (2002) High-resolution electrophysiological assessment of human embryonic stem cell-derived cardiomyocytes: a novel in-vitro model for the study of conduction. *Circ Res.* 91, 659–661.

Klug MG, Soonpaa MH, Koh GY, Field LJ (1996) Genetically selected cardiomyocytes from differentiating embronic stem cells form stable intracardiac grafts. *J. Clin. Invest.* 98, 216–224.

Kolossov E, Fleischmann BK, Liu Q, Bloch W, Viatchenko-Karpinski S, Manzke O, Ji GJ, Bohlen H, Addicks K, Hescheler J (1998) Functional characteristics of ES cell-derived cardiac precursor cells identified by tissue-specific expression of the green fluorescent protein. *J. Cell Biol.* 143, 2045–2056.

Krause DS, Theise ND, Collector MI, Henegariu O, Hwang S, Gardner R, Neutzel S, Sharkis SJ (2001) Multi-organ, multi-lineage engraftment by a single bone marrow-derived stem cell. *Cell* 105, 369–377.

Leor J, Aboulafia-Etzion S, Dar A, Shapiro L, Barbash IM, Battler A, Granot Y, Cohen S (2000) Bioengineered cardiac grafts: A new approach to repair the infarcted myocardium? *Circulation* 102, III56-61.

Li RK, Mickle DA, Weisel RD, Mohabeer MK, Zhang J, Rao V, Li G, Merante F, Jia ZQ (1997) Natural history of fetal rat cardiomyocytes transplanted into adult rat myocardial scar tissue. *Circulation* 96, II-179-86; discussion 186–187.

Martin G (1981) Isolation of a pluripotent cell line from early mouse embryos cultured in medium conditioned by teratocarcinoma stem cells. *Proc. Natl Acad. Sci. USA* 78, 7635.

Menasche P, Hagege AA, Scorsin M, Pouzet B, Desnos M, Duboc D, Schwartz K, Vilquin J T, Marolleau JP (2001) Myoblast transplantation for heart failure. *Lancet* 357, 279–280.

Monzen K, Nagai R, Komuro I (2002) A role for bone morphogenetic protein signaling in cardiomyocyte differentiation. *Trends Cardiovasc. Med.* 12, 263–269.

Muller M, Fleischmann BK, Selbert S, Ji GJ, Endl E, Middeler G, Muller OJ, Schlenke P, Frese S, Wobus AM *et al.* (2000) Selection of ventricular-like cardiomyocytes from ES cells in vitro. *Faseb J.* 14, 2540–2548.

Muller-Ehmsen J, Peterson KL, Kedes L, Whittaker P, Dow JS, Long TI, Laird PW, Kloner RA (2002a) Rebuilding a damaged heart: long-term survival of transplanted neonatal rat cardiomyocytes after myocardial infarction and effect on cardiac function. *Circulation* 105, 1720–1726.

Muller-Ehmsen J, Whittaker P, Kloner RA, Dow JS, Sakoda T, Long TI, Laird PW, Kedes L (2002b) Survival and development of neonatal rat cardiomyocytes transplanted into adult myocardium. *J. Mol. Cell Cardiol.* 34, 107–116.

Mummery C, Ward D, van den Brink CE, Bird SD, Doevendans PA, Opthof T, Brutel de la Riviere A, Tertoolen L, van der Heyden M, Pera M (2002) Cardiomyocyte differentiation of mouse and human embryonic stem cells. *J. Anat.* 200, 233–242.

Mummery CL, van Achterberg TA, van den Eijnden-van Raaij A. J, van Haaster L, Willemse A, de Laat SW, Piersma AH (1991) Visceral-endoderm-like cell lines induce differentiation of murine P19 embryonal carcinoma cells. *Differentiation* 46, 51–60.

Nascone N, Mercola M (1995) An inductive role for the endoderm in Xenopus cardiogenesis. *Development* 121, 515–523.

Olson EN (2001) Development. The path to the heart and the road not taken. *Science* 291, 2327–2328.

Orlic D, Kajstura J, Chimenti S, Jakoniuk I, Anderson SM, Li B et al. (2001) Bone marrow cells regenerate infarcted myocardium. *Nature* 410, 701–705.

Parker TG, Schneider MD (1991) Growth factors, proto-oncogenes, and plasticity of the cardiac phenotype. *Annu. Rev. Physiol.* 53, 179–200.

Reinecke H, Zhang M, Bartosek T, Murry CE (1999) Survival, integration, and differentiation of cardiomyocyte grafts: a study in normal and injured rat hearts. *Circulation* 100, 193–202.

Reinlib L, Field L (2000) Cell transplantation as future therapy for cardiovascular disease?: A workshop of the National Heart, Lung, and Blood Institute. *Circulation* 101, E182-7.

Reubinoff BE, Pera MF, Fong CY, Trounson A, Bongso A (2000) Embryonic stem cell lines from human blastocysts: somatic differentiation in vitro. *Nat. Biotechnol.* 18, 399–404.

Richards M, Fong CY, Chan WK, Wong PC, Bongso A (2002) Human feeders support prolonged undifferentiated growth of human inner cell masses and embryonic stem cells. *Nat. Biotechnol.* 20, 933–936.

Schuldiner M, Yanuka O, Itskovitz-Eldor J, Melton DA, Benvenisty N (2000) From the cover: effects of eight growth factors on the differentiation of cells derived from human embryonic stem cells. *Proc. Natl Acad. Sci. USA* 97, 11307–11312.

Scorsin M, Hagege AA, Marotte F, Mirochnik N, Copin H, Barnoux M, Sabri A, Samuel JL, Rappaport L, Menasche P (1997) Does transplantation of cardiomyocytes improve function of infarcted myocardium? *Circulation* 96, II-188-93.

Sepulveda JL, Vlahapoulous S, Iyer D, Belaguli N, Schwartz RJ (2002) Combinatorial expression of GATA4, Nkx2.5, and serum response factor directs early cardiac gene activity. *J. Biol. Chem.* 277, 25775–25782.

Soonpaa MH, Koh GY, Klug MG, Field LJ (1994) Formation of nascent intercalated disks between grafted fetal cardiomyocytes and host myocardium. *Science* 264, 98–101.

Taylor DA, Atkins BZ, Hungspreugs P, Jones TR, Reedy MC, Hutcheson KA, Glower DD, Kraus WE (1998) Regenerating functional myocardium: improved performance after skeletal myoblast transplantation. *Nat. Med.* 4, 929–933.

Thomson JA, Itskovitz-Eldor J, Shapiro SS, Waknitz MA, Swiergiel JJ, Marshall VS, Jones JM (1998) Embryonic stem cell lines derived from human blastocysts. *Science* **282**, 1145–1147.

Thomson JA, Kalishman J, Golos TG, Durning M, Harris CP, Becker RA, Hearn JP (1995) Isolation of a primate embryonic stem cell line. *Proc. Natl Acad. Sci. USA* **92**, 7844–7848.

Wang D, Chang PS, Wang Z, Sutherland L, Richardson JA, Small E, Krieg PA, Olson EN (2001) Activation of cardiac gene expression by myocardin, a transcriptional cofactor for serum response factor. *Cell* **105**, 851–862.

Wang JS, Shum-Tim D, Galipeau J, Chedrawy E, Eliopoulos N, Chiu RC (2000) Marrow stromal cells for cellular cardiomyoplasty: feasibility and potential clinical advantages. *J. Thorac. Cardiovasc. Surg.* **120**, 999–1005.

Xu C, Inokuma MS, Denham J, Golds K, Kundu P, Gold JD, Carpenter MK (2001) Feeder-free growth of undifferentiated human embryonic stem cells. *Nat. Biotechnol.* **19**, 971–974.

Xu C, Police S, Rao N, Carpenter MK (2002) Characterization and enrichment of cardiomyocytes derived from human embryonic stem cells. *Circ. Res.* **91**, 501–508.

Yoo KJ, Li RK, Weisel RD, Mickle DA, Li G, Yau TM (2000) Autologous smooth muscle cell transplantation improved heart function in dilated cardiomyopathy. *Ann. Thorac. Surg.* **70**, 859–865.

Zaffran S, Frasch M (2002). Early signals in cardiac development. *Circ. Res.* **91**, 457–469.

Zhang M, Methot D, Poppa V, Fujio Y, Walsh K, Murry CE (2001) Cardiomyocyte grafting for cardiac repair: graft cell death and anti-death strategies. *J. Mol. Cell Cardiol.* **33**, 907–921.

12. Genetic engineering of human embryonic stem cells

Micha Drukker, Sujoy Kumar Dhara and Nissim Benvenisty

12.1 Introduction

Human embryonic stem (ES) cells were isolated from the inner cell mass (ICM) of blastocyst stage embryos (Thomson *et al.*, 1998; Reubinoff *et al.*, 2000). Under specific conditions, *in vivo* and *in vitro*, these cells are capable of differentiating into cell types of the three embryonic germ layers, namely ectoderm, mesoderm and endoderm. *In vivo*, following injection into severe combined immunodeficient (SCID) mice the cells develop into teratomas comprised of derivatives from many cell types (Thomson *et al.*, 1998; Reubinoff *et al.*, 2000). *In vitro*, when placed in non-adherent culture dishes, the cells form spherical structures termed embryoid bodies (EBs) and undergo spontaneous differentiation (Itskovitz-Eldor *et al.*, 2000). Addition of growth factors during differentiation was shown to direct the commitment of the cells towards specific lineages (Schuldiner *et al.*, 2000). Furthermore, protocols describing *in vitro* enrichment of specific cell types such as neurons (Carpenter *et al.*, 2001; Reubinoff *et al.*, 2001; Schuldiner *et al.*, 2001; Zhang *et al.*, 2001), pancreatic β cells (Assady *et al.*, 2001), cardiomyocytes (Kehat *et al.*, 2001; Kehat *et al.*, 2002; Mummery *et al.*, 2002; Xu *et al.*, 2002a), trophoblasts (Xu *et al.*, 2002b), endothelial (Levenberg *et al.*, 2002) and hematopoietic cells (Kaufman *et al.*, 2001) were published. However, these cultures still contain additional cell types and thus generating pure populations of specific differentiated cell types remains a challenge. Since human ES cells propagate indefinitely in culture without loosing pluripotency and their differentiation may be induced and directed rapidly, they were suggested to serve as an unlimited source of clinically transplantable cells. Yet, in order to achieve this ambitious task, improved differentiation protocols and purification procedures of particular human ES cell derivatives should be developed.

The application of genetic engineering techniques to mouse ES cells has enabled the creation of mice mutated at specific loci in the genome. Thus, these techniques have greatly deepened our knowledge of genes involved in embryology and pathology. Genetic engineering of human ES cells has also become possible with the development of transfection (Eiges *et al.*, 2001; Zwaka and Thomson, 2003) and infection (Pfeifer *et al.*, 2002; Gropp *et al.*, 2003; Ma *et al.*, 2003) techniques. These developments will improve our understanding of the biological properties underlying pluripotency and differentiation of human ES cells, in addition to the ability to create better reagents for cellular transplantation. This chapter presents the recent developments in the context of experimental designs that may now be performed. Moreover, the clinical benefits derived from these methods are described.

12.2 Methods for introduction of DNA into human ES cells

Direct injection of DNA into the pronuclei of *in vitro* fertilized mouse oocytes is a highly efficient method for producing transgenic mice, but the site of DNA integration is random and cannot be preselected. In contrast, propagation of mouse ES cells in culture provides the means to manipulate genetically the cell's genome and to screen for the desired clones prior to the production of transgenic animals. Therefore, many techniques have been developed over the years in order to introduce genetic changes into mouse ES cells. These techniques include DNA introduction by means of transfection, chemical reagents, electroporation and infection by viral vectors. Infection by retroviral-based vectors is the most efficient method but the DNA integrates randomly and the expression cassette tends to be silenced. In contrast, DNA can be targeted to a specific locus using electroporation or transfection techniques and the transgene usually remains active. However, efficiency is usually sacrificed, as the efficiency of infection is generally much higher. The recent isolation of human ES cells provides an extraordinary opportunity to apply the genetic engineering techniques of mouse ES cells to manipulate the genome of human ES cells. This section will focus on the manipulation techniques used to genetically engineer human ES cells.

12.2.1 Transfection

Transfection is the experimental process by which foreign DNA is introduced into cultured cells by physical or biochemical methods. Although various chemical reagents with diverse mechanisms of action have been developed for this purpose, the method of choice always depends on the type of cell line to be transfected. The most popular transfection method for mouse ES cells is electroporation, a process by which application of short electric impulses creates transient small pores in the bilayer membrane, allowing introduction of exogenous DNA into the cell. Although efficient with mouse ES cells (Thomas and Capecchi, 1987), human ES cells did not survive the voltage shock when electroporation conditions similar to those applied for mouse cells were tested (Eiges *et al.*, 2001; Zwaka and Thomson, 2003). However, when the electroporation parameters were modified in accordance to the cell's size and the electroporation was carried out in protein-rich solution, the efficiency of transfection increased by 100-fold (Zwaka and Thomson, 2003).

Other transfection methods that have been put into use in human ES cells include cationic lipid reagents, multi-component lipid-based reagents and linear polyethylenimine (PEI) (Eiges et al., 2001). Cationic lipid reagents form small unilamellar liposomes harboring a positive charge at their periphery. Thus, when mixed with DNA the liposomes attract electrostatically both the negatively charged phosphate backbone of the DNA and the negatively charged surface of cells. In contrast to the reported high efficiency with many other cell types, LipofectAMINE PLUS (Life Technologies), a cationic lipid reagent, failed to give good transfection efficiency in human ES cells (Eiges et al., 2001). Similarly, FuGENE 6 (Roche Applied Science), a multi-component lipid-based transfection reagent that forms complexes with DNA and transports it into the cell, gave relatively poor results. In contrast, the non-liposomal ExGen 500 reagent (Fermentas), which consists of linear PEI molecules, yielded a high rate of transfection when applied to human ES cells (Eiges et al., 2001). This reagent interacts with DNA molecules forming small, stable and diffusible particles that settle on the cell surface once gravity force is applied and enter the cells by endocytosis. The unique property of this reagent is due to its apparent ability to act as a 'proton sponge' thereby buffering endosomal acid, leading to endosome rupture and DNA release. In approximately 5% to 20% of cells so treated, the exogenous DNA was able to reach the nucleus and be transiently expressed (N.B., unpublished). However, in only approximately 10^{-5}–10^{-6} of the transfected cells, the transgene expression was found to be stable during expansion of undifferentiated cells and during differentiation *in vitro* into EBs (Eiges et al., 2001). Thus, it seems that ExGen 500 is a suitable reagent to allow transfection of human ES cells. For a detailed protocol of ExGen 500 transfection of undifferentiated human ES cells, see *Table 12.1*.

12.2.2 Infection

Historically, the first report on genetic manipulation of mouse ES cells showed that retrovirally-derived vectors can infect the cells and the integrated virus (provirus)

Table 12.1: Transfection protocol of human ES cells using ExGen 500

1. Transfer approximately 1.5×10^5 mitomycin-C treated mouse embryonic fibroblasts (MEFs) to each well of 6 well culture plate (Falcon), pre-coated with Gelatin. Incubate over night.
2. Transfer approximately 1.5×10^5 human ES cells to each well and incubate over night.
3. Change culture medium every day for two to three days. The cells should form small colonies.
4. For each well to be transfected, dilute 5 µg of DNA into 300 µl of sterile 150 mM NaCl solution. Vortex gently and spin down briefly.
5. Add 15 µl ExGen 500 (Fermentas) to the DNA solution (not the reverse order) and vortex the solution immediately for 10 s, and then spin down briefly. Incubate at room temperature for 10 min. Note that further optimization of the ExGen 500/DNA ratio may increase transfection efficiency.
6. Replace the culture medium of the human ES cells by 3 ml of fresh medium.
7. Uniformly disperse 300 µl of DNA/ExGen 500 solution in each well. Gently swirl the plate to achieve even distribution of DNA complexes.
8. Centrifuge culture plate, for 5 min at 280 g. Transfer to 37°C incubator for 30 min.
9. Remove culture medium and wash 3 times with PBS. Then add 2 ml of fresh medium.
10. Return plate to incubator. Transient transfection will reach its peak expression after approximately 48 h.

was transmitted through the germ line (Robertson et al., 1986). Despite the high efficiency of infection, simple murine retroviruses have several limitations. Among these are: (10) occasional lack of significant proviral transcription due to *de novo* methylation of the integrated viral sequences [see references in Cherry et al. (2000)]; (2) effects of viral long terminal repeats (LTRs) on the transgene expression, consequently these vectors may not be suitable for delivering transgenes regulated by tissue specific promoters; (3) uncontrolled location and copy number of viral integration events into the genome; and (4) limited size of the transgene to approximately 6 kb. Hence, these drawbacks restrict broader use of retrovirus-based vectors for genetic manipulation of mouse ES cells.

In comparison to simple retroviral vectors, vectors derived from the more complex family of retroviruses, the lentiviruses, seem to have a few advantages. As demonstrated by Pfeifer et al. (2002) these vectors can be used for delivering transgenes to mouse ES cells. Moreover, the transgene was actively expressed in undifferentiated cells as well as in *in vitro* and *in vivo* differentiated cells without apparent silencing effects. When the transduced ES cell lines were injected into mouse blastocysts, viable chimeras were produced and transgene expression was demonstrated in many tissues.

Infection of lentiviral vectors into human ES cells was also successful at very high efficiencies, and expression persisted during proliferation and expansion of the undifferentiated cells (Pfeifer et al., 2002; Gropp et al., 2003; Ma et al., 2003). Moreover, expression of the transgene was maintained during differentiation into hematopoietic precursors (Ma et al., 2003). In order to elaborate further on the factors that may affect infection efficiency by lentiviral vectors, we have performed infection experiments assessing the influence of (1) presence of serum in the culture medium; (2) pre-coating of the culture plates with mouse embryonic fibroblasts (MEFs); and (3) application of centrifugation force. Our results indicate that the presence of serum does not change the efficiency of infection. In contrast, pre-coating of the culture plates with MEFs resulted in higher proportion of infected MEFs than human ES cells. Favorably, centrifugation resulted in better infection rates than controls, suggesting that centrifugation precipitates the viral particles onto the cells, thereby increasing the efficiency of infection. For a detailed protocol of lentiviral infection of undifferentiated human ES cells see *Table 12.2*. It is evident that the use of lentiviral vectors to infect human ES cells has great

Table 12.2: Infection protocol of human ES cells using lentiviral particles

1. Transfer approximately 3×10^5 trypsinized human ES cells in 0.7 ml culture medium to each well of 6 well culture dish (Falcon), pre-coated by Gelatin.
2. Add to it: 300 µl lentiviral particles (10^7 infectious units/ml), 10 µl (0.4 mg/ml) protamine sulfate (final concentration 4 µg/ml) and 30 µl of dNTPs mix, 10 mM each (final concentration 300 µM).
3. Centrifuge culture plate, for 20 min at 1000 *g*. Incubate over night at 37°C.
4. Transfer each well to a new well, pre-coated with MEFs.

Note: Protamine sulfate improves the contact of the viral particles to the cells, and dNTPs improve reverse transcription activity.

potential to serve as an excellent delivery system for transgenesis. However, as these vectors express their transgene constitutively and integrate into the genome randomly, they may be better suited for those experiments in which random integration and high levels of transgene expression are desired.

12.2.3 Transient versus stable integration

A crucial factor in determining the method by which genetic manipulation should be performed is the duration of the experiment and whether transient or stable expression is desired. If only a short expression period is required then achieving transient transfection by one of the aforementioned methods should be adequate. When human ES cells are transfected transiently, the plasmid is generally introduced into the cells in a super-coiled form, since transcription is more efficient in this state than from a linearized plasmid (Weintraub et al., 1986). The expression level usually reaches its peak within 48 h and is expected to be high because multiple copies of DNA are often introduced into each cell. An example of transient transfection of human ES cell colonies, by an enhanced green fluorescent protein (eGFP) expression vector, is presented in *Color Plate 10a*. A protocol describing ExGen 500 mediated transfer of genes to human ES cells is presented in *Table 12.1*.

Stable integration of exogenous or foreign DNA into human ES cells can be carried out either by infection with retroviral vectors or by transfection with plasmids. As noted above (see section 12.2.2), the process of retroviral infection is highly efficient (*Color Plate 10b*) but suffers from a major drawback: the inability to be directed to a specific locus in the genome. In plasmid transfection, the efficiency of stable integration in human ES cells is low, but may allow isolation of targeted clones (Zwaka and Thomson, 2003). Linearization of the plasmid prior to transfection is highly recommended since otherwise the plasmid may break within the transgene while leaving the selection cassette intact, thus leading to the potential formation of resistant clones that do not express a functional transgene. If successful integration occurs leaving undamaged selection and transgene cassettes, stably transfected clones can be isolated as demonstrated for human ES cells in *Color Plate 10c*. Note the uniformity and high expression level of eGFP in the cells.

As integration events of non-viral vectors occur at a very low frequency, selection markers should be used for the isolation of stably expressing clones. Following selection, subculturing positive clones enables screening of the different clones for transgene expression level and integration site. The transfer of individual clones is usually performed by micropipette, with or without pre-incubation of the cells in the presence of proteolytic enzymes such as trypsin or dispase. In order to retain their pluripotency, human ES cells are propagated in direct contact with MEFs or human embryonic fibroblasts (HEFs) (Richards et al., 2002). If selection is applied following transfection one must ensure the feeder cell line is resistant to the selection reagent. Alternatively, human ES cells may be propagated and selected directly in culture medium conditioned previously by MEFs (Xu et al., 2001). A protocol for the establishment of stable transfected clones is presented in *Table 12.3*.

Table 12.3: Establishment of stably transfected human ES cell clones

1. Perform all steps in *Table 12.1*. Apply selection reagent 24 h after transfection.
2. Replace medium containing selection reagent every day for 5–7 days. Massive cell death should be observed in the first 4 days.
3. Approximately one week following transfection, resistant clones of about few hundred cells should be easily noticed using a microscope (magnification ×40).
4. While monitoring the process through microscope (magnification ×40), transfer each independent clone, using a micropipette to one well of 24-well culture plate pre-coated with MEFs and filled with 1ml selection medium. Use separate micropipette for each clone.
5. Change growth medium (with selection reagent) after 24 h. Successfully transferred clones should be attached to the MEFs.
6. Allow the cells to propagate for additional 3 to 4 days. Monitor the development of the clones on a daily basis. Mark the thriving clones.
7. Transfer the clones to 12-well culture plates: wash with 1 ml Phosphate Buffer Saline (PBS); apply 200 µl trypsin EDTA; wait 5 min while gently swirling the plate every minute; add 2 ml fresh medium containing serum and gently collect the cells into 10 ml tube.
8. Centrifuge the cells at 300 *g* for 5 min. Aspirate medium and re-suspend in growth medium containing selection reagent. Plate the cells onto one well of a 12-well culture plate in 1 ml volume.
9. Further expansion should be performed as in step 8. In future passages, the culture should be divided by a factor of 3 up to 4.
10. Make sure to freeze independent clones once the clones have grown to confluent cultures in 2 wells of 6 well culture plate. At this stage selection is no longer required.

Note: The MEFs used during selection and propagation of the clones must be resistant to the selection reagent. If you cannot obtain such, it is possible to propagate the clones in the presence of conditioned medium (human ES cell medium incubated in the presence of MEFs for 24 h), supplemented with selection reagent.

12.3 Alteration of gene expression in human ES cells

One of the most important advances in mammalian genetic research over the past two decades is the production of genetically engineered mice through genetic manipulation of mouse ES cells. The production of genetically altered mice has become routine and standardized at many scientific institutions. This is because mouse ES cells are readily propagated *in vitro*, techniques for their genetic manipulation are relatively simple, and manipulated ES cells can contribute to germ cells thereby allowing the genetic alteration to be continued in numerous offspring. Moreover, gene function and certain developmental processes of the early embryo may also be studied in culture by analyzing the differentiation of ES cells into EBs. The recent isolation of human ES cells is an exciting opportunity to exploit the knowledge collected thus far on genetic manipulation of mouse ES cells, in order to manipulate genetically the human cells for studying developmental processes in cell culture systems.

There are two main categories of genetic manipulation experiments that may be performed for this purpose: (1) over-expression of genes, and (2) silencing the expression of an endogenous gene for functional studies. The over-expression of selected genes of interest can be either constitutive or inducible. These approaches can be applied with the goal of either introducing cellular genes in order to induce differentiation into a certain lineage or tracking cells as they differentiate towards specific cell lineages. This section will outline the experiments performed thus far in human ES cells using these strategies. As these reports are still rare, experimental approaches formally described in mouse ES cells but which could be adapted to human ES cells will also be presented.

12.3.1 Over-expression of genes in human ES cells

The primary therapeutic objective of human ES cell research is to induce their differentiation into clinically beneficial cell types. Although these cells may differentiate *in vitro* into cell types from the three embryonic germ layers (Itskovitz-Eldor *et al.*, 2000), it seems that some cell types are difficult to obtain and other types form only at low frequencies. The knowledge obtained using mouse ES cells suggests that these obstacles may be overcome, at least to some extent, by using either of two general strategies that take advantage of the ease by which ES cells can be genetically manipulated. These include lineage-restricted expression of marker genes for the identification of specific cell types and over-expression of cellular gene(s) that play a major role during differentiation.

Expression of marker genes in ES cells

One of the primary requirements in transplantation of derivatives of human ES cells is to achieve pure populations of the differentiated cells prior to their transfer into a patient. To achieve this goal cells can simply be selected from a heterogeneous population by means of an antibody that recognizes a cell surface protein specific for the desired cells. Unfortunately, such markers are unknown for many cell types. Yet, many intracellular genes are expressed in certain developmental stages and thus can serve as markers. In order to use these markers, the regulatory sequence of a lineage-restricted marker gene is fused to a reporter gene and the transgene is introduced into the cells by transfection (see section 12.2.1). Then, during differentiation the reporter gene is expressed only in a specific cell type. Using certain reporter genes such as eGFP, which can be viewed in living cells, the marked cells may be sorted out using non-damaging means such as fluorescence activated cell sorting (FACS). This principle may also serve to exclude undifferentiated human ES cells from heterogeneous populations of differentiated and undifferentiated cells as previously demonstrated using a transgene of eGFP expressed under the transcriptional control of mouse Rex-1 promoter sequence (Eiges *et al.*, 2001). As Rex-1 is specifically expressed in undifferentiated cells but not in other cells, eGFP expression was visible only in the undifferentiated cells within colonies of human ES cells while differentiated cells, at the periphery of the of the colonies, did not express eGFP. Interestingly after differentiation of these Rex1 promoter-eGFP ES cells, eGFP fluorescence was noted in some cells within EBs, demonstrating that differentiated cultures may still contain undifferentiated cells. This point is of great importance since injection of undifferentiated ES cells into immunocompromised mice leads to the formation of teratomas, which may be a potential problem for patients receiving ES cell based therapies. In order to separate the undifferentiated cells from the differentiated ones, the cells were sorted according to eGFP expression by FACS (Eiges *et al.*, 2001).

Other than expression of visible markers it is also possible to express dominant selection markers under the transcriptional control of tissue specific promoters. This approach was used by Li et al. (1998) to isolate neuronal cells from a mixed population of differentiating mouse ES cells. In this work, one allele of the *sox2* gene was targeted with the βgeo construct (a bicistronic construct expressing the *neo* gene fused to the *lacZ* gene). As *sox2* is expressed only in neuronal cells the genetically altered clones express the *neo* gene, which confers resistance to the G418 drug, solely in neurons. Indeed, most of the surviving cells express many neuro-epithelial markers (Li et al., 1998).

Although it seems that the direct selection approach is highly efficient, it is reasonable to speculate that in some cases the differentiation process involves cross-talk between different types of cells, thus, application of selection in these instances will probably fail.

Over-expression of cellular genes in ES cells

Expression of cellular genes, most importantly transcription factors, in ES cells can serve to direct cell fate during *in vitro* differentiation to a certain lineage. The general design of such an experiment is to introduce a gene involved in normal development for induction of differentiation. For example, transfection of hepatocyte nuclear factor 3α or 3β (HNF3α or HNF3β, also termed FOXA1 and FOXA2, respectively) genes into mouse ES cells induced expression of hepatic and lung markers during differentiation, demonstrating the importance of these genes during endoderm commitment (Levinson-Dushnik and Benvenisty, 1997). For more examples of mouse ES cells forced differentiation see the review by O'Shea (2001). Interestingly, in some instances different expression levels of a given gene may induce differentiation into completely different lineages as demonstrated by Niwa et al. (2000).

12.3.2 Silencing gene expression in ES cells

One of the most powerful tools that biologists can use to understand gene function is the ability to interfere with gene expression. Following the isolation of mouse ES cell lines various techniques were developed for this purpose. The most effective, but also somewhat time consuming, is gene targeting of the gene of interest by a replacement vector that interferes with normal expression. Other means of interfering with the expression of a specific gene include various reagents that work at the post-transcriptional level. These include mainly antisense RNA transcripts of the target gene and double-stranded RNA molecules (dsRNAs), which may have RNA interference (RNAi) effects. As of today, only few of these techniques have been applied to human ES cells although they have been used in mouse ES cell studies. However, since this field is relatively new it is expected that reports demonstrating the use of these techniques in human ES cells will emerge in the near future. Therefore, this section presents experiments performed in mouse ES cells, with comments on their possible use in human ES cells.

Gene targeting in ES cells

Gene targeting by constructs with modified sequences has been used extensively over the years to manipulate the mouse genome. The most common approach is to build a DNA construct containing a genomic fragment of the targeted gene in which a selection-conferring gene or other gene of interest is inserted, such as lacZ providing a lineage tag, usually disrupting gene function. Following linearization, the construct is transfected into ES cells, the cells are selected for drug resistance and the clones are screened for replacement events of the endogenous gene. If homologous recombination occurs, then one of the alleles would be replaced by the exogenous DNA fragment thus leading to an aberrant transcript, or complete loss of the transcript. As most recombination events occur at random, addition of a negative selection marker adjacent to the homologous arm may serve to enhance selection for homologous recombinant clones. This approach known as positive negative selection (PNS) was shown dramatically to enrich homologous recombination events (Mansour et al., 1988).

Another option is to use an endogenous gene for negative selection. The hypoxanthine guanine phosphoribosyltransferase (HPRT) gene is ideal for this purpose as it is X-linked and cells lacking HPRT activity can be selected in HAT medium. Therefore, mutation of the single allele may be recovered, in a very high efficiency, using a disruption vector that contains a positive selection cassette in combination with HAT medium (Thomas and Capecchi, 1987). This methodology was recently used to delete the *HPRT* gene in human ES cells achieving a 50-fold increase in successful homologous recombination due to the use of HAT medium (Zwaka and Thomson, 2003).

A high proportion of gene-targeted clones can be derived using a construct containing a promoter-less positive selection cassette. By this means, the selection marker is expressed only upon integration into a gene, downstream to its promoter sequence (Doetschman et al., 1988). Although as high as 85% efficiency was reported using these constructs (te Riele et al., 1990), it is important to emphasize that this procedure suits only genes which are actively expressed in the cells as otherwise the selection marker will not be expressed. Zwaka and Thomson (2003) used this method recently to target the locus encoding for the OCT4 gene (*POU5F1*) achieving homologous recombination events in up to 40% of the clones.

Another method for gene disruption in mouse ES cells employs the use of another type of constructs termed 'trap constructs' (Stanford et al., 2001). In contrast to the targeting vectors outlined above, these vectors only contain a selection marker, which is expressed upon integration into a gene. This methodology is used in order to target genes at random and thus is not suitable for disruption of specific genes.

Silencing gene expression at the post-transcriptional level

Although gene targeting in mouse ES cells has proved to be an excellent method for elucidating gene function, this procedure may be time-consuming. Other methods that can advance our understanding of gene function are RNAi and antisense. As these methods work in *trans* the constructs can be introduced into cells transiently for the temporary inhibition of gene expression, or stably into the genome either by random transfection or infection for long-term silencing.

A commonly used technique specifically to repress gene expression is to express its cDNA sequence or part of it, in the reverse orientation, thus it was termed antisense. It is relatively simple to design and to carry out, but in many cases efficiency is low. Therefore, only a few experiments have employed the use of antisense constructs in mouse ES cells to date. For instance, using *vav* antisense construct, the *vav* gene was demonstrated to play a major role in *in vitro* hematopoietic differentiation of mouse ES cells (Wulf et al., 1993).

A relatively more efficient system for gene silencing is the RNAi method. In recent years a new mechanism of specific RNA dependent RNA degradation has been revealed. In short, the mechanism of silencing involves introduction of dsRNA oligonucleotides termed short inhibitory RNAs (siRNAs). These fragments are then bound to a nuclease complex thus enabling its specific direction and consequently degradation of homologous mRNA molecule (Hammond et al., 2001). Two recent reports were able to demonstrate the great potential of this technique for interfering with gene expression in mouse cells. These investigations show that several hundred base pairs long dsRNA molecules are able to induce specific RNAi responses in ES cells and embryonic carcinoma (EC) cell lines (Billy et al., 2001; Yang et al., 2001). Although shown to be efficient in mouse ES and EC cells, the introduction of long dsRNA molecules into mammalian cells may also induce interferon-α and interferon-β synthesis leading to dramatic changes in gene expression and cellular processes (Janeway et al., 2001). In order to avoid this response, siRNAs, which do not cause such effects, can be introduced into cells or synthesized *in situ* (within the cell). Because efficient siRNA expression vectors that can be introduced into cells by either transfection or infection were recently constructed (Brummelkamp et al., 2002a,b), it is reasonable to speculate that 'knockdown' gene expression studies in mouse and human ES cells will be undertaken in the near future.

12.4 The potential clinical applications of genetically modified human ES cells

Since the first establishment of human ES cell lines (Thomson et al., 1998; Reubinoff et al., 2000) many hopes have been raised by scientists and the general public on their potential use for cell based therapeutic applications of various human diseases and disabilities. It was estimated that in the US alone approximately 3000 people die every day from diseases that may potentially be cured in the future by human ES cell-differentiated derivatives (Lanza et al., 2001). For this ambitious task to be achieved, genetic engineering techniques may be of extremely high value. This section presents several different genetic strategies that may aid researchers in creating clinically applicable cell cultures.

12.4.1 Genetic engineering for cellular therapy

When considering the use of human ES cells as cellular reagents for broad transplantation use, the cells must fulfil the following requirements: (1) the culture must be of an homogenous nature, that is, contain only cells which are needed to treat a specific pathology; (2) the cells should not be rejected by the patient; and (3) special care should be taken that the cells will not over-proliferate or create tumors, and if possible be regulated after their transplantation. In order to obtain a pure population of a specific cell type, one option may be to introduce a transgene into the cells that contains a marker protein whose expression would be controlled by a tissue specific promoter (see section 12.3.1). The marker can be of an inert nature in the cell context and only aid in identification and separation of a specific cell type, i.e., GFP. Another option for directing differentiation towards specific cell type is to over-express a key-regulating gene that has a direct role in differentiation (see section 12.3.1).

Silencing of endogenous gene expression may also aid to produce immunologically tolerated differentiated cells. As recently shown, differentiated human ES cells may express high levels of class I major histocompatibility complex (MHC-I) proteins that might cause rejection of the cells upon transplantation (Drukker *et al.*, 2002). Therefore, deleting genes that encode for these proteins in human ES cell lines may enable their use in non-MHC-I matched patients.

The principle of negative selection may be important for safe transplantation. A major risk in transplantation of ES cell-derived differentiated cell populations is the possible presence of contaminating undifferentiated ES cells and their potential for generating teratomas. For instance, transplantation of partially differentiated mouse ES cells into a rat model of Parkinson's disease resulted in lethal teratoma formation in 20% of the rats although in 56% of the rats, dopaminergic neurons were developed (Bjorklund *et al.*, 2002). In order to reduce the risk of teratoma formation we developed transgenic human ES cell line which expresses eGFP under the control of the murine Rex-1 promoter which is expressed in pluripotent cells (Eiges *et al.*, 2001) (see section 12.3.1). Thus, depletion of eGFP-expressing cells prior to transplantation may reduce the risk of teratoma formation. In other instances even uncontrolled proliferation of cells without overt teratoma formation may also be hazardous to patients (Freed *et al.*, 2001). A general strategy aimed to kill specifically the transplanted cells is to introduce a suicide gene into the human ES cell line, which can be later activated. Therefore, we have recently demonstrated in our laboratory that human ES cells expressing the herpes simplex thymidine kinase (HSV-*tk*) gene can be eliminated at will. Following transplantation of HSV-tk^+ human ES cells into SCID mice, teratomas were formed but oral administration of the thymidine analog (ganciclovir) was sufficient to stop tumor growth (Schuldiner *et al.*, 2003). Thus, this methodology may be used to gain control of cell growth for those cases in which the transplanted cells may be proliferating uncontrollably or causing harmful symptoms.

12.4.2 Genetic engineering in nuclear transplantation therapy

As mentioned above, in order to carry out successful therapy the issue of rejection of human ES cell should be solved. As human ES cells were shown to express MHC-I molecules and that this expression can be highly induced by differentiation and by interferon γ (Drukker *et al.*, 2002), the cells will most probably be rejected by the patient's immune system. Overcoming this problem may be achieved by several ways, including administration of immunosuppressive agents, altering expression of MHC proteins by genetic manipulation (see section 12.4.1) or by the somatic nuclear transfer technique. Somatic cell nuclear transfer entails the transfer of a somatic cell nucleus into an enucleated oocyte, resulting in a nuclear transfer-derived blastocyst from which a fully compatible ES cell line may be established. This procedure termed therapeutic cloning was demonstrated recently in mice (Munsie et al., 2000; Wakayama et al., 2001), where not only ES cell lines were established, but the novel ES cell lines were also shown to differentiate into various cell types including dopaminergic and serotoninergic neurons (Munsie et al., 2000). Thus, these new cell lines may serve to be excellent reagents for transplantation since they are genetically identical to the nuclear donor. Moreover, somatic nuclear transfer may also take care of inherited disorders as was recently shown by Rideout et al. (2002) who were able to correct an inherited gene defect, which causes immunodeficiency, in an established line of mice. In short, nuclei from somatic cells from immunodeficient Rag 2-/- mice were transferred into enucleated oocytes and ES cell lines were derived. Then, the genetic defect underlying immunodeficiency was corrected using homologous recombination and the 'modified' cells were induced to differentiate to produce hematopoietic progenitors. When transplanted back into immunodeficient mice the cells contributed to lymphoid and myeloid lineages leading to correction of pathology. This report was the first to show that nuclear transplantation therapy is possible for inherited disorders.

12.5 Conclusions

Human ES cells may offer a number of advantages over other stem cells in regard to their use in transplantation therapies. These include their capacity to propagate readily in culture in high numbers while remaining undifferentiated, the relative ease by which they may be induced to differentiate and genetically modified. As demonstrated in mouse ES cells, it is reasonable to speculate that in the near future improved protocols for the production of specific differentiated cell types from human ES cells will be developed. It is very likely that genetic manipulation methods will help to achieve this task. These methods will be used to isolate specific cell types by expression of marker genes or by directing their differentiation towards specific lineages. Also, these techniques will enable safer transplantation either by deleting genes which may cause immune rejection or by introduction of suicide genes which may aid in eliminating cells that over-proliferate following transplantation. Moreover, genetic manipulation techniques may aid to understand gene function in ES cells through the action of inhibitory molecules such as siRNAs or by the more customary methods of gene targeting.

References

Assady S, Maor G, Amit M, Itskovitz-Eldor J, Skorecki KL, Tzukerman M (2001) Insulin production by human embryonic stem cells. *Diabetes* 50, 1691–1697.
Billy E, Brondani V, Zhang H, Muller U, Filipowicz W (2001) Specific interference with gene expression induced by long, double-stranded RNA in mouse embryonal teratocarcinoma cell lines. *Proc. Natl Acad. Sci. USA* 98, 14428–14433.

Bjorklund LM, Sanchez-Pernaute R, Chung S, Andersson T, Chen IY, McNaught KS, Brownell AL, Jenkins BG, Wahlestedt C, Kim KS, Isacson O (2002) Embryonic stem cells develop into functional dopaminergic neurons after transplantation in a Parkinson rat model. *Proc. Natl Acad. Sci. USA* **99**, 2344–2349.

Brummelkamp TR, Bernards R, Agami R (2002a) Stable suppression of tumorigenicity by virus-mediated RNA interference. *Cancer Cell* **2**, 243–247.

Brummelkamp TR, Bernards R, Agami R (2002b) A system for stable expression of short interfering RNAs in mammalian cells. *Science* **296**, 550–553.

Carpenter MK, Inokuma MS, Denham J, Mujtaba T, Chiu CP, Rao MS (2001) Enrichment of neurons and neural precursors from human embryonic stem cells. *Exp. Neurol.* **172**, 383–397.

Cherry SR, Biniszkiewicz D, van Parijs L, Baltimore D, Jaenisch R (2000) Retroviral expression in embryonic stem cells and hematopoietic stem cells. *Mol. Cell Biol.* **20**, 7419–7426.

Doetschman T, Maeda N, Smithies O (1988) Targeted mutation of the Hprt gene in mouse embryonic stem cells. *Proc. Natl Acad. Sci. USA* **85**, 8583–8587.

Drukker M, Katz G, Urbach A, Schuldiner M, Markel G, Itskovitz-Eldor J, Reubinoff B, Mandelboim O, Benvenisty N (2002) Characterization of the expression of MHC proteins in human embryonic stem cells. *Proc. Natl Acad. Sci. USA* **99**, 9864–9869.

Eiges R, Schuldiner M, Drukker M, Yanuka O, Itskovitz-Eldor J, Benvenisty N (2001) Establishment of human embryonic stem cell-transfected clones carrying a marker for undifferentiated cells. *Curr. Biol.* **11**, 514–518.

Freed CR, Greene PE, Breeze RE, Tsai WY, DuMouchel W, Kao R *et al.* (2001) Transplantation of embryonic dopamine neurons for severe Parkinson's disease. *N. Engl. J. Med.* **344**, 710–719.

Gropp M, Itsykson P, Singer O, Ben-Hur T, Reinhartz E, Galun E, Reubinoff BE (2003) Stable genetic modification of human embryonic stem cells by lentiviral vectors. *Mol. Ther.* **7**, 281–287.

Hammond SM, Caudy AA, Hannon GJ (2001) Post-transcriptional gene silencing by double-stranded RNA. *Nat. Rev. Genet.* **2**, 110–119.

Itskovitz-Eldor J, Schuldiner M, Karsenti D, Eden A, Yanuka O, Amit M, Soreq H, Benvenisty N (2000) Differentiation of human embryonic stem cells into embryoid bodies comprising the three embryonic germ layers. *Mol. Med.* **6**, 88–95.

Janeway CA, Travers P, Walport M, Shlomchik M (2001) *Immunobiology*. Garland Publishing, New York, pp. 81–82.

Kaufman DS, Hanson ET, Lewis RL, Auerbach R, Thomson JA (2001) Hematopoietic colony-forming cells derived from human embryonic stem cells. *Proc. Natl Acad. Sci. USA* **98**, 10716–10721.

Kehat I, Kenyagin-Karsenti D, Snir M, Segev H, Amit M, Gepstein A, Livne E, Binah O, Itskovitz-Eldor J, Gepstein L (2001) Human embryonic stem cells can differentiate into myocytes with structural and functional properties of cardiomyocytes. *J. Clin. Invest.* **108**, 407–414.

Kehat I, Gepstein A, Spira A, Itskovitz-Eldor J, Gepstein L (2002) High-resolution electrophysiological assessment of human embryonic stem cell-derived cardiomyocytes: a novel *in vitro* model for the study of conduction. *Circ. Res.* **91**, 659–661.

Lanza RP, Cibelli JB, West MD, Dorff E, Tauer C, Green RM (2001) The ethical reasons for stem cell research. *Science* **292**, 1299.

Levenberg S, Golub JS, Amit M, Itskovitz-Eldor J, Langer R (2002) Endothelial cells derived from human embryonic stem cells. *Proc. Natl Acad. Sci. USA* **99**, 4391–4396.

Levinson-Dushnik M, Benvenisty N (1997) Involvement of hepatocyte nuclear factor 3 in endoderm differentiation of embryonic stem cells. *Mol. Cell Biol.* **17**, 3817–3822.

Li M, Pevny L, Lovell-Badge R, Smith A (1998) Generation of purified neural precursors from embryonic stem cells by lineage selection. *Curr. Biol.* **8**, 971–974.

Ma Y, Ramezani A, Lewis R, Hawley RG, Thomson JA (2003) High-level sustained transgene expression in human embryonic stem cells using lentiviral vectors. *Stem Cells* **21**, 111–117.

Mansour SL, Thomas KR, Capecchi MR (1988) Disruption of the proto-oncogene int-2 in mouse embryo-derived stem cells: a general strategy for targeting mutations to non-selectable genes. *Nature* **336**, 348–352.

Mummery C, Ward D, van den Brink CE, Bird SD, Doevendans PA, Opthof T, Brutel de la Riviere A, Tertoolen L, van der Heyden M, Pera M (2002) Cardiomyocyte differentiation of mouse and human embryonic stem cells. *J. Anat.* **200**, 233–242.

Munsie MJ, Michalska AE, O'Brien CM, Trounson AO, Pera MF, Mountford PS (2000) Isolation of pluripotent embryonic stem cells from reprogrammed adult mouse somatic cell nuclei. *Curr. Biol.* **10**, 989–992.

Niwa H, Miyazaki J, Smith AG (2000) Quantitative expression of Oct-3/4 defines differentiation, dedifferentiation or self-renewal of ES cells. *Nat. Genet.* **24**, 372–376.

O'Shea KS (2001) Directed differentiation of embryonic stem cells: genetic and epigenetic methods. *Wound Repair Regen.* **9**, 443–459.

Pfeifer A, Ikawa M, Dayn Y, Verma IM (2002) Transgenesis by lentiviral vectors: lack of gene silencing in mammalian embryonic stem cells and preimplantation embryos. *Proc. Natl Acad. Sci. USA* **99**, 2140–2145.

Reubinoff BE, Pera MF, Fong CY, Trounson A, Bongso A (2000) Embryonic stem cell lines from human blastocysts: somatic differentiation *in vitro*. *Nat. Biotechnol.* **18**, 399–404.

Reubinoff BE, Itsykson P, Turetsky T, Pera MF, Reinhartz E, Itzik A, Ben-Hur T (2001) Neural progenitors from human embryonic stem cells. *Nat. Biotechnol.* **19**, 1134–1140.

Richards M, Fong CY, Chan WK, Wong PC, Bongso A (2002) Human feeders support prolonged undifferentiated growth of human inner cell masses and embryonic stem cells. *Nat. Biotechnol.* **20**, 933–936.

Rideout WM, 3rd, Hochedlinger K, Kyba M, Daley GQ, Jaenisch R (2002) Correction of a genetic defect by nuclear transplantation and combined cell and gene therapy. *Cell* **109**, 17–27.

Robertson E, Bradley A, Kuehn M, Evans M (1986) Germ-line transmission of genes introduced into cultured pluripotential cells by retroviral vector. *Nature* **323**, 445–448.

Schuldiner M, Yanuka O, Itskovitz-Eldor J, Melton DA, Benvenisty N (2000) Effects of eight growth factors on the differentiation of cells derived from human embryonic stem cells. *Proc. Natl Acad. Sci. USA* **97**, 11307–11312.

Schuldiner M, Eiges R, Eden A, Yanuka O, Itskovitz-Eldor J, Goldstein RS, Benvenisty N (2001) Induced neuronal differentiation of human embryonic stem cells *Brain Res.* **913**, 201–205.

Schuldiner M, Itskovitz-Eldor J, Benvenisty N (2003) Selective ablation of human embryonic stem cells expressing a 'suicide' gene. *Stem Cells* **21**, 257–265.

Stanford WL, Cohn JB, Cordes SP (2001) Gene-trap mutagenesis: past, present and beyond. *Nat. Rev. Genet.* **2**, 756–768.

te Riele H, Maandag ER, Clarke A, Hooper M, Berns A (1990) Consecutive inactivation of both alleles of the pim-1 proto-oncogene by homologous recombination in embryonic stem cells. *Nature* **348**, 649–651.

Thomas KR, Capecchi MR (1987) Site-directed mutagenesis by gene targeting in mouse embryo-derived stem cells. *Cell* **51**, 503–512.

Thomson JA, Itskovitz-Eldor J, Shapiro SS, Waknitz MA, Swiergiel JJ, Marshall VS, Jones JM (1998) Embryonic stem cell lines derived from human blastocysts. *Science* **282**, 1145–1147.

Wakayama T, Tabar V, Rodriguez I, Perry AC, Studer L, Mombaerts P (2001) Differentiation of embryonic stem cell lines generated from adult somatic cells by nuclear transfer. *Science* **292**, 740–743.

Weintraub H, Cheng PF, Conrad K (1986) Expression of transfected DNA depends on DNA topology. *Cell* **46**, 115–122.

Wulf GM, Adra CN, Lim B (1993) Inhibition of hematopoietic development from embryonic stem cells by antisense vav RNA. *Embo J.* **12**, 5065–5074.

Xu C, Inokuma MS, Denham J, Golds K, Kundu P, Gold JD, Carpenter MK (2001) Feeder-free growth of undifferentiated human embryonic stem cells. *Nat. Biotechnol.* **19**, 971–974.

Xu C, Police S, Rao N, Carpenter MK (2002a) Characterization and enrichment of cardiomyocytes derived from human embryonic stem cells. *Circ. Res.* **91**, 501–508.

Xu RH, Chen X, Li DS, Li R, Addicks GC, Glennon C, Zwaka TP, Thomson JA (2002b) BMP4 initiates human embryonic stem cell differentiation to trophoblast. *Nat. Biotechnol.* **20**, 1261–1264.

Yang S, Tutton S, Pierce E, Yoon K (2001) Specific double-stranded RNA interference in undifferentiated mouse embryonic stem cells. *Mol. Cell Biol.* **21**, 7807–7816.

Zhang SC, Wernig M, Duncan ID, Brustle O, Thomson JA (2001) *In vitro* differentiation of transplantable neural precursors from human embryonic stem cells. *Nat. Biotechnol.* **19**, 1129–1133.

Zwaka TP, Thomson JA (2003) Homologous recombination in human embryonic stem cells. *Nat. Biotechnol.* **21**, 319–321.

13. ES cells for transplantation: coping with immunity

J. Andrew Bradley, Eleanor M. Bolton and Roger A. Pedersen

13.1 Introduction

The capacity of pluripotent human stem cells to differentiate into diverse functional cell types has generated excitement about their potential use as a source of cells for transplantation in the treatment of a wide range of diseases. However, their exogenous origin confers upon pluripotent stem cells a status comparable to that of unrelated organ donors. This raises the question of how to deal with the disparity between the immune identity of the donor tissue and that of its intended recipient. Several distinct strategies may be envisioned for resolving this 'immunity obstacle' to the therapeutic uses of pluripotent stem cells. We and others have recently reviewed the available options (Bradley et al., 2002; Drukker and Benvenisty, 2004) and they are listed in Table 13.1.

In the first instance, it should be possible to utilize the inherent genetic diversity of the sources from which the pluripotent stem cells were derived (donated embryos, in the case of ES cells) to effect a tissue match, as is currently done with kidney and bone marrow transplantation. In the case of pluripotent stem cells derived from surplus embryos generated in therapies for infertility, the genetic identity of the cells will specifically reflect that of the parental gametes and should generally resemble the population to which the infertile couple belonged. As detailed in section 13.3.1, the donor population can be modeled mathematically to provide an estimate of genetic diversity of stem cells derived from such sources.

Table 13.1: Strategies for overcoming the immune barrier to stem cell transplantation.

- Transplantation to immune privileged sites (e.g., central nervous system, thymus)
- HLA matching (stem cell banking and oocyte manipulation)
- Genetic modification of stem cells
- Non-specific immunosuppression
- Mixed hematopoietic chimerism

While their embryonic origin confers upon ES cells an inherent genetic diversity, it also limits their capacity for closely matching a prospective recipient. This has focused attention on the alternative strategy of utilizing oocyte micromanipulation, developed initially for domestic laboratory species, to generate matching tissues through somatic cell nuclear transfer. Although it has been demonstrated as feasible in animal models, this approach had, until recently, remained hypothetical for human oocytes. The evidence for derivation of a human ES cell line (SCNT-hES-1) following somatic cell nuclear transfer (Hwang et al., 2004) prompts a reappraisal of how this technology can contribute to tissue matching. An additional way in which human oocytes could be used to enhance the degree of immune matching is through parthenogenesis (Lin et al., 2003). This approach involves activation of oocytes using electrical or chemical stimulation, rather than fertilization by a sperm, and it results in an embryo with strictly maternal inheritance. Derivation of parthenogenetic ES cells with substantially similar properties to those derived from zygotes has been demonstrated in mice and non-human primates, thus justifying a consideration of how this approach could be adopted for achieving clinically useful materials. These strategies are considered together in section 13.3.2.

Because they originate (and are maintained) in culture, pluripotent stem cells offer opportunities for intervention that would not exist with solid organs or even bone marrow stem cells. Human ES (hES)cells can be modified using gain- and loss-of-genetic function strategies (Vallier et al., 2004; see also Chapter 12). The differentiated progeny of ES cells could also be used as vehicles to deliver gene products to a local tissue environment, with the aim of promoting graft acceptance (section 13.3.3).

Finally, the prospective use of stem cells in transplantation medicine evokes new opportunities to use classical methods, particularly immunosuppression, to deal with mismatch between donor and recipient immune identities. Advance knowledge of the immune identity of the stem cell-derived donor cells provides novel opportunities to intervene in the recipient's immune response, including strategies for induction of transplant tolerance through mixed hematopoietic chimerism and other means of modifying the recipient's response to imperfectly matched donor tissue (section 13.4).

Taken together, the diverse strategies available for coping with the immunity makes the immune matching problem seem less like an obstacle than an opportunity to advance our understanding, both of the developmental properties of pluripotent stem cells as unique biological entities, and also of the role during *in vitro* development of human genes that are essential for *in vivo* immune function.

13.2 Immune profile

The strength of the immune response to allogeneic tissue is very considerable and not to be underestimated. Countless experimental studies in the field of transplant immunology have shown that organs, tissue and cells transplanted between genetically disparate individuals (allografts) provoke a powerful immune response that invariably results in complete graft rejection within one or two weeks. An exception to this general rule applies to allografts placed in 'privileged sites' within the recipient, such as the anterior chamber of the eye, brain, thymus, or testis. Grafts implanted into these special sites enjoy a variable degree of protection from immune injury, even in the absence of exogenous immunosuppression (Suter et al., 2003).

Tissues derived from ES cells would likely, in most circumstances, suffer a similar fate to conventional allografts unless steps were taken to circumvent the rejection response, with the possible exception of hES cell-derived neuronal tissue transplanted into the brain or spinal cord or perhaps the transplantation of hES cell-derived retinal tissue into the eye.

Graft rejection is triggered by the recognition of antigens on the cell surface of the transplanted tissues. These 'transplantation' or histocompatibility antigens are the result of genetic polymorphism in the human population creating antigenic differences between the donor and recipient. Three categories of histocompatibility antigens give rise to rejection: these are, in order of importance, blood group antigens (ABO in humans), major histocompatibility complex (MHC) antigens and minor histocompatibility (mHC) antigens.

13.2.1 ABO blood group antigens

ABO blood group antigens result from structural polymorphisms in carbohydrate residues on glycolipids. They need to be considered because they are expressed not only on the surface of red blood cells but also on most endothelial and epithelial cells. Individuals who do not inherit a particular ABO blood group antigen develop cross-reactive antibodies to those antigens through exposure during infancy to normal intestinal bacteria with 'blood group like' antigens on their cell surface (Springer and Horton, 1969). Prior expression during fetal development of a particular blood group antigen (including blood group H antigen, expressed by all individuals) confers immunological tolerance towards it and persistent exposure to bacterial antigens is unable to stimulate a cross-reactive antibody response. As a result, blood group O individuals who lack group A and B antigens have naturally occurring antibodies to both antigens, whereas blood group AB individuals ('universal recipients') have neither anti-A nor anti-B antibodies. Likewise, blood group A individuals have naturally occurring antibodies to blood group B antigens and blood group B individuals have anti-A antibodies. If a recipient were to receive an ABO blood group incompatible graft, preformed natural antibodies would bind to the graft and likely mediate rapid complement dependent graft destruction (Paul and Baldwin, 1987; Cooper, 1990) . For this reason an important rule for all types of clinical transplantation is to avoid ABO blood group incompatibility. This includes tissue transplants such as corneal and pancreatic islet transplants, as well as bone marrow and solid organ transplants (Clayton et al., 1993; Borderie et al., 1997). It will probably be mandatory, and also fairly straightforward, to apply a similar rule for stem cell-derived transplants.

13.2.2 MHC antigens

For tissue allografts that are blood group compatible, the graft rejection response is directed predominantly against polymorphic cell surface glycoprotein molecules encoded by the major histocompatibility complex (MHC). This cluster of closely linked genes spans around 4000 kilobases of DNA and is the most highly polymorphic region of the genome. It is termed H-2 in mice and is situated on chromosome 17, whereas in humans it is known as the human leukocyte antigen (HLA) system and is located on the short arm of chromosome 6 (*Figure 13.1*). The genes within the MHC comprise three regions, known as class I, class II and class III; the most important transplant antigens are all encoded by distinct genes in the class I and class II regions. The class III region, which is situated between the class I and class II regions, encodes various proteins that are of immunological importance but none are major transplant antigens. Genes in the class I region of the MHC encode the heavy chains of the three classical class I MHC molecules (designated H-2K, H2-D and H-2L in mice and, in order of discovery, HLA-A, HLA-B and HLA-C in humans) along with other non-classical class I molecules (such as HLA-E, HLA-F and HLA-G in humans). The latter are less important as transplant antigens. The MHC class II region encodes the class II molecules (H-2A and H-2E in mice and HLA-DR, HLA-DP, and HLA-DQ in humans). The HLA gene products which are of most relevance to human organ and tissue transplantation are the HLA class I gene products HLA-A and HLA-B and the HLA class II gene product HLA-DR. These three loci or their products are those that are routinely typed and attempts made to match for them to maximize graft survival after kidney and bone marrow transplantation.

The reason that MHC class I and class II antigens are such powerful transplant antigens is because of their physiological role in the presentation of peptide antigens to T cells. The fact that they are highly polymorphic and abundantly expressed also contributes. Both types of MHC molecule have a broadly similar three-dimensional structure (*Figure 13.2*). This comprises a deep groove walled by two parallel alpha helices and floored by a beta-pleated sheet which enables effective binding and presentation of linear antigenic peptide for surveillance by the T-cell receptor. This three-dimensional structure is achieved in a different way by class I and class II MHC molecules. MHC class I molecules comprise a polymorphic alpha or heavy chain which associates non-covalently at the cell surface with a smaller non-polymorphic chain, β_2-microglobulin, encoded on chromosome 15 in humans. MHC class II molecules on the other hand comprise two polymorphic polypeptide chains designated alpha and beta, the membrane distal domains of which form the peptide binding cleft.

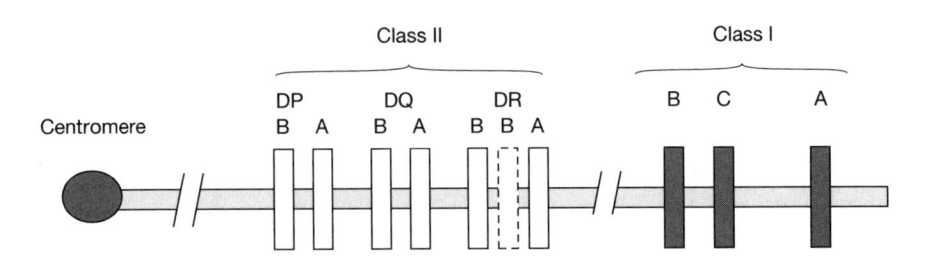

Figure 13.1: Genomic organization of human leukocyte antigen (HLA) loci on chromosome 6.

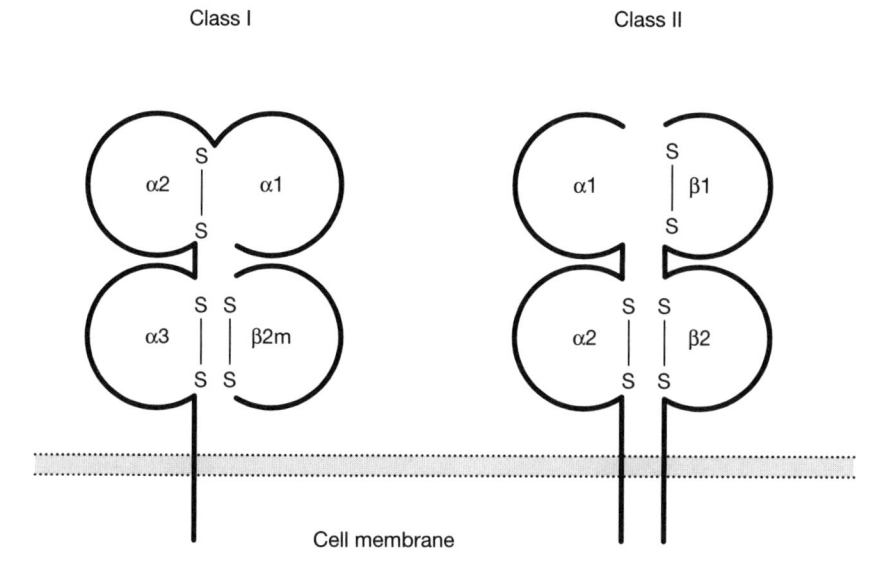

Figure 13.2: Structure of major histocompatibility complex (MHC) class I and class II molecules.

The tissue distribution of MHC class I and class II molecules differs markedly. In humans, HLA class I molecules are present on most nucleated cells and are expressed most strongly by hematopoietic cells. The distribution of HLA class II molecules is much more restricted. HLA class II is expressed strongly on only a subset of hematopoietic cells (dendritic cells, B lymphocytes and macrophages) and thymic epithelial cells. Other cell types express little or no HLA class II constitutively, although expression is readily inducible on many cell types by exposure to the cytokine interferon-γ.

13.2.3 Minor histocompatibility antigens

Minor histocompatibility antigens represent a further immunological barrier to successful transplantation. They are peptides derived from polymorphic donor proteins expressed by the donor and most allografts will express multiple minor antigens. Their name in the context of transplantation is potentially misleading because although less potent transplant antigens than MHC antigens, they are capable of initiating graft rejection (Simpson *et al.*, 2001).

13.2.4 Alloantigen recognition and the rejection response

MHC proteins expressed by transplanted tissue trigger a graft rejection response by two distinct allorecognition pathways, called the 'direct' and 'indirect' pathways (Game and Lechler, 2002; Heeger, 2003) (*Figure 13.3*). Both allorecognition pathways require the services of 'professional' antigen presenting cells (APC), typically myeloid dendritic cells, that express a high density of class I and class II MHC molecules, along with the full range of co-stimulatory molecules needed to trigger T-cell activation. In the direct pathway, intact allogeneic donor MHC molecules

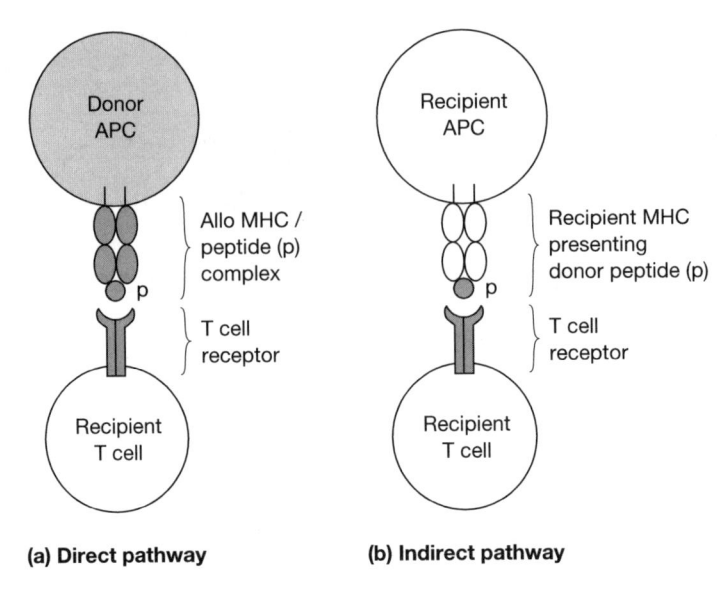

Figure 13.3: Alloantigen recognition pathways.
(A) In the direct pathway, the recipient T-cell receptor recognizes allogeneic major histocompatibility complex (MHC) antigens and naturally occurring endogenous peptide on the donor antigen presenting cells (APC). (B) In the indirect pathway, the T-cell receptor recognizes donor alloantigen after it has been taken up by a recipient APC and presented as an antigenic peptide bound to recipient MHC.

(and their bound peptides) expressed on the surface of donor APC from the grafted tissue stimulate recipient T cells directly. Normal adult tissues are rich in dendritic cells and so when an organ or tissue transplant is undertaken, large numbers of donor APC are transferred with the graft and are important in initiating rejection. During the first few days after transplantation, donor APC migrate from the graft to the lymph nodes and spleen of the recipient where they present intact allogeneic MHC to alloreactive host T and B lymphocytes. Grafts derived *in vitro* from ES cells will not contain dendritic cells, unless the ES cells have been induced to differentiate along the hematopoietic pathway, and as a result the direct allorecognition pathway may not become fully operational.

In the indirect pathway, donor MHC antigens that have been released or shed from the graft are captured by recipient APCs, and processed by endosomes and then presented at the recipient APC cell surface as antigenic peptides in the binding cleft of class II molecules. Minor HC antigens are all presented as peptides via the indirect pathway. This pathway is not dependent on donor dendritic cells and is likely to be the dominant route for triggering alloimmunity for most tissue grafts derived from ES cells.

T cells, after direct or indirect activation, undergo clonal expansion and differentiate into regulatory and effector cells. Allograft rejection is a T-cell dependent phenomenon, and is mediated by cytotoxic T cells, non-specific effectors (including macrophages), and alloantibody. natural killer (NK) cells and eosinophils may also play a role (Le Moine *et al.*, 2002; Le Moine and Goldman, 2003).

13.2.5 HLA expression by hES cells and their derivatives

The cell surface expression of human ES cells and their differentiated derivatives was recently described by Drukker and colleagues (Drukker *et al.*, 2002). The two human ES cell lines examined (H9 and H13) both expressed detectable but very low levels of the classical HLA class I molecules (HLA-A, -B, -C). After *in vitro* differentiation of human ES cells into embryoid bodies (EBs) a two- to four-fold increase in HLA class I expression was observed. Human teratoma cell lines that had been derived from the H9 line (representing *in vivo* differentiation of human ES cells) showed an eight- to ten-fold increase in HLA class I expression. It was notable, however, that even after *in vitro* or *in vivo* differentiation, expression of HLA class I antigen by ES cell lines was relatively modest and lower than that observed in somatic cell lines such as HeLa cells or monocytes. The non-classical HLA-G molecule was not detectable in either ES cells or their differentiated products.

In contrast to the classical HLA class I antigens, expression of HLA class II antigen was not detectable in either the undifferentiated or differentiated human ES cells. The addition of interferon-γ to the culture medium of undifferentiated human ES cells or teratoma cells led to a marked upregulation of HLA class I but did not induce detectable cell surface expression of HLA class II molecules. Nevertheless, there is no *a priori* reason to suppose that the fully differentiated products of hES cells will not be capable of expressing HLA class II when exposed to interferon-γ *in vitro* or *in vivo*.

Drukker and colleagues (Drukker *et al.*, 2002) also showed that although human ES cells expressed relatively low levels of class I HLA they were not unduly sensitive to lysis by natural killer (NK) cells. The human ES cells examined did not express receptors for the various NK cell ligands examined (Nkp30, Nkp44 Nkp46 and CD16) and hence likely escaped effective recognition by NK cells.

Draper *et al.* (2002) showed that another human ES line (H7) also expressed detectable levels of HLA class I on the cell surface which was reduced slightly when differentiation was induced. Exposure to interferon-γ induced strong expression of HLA class I in both the undifferentiated and differentiated cell cultures. In sum, hES cells and their differentiated progeny are able to express HLA cell surface antigens to a level sufficient to manifest their immune identity, thus necessitating a strategy for immune matching in order to achieve graft acceptance following transplantation.

13.3 Strategies for matching donor and recipient

13.3.1 HLA matching

Reducing the HLA mismatch between donor and recipient would undoubtedly reduce the immunological barrier to stem-cell transplantation. One potential way of achieving this would be to create a bank of hES cell lines so that for each potential recipient the hES cell line with the best HLA match could be selected for transplantation.

Experience in kidney transplantation has shown that the most important HLA molecules to match are HLA-A, HLA-B and HLA-DR and these three loci are those that are routinely typed for and considered when attempting to match a cadaveric donor kidney to the most suitable potential recipient. Data from national and international kidney transplant registries (Opelz et al., 1999) show that there is a progressive increase in renal allograft survival as the HLA-A, -B and -DR mismatch grade decreases from 6 (complete mismatch) to zero (complete match). It is important to note however that the difference in graft survival between a well matched graft and a poorly matched graft is relatively modest (10% graft survival difference at 5 years) (Opelz et al., 1999).

Functional allelic variants of over 250 HLA-A alleles, 500 HLA-B alleles and 300 HLA-DR alleles have now been revealed by DNA typing. This degree of polymorphism poses a considerable challenge to HLA matching. Consequently it is possible in practice to obtain a zero HLA-A, -B, -DR mismatch in only a minority of kidney graft recipients even with a national organ sharing program. The aim of the UK organ sharing program is to obtain a favorable match for most kidneys, which is defined as a zero mismatch at HLA-DR and no more than 1 mismatch at HLA-A and/or -B. Constructing a bank of hES cells with a view to HLA matching of therapeutic hES cell tissues is a practical approach to reducing the immunological barrier to transplantation, and the recently established UK Stem Cell Bank will enable accomplishment of this objective. The size of stem cell bank required to match the HLA genotype of hES cell lines to potential recipients depends on the degree of HLA matching required and the proportion of the total recipient pool in which a match is sought. We have undertaken a computer simulation to determine how large a hES cell bank would need to be to make HLA matching a practical approach for minimizing HLA disparity between hES donor type and potential recipients (Taylor CJ, Pedersen RA, Bolton EM Bradley JA, unpublished). To generate a set of HLA types representative of a 'random' population from which hES cell donors might be derived, we used data from a series of 1500 consecutive cadaveric organ donors reported to the UK transplant service during a two-year period. Over 6000 patients registered on the UK kidney transplant waiting list were then used to determine the likelihood of obtaining a blood group and HLA matched cell donor for patients in the recipient pool. Assuming the need for ABO blood group compatibility and accepting that each hypothetical (stem cell) donor could be used for an unlimited number of recipients, our analysis revealed that a donor cohort of 250 would provide a zero HLA (HLA-A, -B, and -C) mismatch for 20% of potential recipients and a favorable HLA match for almost 80%. Further increasing the size of the donor pool beyond 250 conferred very little additional benefit for improved HLA matching. This analysis suggests that a cell bank will allow a reasonable degree of HLA matching for most potential recipients but would provide relatively few with a zero HLA mismatch. If hES-cell-derived tissue does not contain class II expressing hematopoietically derived cells, and if inflammatory cytokines are unable to induce expression of class II HLA on tissue that differentiates from human ES cells, then matching for class II HLA molecules may not be important. However, in our simulation, DR matching was achieved in over 90% of recipients and HLA-A and –B mismatches accounted for most of the residual HLA disparity. This analysis reveals a productive strategy for immune matching, consisting of maximizing the degree of HLA similarity through generating and identifying hES cell lines that match a substantial fraction of the recipient population.

Criteria for matching donor and recipient for bone marrow transplantation are much more stringent than for organ grafting because bone marrow grafts are more susceptible to rejection and contain functional T cells able to respond to incompatibilities in the host, resulting in potentially fatal graft-versus-host-disease. Such a high level of hES cell matching only becomes necessary if hES cells are to be used for reconstituting hematopoietic tissue in patients with immunodeficiency diseases, in which case a perfect match, such as might be obtained by nuclear transfer (therapeutic cloning) or from parthenogenetically-derived embryos may be desirable.

13.3.2 Oocyte manipulation strategies for immune matching

Successes in carrying out somatic cell nuclear transfer in domestic and laboratory species have encouraged hopes for the potential for combined oocyte micromanipulation and stem cell derivation studies with human material. The procedure of somatic cell nuclear transfer (also known as genome replacement or therapeutic cloning) involves transplanting the genome of a somatic cell in place of the chromosomes of an oocyte which is then activated to proceed through early development to the blastocyst stage. The resulting embryo has a genetic identity consisting of the nuclear DNA of the somatic cell, and the mitochondrial DNA of the oocyte. When such studies were carried out with mouse oocytes, several groups were able to generate ES cells from the resulting blastocysts. The resulting stem cell lines had similar properties of cell surface and molecular markers, pluripotency and germ line contribution, to ES cells derived from zygotes (Munsie *et al.*, 2000; Kawase *et al.*, 2000; Wakayama *et al.*, 2001; Hochedlinger and Jaenisch, 2002; Rideout *et al.*, 2002). Mouse ES cells, obtained through somatic cell nuclear transfer, differentiated into hematopoietic stem cells that were capable of multilineage, long-term engraftment, thereby alleviating a genetic disease in the recipient mouse (Rideout *et al.*, 2002). These studies have thus established proof in principle of the relevance of the oocyte genome replacement approach for matching stem cell genetic identity to the recipient immune system.

Until recently, the human relevance of the foregoing studies had remained hypothetical. However, two recent studies have extended the genome replacement approach to human materials, with apparent success in generating pluripotent stem cell lines. In the study of Chen *et al.* (2003), human fibroblast nuclei were transferred to enucleated rabbit oocytes, which were induced to develop by electrical activation. After developing embryos reached the blastocyst stage, ES cells were derived using approaches similar to those used previously with human embryos (Thomson *et al.*, 1998). Four of the stem cells lines isolated using this approach were cultured for more than 25 passages, maintaining pluripotency, appropriate cell surface marker gene expression, and a normal (human) karyotype. Interestingly, these cell lines had human nuclear DNA identical to the nuclear donor, in combination with rabbit mitochondrial DNA, reflecting the distinct species of origin of their nuclei and cytoplasm. However, such cross-species compatibility between nuclear and mitochondrial function is surprising, in view of the limited extent of such compatibility even between primate species (Kenyon and Moraes, 1997). Therefore, further studies are needed to validate such an interspecies approach, in particular to determine whether surviving pluripotent stem cells are heteroplasmic for donor cell-derived human mitochondria, in addition to those derived from the oocyte, as has been observed in some cases of bovine nuclear transfer (Takeda et al., 2003). In any case, pluripotent stem cells derived in this manner would have to be treated as xenografts if their progeny were used for human transplantation.

Another recent report has simplified the approach to human somatic cell nuclear transfer, at least conceptually, by demonstrating the feasibility of deriving a pluripotent stem cell line from a manipulated human oocyte (Hwang et al., 2004). In this case, a single pluripotent human stem cell line (SCNT-hES-1) was generated, manifesting the expected ES cell properties of cell surface markers and other molecular features, pluripotency, and karyotypic stability. Interestingly, all the embryos that developed to the blastocyst stage were derived from oocytes whose genome was replaced with an autologous somatic nucleus (i.e., a cumulus cell from the same woman who donated the oocyte). Whether the use of an autologous human genome will be necessary for such derivations in general remains to be determined. In any case, it is unclear what biological mechanism would underlie such a requirement. Nevertheless, the demonstration provides proof-of-principle for the widely heralded potential of somatic cell nuclear transfer in humans. Therefore, it compels a more critical assessment of the practical contribution of this approach as a strategy for immune matching.

Despite its appeal as the best possible means of achieving an exact match between the immune identity of the donor cells and the intended recipient, the sequence of procedures (somatic cell nuclear transfer, followed by pluripotent stem cell derivation, and then directed differentiation into a clinically useful tissue), is too logistically complex for widespread use in generating individually customized tissues for transplantation. So, how could the strengths of the somatic cell nuclear transfer technology best be harnessed to enhance solutions to the immune matching problem? The genome replacement approach could be used to accelerate the generation of pluripotent stem cell lines that are homozygous for rare HLA haplotypes. This is a powerful strategy because the reduced genetic diversity in an HLA homozygous haplotype means that it can serve as a matched donor for a larger fraction of the population than a heterozygous donor. For example, the haplotype HLA-A2, HLA-B44 and HLA-DR4 in the homozygous state would be a zero HLA mismatch for ~7% of the UK recipient pool. However, homozygosity of this particular haplotype occurs in less than 1% of the Caucasian population, so a substantial number of ES cell lines would probably have to be generated from surplus embryos before encountering a single one that is homozygous for this haplotype by chance alone. By using somatic cell nuclear transfer with a homozygous HLA haplotype nuclear donor, it would be possible to achieve matching far more effectively than by relying on chance alone. This application of the technology would of course raise the ethical issues involved in identifying and securing the consent of the rare individuals bearing the homozygous HLA haplotypes, and of generating embryos for the express purpose of deriving stem cell lines carrying the specified genotypes. However, the utility of applying the genome replacement approach in this way is obvious and, if successful, would lead to generation of a series of pluripotent human stem cell lines with a degree of universality in their donor potential. Each such line would enable matching a predictable fraction (e.g., 5–10%) of the recipient population.

A second oocyte manipulation approach, parthenogenesis, appeals for similar reasons, namely that it could be used to generate pluripotent stem cell lines with a high probability of homozygosity for HLA haplotypes. The homozygosity arises from the fact that the oocyte's first meiotic division separates homologous chromosomes, leaving the two daughter chromatids of a single parental chromosome in the egg. If recombination has not disrupted the HLA region, then the resulting oocyte will be homozygous if (as typically is the case) it is prevented from extruding the second polar body following parthenogenetic activation (Lin et al., 2003). The demonstration that parthenogenetically activated oocytes of Cynomolgus monkeys can give rise to pluripotent stem cells (Cibelli et al., 2002; Vrana et al., 2003) suggests that similar cell lines could be generated from human parthenogenetic embryos (Taylor and Braude, 1994; Santos et al., 2003). Parthenogenetically-derived mouse ES cell lines were found to contribute to most tissues in chimeric mice, with the exception of skeletal muscle and testis, where contribution was minimal (Allen et al., 1994). Their origin from exclusively maternal genes means that parthenogenetic stem cells might express atypical levels of imprinted genes (Szabo and Mann, 1994). However, this should not be taken to imply that their epigenetic status is inherently unstable, only that their parentage is atypical, as compared with zygotes. In any case, the epigenetic status of human pluripotent stem cells, whatever their origins, would need thorough analysis prior to their use in clinical applications (P.J. Rugg-Gunn and R.A. Pedersen, unpublished observations). In sum, the possibility of generating human pluripotent stem cell lines from parthenogenetically activated oocytes further enriches the potential for immune matching through the generation of homozygous HLA haplotypes, thus providing 'limited universality' as a clear alternative to individually customized pluripotent stem cell lines.

13.3.3 Genetic manipulation of hES cells

The amenability of hES cells to genetic alteration has been discussed elsewhere in this book. Genetic alteration of hES cells is desirable and, in principle, achievable in order to 'fast track' the developmental pathways involved in differentiation of stem cells to specialized tissues that may be used for tissue repair and regeneration. It is also a viable approach for altering the immunogenicity of hES cells in order to overcome the immunological barriers to transplantation. This is an approach that has been pursued actively in rodent and higher animal models of tissue transplantation in an attempt to achieve prolonged or indefinite graft survival without recourse to immunosuppressive drugs. The primary and most obvious targets for alteration are the MHC class I and class II molecules expressed on the cell surface. Because of the highly polymorphic nature of these molecules, modification of the molecules themselves to reduce their immunogenicity may be less achievable than prevention of their cell surface expression. Knocking out individual MHC class I and class II genes would be experimentally complex because there are so many of them. A better approach would be to target a common protein that influences expression of the molecules on the cell surface. Two such molecules are critical for the final folding of MHC molecules, giving them their unique and individual recognition elements. These are, for class I, the common protein β2 microglobulin and the peptide bound in the peptide cleft formed by the α1 and α2 domains of the class I heavy chain protein. Both of these are integral to the correct assembly of the class I molecule within the cell and for its insertion as a stable molecule on the cell surface. For class II, correct folding and membrane insertion requires intracellular assembly in the presence of the invariant chain which acts as a chaperone for the eventual insertion of exogenously derived, immunogenic peptides in the cleft formed by the α1 domain of the class II α chain and the β1 domain of the class II β chain.

There are a few key components to be targeted in order to prevent correct assembly of the class I and class II molecules. Candidate genes are those that encode, for class I, the $β_2$-microglobulin, TAP and tapasin proteins regulating class I folding and peptide assembly and, for class II, the genes encoding the invariant chain (Ii) and the class II transactivator (CIITA) proteins regulating class II expression.

Genetic alteration of mouse ES cells has become a commonplace methodology for the creation of new strains of transgenic or knockout mice. Transplantation studies have made use of both MHC class I-deficient mice created by disruption, through targeted mutation, of the $β_2$-microglobulin ($β_2$m) gene (Zijlstra et al., 1990; Koller et al., 1990) and class II-deficient mice which have a disrupted class II A^b β-chain gene (Grusby et al., 1991). The lack of a functional $β_2$m gene results in absence of H-2K expression detectable by flow cytometry together with a 20-fold reduction in H-2D expression and very few remaining peripheral CD8 T lymphocytes. The class II A^b β-chain gene disruption prevents normal inducible class II expression on thymic or other epithelium, or on APCs; these mice are also devoid of Eα,β class II through a natural mutation, and there are very few remaining peripheral CD4 T lymphocytes. Skin grafts deficient in expression of either class I or class II target molecules are not, however, fully protected from rejection by immunologically intact allogeneic hosts, although MHC class I-deficient grafts are rejected more slowly. Mice lacking both MHC class I and class II molecules have also been created by crossing these two strains (Grusby et al., 1993). They display a surprisingly healthy phenotype relying, presumably, on their intact innate immune system, and are able to breed normally. Nevertheless, skin and heart grafts from class I- and II-deficient mice are quickly rejected when transplanted into normal allogeneic recipients (Lee et al., 1997; Pollak and Blanchard, 2000). This was an unexpected observation since, although the double knockouts still express small amounts of MHC, it was felt that these levels were insufficient either to stimulate or to act as targets for a direct immune response. Rejection is attributed to the highly effective indirect pathway of allorecognition that empowers recipient immune cells continually to respond to the very low but persistent residual levels of MHC expression on donor tissue. Minor histocompatibility antigens within donor cells may also be cross-presented by residual class I molecules and act as a focus for recipient CD8 effector cells (Valujskikh *et al.*, 2002; Bradley, 2004; He and Heeger, 2004). A more robust deletion of cell surface MHC class I expression has been achieved through crossing H-2K^b-/- mice with H-2D^b-/- mice, each bearing a targeted deletion of the respective class I gene (Chen *et al.*, 1996). Cell surface expression of MHC class I is not detectable in the resulting K^{b}D^b-/- mice although their few remaining CD8 T cells appear to be fully functional; this particular model has yet to be used in transplantation studies. It is apparent from the preceding description, however, that targeted deletion of both HLA class I and class II locus genes in hES cells would present a considerable technical challenge.

Over the past two decades, genetic analysis of patients with the spectrum of severe immunodeficiency diseases known collectively as bare lymphocyte syndrome, together with studies of transgenic mice, have improved our understanding of the regulatory pathways controlling MHC expression. Patients with bare lymphocyte syndrome suffer from repeated and life-threatening bacterial, viral and fungal infections attributed to an absence of HLA class I and class II antigen expression and reduced numbers of CD4 and CD8 T lymphocytes. Patients at the more serious end of the disease spectrum frequently have deletions of the RFX5 and CIITA genes that are now known to be pivotal for regulation of class II expression (Fruh *et al.*, 1995). It has become clear that defects in RFX genes are also associated with reduced MHC class I expression suggesting a common regulatory pathway for MHC class I and class II expression on different cell types not restricted to lymphocytes alone. It has recently been shown that a group of conserved regulatory elements, termed the SXY module, comprises a promoter region that controls MHC class II expression and regulates expression of classical class I molecules (Rousseau *et al.*, 2004). This offers potential as a novel target for genetic manipulation aimed at achieving complete deletion of MHC molecules.

An important observation that emerged from the study of K^{b}D^b-/- mice was that class I negative splenic lymphoblasts are highly susceptible to NK cell lysis (Chen *et al.*, 1996). This illustrates a major drawback of attempts to avoid immunological rejection through creation of 'universal donor' hES cells lacking MHC class I target antigens: the innate immune response has mechanisms for combating perceived threats from potentially dangerous foreign invaders. Some of the most aggressive malignancies and successful pathogenic viruses are those that evade the innate and adaptive immune response by their lack of MHC antigen expression (tumors) or by their ability to hijack class I-peptide loading systems (viruses such as CMV), thereby downregulating cell surface MHC molecules (Storkus *et al.*, 1989; Matsuda *et al.*, 1994; Bellgrau *et al.*, 1995; Strand *et al.*, 1996; Zeidler *et al.*, 1997; Swenson *et al.*, 1998). Consequently, NK cells have evolved a second line of host defense against these survival strategies which involves recognition of the absence of class I molecules. NK cells acknowledge, and are disarmed by, the presence of cell surface class I molecules but they are ruthless killers of target cells that have no class I identity (Qin *et al.*, 1996; Lewandoski, 2001). hES cells depleted of all class I identity would, unfortunately, fall into the category of NK-susceptible cells and hence be at risk of rejection by NK cell-dependent effector mechanisms. Another matter of concern is that a lack of class I molecules would be likely to permit stem cell-derived tissues to harbor and amplify viral particles since there is no class I molecule to present viral peptide and to act as target for virus-specific cytotoxic T cells (Grey et al., 1999).

An alternative strategy to the complete ablation of MHC class I identity, as described above, might instead be to acknowledge histoincompatibility and induce a state of immunological tolerance with a small series of common haplotype hES cell lines (best matches) that have been appropriately genetically modified. We can look at pregnancy as an example where a semi-histoincompatible allograft, the fetus, induces a state of specific immunological tolerance in the mother, attributable to expression by the placenta and trophoblast of a range of disparate protective genes. Two molecules that are thought to exert a protective effect towards the fetus have been studied extensively in recent years. The first of these, HLA-G, is a non-classical MHC class I antigen expressed at high levels in the placenta. It is relatively non-polymorphic and functions as a common ligand for the killer inhibitory receptors of NK cells that are found in high numbers in the pregnant uterus (Le Bouteiller and Blaschitz, 1999). The developing conceptus, which expresses very low levels of classical MHC class I, is thus protected from attack by NK cells. HLA-G therefore offers potential as a protective molecule for expression on hES cells, although it remains possible that the protection afforded to the fetus by placental expression of HLA-G is regulated by the fetal environment, outside of which HLA-G retains strong immunogenicity (Horuzsko et al., 1997; van der Meer et al., 2004). The second of these molecules thought to be protective for the developing fetus is indoleamine 2,3-dioxygenase (IDO) which is produced in the trophoblast. IDO is normally produced by macrophages and dendritic cells to downregulate T-cell responses *in vitro* through its catabolic effect on the essential amino acid, tryptophan (Munn et al., 1999). Cell lines and transgenic mice that overexpress IDO elicit reduced alloimmune responses both *in vitro* and *in vivo*, highlighting its potential as a 'protective' molecule for genetically modifying hES cells.

A number of other molecules have been described that contribute to the maintenance of a state of immunological privilege characteristic of a range of tissues including the eye, testis, brain, the developing fetus and certain malignant tissues. For example, it is thought that high expression of the anti-apoptotic protein, Fas ligand (FasL) in the trophoblast confers fetal protection, possibly by inducing apoptosis of activated CD95+ maternal lymphocytes (Uckan et al., 1997). Other putative anti-apoptotic molecules include the A20 zinc finger protein induced by high levels of TNF-α in endothelial cells, bcl-2 and bcl-xL. Proof of this principle has been shown in both organ transplantation studies where gene transfer of anti-apoptotic molecules has protected allogeneic tissues from immunological rejection (Ke et al., 2000) and in neuronal cell transfer studies where bcl-xL transduced mouse ES cells were protected from neurotoxin induced cell death on transfer to the striatum of cyclosporine-treated rats (Shim et al., 2004). An alternative choice of molecule might be one that confers a protective milieu, such as the anti-inflammatory cytokine, interleukin 10 which is secreted in the placenta (Fairchild and Waldmann, 2000; Moreau et al., 1999). In summary, while some of the apparent targets for genetic alteration may not, after all, prove to be effective in reducing the immunogenicity of stem cell-derived grafts, other novel targets have potential and deserve to be examined in greater detail.

13.4 Strategies for preventing allograft rejection
13.4.1 Immunosuppressive therapy

A wide choice of immunosuppressive agents is now available and there is little doubt that these could be used successfully to overcome, at least at the phenotypic level, the immunological barrier to the use of hES-cell-derived tissue for clinical transplantation (*Table 13.2*). This view is based on the knowledge that these agents are used to prevent the rejection of different types of organ transplantation in clinical transplantation with increasing effectiveness. Even if no attempt were made to match the HLA antigens of hES-cell-derived tissue with its intended recipient there is no reason to doubt the ability of currently available agents effectively to control acute rejection. After all, living unrelated donor kidney transplants are usually poorly matched for HLA antigens, yet with modern immunosuppressive therapy, graft survival is as good as or better than for well matched cadaveric kidney transplants. Moreover, no attempt is made to match for HLA when undertaking heart or liver transplants, yet the results are generally very good. The one-year graft survival after solid organ transplantation is currently around 85% and five-year graft survival is around 70%. Irreversible acute graft rejection has become an uncommon cause of graft loss and the emphasis now in solid organ transplantation is on minimizing the side effects of immunosuppressive treatment by careful choice of immunosuppressive agents and managing risk factors. Modern immunosuppressive therapy has also enabled the introduction of human pancreatic islet transplantation and again no attempt is made to use HLA matched tissue (Shapiro *et al.*, 2000). Indeed, most recipients receive multiple islet transplants from different donors with no regard to HLA matching.

While immunosuppressive agents prevent acute graft rejection, this is achieved at a considerable price in terms of unwanted side effects. Such side effects are also a long-term problem, since immunosuppressive therapy must be continued indefinitely to prevent rejection. The non-specific immunosuppression that is an inevitable consequence of current immunosuppressive agents increases the risk of

Table 13.2: Immunosuppressive agents.

- **Calcineurin inhibitors**
  - Cyclosporine
  - Tacrolimus
- **Corticosteroids**
- **Antiproliferative agents**
  - Azathioprine
  - Mycophenolate mofetil
- **Sirolimus**
- **Biological agents**
  - Anti-CD25
  - Anti-CD3
  - Anti-CD52
  - Anti-lymphocyte globulin/serum

both infection and malignancy. In addition, immunosuppressive agents cause a wide range of potentially serious and sometimes distressing agent-specific side effects that may increase mortality and reduce the quality of life after transplantation. A significant number of recipients fail to adhere to their prescribed immunosuppression after solid organ transplantation, thereby risking graft failure (Nevins and Matas, 2004).

Whether the side effects of immunosuppressive therapy can be justified for hES-cell-derived tissue will depend on the likely benefit of the tissue transplanted. For example, if the transplant were a life-saving transplant of cardiac tissue, the justification for immunosuppressive therapy would be easy to argue. On the other hand, if the transplant were of insulin producing tissue to treat diabetes mellitus the benefit over existing insulin replacement therapy might be more marginal.

The approach to using immunosuppressive agents to prevent rejection of hES-cell-derived tissue would likely be similar to that used to prevent organ allograft rejection. If so a number of general principles can be stated. First, effective immunosuppression would be achieved using a combination of available agents as triple or quadruple therapy. Secondly, the levels of immunosuppression required to prevent rejection would probably be highest during the first few months after the transplant, and thereafter it may be possible to reduce the level of immunosuppression. Thirdly, immunosuppression would have to be continued indefinitely to ensure continued graft survival. Bearing these principles in mind it is instructive briefly to review the available immunosuppressive agents used in clinical transplantation and their specific side effects.

Immunosuppressive agents can be broadly categorized into five groups (*Table 13.2*).

Calcineurin antagonists

Two calcineurin antagonists are available, namely cyclosporine and tacrolimus, and one or other of them is used as the mainstay of most immunosuppressive drug regimens after solid organ transplantation (Kaufman et al., 2004). Although cyclosporine and tacrolimus are very different in their physical structure, they exert their principal immunosuppressive effects through the same mechanism. The two drugs bind to specific receptors known as immunophilins in the cytoplasm of lymphocytes. Cyclosporine binds to cyclophilin and tacrolimus binds to FK-binding protein (FKBP). The drug and immunophilin complex then binds to and inhibits the enzyme calcineurin which is a calcium/calmodulin-dependent phosphatase. Calcineurin is responsible in activated lymphocytes for facilitating the translocation of nuclear factor of activated T cells (NFAT) from the cytoplasm into the cell nucleus where it acts by increasing the transcription of IL-2 and other T-cell growth factors. By blocking the production of T-cell growth factors calcineurin inhibitors prevent the clonal expansion of both helper and cytotoxic T lymphocytes. Cyclosporine and tacrolimus are both administered twice daily and their dose is adjusted on the basis of the blood levels of the drugs which must be monitored to allow optimal therapy, especially during the initial phase of immunosuppression. Not surprisingly, in view of their similarity in mode of action, cyclosporine and tacrolimus share similar side effects. Like other immunosuppressive drugs they increase the susceptibility to infection and certain types of malignancy. Both agents are also nephrotoxic, which is probably their biggest weakness, and they both cause hypertension. Cyclosporine may cause cosmetic side effects, notably hirsutism and hyperplasia of the gums, whereas tacrolimus may cause diabetes and neurological symptoms. Nevertheless, the availability of calcineurin antagonists revolutionized solid organ transplantation and is likely to play a role in stem cell-based therapies as well.

Corticosteroids

Corticosteroids have potent anti-inflammatory and immunosuppressive properties and are an integral component of most immunosuppressive regimens after solid organ transplantation (Kaufman *et al.*, 2004). They exert their effects through multiple pathways and have numerous and well known side effects that include hypertension, diabetes, osteoporosis, peptic ulceration and cushingoid features.

Antiproliferative agents

The antiproliferative agent azathioprine has been used as an important component of immunosuppressive regimens since the early days of kidney transplantation. Azathioprine interferes with purine synthesis and inhibits lymphocyte proliferation. It is still widely used but it has now been replaced in many transplant units by the newer antiproliferative drug mycophenolate mofetil which has a more selective effect on lymphocyte proliferation. Mycophenolate is a pro-drug and is converted after ingestion to its active form, namely mycophenolic acid. It is a non-competitive and reversible inhibitor of inosine monophosphate dehydrogenase which is a rate limiting step in *de novo* purine synthesis.

The side effects of these agents include gastrointestinal symptoms, leukopenia and thrombocytopenia.

Sirolimus

Sirolimus is a relatively new and potent immunosuppressive agent (Dupont and Warrens, 2003). It exerts its principal immunosuppressive effect by inhibiting an intracellular kinase called mTOR (mammalian target of rapamycin). By interfering with the downstream signaling events from the IL-2 receptor and other cytokine receptors it inhibits lymphocyte proliferation. Side effects include hyperlipidemia, thrombocytopenia, leukopenia, elevated liver function tests, bone pain, and impaired wound healing.

Biological agents

Several different polyclonal and monoclonal antibody preparations are available and one or other of these is commonly used in the immediate post-transplant period as a temporary component of the immunosuppressive regimen. The most widely used agents are humanized or chimeric monoclonal antibodies to the IL-2 receptor and polyclonal anti-lymphocyte preparations (Webster *et al.*, 2004). The latter may also be used in an attempt to reverse acute rejection episodes which are refractory to treatment with corticosteroids.

New immunosuppressive agents

A number of promising novel immunosuppressive agents are currently in different stages of clinical development (Alsina and Grinyo, 2003). These include FTY720, a synthetic molecule that interferes with lymphocyte homing and causes migration of lymphocytes from the peripheral blood into lymph nodes and Peyer's patches of the intestine (Aki and Kahan, 2003). Biological agents in development include engineered antibodies and fusion proteins that block the delivery of essential co-stimulatory activity to lymphocytes by APC (Schuler *et al.*, 2004). The impending availability of such agents will further enhance the role of immunosuppressive agents in cell transplant therapies.

Side effects of non-specific immunosuppressive therapy

Immunosuppressive therapy increases the risk of infection and malignancy. The risk of viral and fungal infection is particularly high during the first few months after organ transplantation when higher doses of immunosuppressive drugs are given (Fishman and Rubin, 1998; Simon and Levin, 2001). Cytomegalovirus (CMV) infection is one of the most common problems and causes symptomatic disease in up to one third of recipients after organ transplantation. CMV disease may result from reactivation of latent virus or from a primary infection, often after transmission from the donor organ, and can be severe and even life threatening. Transplantation recipients are at particular risk of developing skin cancer, lymphoma and urogenital malignancy although there is also an overall (two fold) increase in most of the common solid malignancies (Penn, 2000; Feng *et al.*, 2003). The majority of skin cancers seen are squamous cell carcinomas, in contrast to the general population where basal cell carcinomas are more common (Euvrard *et al.*, 2003). The risk of skin cancer increases with time, and approximately half of transplant recipients develop a squamous cell carcinoma by twenty years after transplantation. Post-transplant lymphoproliferative disease (PTLD) is the second most common malignancy and may affect up to 5% of adults and 10% of pediatric transplant recipients (Shroff and Rees, 2004) . Most cases of PTLD are associated with EBV infection and the incidence of the disease increases with the amount of immunosuppressive therapy administered. While the physiological benefits of normalized organ function following transplantation are undeniable, the persistence of side effects of long-term immunosuppression remain, emphasizing the desirability of improving recipient tolerance to engraftment whenever a high degree of matching cannot be achieved.

13.4.2 Mixed hematopoietic chimerism

Studies in experimental animals have shown that engrafting allogeneic hematopoietic stem cells may, under appropriate conditions, lead to a state of stable long-term mixed hematopoietic chimerism where donor and recipient hematopoietic stem cells coexist in the absence of either graft rejection or graft versus host disease (Sykes, 2001). A key feature of this stable chimeric state is that the recipient thymus becomes populated by both donor and recipient APC (*Figure 13.4*). Thymocytes of recipient or donor origin with strong affinity for recipient or donor alloantigens are deleted, as part of the natural selection process of T cells in the thymus. As a result of this negative selection process, the chimeric recipient displays specific immunological tolerance to the alloantigens expressed by the donor hematopoietic cells. Such recipients will, therefore, accept a tissue or organ graft from the hematopoietic

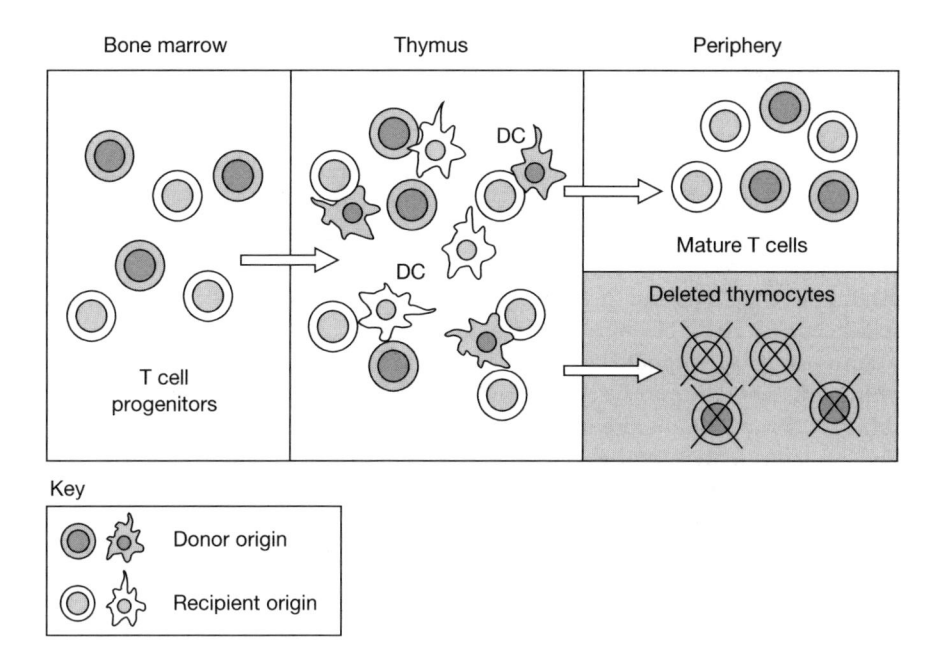

Figure 13.4: Mixed hematopoietic chimerism.
Donor hematopoietic cells populate the host thymus with dendritic antigen presenting cells (APC). Donor and recipient T-cell progenitor cells from the bone marrow enter the thymus and during their maturation are exposed to donor and recipient antigens. Thymocytes that react strongly with donor or recipient alloantigens on dendritic APC are eliminated as part of the natural negative selection process. Residual thymocytes are exported to the periphery as mature T cells. The net result is a state of mutual immunological tolerance between donor and recipient.

stem cell donor with no need for further immunosuppression and retain the ability to mount a normal immunological response against pathogens or tumor cells.

The concept of mixed hematopoietic chimerism is of particular relevance to ES cells because it takes advantage of the possibility that hES cells could be differentiated into hematopoietic stem cells (Kaufman et al., 2001; Chadwick et al., 2003). After stable chimerism had been established the recipient could then be grafted with a therapeutic tissue graft, for example, of pancreatic islets or cardiac myocytes derived, from the same hES cell line without the need for any additional immunosuppressive therapy. This strategy has become a step nearer with the recent demonstration of stable ES-cell derived T cells in MHC mismatched mice, although this has yet to be achieved with hES cells after hematopoietic reconstitution using normal, non-transformed allogeneic ES cells (Burt et al., 2004). Mixed hematopoietic chimerism was originally achieved in experimental animals by a severe conditioning regimen comprising lethal whole body irradiation and/or potent cytotoxic drugs to completely ablate the recipient hematopoietic system, followed by reconstitution with a mixture of autologous and allogeneic hematopoietic stem cells (Ildstad and Sachs, 1984) (Ildstad et al., 1985). More recently, however, a variety of less severe non-myeloablative regimens have been used to achieve successful mixed hematopoietic chimerism. These include the use of depleting antibodies to CD4 and CD8 plus thymic irradiation and conventional immunosuppressive drugs and, most recently, the use of co-stimulatory blockade (Durham *et al.*, 2000; Wekerle *et al.*, 2000).

ES cells may be able to induce donor specific tolerance even in the absence of recipient preconditioning. Fandrich *et al.* (2002 showed that rat ES-cell-like (RESC) lines induced stable mixed chimerism after injection into the portal vein of fully allogeneic recipients. There was no requirement in these studies for any preconditioning regimen and the animals displayed donor specific tolerance without evidence of GVHD. They were shown to accept a heart graft from the same donor strain as the RESC without any evidence of graft rejection, but to reject heart grafts from a different rat strain. The authors suggested that the RESC in their study were able to engraft in the absence of any recipient preconditioning because they lacked co-stimulatory activity and would not trigger a destructive T-cell response in the recipient and second, that they expressed FasL which might trigger self-destruction of any alloreactive recipient T cells that were activated.

These findings are of considerable interest although there is no evidence that human ES cells would behave in a similar way to induce donor specific tolerance. In any event, the direct injection of undifferentiated hES cells would be too high risk because it might lead to the formation of teratomas. However, these results encourage the development of *in vitro* differentiation of human hematopoietic stem cells that could be used in the clinical induction of such mixed chimerism.

13.5 Concluding comments

Current knowledge about the immunogenicity of human ES cells and their differentiated derivatives makes it clear that they do have an immune identity. Moreover, it is apparent that successful strategies for dealing with mismatch between donor and recipient immune identities are more likely to emerge from accommodating the donor immune identity, rather than eliminating it. These conclusions place an emphasis on maximizing the overall genetic diversity of pluripotent stem cell lines available for differentiation into transplantable tissues, a strategy similar to the transplant registries already in use for solid organs and hematopoietic stem cells. However, because they originate from early developmental stages, ES cells provide unique opportunities for expanding the 'immune pool' that could not be achieved with cadaveric solid organ and living donor bone marrow material. The manipulation of human oocytes to replace the gamete genomes with that of a somatic cell (genome replacement) has the potential to confer upon pluripotent stem cells an immune identity that is not just individually customized, but is more broadly applicable for transplantation. Similarly, derivation of pluripotent stem cells from parthenogenetically activated oocytes has the potential to contribute to immune matching in ways that have not been widely appreciated. In both these approaches, the entities emerging from oocyte manipulation have been scarcely studied in human material. Therefore, they represent unprecedented opportunities to gain insights into somatic cell genome reprogramming and the role of epigenetic modifications in the context of the human genome. Moreover, the exploration of specific immune system gene functions in an *in vitro* developmental context provides a potential source of major insights into human functional genomics.

Such opportunities emphasize the importance of expanding the *in vitro* developmental repertoire of controlled differentiation, particular along lineage pathways that can be potentially useful in clinical transplantation. This exhortation seems to apply particularly to cell lineages that could be clinically effective as single purified cell types, such as dopamine neurons (for Parkinson's), pancreatic beta cells (for diabetes) and hematopoietic stem cells. In sum, the strategies available for coping with immunity appear to be sufficiently diverse and robust to justify the claim that pluripotent stem cells and their derivatives do not pose new problems, but rather offer new solutions to the long-standing challenge of matching exogenous, transplantable tissues to their intended recipients.

References

Aki FT, Kahan BD (2003) FTY720: A new kid on the block for transplant immuno-suppression. *Expert Opin. Biol. Ther.* 3, 665–681.

Allen ND, Barton SC, Hilton K, Norris ML, Surani MA (1994) A functional analysis of imprinting in parthenogenetic embryonic stem cells. *Development* 120, 1473–1482.

Alsina J, Grinyo JM (2003) New immunosuppressive agent: expectations and controversies. *Transplantation* 75, 741–742.

Bellgrau D, Selawry H, Moore J, Franzusoff A, Duke RC (1995) A role for CD95 ligand in preventing graft rejection. *Nature* 377, 630–632.

Borderie VM, Lopez M, Vedie F, Laroche L (1997) ABO antigen blood-group compatibility in corneal transplantation. *Cornea* 16, 1–6.

Bradley JA (2004) A roundabout route to rejection: the contribution of cross-primed CD8 T cells. *Am. J. Transplant* 4, 675–677.

Bradley JA, Bolton EM, Pedersen RA (2002) Stem cell medicine encounters the immune system. *Nat. Rev. Immunol.* 2(11), 859–871.

Burt, RK, Verda, L, Kim, DA, Oyama, Y, Luo, K, Link, C (2004) embryonic stem cells as an alternate marrow donor source: engraftment without graft-versus-host disease. *J. Exp. Med.* 199, 895–904.

Chadwick K, Wang L, Li L, Menendez P, Murdoch B, Rouleau A, Bhatia M (2003) Cytokines and BMP-4 promote hematopoietic differentiation of human embryonic stem cells. *Blood* 102, 906–915.

Chen HL, Gabrilovich D, Tampe R, Girgis KR, Nadaf S, Carbone DP (1996) A functionally defective allele of TAP1 results in loss of MHC class I antigen presentation in a human lung cancer. *Nat. Genet.* 13, 210–213.

Chen Y, He ZX, Liu A, Wang K, Mao WW, Chu JX *et al.* (2003) Embryonic stem cells generated by nuclear transfer of human somatic nuclei into rabbit oocytes. *Cell Res.* 13, 251–263.

Cibelli JB, Grant KA, Chapman KB, Cunniff K, Worst T, Green HL *et al.* (2002) Parthenogenetic stem cells in nonhuman primates. *Science* 295, 819.

Clayton HA, Swift SM, James RF, Horsburgh T, London NJ (1993) Human islet transplantation – is blood group compatibility important? *Transplantation* 56, 1538–1540.

Cooper DK (1990) Clinical survey of heart transplantation between ABO blood group-incompatible recipients and donors. *J. Heart Transpl.* 9, 376–381.

Draper JS, Pigott C, Thomson JA, Andrews PW (2002) Surface antigens of human embryonic stem cells: changes upon differentiation in culture. *J. Anat.* 200 (Pt 3), 249–258.

Drukker M, Benvenisty N (2004) The immunogenicity of human embryonic stem-derived cells. *Trends Biotechnol.* 22, 136–141.

Drukker M, Katz G, Urbach A, Schuldiner M, Markel G, Itskovitz-Eldor J, Reubinoff B, Mandelboim O, Benvenisty N (2002) Characterization of the expression of MHC proteins in human embryonic stem cells. *Proc. Natl Acad. Sci. USA* 99, 9864–9869.

Dupont P, Warrens AN (2003) The evolving role of sirolimus in renal transplantation. *Qjm* 96, 401–409.

Durham MM, Bingaman AW, Adams AB, Ha J, Waitze SY, Pearson TC, Larsen CP (2000) Cutting edge: administration of anti-CD40 ligand and donor bone marrow leads to hemopoietic chimerism and donor-specific tolerance without cytoreductive conditioning. *J. Immunol.* 165, 1–4.

Euvrard S, Kanitakis J, Claudy A (2003) Skin cancers after organ transplantation. *N. Engl. J. Med.* 348, 1681–1691.

Fairchild PJ, Waldmann H (2000) Dendritic cells and prospects for transplantation tolerance. *Curr. Opin. Immunol.* 12, 528–535.

Fandrich F, Lin X, Chai GX, Schulze M, Ganten D, Bader M et al. (2002) Preimplantation-stage stem cells induce long-term allogeneic graft acceptance without supplementary host conditioning. *Nat. Med.* 8, 171–178.

Feng S, Buell JF, Chari RS, DiMaio JM, Hanto DW (2003) Tumors and transplantation: The 2003 Third Annual ASTS State-of-the-Art Winter Symposium. *Am. J. Transplant.* 3, 1481–1487.

Fishman JA, Rubin RH (1998) Infection in organ-transplant recipients. *N. Engl. J. Med.* 338, 1741–1751.

Fruh K, Ahn K, Djaballah H, Sempe P, van Endert PM, Tampe R, Peterson PA, Yang Y (1995) A viral inhibitor of peptide transporters for antigen presentation. *Nature* 375, 415–418.

Game DS, Lechler RI (2002) Pathways of allorecognition: implications for transplantation tolerance. *Transpl. Immunol.* 10, 101–108.

Grey ST, Arvelo MB, Hasenkamp W, Bach FH, Ferran C (1999) A20 inhibits cytokine-induced apoptosis and nuclear factor kappaB-dependent gene activation in islets. *J. Exp. Med.* 190, 1135–1146.

Grusby MJ, Auchincloss HJr, Lee R, Johnson RS, Spencer JP, Zijlstra M, Jaenisch R, Papaioannou VE, Glimcher LH (1993) Mice lacking major histocompatibility complex class I and class II molecules. *Proc. Natl Acad. Sci. USA* 90, 3913–3917.

Grusby MJ, Johnson RS, Papaioannou VE, Glimcher LH (1991) Depletion of CD4+ T cells in major histocompatibility complex class II-deficient mice. *Science* 253, 1417–1420.

He C, Heeger PS (2004) CD8 T cells can reject major histocompatibility complex class I-deficient skin allografts. *Am. J. Transplant.* **4**, 698–704.

Heeger PS (2003) T-cell allorecognition and transplant rejection: a summary and update. *Am. J. Transplant* **3**, 525–533.

Hochedlinger K, Jaenisch R (2002) Monoclonal mice generated by nuclear transfer from mature B and T donor cells. *Nature* **415**, 1035–1038.

Horuzsko A, Antoniou J, Tomlinson P, Portik-Dobos V, Mellor AL (1997) HLA-G functions as a restriction element and a transplantation antigen in mice. *Int. Immunol.* **9**, 645–653.

Hwang WS, Ryu YJ, Park JH, Park ES, Lee EG, Koo JM et al. (2004) Evidence of a pluripotent human embryonic stem cell line derived from a cloned blastocyst. *Science* **303**, 1669–1674.

Ildstad ST, Sachs DH (1984) Reconstitution with syngeneic plus allogeneic or xenogeneic bone marrow leads to specific acceptance of allografts or xenografts. *Nature* **307**, 168–170.

Ildstad ST, Wren SM, Bluestone JA, Barbieri SA, Sachs DH (1985) Characterization of mixed allogeneic chimeras. Immunocompetence, in vitro reactivity, and genetic specificity of tolerance. *J. Exp. Med.* **162**, 231–244.

Kaufman DB, Shapiro R, Lucey MR, Cherikh WS, Bustami R, Dyke DB (2004) Immunosuppression: practice and trends. *Am. J. Transplant.* **4** (Suppl 9), 38–53.

Kaufman DS, Hanson ET, Lewis RL, Auerbach R, Thomson JA (2001). Hematopoietic colony-forming cells derived from human embryonic stem cells. *Proc. Natl Acad. Sci. USA* **98**, 10716–10721.

Kawase E, Yamazaki Y, Yagi T, Yanagimachi R, Pedersen RA (2000) Mouse embryonic stem (embryonic stem) cell lines established from neuronal cell-derived cloned blastocysts. *Genesis* **28**, 156–163.

Ke B, Coito AJ, Kato H, Zhai Y, Wang T, Sawitzki B, Seu P, Busuttil RW, Kupiec-Weglinski JW (2000) Fas ligand gene transfer prolongs rat renal allograft survival and down-regulates anti-apoptotic Bag-1 in parallel with enhanced Th2-type cytokine expression. *Transplantation* **69**, 1690–1694.

Kenyon L, Moraes CT (1997) Expanding the functional human mitochondrial DNA database by the establishment of primate xenomitochondrial cybrids. *Proc. Natl Acad. Sci. USA* **94**, 9131–9135.

Koller BH, Marrack P, Kappler JW, Smithies O (1990) Normal development of mice deficient in beta 2M, MHC class I proteins, and CD8+ T cells. *Science* **248**, 1227–1230.

Le Bouteiller P, Blaschitz A (1999) The functionality of HLA-G is emerging. *Immunol. Rev.* **167**, 233–244.

Le Moine A, Goldman M (2003) Non-classical pathways of cell-mediated allograft rejection: new challenges for tolerance induction? *Am. J. Transplant.* **3**, 101–106.

Le Moine A, Goldman M, Abramowicz D (2002) Multiple pathways to allograft rejection. *Transplantation* **73**, 1373–1381.

Lee RS, Grusby MJ, Laufer TM, Colvin R, Glimcher LH, Auchincloss H Jr (1997) CD8+ effector cells responding to residual class I antigens, with help from

CD4+ cells stimulated indirectly, cause rejection of 'major histocompatibility complex-deficient' skin grafts. *Transplantation* 63, 1123–1133.

Lewandoski M (2001) Conditional control of gene expression in the mouse. *Nat. Rev. Genet.* 2, 743–755.

Lin H, Lei J, Wininger D, Nguyen MT, Khanna R, Hartmann C, Yan WL, Huang SC (2003) Multilineage potential of homozygous stem cells derived from metaphase II oocytes. *Stem Cells* 21, 152–161.

Matsuda M, Salazar F, Petersson M, Masucci G, Hansson J, Pisa P, Zhang QJ, Masucci MG, Kiessling R (1994) Interleukin 10 pretreatment protects target cells from tumor- and allo-specific cytotoxic T cells and downregulates HLA class I expression. *J. Exp. Med.* 180, 2371–2376.

Moreau P, Adrian-Cabestre F, Menier C, Guiard V, Gourand L, Dausset J, Carosella ED, Paul P (1999) IL-10 selectively induces HLA-G expression in human trophoblasts and monocytes. *Int. Immunol.* 11, 803–811.

Munn DH, Shafizadeh E, Attwood JT, Bondarev I, Pashine A, Mellor AL (1999) Inhibition of T cell proliferation by macrophage tryptophan catabolism. *J. Exp. Med.* 189, 1363–1372.

Munsie MJ, Michalska AE, O'Brien CM, Trounson AO, Pera MF, Mountford PS (2000) Isolation of pluripotent embryonic stem cells from reprogrammed adult mouse somatic cell nuclei. *Curr. Biol.* 10, 989–992.

Nevins TE, Matas AJ (2004) Medication noncompliance: another iceberg's tip. *Transplantation* 77, 776–778.

Opelz G, Wujciak T, Dohler B, Scherer S, Mytilineos J (1999) HLA compatibility and organ transplant survival. Collaborative Transplant Study. *Rev. Immunogenet.* 1, 334–342.

Paul LC, Baldwin WM 3rd (1987) Humoral rejection mechanisms and ABO incompatibility in renal transplantation. *Transplant. Proc.* 19, 4463–4467.

Penn I (2000) Post-transplant malignancy: the role of immunosuppression. *Drug Saf.* 23, 101–113.

Pollak R, Blanchard JM (2000) Organ donor or graft pretreatment to prolong allograft survival: lessons learned in the murine model. *Transplantation* 69, 2432–2439.

Qin L, Chavin KD, Ding Y, Tahara H, Favaro JP, Woodward JE, Suzuki T, Robbins PD, Lotze MT, Bromberg JS (1996) Retrovirus-mediated transfer of viral IL-10 gene prolongs murine cardiac allograft survival. *J. Immunol.* 156, 2316–2323.

Rideout WM 3rd, Hochedlinger K, Kyba M, Daley GQ, Jaenisch R (2002) Correction of a genetic defect by nuclear transplantation and combined cell and gene therapy. *Cell* 109, 17–27.

Rousseau P, Masternak K, Krawczyk M, Reith W, Dausset J, Carosella ED, Moreau P (2004) In vivo, RFX5 binds differently to the human leucocyte antigen-E, -F, and -G gene promoters and participates in HLA class I protein expression in a cell type-dependent manner. *Immunology* 111, 53–65.

Santos TA, Dias C, Henriques P, Brito R, Barbosa A, Regateiro F, Santos AA (2003) Cytogenetic analysis of spontaneously activated noninseminated oocytes and parthenogenetically activated failed fertilized human oocytes –

implications for the use of primate parthenotes for stem cell production. *J. Assist. Reprod. Genet.* **20**, 122–130.

Schuler W, Bigaud M, Brinkmann V, Di Padova F, Geisse S, Gram H *et al.* (2004) Efficacy and safety of ABI793, a novel human anti-human CD154 monoclonal antibody, in cynomolgus monkey renal allotransplantation. *Transplantation* **77**, 717–726.

Shapiro AM, Lakey JR, Ryan EA, Korbutt GS, Toth E, Warnock GL, Kneteman NM, Rajotte RV (2000) Islet transplantation in seven patients with type 1 diabetes mellitus using a glucocorticoid-free immunosuppressive regimen. *N. Engl. J. Med.* **343**, 230–238.

Shim JW, Koh HC, Chang MY, Roh E, Choi CY, Oh YJ, Son H, Lee YS, Studer L, Lee SH (2004) Enhanced in vitro midbrain dopamine neuron differentiation, dopaminergic function, neurite outgrowth, and 1-methyl-4-phenylpyridium resistance in mouse embryonic stem cells overexpressing Bcl-XL. *J. Neurosci.* **24**, 843–852.

Shroff R, Rees L (2004) The post-transplant lymphoproliferative disorder – a literature review. *Pediatr. Nephrol.* **19**, 369–377.

Simon DM, Levin S (2001) Infectious complications of solid organ transplantations. *Infect. Dis. Clin. North Am.* **15**, 521–549.

Simpson E, Scott D, James E, Lombardi G, Cwynarski K, Dazzi F, Millrain JM, Dyson PJ (2001) Minor H antigens: genes and peptides. *Eur. J. Immunogenet.* **28**, 505–513.

Springer GF, Horton RE (1969) Blood group isoantibody stimulation in man by feeding blood group-active bacteria. *J. Clin. Invest.* **48**, 1280–1291.

Storkus WJ, Alexander J, Payne JA, Dawson JR, Cresswell P (1989) Reversal of natural killing susceptibility in target cells expressing transfected class I HLA genes. *Proc. Natl. Acad. Sci. USA* **86**, 2361–2364.

Strand S, Hofmann WJ, Hug H, Muller M, Otto G, Strand D, Mariani SM, Stremmel W, Krammer PH, Galle PR (1996) Lymphocyte apoptosis induced by CD95 (APO-1/Fas) ligand-expressing tumor cells – a mechanism of immune evasion? *Nat. Med.* **2**, 1361–1366.

Suter T, Biollaz G, Gatto D, Bernasconi L, Herren T, Reith W, Fontana A (2003) The brain as an immune privileged site: dendritic cells of the central nervous system inhibit T cell activation. *Eur. J. Immunol.* **33**, 2998–3006.

Swenson KM, Ke B, Wang T, Markowitz JS, Maggard MA, Spear GS, Imagawa DK, Goss JA, Busuttil RW, Seu P (1998) Fas ligand gene transfer to renal allografts in rats: effects on allograft survival. *Transplantation* **65**, 155–160.

Sykes M (2001) Mixed chimerism and transplant tolerance. *Immunity* **14**, 417–424.

Szabo P, Mann JR (1994) Expression and methylation of imprinted genes during in vitro differentiation of mouse parthenogenetic and androgenetic embryonic stem cell lines. *Development* **120**, 1651–1660.

Takeda K, Akagi S, Kaneyama K, Kojima T, Takahashi S, Imai H, Yamanaka M, Onishi A, Hanada H (2003) Proliferation of donor mitochondrial DNA in nuclear transfer calves (*Bos taurus*) derived from cumulus cells. *Mol. Reprod. Dev.* **64**, 429–437.

Taylor AS, Braude PR (1994) The early development and DNA content of activated human oocytes and parthenogenetic human embryos. *Hum. Reprod.* 9, 2389–2397.

Thomson JA, Itskovitz-Eldor J, Shapiro SS, Waknitz MA, Swiergiel JJ, Marshall VS, Jones JM (1998) Embryonic stem cell lines derived from human blastocysts. *Science* 282, 1145–1147.

Uckan D, Steele A, Cherry A, Wang BY, Chamizo W, Koutsonikolis A, Gilbert-Barness E, Good RA (1997) Trophoblasts express Fas ligand: a proposed mechanism for immune privilege in placenta and maternal invasion. *Mol. Hum. Reprod.* 3, 655–662.

Vallier L, Rugg-Gunn PJ, Bouhon IA, Andersson FK, Sadler AJ, Pedersen RA (2004) Enhancing and diminishing gene function in human embryonic stem cells. *Stem Cells* 22, 2–11.

Valujskikh A, Lantz O, Celli S, Matzinger P, Heeger PS (2002) Cross-primed CD8(+) T cells mediate graft rejection via a distinct effector pathway. *Nat. Immunol.* 3, 844–851.

van der Meer A, Lukassen HG, van Lierop MJ, Wijnands F, Mosselman S, Braat DD, Joosten I (2004) Membrane-bound HLA-G activates proliferation and interferon-gamma production by uterine natural killer cells. *Mol. Hum. Reprod.* 10, 189–195.

Vrana KE, Hipp JD, Goss AM, McCool BA, Riddle DR, Walker SJ *et al.* (2003) Nonhuman primate parthenogenetic stem cells. *Proc. Natl Acad. Sci. USA* 100 (Suppl 1), 11911–11916.

Wakayama T, Tabar V, Rodriguez I, Perry AC, Studer L, Mombaerts P (2001) Differentiation of embryonic stem cell lines generated from adult somatic cells by nuclear transfer. *Science* 292, 740–743.

Webster AC, Playford EG, Higgins G, Chapman JR, Craig JC (2004) Interleukin 2 receptor antagonists for renal transplant recipients: a meta-analysis of randomized trials. *Transplantation* 77, 166–176.

Wekerle T, Kurtz J, Ito H, Ronquillo JV, Dong V, Zhao G, Shaffer J, Sayegh MH, Sykes M (2000) Allogeneic bone marrow transplantation with co-stimulatory blockade induces macrochimerism and tolerance without cytoreductive host treatment. *Nat. Med.* 6, 464–469.

Zeidler R, Eissner G, Meissner P, Uebel S, Tampe R, Lazis S, Hammerschmidt W (1997) Downregulation of TAP1 in B lymphocytes by cellular and Epstein-Barr virus-encoded interleukin-10. *Blood* 90, 2390–2397.

Zijlstra M, Bix M, Simister NE, Loring JM, Raulet DH, Jaenisch R (1990) Beta 2-microglobulin deficient mice lack CD4-8+ cytolytic T cells. *Nature* 344, 742–746.

14. Clinical applications for human ES cells

Timothy J. Kamp and Jon S. Odorico

14.1 Introduction

Most human diseases are characterized by the dysfunction and/or progressive loss of particular populations of cells in the body. Although cellular dysfunction may in some cases be treated with pharmacological therapy or other interventions, in many diseases the loss of specialized cell types marks an irreversible and incurable state. The dropout of dopaminergic neurons in the substantia nigra in Parkinson's disease, the necrosis of cardiac muscle in the setting of a myocardial infarction, and the loss of pancreatic islet β cells in type I diabetes mellitus are some of the best known examples. In all of these conditions and many more, the promise of replacing damaged, diseased, or missing cells with new functional cells holds tremendous promise. While whole organ transplantation has made major advances in the treatment of end-stage organ failure, the scarcity of donor organs limits the widespread application of this therapeutic approach. These factors have contributed to the growing interest in cell-based therapies or cellular transplantation. A new era of regenerative medicine is emerging. It is in this area of future therapy that human embryonic stem (hES) cells have been a focus of interest.

From the time of the initial isolation of hES cells in 1998 (Thomson *et al.*, 1998), the promise for therapeutic applications arising from these cells was recognized. This promise comes from the fact that hES cells are pluripotent and thus capable of differentiating into multiple different human cell types. Derivatives of all three embryonic germ cell layers have been identified in teratomas formed from hES cells injected into immunoincompetent mice wherein a variety of highly differentiated and even complex tissue structures can be found (*Figure 14.1*). Thus, the ability of hES cells to form potentially any cell type in the body has garnered widespread interest in the potential utility of this source of cells for clinical applications.

Human ES cells can be maintained in culture indefinitely. The cells possess a remarkable property of self-renewal, and have high levels of telomerase activity

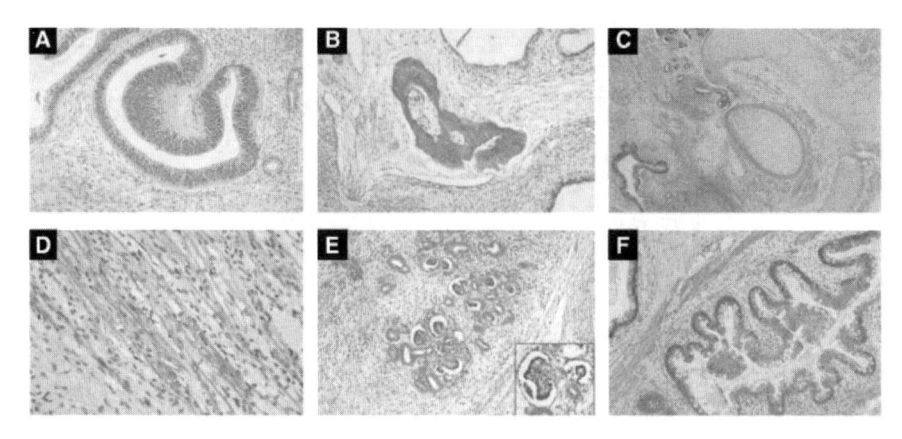

Figure 14.1: Tissue derivatives of all three embryonic germ layers differentiated from human ES cells *in vivo*. Human ES cells injected into immunocompromised mice form benign teratomas. Present within these teratomas are advanced derivatives of ectoderm, such as (**A**) neural epithelium (100×), of mesoderm, such as (**B**) bone (100×), (**C**) cartilage (40×), (**D**) striated muscle (200×), and (**E**) fetal glomeruli and renal tubules (100×; insert, 200×), and of endoderm, such as (**F**) gut (40×). To some degree micro-architectural tissue relationships of complex organs can be reproduced in human ES cell teratomas. H1, H7C, H9, H13, and H14 cell lines, which produced the above teratomas, exhibit a similar range of differentiation. All photomicrographs are of hematoxylin- and eosin-stained sections.

which contributes to their long-term stability (Thomson *et al.*, 1998). Published reports have described cells in culture continuously for 18 months, though higher passage cells may develop karyotypic abnormalities (Draper *et al.*, 2004; Pera, 2004; Cowan *et al.*, 2004). Thus, because hES cells can be extensively expanded in culture, they represent a potentially limitless source of cells for clinical applications.

More than two decades of research using mouse ES cells has provided a strong foundation for the ongoing studies using hES cells. If the results obtained with mouse ES cells in a variety of rodent models hold true with hES cells, then there is considerable hope that hES cells will find utility in clinical applications. In addition, a variety of other sources of donor cells for use in cell-based therapies are actively being investigated and in some cases have reached beginning clinical trials (Couzin and Vogel, 2004; Kondziolka *et al.*, 2000; Menasche *et al.*, 2001; Mazzini *et al.*, 2003; Freed *et al.*, 2001). However, investigations into the potential utility of hES cells in treating human disease are just beginning. The present chapter will describe the major goals that must be achieved and the current progress in the realization of the tremendous potential of hES cells for clinical medicine.

14.2 Goals for bringing hES cell-based therapy to clinical practice

Figure 14.2 shows some of the goals that must be achieved for hES cells to reach clinical applications. Significant basic research and preclinical research in animal models will be essential. While the specifics of each goal will vary with the particular clinical application, the general outlined goals should hold. Furthermore, most of the described goals will hold true for other donor cell types used in transplantation and tissue engineering.

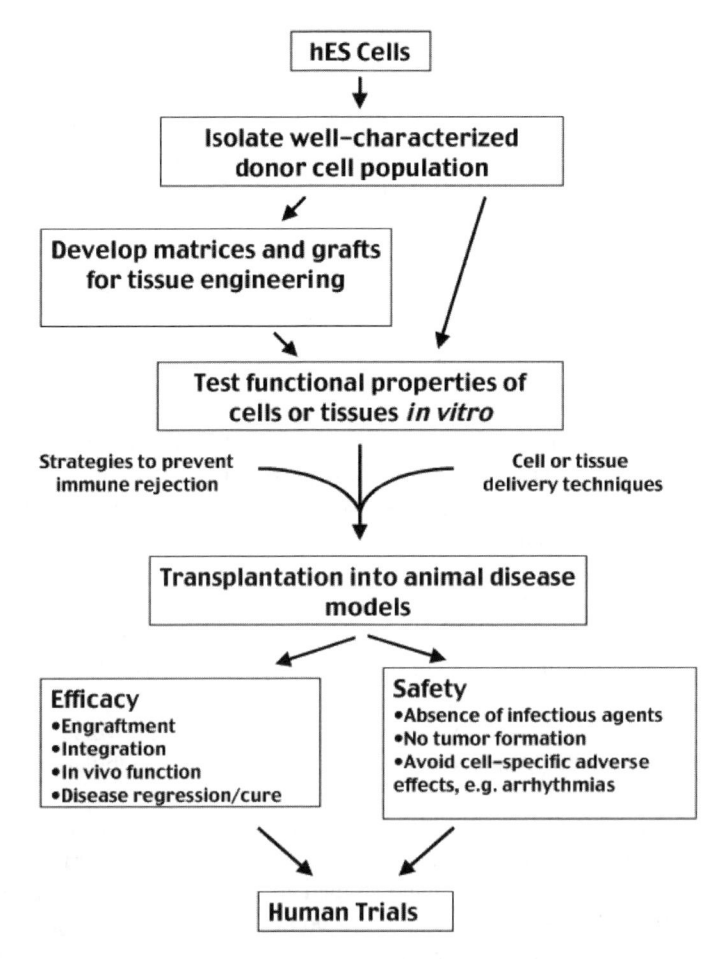

Figure 14.2: Goals for bringing hES cells to clinical applications.

14.2.1 Isolation of well-characterized donor cells in adequate numbers

The first challenge is to isolate a well-characterized donor cell population in adequate numbers to make therapy feasible. In the case of hES cell-based therapies, this will first require the ability to expand hES cells to adequate numbers using a well-defined cell line. The care needed for routine culture of hES cells with existing technology makes expansion of cell lines time consuming and not easily amenable to automation. In addition, many additional human ES cell lines would likely be necessary to scale-up for clinical applications. More hES cell lines would provide reasonable assurances that low passage lines were used that are more likely to be free of karyotypic abnormalities and to have retained their full pluripotency. Considerations related to scale-up and cell banking are beyond the scope of this chapter and are dealt with in Chapter 15.

Establishing defined feeder-free, serum-free conditions that support the derivation and propagation of hES cells would simplify culture protocols.

Although leukemia inhibitory factor (LIF) alone has been successful in maintaining an undifferentiated state in feeder-free culture of mouse ES cells, LIF does not exert a comparable effect in hES cells. Basic fibroblast growth factor (bFGF) is effective in promoting self-renewal in hES cells, but not alone, as embryonic fibroblast conditioned media is also required. Thus, it is generally believed that another self-renewal factor for hES cells must exist and this is an active area of investigation (Amit *et al.*, 2004; Lim and Bodnar, 2002).

Obtaining adequate numbers of undifferentiated hES cells is a starting point for cell banking and for producing a therapeutic product. However, undifferentiated hES cells may not be the ideal cell population for transplantation. The capacity of hES cells to form teratomas suggests that some degree of differentiation to a precursor or progenitor stage or even a terminally differentiated cell type may be desirable. The ideal donor cell type for transplantation will vary depending on the specific clinical application. Progenitor cells will have the advantage of being able to undergo continued cell division in response to stimuli and to develop potentially into multiple cell types necessary to regenerate functional tissues in some cases. However, this must be balanced by the potential of these cells to proliferate or revert to an undifferentiated ES cell phenotype with the undesirable possibility for tumor formation. In addition, these progenitor cells may not be able to differentiate *in vivo* into the fully functional mature cell types needed. In contrast, transplanting more terminally differentiated cell types will provide the appropriate cell types, but cell division will likely be more limited, potentially impacting the ability of the transplanted cells to regenerate the desired tissue. It will likely require significant experimental work to determine the ideal donor cell population for each application.

Producing a robust hES cell-derived donor population requires a reproducible procedure to produce progenitor or differentiated cells for therapy. Several approaches are possible, but the longest studied and most widely used at present is the embryoid body (EB) system. Embryoid bodies are the result of spontaneous differentiation of aggregates of ES cells undergoing spontaneous differentiation when culture conditions no longer favor undifferentiated propagation of ES cells, and this differentiation can be partially directed toward certain lineages by optimizing culture conditions. Some features of the process mimic embryogenesis with derivatives from all three germ layers present, but the EB is highly disorganized and variable in its composition. The EB system allows the inductive interactions among different cell types that are essential for the genesis of certain cells. For example, the endodermal precursors in EBs may be required for the efficient generation of pancreatic lineage cells (unpublished observations, JSO). The EB system has been widely used with mouse ES cells, and with some modification has been adapted to human ES cells. While mouse ES cells were separated into essentially single cells and cultured as hanging drops reproducibly to start the process, human ES cells do not readily form EBs under similar conditions (TJK, JSO unpublished observations). Instead, hES cell colonies are lightly digested to clumps of cells that are then cultured in suspension to allow EB formation. Because cell clumps can be cultured in suspension, techniques for growing multiple EBs using bioreactors have been described for mouse ES cells and likely would be possible for human ES cells (Zandstra et al., 2003). However, the major disadvantage of the EB system is that it produces a highly heterogeneous population of cells which would then likely need to undergo some form of separation to obtain relatively pure populations of progenitor cells for further differentiation or transplantation. Thus, at present this represents an inefficient process for most desired cells types.

A process of directed differentiation without EB formation has the appeal of being more reproducible and efficient than EB formation. Taking advantage of known signals from normal development, it may be possible to differentiate ES cells to a desired cell type. In this case, appropriate growth factors and regulatory molecules could be sequentially applied to a culture of ES cells, thereby driving differentiation in a stepwise fashion into a desired cell type. A strategy of sequential addition of these factors mimicking the temporal sequence of normal development may be necessary. For example, the early addition of bone morphogenetic protein 4 (BMP-4) to human ES cells can result in remarkably homogeneous cultures of trophoblast cells (Xu et al., 2002). In mouse ES cells, addition of ascorbic acid to the culture medium has been suggested strongly to favor cardiac differentiation (Takahashi et al., 2003). Likewise, compounds favoring neural differentiation of mouse ES cells have been described (Ding et al., 2003; Wichterle et al., 2002). In addition, the context of the signals driving lineage-specific differentiation may be important, such as the nature of neighboring cells of other lineages, the matrix substrate, or whether cells are provided a three-dimensional environment to grow (Levenberg et al., 2003).

Because development of many cell types requires cell-to-cell interactions and spatially integrated patterns of inductive signaling, an alternative strategy is to co-culture ES cells with cells or embryonic tissues that will stimulate their differentiation to a particular cell type. An example of this principle is the differentiation of hematopoietic derivatives from ES cells when they are cultured on bone marrow stromal cell lines such as mouse S17 cells (Kaufman et al., 2001). In normal cardiac development, inductive signaling from adjacent endoderm contributes to lateral mesoderm developing into cardiac muscle. Based on this normal developmental interaction, the endodermal-like cell line END-2 was co-cultured with hES cells, which promoted cardiogenesis (Mummery et al., 2003).

All of the above strategies, including using EBs and directing differentiation with cytokines, cell lines or embryonic tissues, may be considered complementary and not mutually exclusive. For certain donor cell types, it may be optimal to use a combination of these approaches for the most efficient production of derivative cells. Formation of neuroectoderm from ES cells is an example where a combination of EB formation with carefully controlled culture conditions with the inclusion of FGF2 or retinoic acid produces highly enriched cultures of neuronal cells (Zhang et al., 2001; Kim et al., 2002; Wichterle et al., 2002). Ultimately, a more comprehensive understanding of the developmental signals regulating cell lineage specification will help guide the establishment of protocols for the efficient differentiation of an increasing number of specific cell lineages.

Many of the cell differentiation strategies described above will result in mixed populations of cells including the desired donor cell population. Techniques have been developed to isolate distinct cell populations from mixed populations in EBs or ES cell differentiation cultures. Some strategies depend on either transgenes or gene trapping, where a cell lineage can be selected or isolated based on activation of a cell type specific promoter. Using a cardiac specific promoter (α-myosin heavy chain) driving a neomycin resistance gene, Klug *et al.* (1996) first demonstrated efficient selection of cardiomyocytes derived from mouse EBs. Thus, in the presence of selection with geneticin only the cardiomyocytes survive, resulting in significantly enriched populations. A related approach is to express a marker protein such as green fluorescent protein (GFP) driven by a cell-type specific promoter. Fluorescent cells can be obtained by fluorescence-activated cell sorting (FACS). Alternatively, by taking advantage of unique cell surface antigens it may be possible to sort cells using specific antibodies as has been readily done with hematopoietic cells. In addition to FACS sorting, magnetic beads offer another immunoaffinity approach to separate cell populations. Although our current state of knowledge of developmental mechanisms limits our ability efficiently to direct differentiation of hES cells into most lineages using a combination of strategies, including cytokine induction and lineage selection, it should be possible to isolate well-characterized cell populations for transplantation.

14.2.2 Test functional properties of cells or tissues in vitro

Many of the methods used to identify and isolate desired cell populations rely on following the expression of various marker proteins or genes. However, a more complete characterization of the donor cell population requires an assessment of function. Because the function of specialized cell types depend on the coordinated interaction of hundreds to thousands of proteins, it is difficult to predict on the basis of measuring expression of a limited number of proteins whether a cell will exhibit a particular function similar to that of a mature adult cell. What functional assay is chosen is obviously dependent on the cell type in question. For hES cell-derived pancreatic beta cells, insulin secretion could be measured in response to provocative stimuli such as high glucose and other secretagogues in either static incubation assays or dynamic perfusion assays. For various neuronal populations, specific neurotransmitter release assays and methods for measuring action potentials will indicate functional integrity. If cardiomyocytes are the desired population, then some characterization of the contractile and electrophysiological properties of these cells will be essential.

Another consideration is the possibility that a subset of cells in a purified population may have not completed their development and, as a result, they may not be terminally differentiated. A particular differentiation protocol may result in a mixed population of cells some of which are maturely functioning while others are poorly or non-functional despite all cells expressing a particular marker protein. Or, a subset of the selected population of cells may be undergoing apoptosis. Therefore, it may be important to assess function at a single cell level to determine what fraction of the population is maturely functional and viable. Assays for measuring the viability and electrophysiology of single cells can provide such information. An advantage of having the donor cells in culture is that they will be amenable to a wide armamentarium of functional assays dictated by the cell type.

Functional assays for progenitor cells will be essential for these earlier stage cells, but precursor or progenitor cells are not anticipated to exhibit mature functional properties. In the case of hematopoietic stem cells, colony forming assays in specialized three-dimensional culture conditions using methylcellulose and transplantations into irradiated mice provide important assays to determine the clonogenic and differentiative potential of hematopoietic precursors at various stages (Kaufman *et al.*, 2001). *In vitro* and *in vivo* clonogenic assays will need to be developed for progenitors of other lineages, such as pancreatic islet and neural lineages. After demonstrating robust function *in vitro*, it will be critical to determine whether cells have the desired functional effect when transplanted.

14.2.3 Efficient and safe cell delivery approaches

Cell transplantation will require delivery strategies that can implant viable cells in a way that favors their survival and integration. The particular application will dictate the ideal delivery method. Many catheter-based strategies are being evaluated for cell therapy-based applications. Catheters can be safely inserted into patients and guided to the target organ for cell delivery either into the vasculature feeding the organ or directly into the parenchyma. Specialized catheters for efficient cell delivery are actively being developed. For some applications direct surgical delivery of cells may be optimal, and for others, the surgical implantation of tissues engineered *ex vivo* will be appropriate. Engineered tissue constructs have already been widely used for bone, joint, and skin reconstructive surgery employing cells derived from human donors. For some applications, such as islet transplantation, the site is not as critical as it would be for other applications, such as transplantation of dopaminergic neurons where precise tissue localization of the cellular transplant may be necessary to achieve optimal function. A simple protocol for transplantation of human pancreatic islet cells into the portal vein and embolized into the liver has been established (Shapiro *et al.*, 2000). This technique generally shows good success and likely represents a reasonable approach to deliver any hES cell-derived islet cells. Alternatively, if production and secretion of a soluble molecule is the desired effect, it may not be necessary for cells to integrate into existing tissue at all. This concept has been applied in the creation of a cell holder using a modified inferior vena cava filter in which the cells remain in the blood circulation but are encapsulated and easily removable (US patent number 6,716,208, Nephros Therapeutics, Inc.).

14.2.4 Efficacy in animal models of disease

The major tests of hES cell-based therapies will come with testing of these therapies in animal models of disease. Each particular application will need a specific animal model, e.g., 6-hydroxy dopamine-induced Parkinson's disease, streptozotocin-induced diabetes, a surgical myocardial infarction model, etc. These models can involve small animals such as humanized mouse models or larger animals such as non-human primates. Perhaps the most robust pre-clinical evidence for the efficacy of ES cell-based therapy will come from studies using non-human primates and primate ES cells for therapy.

Ultimately, the goal of hES cell-based therapies is to recover stable normal functioning of the damaged or diseased target organ system following cell or tissue transplantation. However, to reach that final goal, it will likely be necessary to optimize multiple steps along the way. First, the transplanted cells or tissue constructs will need to engraft into the recipient. The cells must be able to survive where transplanted and it will often be desirable for them actively to grow and divide following transplantation, as would be the case in the transplantation of ES cell-derived lineage-restricted progenitor cells. Secondly, the donor cells will typically need functionally to integrate into the host tissue. In the case of the heart, this means forming an electrical connection with neighboring heart cells and become part of the electrical syncitium. In the case of regeneration of neural tissue, the regenerated or engrafted neurons will need to form appropriate synapses with neighboring cells. Next, even if the cells engraft and integrate into the host tissue, they will still need to exhibit appropriate functional properties ultimately to make the therapy successful. Finally, if the above steps are successful, the last goal will be to maintain stable graft function over time to provide a durable and potentially curative treatment.

Understanding potential difficulties with the tested cell therapy will require specialized approaches and techniques as appropriate for the given application. A critical feature of these pilot studies will be to discriminate between donor cells and native recipient tissue. This discrimination could potentially be accomplished by taking advantages of differences in certain surface antigens, e.g., HLA haplotypes, or purposely using male donor cells in females to allow detection of Y chromosomes by *in situ* fluorescence as a marker of donor cells. Alternatively, transplanted hES cells can be genetically modified to express a unique protein such as green fluorescent protein (GFP), beta galactosidase or other markers to allow tracking of the transplanted cells. Tissues can then be obtained for histological evaluation to determine the presence of donor cells. By using marker proteins such as GFP, it will be theoretically possible to identify living donor cells after transplantation, which could be re-isolated and functionally characterized. As transplantation experiments move forward, *in vivo* non-invasive imaging techniques to identify donor cells in the intact animal and track them will become increasingly important. Success in labeling donor cells with dense iron particles has allowed applications using magnetic resonance imaging (MRI) to track donor cells in living animals (Bulte *et al.*, 2001).

As scientific progress is made in the derivation of purified populations of functional cell types from human ES cells for testing in appropriate animal models, significant work can and should go on in demonstrating efficacy of mouse ES cell-derived tissues. For many initial studies with murine and human ES cell-derived tissues, the severe combined immunodeficient (SCID) mouse may be the workhorse model.

However, before human trials are initiated, large animal models, and in particular, models in the non-human primate will be desirable. For some applications this will be possible, but for others, surrogate models involving surgically or chemically induced disease will only be possible. Indeed, better large animal models more closely approximating certain human diseases are desperately needed. For example, a model of autoimmune diabetes is the non-obese diabetes (NOD) mouse, and reports of experimental cures in this model abound in the scientific literature, but no reliable large animal or primate autoimmune diabetes models exist in which to test these murine cures.

14.2.5 Avoidance of immune rejection

Just as in solid organ transplantation, immune-mediated rejection represents a major challenge to the success of cellular transplantation. Since normal adult somatic cells express major histocompatibility complex (MHC) antigens, it is predictable that mismatches between donor and recipient will result in immune-mediated destruction of transplanted cells or tissues, except perhaps in the case of inoculation into immunologically privileged sites such as the central nervous system. Likewise, because ES cell-based therapy is envisioned to generate normal somatic cells, it might be expected that ES cell-derived tissues would be immunogenic as well, though some data suggests they may be less immunogenic than adult tissues, and genetically matched human tissues could possibly one day be derived through therapeutic cloning.

Basic and applied research in immunobiology and the development of new immunosuppressive medications will be applicable to preventing rejection of ES cell-derived tissues after transplantation. Immunosuppressive drugs and biological agents have become the mainstay in clinical organ transplantation to prevent allograft rejection and these drugs could be used in early ES cell-based therapy trials. Unfortunately, systemic immunosuppression heightens the risk of infection and malignancy, and these complications are major causes of death after transplantation. Therefore, there is ongoing research to find less toxic and more effective ways to prevent rejection of transplanted organs or tissues. A number of strategies for preventing allograft rejection while avoiding non-specific chronic immunosuppression are in various stages of experimental or clinical development. Induction of specific tolerance to the donor organ has been achieved in animals by combining lymphocyte depletion and inoculation of donor strain cells. In humans, however, success with this approach has proved more elusive, except in a few rare cases (Butcher et al., 1999; Sorof et al., 1995; Spitzer et al., 1999). It is clear that it will be necessary to reduce the toxicity of the conditioning regimen in order to achieve wide application. Blockade of signaling through co-stimulatory molecules and/or induction of chimerism (Wekerle et al., 2000, 2001) or induction of immunoregulatory T cells are effective in promoting donor-specific unresponsiveness in rodent models, but significant work needs to be done before these strategies will be applied consistently in human allotransplantation (Lee et al., 2003). The extensive investigations in bone marrow and solid organ transplant immunology will certainly provide a strong foundation for understanding and avoiding immune rejection of stem cell-derived transplants. In addition, the strategies developed to prevent solid organ rejection will likely show efficacy in cellular transplantation.

Stem cell-based therapies will also present new opportunities to avoid rejection compared with solid organ transplants. Cells may be manipulated in culture with the goal of escaping recognition and/or avoiding destruction by immunological mechanisms. For example, one might purposefully overexpress molecules that would promote apoptosis of lymphocytes or overexpress immunosuppressive cytokines, or alternatively delete MHC antigens. For example, one strategy would be to develop a 'universal' donor cell phenotype through genetic manipulation, thereby generating a cell line that failed to express any MHC antigens. Unfortunately, such cells may be still susceptible to natural killer cell-mediated lysis, or minor antigens may stimulate a rejection response via indirect antigen presentation (Herberman, 1986). Not only can the MHC antigens be manipulated in ES cells, but genes encoding other molecules involved in the immune response could also be altered. Numerous examples of such approaches for transplants of somatic cell populations have been reported and could be applied to human ES cell-derived differentiated cell populations.

ES cell derivatives may elicit reduced allogeneic immune responses because of lower levels of MHC antigen expression and therefore may be less susceptible to immune-mediated rejection. Differentiated ES cells generally express low levels of Class I MHC antigens and undetectable levels of Class II MHC antigens (Drukker et al., 2002; Li et al., 2004; Tian et al., 1997; unpublished observations JSO). Purified cell populations of a particular lineage when isolated from ES cells may be devoid of, or have significantly reduced numbers of antigen presenting cells (i.e., dendritic cells or B cells) depending on the selection methods. These cell types, which normally circulate through tissues and organs of living individuals, are generally considered to mediate direct allorecognition in tissue and solid organ transplant recipients. Whether these factors would contribute to a significantly reduced rejection response and a reduced requirement for systemic immunosuppression are questions that remain to be answered.

A direct approach to minimizing immune rejection of human ES cell-derived transplants is to develop large banks of cell lines derived from genetically different embryos that would represent a broad spectrum of the MHC and minor histocompatibility antigens. However, the number of different ES cell lines needed would likely be in the tens of thousands only to obtain lines with partial matches, and this may not be feasible in the short term (reviewed in more detail in Chapter 13). Alternatively, selective genetic modification of ES cells, such as making the cells homozygous for many of the human leukocyte antigens (HLA), could simplify tissue matching and allow a much smaller bank of needed cell lines.

The prospect of using somatic cell nuclear transfer technology to develop human ES cells lines that could be used to create customized stem cell therapies for individual patients has also been proposed and would minimize, if not eliminate, the risk of immunological rejection of the stem cell transplant (section 14.5). In this case, because the nuclear material for transfer would be obtained from a somatic cell of a prospective recipient, the newly derived ES cell line and recipient would be genetically and immunologically matched (Lanza et al., 2002). No immunosuppression would be needed for such an 'autologous' transplant. Recently, Hwang and colleagues demonstrated the isolation of a pluripotent human ES cell line from a cloned blastocyst, which was produced through the transfer of cumulus cell nuclei into enucleated ova from the same donor (Hwang et al., 2004). The inefficiency and cost of using this technique to produce a new ES cell line for each potential patient makes this approach currently impractical. Therefore, there are a number of promising strategies to overcoming immune rejection of human ES cell-derived transplants, but at this stage most are simply hypotheses and limited testing of these strategies has been performed.

14.2.6 Safety

Safety considerations will be of major importance in bringing ES cell-related therapies to clinical application. There are several major areas of safety concern including the potential for transmission of infectious diseases, the possibility of tumor formation, and a myriad of tissue type-specific adverse outcomes. To avoid or lessen the risk of transmission of infectious agents, specific procedures and quality control tests are currently utilized in the setting of bone marrow transplantation and islet transplantation. Handling the cells in strict accordance with current Good Manufacture Practice (cGMP) guidelines for biological therapies will be essential. Screening cell lines for human pathogenic agents will also be essential (see Chapter 15).

The possibility of transmitting pathogens of other animal species to the recipient (zoonoses) through exposure to animal serum or tissue used during the production process is a particular concern. In the case of hES cells, the use of feeder cells derived from mouse embryos is a significant limitation, and could be the source of zoonotic pathogens, such as endogenous retroviruses or potentially oncogenic viruses. Even in the setting of pig-to-human or non-human primate xenotransplantation, where there is direct, intentional contact of tissues in a heavily immunosuppressed individual, the precise risk is not known and there are differing opinions as to the magnitude of the threat for human recipients of pig xenografts (Cunningham *et al.*, 2001; Elliott *et al.*, 2000; Fishman, 2001; Michaels *et al.*, 2001; Paradis *et al.*, 1999; Weiss, 1999, 2003). To determine whether similar zoonotic infections are a risk to recipients of hES cell-derived transplants, we will first need to make an assessment as to which viruses should be monitored in cell lines and transplant recipients, then develop and test accurate assays. Although the threat of zoonotic infections in pig xenograft recipients is real, it is interesting that follow-up studies in these patients have rarely detected transmission of endogenous retroviruses and therefore the problem may constitute a mostly theoretical hazard (Cunningham *et al.*, 2001; Elliott *et al.*, 2000; Michaels *et al.*, 2001; Paradis *et al.*, 1999). Nonetheless, because of safety concerns, existing and future lines exposed to animal tissues will likely need to be tested rigorously for a panel of infectious agents and possibly endogenous retroviruses (see Chapter 15).

To avoid the risk of zoonoses entirely it would be necessary to derive and maintain lines in the absence of animal cells and serum. Recent efforts have demonstrated alternative ways of propagating human ES cells using human cells as feeders such as human mesenchymal stem cells or human fibroblasts (Cheng *et al.*, 2003; Richards *et al.*, 2002). The use of human feeder cells avoids the concern of transmitting animal pathogens or developing a zoonotic infection, but the feeder cells would still need extensive characterization to document the absence of infectious agents. Consequently, methods to culture the human ES cells in the absence of feeder cells have been an important focus. Although success in growing the cells in the absence of feeder cells has been demonstrated, the required use of conditioned media from feeder cells, which could still potentially result in contamination, may still be a concern (Xu *et al.*, 2001). Thus, the derivation and prolonged propagation of hES cells completely free of animal products is an important goal.

The second major safety concern is the possibility of tumor formation from the transplanted cells. This is particularly a concern when relatively undifferentiated cells are transplanted. For example, a defining feature of ES cells is the ability to form teratomas when transplanted into adult immunodeficient animals. However, teratomas are benign tumors and ES cells have not yet been documented to form malignant teratocarcinomas. Furthermore, transplanting differentiated progeny from mouse ES cells does not result in tumor formation at least in short-term studies (Klug *et al.*, 1996; McDonald *et al.*, 1999; Soria *et al.*, 2000). In order to safeguard against a very small number of undifferentiated cells, negative selection strategies could also be employed during the process of deriving the differentiated cell type of interest for transplantation. For example, cell sorting or sensitivity to a toxin could potentially remove cells that either reverted to, or maintained an undifferentiated phenotype and expressed Oct 4 or other specific gene(s) not expressed in the mature cell type of interest. An additional concern would be the possibility that the ES cells became transformed after multiple passages. Once transformed, even a terminally differentiated population devoid of undifferentiated stem cells might still be tumorigenic. Avoiding tumor formation will require careful and detailed characterization of each hES cell line and the differentiated populations destined for transplantation. In addition, transplants of hES cell derivatives into immunodeficient mice with long-term follow-up will be important in excluding significant tumorigenicity and ensuring that the transplanted cells are safe. Perhaps the most rigorous proof of lack of tumor formation for a particular transplant strategy will come from the use of non-human primate ES cells in the same species disease model. Measures to safeguard against tumor formation or kill tumorigenic cells once they arise may ultimately be necessary.

In addition to the general risks associated with any type of ES cell-based therapy as just described, there are application-specific risks. For example, cell therapy in heart disease may be associated with the risk of inducing life-threatening arrhythmias. Therapy with stem cells for Parkinson's disease or other CNS diseases may be associated with an increased risk of seizures. Thus, site and application-specific patient safety issues will need to be considered.

Lessons from the gene therapy trials have suggested that premature clinical trials without adequate animal testing can result in major setbacks and excessive patient risks. Therefore, it will be prudent to test cell-based therapies using ES cell-derived donor cells extensively in animal models. An attractive pre-clinical, large animal model for safety testing will likely be non-human primates using ES cell lines derived from these species. It may also be possible to test the hES cell lines destined for therapy themselves in certain animal models such as 'humanized' mouse models or potentially in non-human primate models.

14.2.7 Clinical trials

The final step in bringing hES cell-derived donor cells or tissues to therapeutic use will be the careful evaluation in clinical trials. Phase 1 safety studies with adequate duration of follow-up will be necessary. Next randomized, blinded trials with appropriate control groups will be essential to delineate the benefit of treatment given the potential for powerful placebo effects. In addition, it is possible that some observed benefits might be due to the delivery procedure, associated growth factors in media, or other related confounding factors rather than due to the transplanted cells themselves. Rigorous clinical trials will also allow clear reporting of adverse effects to optimize safety. In addition, long-term follow-up will be necessary to confirm lasting benefit as well as to exclude late adverse effects such as malignancy.

14.3 Tissue engineering with hES cells

Tissue engineering can be broadly defined as the application of the principles of life sciences and engineering to generate biological substitutes to replace or support the function of defective or injured body parts (www.ptei.org/about_te/index.html). Operationally, this might entail combining one or more different cell types with one or more different polymer matrices to form a defined shape and perform a specific function. A classical example of a biological product engineered as a tissue/scaffold composite already in clinical use is the construction of an artificial skin substitute generated by seeding human foreskin keratinocytes onto a dermal layer composed of human fibroblasts in a bovine Type I collagen matrix (Apligaf®, Organogenesis, Inc. Canton, MA, USA). Another example of tissue engineering to produce new blood vessels is the work of Niklason *et al.* (1999) who have expanded aortic smooth muscle cells and endothelial cells obtained from a vascular biopsy and seeded them into a biodegradable polymer tubular scaffold using a pulsatile-flow bioreactor. While some tissue engineered constructs are far along in pre-clinical development or even in clinical use, others, such as bioengineered kidney and heart tissue, lag behind in development. Still, remarkable progress has been made over the last decade in the field of tissue engineering.

One technical aspect that hinders development of some tissue engineering constructs and/or their eventual scale-up for clinical use is the limited availability of human cells. Each application requires a large, readily available source of functional human cells having defined specificity. Human ES cells offer a potentially limitless source of cells for tissue engineering applications (Koh and Atala, 2004a). Going one step further, recent work demonstrating the ability to generate genetically identical or autologous human ES cells through therapeutic cloning (see below) now establishes the prospect of producing cells for these applications that are both unlimited in number and not susceptible to rejection in the recipient (Hwang *et al.*, 2004; Koh and Atala, 2004b).

The efficient differentiation of human ES cells into defined cell lineages in culture will be critical to the application of ES cells to tissue engineering. Currently, *in vitro* culture systems for ES cell differentiation and tissue engineering are rather crude approximations of the complex biochemical and physical environments that are experienced by cells during embryonic organ development and adult reparative processes *in vivo*. Commonly employed two-dimensional culture dishes have limitations. Optimal conditions for tissue remodeling and human ES cell differentiation may ultimately depend on three-dimensional cell–cell and cell–matrix interactions that can be provided by biodegradable polymer scaffolds. Bioactive molecules, cytokines, and growth factors can be delivered locally to the cells by the co-administration of the molecule and cells to the polymer scaffolds establishing a favorable microenvironment for differentiation (Leach and Schmidt, 2005; Newman and McBurney, 2004). Alternatively, polymer scaffolds may be engineered to be able to bind to and directly deliver the bioactive molecules of interest (Hartgerink *et al.*, 2001, 2002). Only initial pilot experiments have been conducted with human ES cells and 3D scaffolds, however (Levenberg *et al.*, 2003). This research area will likely see increased activity over the coming years as more knowledge about the factors and signals necessary for efficient differentiation of human ES cells is gained, and more sophisticated polymer scaffolds are created. The combination of human ES cells and tissue engineering in the production of organ and tissue replacement therapies is a logical experimental strategy that should see greater focus in the future.

14.4 Cell-based therapy for delivery of bioactive molecules

Many genetic diseases result from the failure to produce a single functional protein and would be amenable to treatment with cells engineered to produce that protein. Because ES cells are relatively permissive for stable genetic modification, they could function as an effective vehicle for delivery of bioactive gene products either to correct a genetically inherited disease or deliver a therapeutic protein.

14.4.1 Nexus of cellular transplantation and gene therapy for inherited human disorders

Many gene therapy protocols have focused on delivering the vector directly to the patient with the hope that one or more cell types in the recipient would express the vector and the gene of interest. This approach, termed *in vivo* gene therapy, is associated with potentially severe toxicities directly related to the viral vectors themselves (Kaiser, 2003; Romano *et al.*, 1998; Somia and Verma, 2000). Other practical and safety considerations include the inability exactly to control gene dosing, accidental insertional mutagenesis, compromised function in some host cells, stimulated immunity to viral proteins on host cells, and viral transport to ectopic sites causing undesirable effects. An alternative strategy is to engineer cells *ex vivo* to express a protein of interest, and then transplant the engineered cells to the host or patient. Because the recipient is not directly exposed to large titers of the viral vector, the *ex vivo* strategy may avoid the toxicities and immune stimulation associated with direct delivery. This approach would be especially attractive if the situation existed such that the 'cellular vehicle' for gene delivery is autologous or isogenic to the prospective recipient. An example as applied to the treatment of Type I, insulin-deficient diabetes might first entail harvesting liver cells from a diabetic patient, then subjecting the liver cells in culture to gene therapy with the insulin gene to allow them to express insulin, and then transplant the autologous genetically modified cells to the same diabetic patient (Alam and Sollinger, 2002). This would have the advantage of avoiding rejection and the requirement for long-term immunosuppression, but the genome of somatic cells is not readily and stably modified, and it is usually challenging to expand primary cells in culture over many passages.

ES cells possess several properties that are useful in a combined gene and cell therapy strategy. Stable genetic modification in mouse and human ES cells is straightforward using a variety of methods including homologous recombination (see Chapter 12, Zwaka and Thomson, 2003). Also, because ES cells can be repeatedly expanded and differentiated in culture, they could provide a ready source of cells for continued therapy as might be needed in the case of repeated treatments in patients. Though permissive for stable genetic manipulation and capable of many-fold expansion *in vitro*, undifferentiated ES cells themselves may not serve as an effective transplantable cellular vehicle for gene therapy primarily because they may be tumorigenic and would not be autologous. However, it now appears feasible to make mouse or human ES cells isogenic to a prospective recipient by somatic cell nuclear transfer (Hwang *et al.*, 2004; Munsie *et al.*, 2000; Rideout *et al.*, 2002).

Recently, a combinatorial cell and gene therapy strategy employing ES cells was elegantly shown by Rideout *et al.* (2002) to correct a genetic defect in mice. In their experiment, *Rag2* -/- immunodeficient mice donated cells, which were used for nuclear transfer to create *Rag2*-/- cloned blastocysts from which isogenic *Rag2*-/- ES cells were generated by standard derivation procedures. The *Rag2* deficiency was corrected by homologous recombination, taking advantage of the ease of stably manipulating the genome in ES cells. Once the genetic defect was corrected, ES cells were then differentiated into hematopoietic derivatives *in vitro* and transplanted back into the *Rag2*-/- mice from which the donated cells were obtained. Ultimately, the immune system in these mice was partially restored (Rideout *et al.*, 2002). Thus, human ES cells and their differentiated progeny provide an opportunity to repair genetic defects within stem cells to treat an inherited disease. Because human ES cells can be differentiated into many cell types, the approach demonstrated by Rideout *et al.* may be broadly applicable to a variety of human genetic diseases that can be corrected by cellular transplantation.

14.4.2 Continuous local or systemic delivery of bioactive molecules by ex vivo gene therapy using stem cells

Just as the combination of gene therapy and ES cell-based therapy may be used to correct a genetic defect, ES cells or other types of stem cells could provide a convenient vehicle to deliver a therapeutic protein or other bioactive molecule. For some diseases or applications it would be desirable to have continuous delivery of the bioactive substance over a long period of time. In this setting, current medical standard of care dictates the use of long-term indwelling intravenous catheters or mini-pumps for this purpose. However, these procedures are commonly associated with infectious and thrombotic complications. In the case of direct delivery of a molecule to the brain, an indwelling catheter or pump is even more problematic. Thus, alternative methods for continuous local administration of a therapeutic protein are being investigated. Cells modified *ex vivo* to produce and secrete a therapeutic molecule and then transplanted provide a potential means to deliver a drug continuously until the cell dies or production is turned off. Depending on the site of integration of the modified cells, either local delivery within a specific organ or systemic delivery may be achieved.

Stem cells may be ideal cells for *ex vivo* gene therapy because they can be genetically modified and then cultured and expanded. Stem cells of human origin, such as human ES cells, could be used for clinical application in this regard, but much research must be done to demonstrate proof of concept in animals first. In this vein, rodent neural stem cells were recently modified *ex vivo* to express glial cell line-derived neurotrophic factor (GDNF) and cells secreting GDNF were transplanted into the substantia nigra in a rat model of Parkinson's disease (Ostenfeld *et al.*, 2002). They found that GDNF delivered by neurospheres enhanced the survival of co-transplanted primary dopamine neurons whereas neurospheres delivering GFP did not. ES cell derivatives could also function in the same way as neural stem cells in the aforementioned application as ES cells are amenable to stable genetic manipulation to express a transgene without the need for viral vectors, either constitutively, or inducibly for dose control. Moreover, the genetically modified ES cells could be differentiated *in vitro* into a desired cell type depending on the application. Applications of human ES cells in neural diseases would be particularly attractive because of the ease of differentiating the cells into this lineage and the relatively immunologic privilege of the central nervous system afforded by the blood–brain barrier. Yet, such clinical applications are largely hypothetical at this point and will first require much basic research.

To allow stem cells to secrete a protein product into the systemic circulation and to protect the host against potential tumor formation, the use of encapsulation devices have been proposed by some. Using other cell types to treat a variety of diseases including diabetes, liver failure, renal failure, and sepsis among others, investigators have studied many different encapsulation devices, both implantable and extra-corporeal, in animal and early human clinical trials (Baccarani *et al.*, 2004; Demetriou *et al.*, 2004; de Vos *et al.*, 2002; Efrat, 2002; Fissell *et al.*, 2001, 2003; Humes *et al.*, 1999, 2003; Maguire *et al.*, 2000). Perhaps the most widely studied disease application for encapsulation technology is the encapsulation of insulin producing tissue for the treatment of Type I diabetes, in which numerous different implantable devices have been tested, including implantable bioartificial devices containing functional cells that are then anastomosed to blood vessels, implantable devices that allow diffusion of molecules from cells within the device, and biodegradable or non-biodegradable micro- and macro-capsules. With these devices, cell lines (from insulinomas) that have tumor potential, or primary cells from other species (pig or primate isolated islets of Langerhans) that would stimulate a vigorous immune reaction are encapsulated *ex vivo* in a fixed, enclosed space surrounded by a semi-permeable membrane to allow oxygen and nutrient influx, while permitting efflux of waste products and bioactive molecules. These devices also have the advantage of restricting access to the cellular elements of the immune system, thereby theoretically preventing immune mediated damage. Unfortunately, many encapsulation devices fail in practice adequately to preserve long-term function of the contained cells. Damage to cells may be mediated in part by inflammatory cytokines small enough to gain access through the pores, or by cellular components of the immune system that gain access through microscopic defects in the capsules. In addition, devices that are implanted into the peritoneum or subcutaneous tissue may suffer over time from the accumulation of fibrous tissue around the device, which ultimately restricts the supply of nutrients or diffusion of bioactive molecules. Other devices, such as those that are connected to the circulatory system, are limited by the need for surgery for implantation, and risks of vascular compromise and thrombosis as well as risks of foreign-body type infection. On the other hand, larger implantable encapsulation devices may provide an opportunity to refill the chamber with additional doses of cells if the first dose is exhausted. Newer generation encapsulation devices and technologies are continuing to evolve with increasing promise of clinical utility by allowing regulated delivery of biomolecules while protecting cells from immune mediated damage on the one hand and protecting the patient from tumor formation on the other.

14.5 Somatic cell nuclear transfer and hES cells

While hES cells open up new possibilities for cell replacement therapies, hES cell-derived tissues would likely be susceptible to immunological attack by the recipient of the transplant unless the ES cell line and recipient were genetically matched. The closely timed reports of the first cloned mammal, Dolly the sheep (Wilmut et al., 1997), and of the derivation of human ES cells (Thomson et al., 1998) immediately suggested to many in the stem cell field the prospect of combining these technologies to get around the problem of immunological rejection of ES cell-derived tissue transplants (Gurdon and Colman, 1999; Lanza et al., 1999; Solter, 1998; Solter and Gearhart, 1999). Thus, the concept of therapeutic cloning, or the generation of genetically identical ES cell lines through somatic cell nuclear transfer (SCNT), was born. To distinguish it from *reproductive* cloning, which describes the use of SCNT to generate genetically identical viable offspring, the term *therapeutic* cloning is now commonly used. In essence, therapeutic cloning as first described by Solter (1998) entails the *in vitro* electrofusion of a somatic cell, initially obtained through a tissue biopsy or blood sample, and an enucleated oocyte to produce an embryo genetically matched to the somatic cell donor, except for some oocyte-derived mitochondrial antigens. From this genetically matched embryo, subsequently grown in culture to the blastocyst stage, an ES cell line would be derived. From this isogenic ES cell line, differentiated cell and tissue types would be generated through a variety of *in vitro* protocols. The resulting tissue should theoretically be compatible, if not genetically identical, to the individual from whom the somatic cell was obtained. A hypothetical scheme for generating autologous human ES cells through therapeutic cloning for treating patients is shown in *Color Plate 11*. It would be anticipated that the autologous human ES cell-derived tissues would consequently be protected from immunological rejection when transferred back into the somatic cell donor, and would be accepted without the need for immunosuppressive medications. Notwithstanding the negative connotations connected to human therapeutic cloning and the ethical implications inherent in the procedure, research in this area has progressed because of the potential impact on human medicine and what it can teach us about basic biological properties ranging from imprinting to X chromosome inactivation to plasticity in human embryos.

In recent years investigators have begun to address the feasibility of key elements of this hypothetical scheme upon which Solter initially claimed 'there are no theoretical obstacles' (Solter, 1998). Shortly after the description of SCNT using an adult somatic cell to derive Dolly, reports of NT-derived mice and calves emerged (Cibelli et al., 1998a,b; Wakayama et al., 1998) which demonstrated that this was not a rare, aberrant phenomena unique to sheep and could be reproduced in other mammalian species. Yet the efficiency of SCNT remains low and is critically dependent on the cell type, which is used as the nuclear source. ES cells and cumulus cells appear to have the highest efficiency. Further research will be necessary to help answer basic questions, such as why are some cells easier to reprogram than others, and what factors in the oocyte cytoplasm contribute to reprogramming and resetting telomere lengths? As insights are gained, methods to increase the efficiency or artificially reprogram nuclei may be devised.

Until recently a major question was: Could this be done in humans, and if so, how many human oocytes would be needed and from whom would they come? The landmark report by Hwang and colleagues demonstrates that autologous human ES cell lines can be created using SCNT with the donor's own cumulus cell as the nuclear donor (Hwang et al., 2004). However, they also report that this process is currently inefficient and not practical on a large scale. Whereas human ES cell lines can be derived from isolated inner cell masses at an efficiency of approximately 30%, the frequency of deriving a cloned human ES cell line is estimated to be approximately 5% (1 cell line from 20 isolated inner cell masses) (Hwang et al., 2004; Lanzendorf et al., 2001; Reubinoff et al., 2000; Thomson et al., 1998). In addition, in order to generate a single cloned human ES cell line, it was necessary to collect 242 oocytes from 16 volunteers. Thus, practically speaking, nuclear reprogramming in all mammalian species remains an inefficient process and the idea of producing genetically matched human ES cell lines on a scale needed to provide custom cell lines for many thousands of individuals in need is far from reality.

One of the key practical and ethical challenges in the scheme is the source of human oocytes. Female donors in the Hwang study needed to undergo ovarian stimulation protocols that are painful and can have serious side effects. Still, there are significant limitations in the number of eggs available at any one time. An alternative source of oocytes could potentially be from human ES cells themselves if oocytes derived from human ES cells were shown to possess nuclear reprogramming functions of similar efficiency to that of normal oocytes (Hubner et al., 2003). Alternatively, since cross-species nuclear reprogramming may be possible, oocytes could be recovered from other species (Dominko et al., 1999). Although this may provide a ready source of eggs, it raises additional ethical questions. Ultimately, a greater understanding of the biology of nuclear reprogramming will be necessary in order to refine the technology before it can be applied on a routine basis.

If SCNT-derived ES cells lost or lacked an ability to differentiate into many somatic cell types, they would not be useful despite being matched to the recipient. Studies now indicate that SCNT-derived ES cells are pluripotent. The broad differentiative capacity during development *in vivo* of SCNT-derived ES cell lines was confirmed by Cibelli and colleagues (Cibelli *et al.*, 1998b). Similar studies extending these findings to mice were later reported (Kawase *et al.*, 2000; Munsie *et al.*, 2000; Wakayama *et al.*, 2001). More recently, the Hwang *et al.* (2004) study showed that SCNT-derived human ES cell lines do indeed have pluripotential *in vitro* and *in vivo*. Importantly, SCNT-derived mouse ES cell lines are able to differentiate into the hematopoietic lineage under the correct conditions and provide cells capable of reversing a genetic defect in the nuclear donor (Rideout *et al.*, 2002).

The presence of oocyte-derived mitochondria and mitochondrial DNA encoded proteins in NT cells has raised questions about the histocompatibility of the cloned cells (Evans *et al.*, 1999; Lanza *et al.*, 2002). It has been suggested that non-self mitochondrial antigens can be presented to the host immune system and stimulate a T-cell response based on minor histocompatibility antigenic differences (Davies *et al.*, 1991). To address this, Lanza and colleagues transplanted fetal tissues (cardiac, skeletal muscle, and renal promordia) from cloned calves into non-immunosuppressed adult cows that were the nuclear donors (Lanza *et al.*, 2002). They showed that these grafts survived for up to 12 weeks without significant loss of cell viability and with fewer infiltrating cells than allogeneic control grafts. They also demonstrated that the cloned fetal tissues elicited reduced delayed-type hypersensitivity responses in the nuclear donor animals compared with allogeneic cells and the frequency of T cells secreting interferon γ in mixed lymphocyte cultures was significantly lower when cloned renal cells or nuclear donor fibroblasts were used as stimulators compared with fully allogeneic stimuli. These data suggest that SCNT-derived tissues may be immunologically protected and that mitochondrial-encoded proteins from the oocyte donor may not pose a substantial impediment to transplantation, although further immunological studies on such recipients and longer follow-up of functional graft survival are warranted.

In summary, recent proof of concept studies demonstrate that autologous ES cells can be generated from human embryos; they are pluripotent and based on animal studies, may be less susceptible to immunological rejection. However, many scientific and ethical hurdles lay ahead before therapeutic cloning could be applied for treating patients. As would be the case for allogeneic ES cell-derived therapies, autologous ES cell-derived therapies could be safety and efficacy tested in non-human primate pre-clinical models.

14.6 Progress and promise in disease-specific cell therapies

Encouraging results from a number of animal models of disease treated with ES-derived cells have fueled the interest in human ES cells for therapeutic applications. The vast majority of work to date has been done using non-human ES cells. Most of the work has used small animal models such as mouse and rat, and little has been done in non-human primates or other relevant large animal models. This represents a highly active area of investigation, so the frontier is rapidly changing. This section will only highlight a handful of representative applications for hES cell therapy, which currently have made the most progress in pilot studies, but the range of potential uses of ES cells in clinical therapies is acknowledged to be far greater (*Figure 14.3*).

14.6.1 Hematologic disorders

Multiple clinical applications may arise from the use of human ES cells to produce hematopoietic stem cells or defined blood products. For the many hematological or other malignancies that are treated with bone marrow transplantation, having an unlimited supply of donor stem cells to reconstitute hematopoiesis would be a major advance. In addition, the ability to produce *in vitro* supplies of blood products such as red blood cells and platelets would revolutionize transfusion medicine. It was recognized from the first experiments demonstrating embryoid body formation from mouse ES cells that hematopoietic cells were present in the embryoid bodies (Doetschman *et al.*, 1985); however, the ES cells forming blood elements largely recapitulate yolk sac or primitive hematopoiesis (Keller *et al.*, 1993). In contrast, the hematopoiesis that occurs in the adult bone marrow, referred to as definitive hematopoiesis, is distinguished by the presence adult isoforms of β-globin in red blood cells and most importantly by the ability to engraft and repopulate the marrow following lethal irradiation. Only recently have efforts succeeded in obtaining adult or definitive hematopoiesis from mouse ES-derived cells. An initial study used the chronic myeloid leukemia-associated BCR/ABL protein to transform ES cells and demonstrated the ability to engraft lethally irradiated mice, but these animals had leukemia (Perlingeiro *et al.*, 2001). An alternative strategy used controlled expression of the homeobox gene HoxB4 in developing EBs cultured on OP9 stromal cells to produce cells exhibiting the

Neurodegenerative
Parkinson's disease
spinal cord injury
demyelinating diseases
peripheral nerve injuries

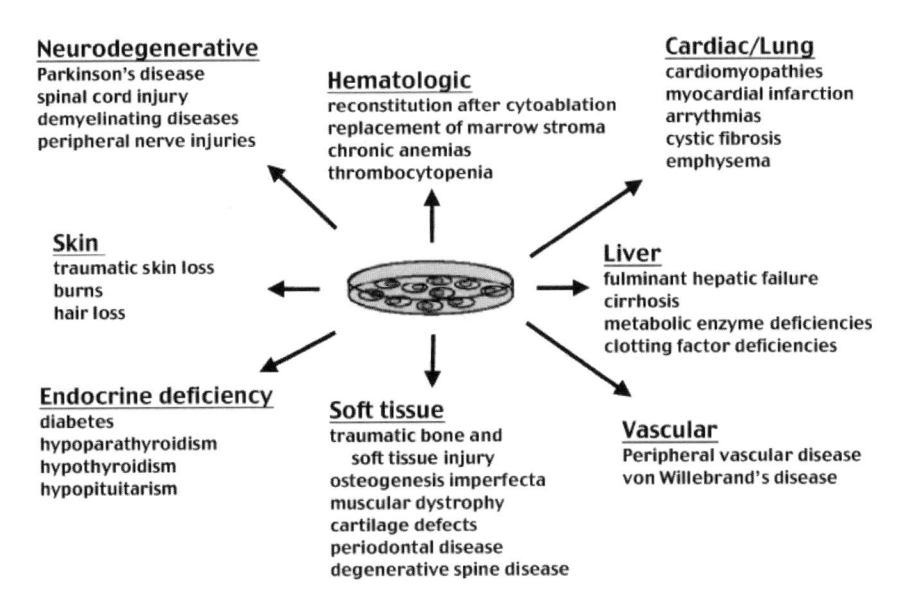

Hematologic
reconstitution after cytoablation
replacement of marrow stroma
chronic anemias
thrombocytopenia

Cardiac/Lung
cardiomyopathies
myocardial infarction
arrythmias
cystic fibrosis
emphysema

Skin
traumatic skin loss
burns
hair loss

Liver
fulminant hepatic failure
cirrhosis
metabolic enzyme deficiencies
clotting factor deficiencies

Endocrine deficiency
diabetes
hypoparathyroidism
hypothyroidism
hypopituitarism

Soft tissue
traumatic bone and
soft tissue injury
osteogenesis imperfecta
muscular dystrophy
cartilage defects
periodontal disease
degenerative spine disease

Vascular
Peripheral vascular disease
von Willebrand's disease

Figure 14.3: Human conditions potentially treatable with human ES cell derivatives. This incomplete list highlights some of the human diseases undergoing active research for stem cell-based therapies, but potentially many other conditions will be added to the list as research progresses.

definitive hematopoiesis stem cell phenotype (Kyba et al., 2002). These cells could reconstitute hematopoiesis without producing leukemia in lethally irradiated mice, which provides strong evidence for definitive hematopoiesis. Human ES cells have only recently been investigated for their potential to form blood, and clear evidence has been provided that like mouse ES cells, human ES cells form primitive hematopoietic precursor cells in EBs (Kaufman et al., 2001). Therefore, it seems reasonable to hope that with manipulation, definitive hematopoietic cells will be derived from human ES cells, providing a powerful new source of donor cells for bone marrow transplantation. In addition, progress in forming isolated blood elements has been made, for example, with the production of functional platelets from murine ES cells (Fujimoto et al., 2003). These experiments represent a promising beginning to realizing the potential of hES cells in hematological-related disorders.

14.6.2 Neurological diseases

Cell-based therapies have been the subject of much investigation for the treatment of a variety of different neurological diseases. Studies using cells derived mostly from mouse ES cells have been tested in different animal models of disease with encouraging results in proof-of-principle studies. Parkinson's disease has proven a popular target for cell-based therapy, given the specific loss of dopaminergic neurons in the substantia niagra. In a rat model of Parkinson's disease produced by 6-hydroxydopamine, both dopaminergic neurons differentiated from mouse genetically-engineered ES cells and undifferentiated ES cells have been successfully transplanted and form functional, integrated dopaminergic neurons with resulting improvement in the animals' behavior (Bjorklund et al., 2002; Kim et al., 2002). There is also promise for treating demyelinating diseases with oligodendrocytes or astrocytes derived from ES cells in mouse and rat models (Brustle et al., 1999; Liu et al., 2000). Likewise, treating a rat spinal cord injury with neurally differentiated ES cells resulted in significantly better outcomes than placebo (McDonald et al., 1999). Mouse ES cells have recently been demonstrated to undergo directed differentiation into motor neurons which suggests potential promise for motor neuron diseases such as amyotrophic lateral sclerosis (Wichterle et al., 2002). Even blindness may become a target for ES cell therapy as transplanted ES cells prevented retinal degeneration in a rat model (Schraermeyer et al., 2001). Most of the studies in neurological diseases have been proof-of-concept, and longer term studies in larger animal models are needed to assess the benefit and risks of these therapies. Most of the studies have used mouse ES cells, but recent data demonstrate that human ES cells are also capable of *in vitro* differentiation into neural precursors that can be transplanted into rat brains and mature into neurons, astrocytes, and oligodendrocytes (Reubinoff et al., 2001; Zhang et al., 2001). At this time only one clinical trial using human embryonic-like cells for therapy has been reported (Kondziolka et al., 2000). Neurons from an embryonal carcinoma cell line were transplanted into patients with large basal ganglia strokes in a phase 1 clinical trial. No adverse effects were detected and transplanted cells have been detected up to 27 months following transplant (Nelson et al., 2002).

14.6.3 Endocrine deficiencies

A number of endocrinopathies characterized by a loss of physiological hormone production could theoretically be treated with stem cell-based therapies; however, type I diabetes mellitus has been the focus of most attention to date. Type I diabetes mellitus is due to the autoimmune-mediated destruction of β cells within the islets of Langerhans of the pancreas, and results in the absence of responsiveness to glucose and insulin secretion. Work with mouse ES cells has demonstrated that murine pluripotent stem cells can differentiate into insulin producing cells similar to pancreatic β cells (Blyszczuk et al., 2003; Hori et al., 2002; Kahan et al., 2003; Lumelsky et al., 2001; Soria et al., 2000). For some of these experiments, there is concern that immunohistochemical identification of insulin-staining cells has been accurate (Rajagopal et al., 2003). Clearly, more comprehensive characterization of the phenotypical markers and functions of the cells should be performed before they are called definitive, adult β cells. In some of the studies transplantation of the ES-derived insulin producing cells into streptozotocin-induced diabetic mice has improved glycemic control, suggesting that appropriately functioning β-like cells are present (Blyszczuk et al., 2003; Hori et al., 2002; Soria et al., 2000). Human ES cells may also be capable of differentiating into insulin producing cells though concerns regarding insulin uptake from the media may still exist based on a high insulin content in the supplemented medium (Assady et al., 2001; Segev et al., 2004).

To date, it has proved difficult to direct pancreatic lineage differentiation from ES cells, but it is likely that understanding normal developmental signals will deliver clues to develop more efficient *in vitro* differentiation protocols. Progress in isolated allogeneic islet transplant procedures in humans performed by injecting the islet cell suspension into the portal vein have demonstrated the feasibility of treating human diabetes with cell therapy (Shapiro et al., 2000). Similar immunosuppressive protocols and technical procedures for cell delivery will likely prove useful when human ES cell-derived islets cells are tested in the future.

14.6.4 Heart disease

Given the high prevalence of heart disease, which remains the number one killer in western societies, there are multiple potential applications that could dramatically alter patient care. Many forms of heart disease are due to the loss of functional cardiomyocytes due to various insults, which can cause necrosis or apoptosis of myocytes. These conditions therefore may be ideally suited for cell replacement therapy potentially using ES-derived cells. For example, myocardial infarction results in a sudden loss of working myocardium that is replaced by non-functional scar tissue, and this can set in place progressive remodeling and an associated decline in heart function leading to congestive heart failure. Ideally myocardial infarction is treated by rapid reperfusion before substantial loss of muscle occurs, but in a significant number of patients it is either too late for reperfusion or reperfusion is not possible. In these patients, cell-based therapy or tissue engineered constructs could theoretically regenerate or repair infarcted myocardium and thus reduce the number of patients progressing to congestive heart failure.

The feasibility of using ES-derived cells for transplantation to the heart was first demonstrated in 1996 with an enriched population of mES cell-derived cardiomyocytes transplanted into normal mouse hearts (Klug et al., 1996). A number of animal studies have used a variety of donor cell types to test the effect of cell therapy in injured cardiac muscle with promising early results (Couzin and Vogel, 2004). Results using skeletal myoblasts, hematopoietic stem cells, mesenchymal stem cells and other cell types have demonstrated improvement in cardiac function and sometimes survival in short-term studies; however, little data are yet available using ES cells as donor cells. Recently, in a rat myocardial infarction model, undifferentiated mouse ES cells were transplanted and were shown to form cardiomyocytes with an associated improvement in left ventricular function (Behfar et al., 2002). Other forms of heart disease marked by loss of functional cardiomyocytes, such as in many cardiomyopathies, may also provide a target for cell-based therapies. In the case of sinus node dysfunction, which is the most common cause for pacemaker implantation, a biopacemaker could potentially be engineered from ES-derived nodal cardiomyocytes. Therefore, cell-based therapy using ES cells in heart disease holds significant promise, but research is only beginning.

14.6.5 Other potential disease applications

Bone tissue engineering applications

Bone grafting is sometimes necessary to repair large bone defects created by trauma, disease or surgery. The source of bone is usually cadaver bone (allogeneic), but in some circumstances it may be obtained through a biopsy from the patient themselves (autologous). These currently available sources have some pitfalls: cadaveric bone may be susceptible to immunological rejection and may be associated with rare transmission of infectious diseases, while in the case of autologous bone transplantation, limited amounts of tissue are available and the biopsy procedure itself can be quite painful. Alternative, 'off-the-shelf' sources of bone could include engineered products derived by differentiating mesenchymal stem cells (MSCs) and/or ES cells into bone tissue *in vitro* (Buttery et al., 2001; Pittenger et al., 1999; Sottile et al., 2003). Sottile et al. (2003) also showed that human ES cells activate osteogenic markers such as osteopontin, bone sialoprotein, in response to factors that promote osteogenesis. Under these conditions, human ES cells are able to differentiate into cells resembling osteoblasts and nodules that exhibit mineralization with hydroxyapatite. Refining optimal growth and differentiation conditions for human ES cell bone generation in simple tissue engineering constructs and demonstrating efficacy in small animal models of disease is an important next step. Early human clinical trials for a human ES cell-derived bone tissue are clearly a number of years away. But, OsteoCel™, a human MSC-derived biocompatible matrix product produced by Osiris Therapeutics, Inc., has already entered Phase I human trials. Lessons learned from pre-clinical and clinical studies with human MSCs would be relevant to human ES cell-derived bone regeneration as the development of an effective 'off-the-shelf' therapy for bridging bone defects moves forward.

Pulmonary disease

Cystic fibrosis, emphysema, and chronic bronchitis are among many lung diseases that are characterized by diminished lung function, in some cases so reduced that permanent oxygen or ventilatory support is required. Under these circumstances, lung transplantation may be the only option for restoring lung function. A recent report suggests the possibility that murine ES cells can be induced to differentiate into distal lung epithelial cells (alveolar Type II pneumocytes) (Rippon *et al.*, 2004). Human ES cells may have similar potential. Therefore, strategies to repair or regenerate injured lung tissue using derivatives of hES cells are under investigation.

14.6 Future

Although numerous scientific hurdles remain before human ES cell-derived tissues and cells enter their first clinical trials, significant progress has been achieved in identifying functional specialized cell types and in refining *in vitro* differentiation protocols. These are small, first steps, but advances in solid organ transplantation, cellular therapeutics, and tissue engineering will pave the way for future progress. Important short-term goals for many therapeutic applications will be to demonstrate reproducible proof-in-principle cures of animal models of human disease with human ES cell-derived tissue. Such cures may quiet some detractors, and advances in the efficiency of nuclear transfer technology will provide some impetus to continue to pursue therapeutic cloning strategies. In addition, the derivation of additional hES cells lines will be essential to assure a diverse range of lines free of karyotypic and genetic abnormalities. There is tremendous promise for hES cell-based therapies, but even if successful clinical applications do not readily emerge, the ongoing research with hES cells will undoubtedly enhance our understanding of basic mechanisms of human development and disease.

Acknowledgments

Karen Heim assisted with manuscript preparation and Joan Kozel created *Color Plate 11*. Will Burlingham critically reviewed parts of the manuscript. Stem cell research in the laboratory of JSO is supported by grants from NIH-DK01-014, the Juvenile Diabetes Research Foundation #1-2001-191 and 1-2004-145, the Roche Organ Transplant Research Foundation # 221283847, and Geron, Inc. TJK received support from NIH PO1 HL47053, NIH R21 HL72089, and Geron, Inc.

References

Alam T, Sollinger HW (2002) Glucose-regulated insulin production in hepatocytes. *Transplantation* **74**, 1781–1787.

Amit M, Shariki C, Margulets V, Itskovitz-Eldor J (2004) Feeder layer- and serum-free culture of human embryonic stem cells. *Biol. Reprod.* **70**, 837–845.

Assady S, Maor G, Amit M, Itskovitz-Eldor J, Skorecki KL, Tzukerman M (2001) Insulin production by human embryonic stem cells. *Diabetes* **50**, 1691–1697.

Baccarani U, Donini A, Sanna A, Risaliti A, Cariani A, Nardo B et al. (2004) First report of cryopreserved human hepatocytes based bioartificial liver successfully used as a bridge to liver transplantation. *Am. J. Transplant.* 4, 286–289.

Behfar A, Zingman LV, Hodgson DM, Rauzier JM, Kane GC, Terzic A, Puceat M (2002) Stem cell differentiation requires a paracrine pathway in the heart. *FASEB J.* 16, 1558–1566.

Bjorklund LM, Sanchez-Pernaute R, Chung S, Andersson T, Chen IY, McNaught KS et al. (2002) Embryonic stem cells develop into functional dopaminergic neurons after transplantation in a Parkinson rat model. *Proc. Natl Acad. Sci. USA* 99, 2344–2349.

Blyszczuk P, Czyz J, Kania G, Wagner M, Roll U, St Onge L, Wobus AM (2003) Expression of Pax4 in embryonic stem cells promotes differentiation of nestin-positive progenitor and insulin-producing cells. *Proc. Natl Acad. Sci. USA* 100, 998–1003.

Brustle O, Jones KN, Learish RD, Karram K, Choudhary K, Wiestler OD, Duncan ID, McKay RD (1999) Embryonic stem cell-derived glial precursors: a source of myelinating transplants. *Science* 285, 754–756.

Bulte JW, Douglas T, Witwer B, Zhang SC, Strable E, Lewis BK et al. (2001) Magnetodendrimers allow endosomal magnetic labeling and in vivo tracking of stem cells. *Nat. Biotechnol.* 19, 1141–1147.

Butcher JA, Hariharan S, Adams MB, Johnson CP, Roza AM, Cohen EP (1999) Renal transplantation for end-stage renal disease following bone marrow transplantation: a report of six cases, with and without immunosuppression. *Clin. Transplant.* 13, 330–335.

Buttery LD, Bourne S, Xynos JD, Wood H, Hughes FJ, Hughes SP, Episkopou V, Polak JM (2001) Differentiation of osteoblasts and in vitro bone formation from murine embryonic stem cells. *Tissue Eng.* 7, 89–99.

Cheng L, Hammond H, Ye Z, Zhan X, Dravid G (2003) Human adult marrow cells support prolonged expansion of human embryonic stem cells in culture. *Stem Cells* 21, 131–142.

Cibelli JB, Stice SL, Golueke PJ, Kane JJ, Jerry J, Blackwell C, Ponce de Leon FA, Robl JM (1998a) Cloned transgenic calves produced from nonquiescent fetal fibroblasts. *Science* 280, 1256–1258.

Cibelli JB, Stice SL, Golueke PJ, Kane JJ, Jerry J, Blackwell C, Ponce de Leon FA, Robl JM (1998b) Transgenic bovine chimeric offspring produced from somatic cell-derived stem-like cells. *Nat. Biotechnol.* 16, 642–646.

Couzin J, Vogel G (2004) Cell therapy. Renovating the heart. *Science* 304, 192–194.

Cowan CA, Klimanskaya I, McMahon J, Atienza J, Witmyer J, Zucker JP et al. (2004) Derivation of embryonic stem-cell lines from human blastocysts. *N. Engl. J. Med.* 350, 1353–1356.

Cunningham DA, Herring C, Fernandez-Suarez XM, Whittam AJ, Paradis K, Langford GA (2001) Analysis of patients treated with living pig tissue for evidence of infection by porcine endogenous retroviruses. *Trends Cardiovasc. Med.* 11, 190–196.

Davies JD, Wilson DH, Hermel E, Lindahl KF, Butcher GW, Wilson DB (1991)

Generation of T cells with lytic specificity for atypical antigens. I. A mitochondrial antigen in the rat. *J. Exp. Med.* **173**, 823–832.

de Vos P, Hamel AF, Tatarkiewicz K (2002) Considerations for successful transplantation of encapsulated pancreatic islets. *Diabetologia* **45**, 159–173.

Demetriou AA, Brown RS Jr, Busuttil RW, Fair J, McGuire BM, Rosenthal P *et al.* (2004) Prospective, randomized, multicenter, controlled trial of a bioartificial liver in treating acute liver failure. *Ann. Surg.* **239**, 660–667.

Ding S, Wu TY, Brinker A, Peters EC, Hur W, Gray NS, Schultz PG (2003) Synthetic small molecules that control stem cell fate. *Proc. Natl Acad. Sci. USA* **100**, 7632–7637.

Doetschman TC, Eistetter H, Katz M, Schmidt W, Kemler R (1985) The in vitro development of blastocyst-derived embryonic stem cell lines: formation of visceral yolk sac, blood islands and myocardium. *J. Embryol. Exp. Morphol.* **87**, 27–45.

Dominko T, Mitalipova M, Haley B, Beyhan Z, Memili E, McKusick B, First NL (1999) Bovine oocyte cytoplasm supports development of embryos produced by nuclear transfer of somatic cell nuclei from various mammalian species. *Biol. Reprod.* **60**, 1496–1502.

Draper JS, Smith K, Gokhale P, Moore HD, Maltby E, Johnson J, Meisner L, Zwaka TP, Thomson JA, Andrews PW (2004) Recurrent gain of chromosomes 17q and 12 in cultured human embryonic stem cells. *Nat. Biotechnol.* **22**, 53–54.

Drukker M, Katz G, Urbach A, Schuldiner M, Markel G, Itskovitz-Eldor J, Reubinoff B, Mandelboim O, Benvenisty N (2002) Characterization of the expression of MHC proteins in human embryonic stem cells. *Proc. Natl Acad. Sci. USA* **99**, 9864–9869.

Efrat S (2002) Cell replacement therapy for type 1 diabetes. *Trends Mol. Med.* **8**, 334–339.

Elliott RB, Escobar L, Garkavenko O, Croxson MC, Schroeder BA, McGregor M, Ferguson G, Beckman N, Ferguson S (2000) No evidence of infection with porcine endogenous retrovirus in recipients of encapsulated porcine islet xenografts. *Cell Transplant.* **9**, 895–901.

Evans MJ, Gurer C, Loike JD, Wilmut I, Schnieke AE, Schon EA (1999) Mitochondrial DNA genotypes in nuclear transfer-derived cloned sheep. *Nat. Genet.* **23**, 90–93.

Fishman JA (2001) Infection in xenotransplantation. *J. Card. Surg.* **16**, 363–373.

Fissell WH, Kimball J, MacKay SM, Funke A, Humes HD (2001) The role of a bioengineered artificial kidney in renal failure. *Ann. NY Acad. Sci.* **944**, 284–295.

Fissell WH, Lou L, Abrishami S, Buffington DA, Humes HD (2003) Bioartificial kidney ameliorates gram-negative bacteria-induced septic shock in uremic animals. *J. Am. Soc. Nephrol.* **14**, 454–461.

Freed CR, Greene PE, Breeze RE, Tsai WY, DuMouchel W, Kao R *et al.* (2001) Transplantation of embryonic dopamine neurons for severe Parkinson's disease. *N. Engl. J. Med.* **344**, 710–719.

Fujimoto TT, Kohata S, Suzuki H, Miyazaki H, Fujimura K (2003) Production of functional platelets by differentiated embryonic stem (ES) cells in vitro. *Blood* **102**, 4044–4051.

Grossman Z, Herbermann RB (1986) Natural killer cells and their role in tumor immunology. *Cancer Res.* 46, 2651–2658.

Gurdon JB, Colman A (1999) The future of cloning. *Nature* 402, 743–746.

Hartgerink JD, Beniash E, Stupp SI (2001) Self-assembly and mineralization of peptide-amphiphile nanofibers. *Science* 294, 1684–1688.

Hartgerink JD, Beniash E, Stupp SI (2002) Peptide-amphiphile nanofibers: a versatile scaffold for the preparation of self-assembling materials. *Proc. Natl Acad. Sci. USA* 99, 5133–5138.

Hori Y, Rulifson IC, Tsai BC, Heit JJ, Cahoy JD, Kim SK (2002) Growth inhibitors promote differentiation of insulin-producing tissue from embryonic stem cells. *Proc. Natl Acad. Sci. USA* 99, 16105–16110.

Hubner K, Fuhrmann G, Christenson LK, Kehler J, Reinbold R, De La Fuente R, Wood J, Strauss JF III, Boiani M, Scholer HR (2003) Derivation of oocytes from mouse embryonic stem cells. *Science* 300, 1251–1256.

Humes HD, Buffington DA, MacKay SM, Funke AJ, Weitzel WF (1999) Replacement of renal function in uremic animals with a tissue-engineered kidney. *Nat. Biotechnol.* 17, 451–455.

Humes HD, Buffington DA, Lou L, Abrishami S, Wang M, Xia J, Fissell WH (2003) Cell therapy with a tissue-engineered kidney reduces the multiple-organ consequences of septic shock. *Crit. Care Med.* 31, 2421–2428.

Hwang WS, Ryu YJ, Park JH, Park ES, Lee EG, Koo JM *et al.* (2004) Evidence of a pluripotent human embryonic stem cell line derived from a cloned blastocyst. *Science* 303, 1669–1674.

Kahan BW, Jacobson LM, Hullett DA, Oberley TD, Odorico JS (2003) Pancreatic precursors and differentiated islet cell types from murine embryonic stem cells: an in vitro model to study islet differentiation. *Diabetes* 52, 2016–2024.

Kaiser J (2003) Gene therapy. Seeking the cause of induced leukemias in X-SCID trial. *Science* 299, 495.

Kaufman DS, Hanson ET, Lewis RL, Auerbach R, Thomson JA (2001) Hematopoietic colony-forming cells derived from human embryonic stem cells. *Proc. Natl Acad. Sci. USA* 98, 10716–10721.

Kawase E, Yamazaki Y, Yagi T, Yanagimachi R, Pedersen RA (2000) Mouse embryonic stem (ES) cell lines established from neuronal cell-derived cloned blastocysts. *Genesis* 28, 156–163.

Keller G, Kennedy M, Papayannopoulou T, Wiles MV (1993) Hematopoietic commitment during embryonic stem cell differentiation in culture. *Mol. Cell Biol.* 13, 473–486.

Kim JH, Auerbach JM, Rodriguez-Gomez JA, Velasco I, Gavin D, Lumelsky N *et al.* (2002) Dopamine neurons derived from embryonic stem cells function in an animal model of Parkinson's disease. *Nature* 418, 50–56.

Klug MG, Soonpaa MH, Koh GY, Field LJ (1996) Genetically selected cardiomyocytes from differentiating embryonic stem cells form stable intracardiac grafts. *J. Clin. Invest.* 98, 216–224.

Koh CJ, Atala A (2004a) Tissue engineering, stem cells, and cloning: opportunities for regenerative medicine. *J. Am. Soc. Nephrol.* 15, 1113–1125.

Koh CJ, Atala A (2004b) Therapeutic cloning applications for organ transplantation. *Transpl. Immunol.* **12**, 193–201.

Kondziolka D, Wechsler L, Goldstein S, Meltzer C, Thulborn KR, Gebel J *et al.* (2000) Transplantation of cultured human neuronal cells for patients with stroke. *Neurology* **55**, 565–569.

Kyba M, Perlingeiro RC, Daley GQ (2002) HoxB4 confers definitive lymphoid-myeloid engraftment potential on embryonic stem cell and yolk sac hematopoietic progenitors. *Cell* **109**, 29–37.

Lanza RP, Chung HY, Yoo JJ, Wettstein PJ, Blackwell C, Borson N *et al.* (2002) Generation of histocompatible tissues using nuclear transplantation. *Nat. Biotechnol.* **20**, 689–696.

Lanza RP, Cibelli JB, West MD (1999) Human therapeutic cloning. *Nat. Med.* **5**, 975–977.

Lanzendorf SE, Boyd CA, Wright DL, Muasher S, Oehninger S, Hodgen GD (2001) Use of human gametes obtained from anonymous donors for the production of human embryonic stem cell lines. *Fertil. Steril.* **76**, 132–137.

Leach JB, Schmidt CE (2005) Characterization of protein release from photocrosslinkable hyaluronic acid-polyethylene glycol hydrogel tissue engineering scaffolds. *Biomaterials* **26**, 125–135.

Lee MK, Moore DJ, Markmann JF (2003) Regulatory CD4+CD25+T cells in prevention of allograft rejection. *Front. Biosci.* **8**, s968–s981.

Levenberg S, Huang NF, Lavik E, Rogers AB, Itskovitz-Eldor J, Langer R (2003) Differentiation of human embryonic stem cells on three-dimensional polymer scaffolds. *Proc. Natl Acad. Sci. USA* **100**, 12741–12746.

Li L, Baroja ML, Majumdar A, Chadwick K, Rouleau A, Gallacher L *et al.* (2004) Human embryonic stem cells possess immune-privileged properties. *Stem Cells* **22**, 448–456.

Lim JW, Bodnar A (2002) Proteome analysis of conditioned medium from mouse embryonic fibroblast feeder layers which support the growth of human embryonic stem cells. *Proteomics* **2**, 1187–1203.

Liu S, Qu Y, Stewert TJ, Howard MJ, Chakrabortty S, Holekamp TF, McDonald JW (2000) Embryonic stem cells differentiate into oligodendrocytes and myelinate in culture and after spinal cord transplantation. *Proc. Natl Acad. Sci. USA* **97**, 6126–6131.

Lumelsky N, Blondel O, Laeng P, Velasco I, Ravin R, McKay R (2001) Differentiation of embryonic stem cells to insulin-secreting structures similar to pancreatic islets. *Science* **292**, 1389–1394.

Maguire PJ, Stevens C, Humes HD, Shander A, Halpern NA, Pastores SM (2000) Bioartificial organ support for hepatic, renal, and hematologic failure. *Crit. Care Clin.* **16**, 681–694.

Mazzini L, Fagioli F, Boccaletti R, Mareschi K, Oliveri G, Olivieri C, Pastore I, Marasso R, Madon E (2003) Stem cell therapy in amyotrophic lateral sclerosis: a methodological approach in humans. *Amyotroph. Lateral Scler. Other Motor Neuron Disord.* **4**, 158–161.

McDonald JW, Liu XZ, Qu Y, Liu S, Mickey SK, Turetsky D, Gottlieb DI, Choi

DW (1999) Transplanted embryonic stem cells survive, differentiate and promote recovery in injured rat spinal cord. *Nat. Med.* 5, 1410–1412.

Menasche P, Hagege AA, Scorsin M, Pouzet B, Desnos M, Duboc D, Schwartz K, Vilquin JT, Marolleau JP (2001) Myoblast transplantation for heart failure. *Lancet* 357, 279–280.

Michaels MG, Jenkins FJ, St George K, Nalesnik MA, Starzl TE, Rinaldo CR, Jr (2001) Detection of infectious baboon cytomegalovirus after baboon-to-human liver xenotransplantation. *J. Virol.* 75, 2825–2828.

Mummery C, Ward-van Oostwaard D, Doevendans P, Spijker R, van den Brink S, Hassink R *et al.* (2003) Differentiation of human embryonic stem cells to cardiomyocytes: role of coculture with visceral endoderm-like cells. *Circulation* 107, 2733–2740.

Munsie MJ, Michalska AM, O'Brien CM, Trounson A, Pera MF, Mountford PS (2000) Isolation of pluripotent embryonic stem cells from reprogrammed adult mouse somatic cell nuclei. *Curr. Biol.* 10, 989–992.

Nelson PT, Kondziolka D, Wechsler L, Goldstein S, Gebel J, DeCesare S *et al.* (2002) Clonal human (hNT) neuron grafts for stroke therapy: neuropathology in a patient 27 months after implantation. *Am. J. Pathol.* 160, 1201–1206.

Newman KD, McBurney MW (2004) Poly(D,L lactic-co-glycolic acid) microspheres as biodegradable microcarriers for pluripotent stem cells. *Biomaterials* 25, 5763–5771.

Niklason LE, Gao J, Abbott WM, Hirschi KK, Houser S, Marini R, Langer R (1999) Functional arteries grown in vitro. *Science* 284, 489–493.

Ostenfeld T, Tai YT, Martin P, Deglon N, Aebischer P, Svendsen CN (2002) Neurospheres modified to produce glial cell line-derived neurotrophic factor increase the survival of transplanted dopamine neurons. *J. Neurosci. Res.* 69, 955–965.

Paradis K, Langford G, Long Z, Heneine W, Sandstrom P, Switzer WM, Chapman LE, Lockey C, Onions D, Otto E (1999) Search for cross-species transmission of porcine endogenous retrovirus in patients treated with living pig tissue. The XEN 111 Study Group. *Science* 285, 1236–1241.

Pera MF (2004) Unnatural selection of cultured human ES cells? *Nat. Biotechnol.* 22, 42–43.

Perlingeiro RC, Kyba M, Daley GQ (2001) Clonal analysis of differentiating embryonic stem cells reveals a hematopoietic progenitor with primitive erythroid and adult lymphoid-myeloid potential. *Development* 128, 4597–4604.

Pittenger MF, Mackay AM, Beck SC, Jaiswal RK, Douglas R, Mosca JD, Moorman MA, Simonetti DW, Craig S, Marshak DR (1999) Multilineage potential of adult human mesenchymal stem cells. *Science* 284, 143–147.

Rajagopal J, Anderson WJ, Kume S, Martinez OI, Melton DA (2003) Insulin staining of ES cell progeny from insulin uptake. *Science* 299, 363.

Reubinoff BE, Pera MF, Fong CY, Trounson A, Bongso A (2000) Embryonic stem cell lines from human blastocysts: somatic differentiation in vitro. *Nat. Biotechnol.* 18, 399–404.

Reubinoff BE, Itsykson P, Turetsky T, Pera MF, Reinhartz E, Itzik A, Ben Hur T

(2001) Neural progenitors from human embryonic stem cells. *Nat. Biotechnol.* **19**, 1134–1140.

Richards M, Fong CY, Chan WK, Wong PC, Bongso A (2002) Human feeders support prolonged undifferentiated growth of human inner cell masses and embryonic stem cells. *Nat. Biotechnol.* **20**, 933–936.

Rideout WM, Hochedlinger K, Kyba M, Daley GQ, Jaenisch R (2002) Correction of a genetic defect by nuclear transplantation and combined cell and gene therapy. *Cell* **109**, 17–27.

Rippon HJ, Ali NN, Polak JM, Bishop AE (2004) Initial observations on the effect of medium composition on the differentiation of murine embryonic stem cells to alveolar type II cells. *Cloning Stem Cells* **6**, 49–56.

Romano G, Claudio PP, Kaiser HE, Giordano A (1998) Recent advances, prospects and problems in designing new strategies for oligonucleotide and gene delivery in therapy. *In Vivo* **12**, 59–67.

Schraermeyer U, Thumann G, Luther T, Kociok N, Armhold S, Kruttwig K, Andressen C, Addicks K, Bartz-Schmidt KU (2001) Subretinally transplanted embryonic stem cells rescue photoreceptor cells from degeneration in the RCS rats. *Cell Transplant.* **10**, 673–680.

Segev H, Fishman B, Ziskind A, Shulman M, Itskovitz-Eldor J (2004) Differentiation of human embryonic stem cells into insulin-producing clusters. *Stem Cells* **22**, 265–274.

Shapiro AM, Lakey JR, Ryan EA, Korbutt GS, Toth E, Warnock GL, Kneteman NM, Rajotte RV (2000) Islet transplantation in seven patients with type 1 diabetes mellitus using a glucocorticoid-free immunosuppressive regimen. *N. Engl. J. Med.* **343**, 230–238.

Solter D (1998) Dolly is a clone – and no longer alone. *Nature* **394**, 315–316.

Solter D, Gearhart J (1999) Biomedicine – Putting stem cells to work. *Science* **283**, 1468–1470.

Somia N, Verma IM (2000) Gene therapy: trials and tribulations. *Nat. Rev. Genet.* **1**, 91–99.

Soria B, Roche E, Berna G, Leon-Quinto T, Reig JA, Martin F (2000) Insulin-secreting cells derived from embryonic stem cells normalize glycemia in streptozotocin-induced diabetic mice. *Diabetes* **49**, 157–162.

Sorof JM, Koerper MA, Portale AA, Potter D, DeSantes K, Cowan M (1995) Renal transplantation without chronic immunosuppression after T cell-depleted, HLA-mismatched bone marrow transplantation. *Transplantation* **59**, 1633–1635.

Sottile V, Thomson A, McWhir J (2003) In vitro osteogenic differentiation of human ES cells. *Cloning Stem Cells* **5**, 149–155.

Spitzer TR, Delmonico F, Tolkoff-Rubin N, McAfee S, Sackstein R, Saidman S, Colby C, Sykes M, Sachs DH, Cosimi AB (1999) Combined histocompatibility leukocyte antigen-matched donor bone marrow and renal transplantation for multiple myeloma with end stage renal disease: the induction of allograft tolerance through mixed lymphohematopoietic chimerism. *Transplantation* **68**, 480–484.

Takahashi T, Lord B, Schulze PC, Fryer RM, Sarang SS, Gullans SR, Lee RT (2003)

Ascorbic acid enhances differentiation of embryonic stem cells into cardiac myocytes. *Circulation* **107**, 1912–1916.

Thomson JA, Itskovitz-Eldor J, Shapiro SS, Waknitz MA, Swiergiel JJ, Marshall VS, Jones JM (1998) Embryonic stem cell lines derived from human blastocysts. *Science* **282**, 1145–1147.

Tian L, Catt JW, O'Neill C, King NJ (1997) Expression of immunoglobulin superfamily cell adhesion molecules on murine embryonic stem cells. *Biol. Reprod.* **57**, 561–568.

Wakayama T, Perry AC, Zuccotti M, Johnson KR, Yanagimachi R (1998) Full-term development of mice from enucleated oocytes injected with cumulus cell nuclei. *Nature* **394**, 369–374.

Wakayama T, Tabar V, Rodriguez I, Perry AC, Studer L, Mombaerts P (2001) Differentiation of embryonic stem cell lines generated from adult somatic cells by nuclear transfer. *Science* **292**, 740–743.

Weiss RA (1999) Xenografts and retroviruses. *Science* **285**, 1221–1222.

Weiss RA (2003) Cross-species infections. *Curr. Top. Microbiol. Immunol.* **278**, 47–71.

Wekerle T, Kurtz J, Ito H, Ronquillo JV, Dong V, Zhao G, Shaffer J, Sayegh MH, Sykes M (2000) Allogeneic bone marrow transplantation with co-stimulatory blockade induces macrochimerism and tolerance without cytoreductive host treatment. *Nat. Med.* **6**, 464–469.

Wekerle T, Kurtz J, Sayegh M, Ito H, Wells A, Bensinger S, Shaffer J, Turka L, Sykes M (2001) Peripheral deletion after bone marrow transplantation with costimulatory blockade has features of both activation-induced cell death and passive cell death. *J. Immunol.* **166**, 2311–2316.

Wichterle H, Lieberam I, Porter JA, Jessell TM (2002) Directed differentiation of embryonic stem cells into motor neurons. *Cell* **110**, 385–397.

Wilmut I, Schnieke AE, McWhir J, Kind AJ, Campbell KH (1997) Viable offspring derived from fetal and adult mammalian cells. *Nature* **385**, 810–813.

Xu C, Inokuma MS, Denham J, Golds K, Kundu P, Gold JD, Carpenter MK (2001) Feeder-free growth of undifferentiated human embryonic stem cells. *Nat. Biotechnol.* **19**, 971–974.

Xu RH, Chen X, Li DS, Li R, Addicks GC, Glennon C, Zwaka TP, Thomson JA (2002) BMP4 initiates human embryonic stem cell differentiation to trophoblast. *Nat. Biotechnol.* **20**, 1261–1264.

Zandstra PW, Bauwens C, Yin T, Liu Q, Schiller H, Zweigerdt R, Pasumarthi KB, Field LJ (2003) Scalable production of embryonic stem cell-derived cardiomyocytes. *Tissue Eng.* **9**, 767–778.

Zhang SC, Wernig M, Duncan ID, Brustle O, Thomson JA (2001) In vitro differentiation of transplantable neural precursors from human embryonic stem cells. *Nat. Biotechnol.* **19**, 1129–1133.

Zwaka TP, Thomson JA (2003) Homologous recombination in human embryonic stem cells. *Nat. Biotechnol.* **21**, 319–321.

15. Production of human embryonic stem cell-derived cellular product for therapeutic use

Ramkumar Mandalam, Yan Li, Sandra Powell, Elisa Brunette and Jane Lebkowski

15.1 Introduction

In the past two decades cell-based therapies have found applications in adoptive immunotherapy, skin care, and hematopoietic stem cell transplantation. More recently, experimental cellular therapies have been evaluated for the treatment of Parkinson's disease (Freed *et al.*, 2001), spinal cord injury (Wirth *et al.*, 2001), cardiovascular disease (Menasche *et al.*, 2003) and diabetes (Shapiro *et al.*, 2000) as these cell-based approaches have the potential of repairing, replacing, restoring and regenerating normal tissue function. Human embryonic stem cells (hESCs) offer a new paradigm for treatment of degenerative diseases due to their unique properties. hESCs have extensive replicative potential and the ability to differentiate into any somatic cell or tissue in the body. These two properties of hESCs enable the design of bulk manufacturing schemes for generating the large numbers of therapeutic cells required for the widespread treatment of Parkinson's disease, cardiovascular disease, diabetes, and many other degenerative diseases for which there are currently no curative therapies. However, as with any other pharmaceutical drug, the hESC-derived cells must undergo pre-clinical safety and efficacy testing and must be produced using qualified and reproducible protocols in a controlled manufacturing environment.

Production of cells for therapeutic application falls into two categories: individualized and bulk. Individualized cell therapies are patient-specific and their production is designed to fulfil the treatment requirements of a single patient. Bulk production of therapeutics is designed to serve multiple patients. The individualized processing/production of cells has mainly involved autologous cells (Huan *et al.*, 1992; Henon, 1993) although patient-specific allogeneic cell production such as cord blood transplantation is also in use today (Laughlin *et al.*, 2001; Rubinstein *et al.*, 1998). Although individualized cell production is more laborious and expensive, it cannot be avoided in some cases due to the limited amount of starting material and replicative potential of many cell types.

To date, the bulk production of cells has focused mainly on generation of fibroblasts, keratinocytes and related cells for use as temporary skin substitutes (Parenteau et al., 2000; Naughton, 2000). Bulk production of cells using pre-qualified reagents and appropriate process controls enables uniform and reproducible generation of multi-dose lots while significantly decreasing costs due to reduced operational and quality expenses compared with individualized patient-specific production. The quality-tested therapeutic cell product can be made available as an 'off-the-shelf' product, thus facilitating timely treatment for both acute and chronic injuries.

In this chapter, development of current Good Manufacturing Practice (cGMP) compliant production processes for generation of hESC-derived differentiated cells is addressed. Specifically, the manufacturing and regulatory requirements including hESC line qualification, raw materials qualification, process development, cell production, cryopreservation and formulation, and product release will be discussed.

15.2 Required properties for an hESC-based cell therapy

hESC-derived therapeutics will be used for the treatment of degenerative diseases with the primary goal of repairing and restoring proper function in the target tissue/organ without compromising safety. The product will be administered to the localized target tissue by injection during a surgical procedure or through a direct delivery device. For example, treatment of spinal cord injury would require direct injection of the cellular product to the lesion site while delivery of cardiomyocytes to patients with heart disease may be accomplished by cardiac catheter delivery. For some applications even systemic delivery may be possible. Since hESCs have the ability to differentiate into any cell/tissue in the body, it is imperative that the differentiation of hESCs yield a defined composition of cells which are safe. Specifically, the hESC-derived product must be free of adventitious agents, have a defined cell composition, display functional stability, have prolonged shelf-life, and require only simple processing for administration to patients at a health care provider's site. To ensure safety of patients, all of the reagents used during production/formulation of cells will have to be tested, qualified and approved according to the guidelines set by governmental regulatory agencies. The final release assays should include complete characterization and quantification of the cell population.

15.3 Qualification of hESCs and raw materials

15.3.1 hESCs

For the production of hESC-based therapeutics, hESCs will serve as the starting cell population for production of differentiated cells. Hence, the cells must be qualified per regulatory guidelines for the production of biologics as described in applicable sections of the US Food and Drug Administration (FDA) Points to Consider (1993), Guidance for Human Somatic Cell Therapy and Gene Therapy (1998), Cell and Gene Therapy Products (US Pharmacopoeia, Chapter 1046) and in Title 21 of the Code of Federal Regulations (CFR), part 1271 (21 CFR Part 1271) document. Briefly these requirements include assessment of donor suitability, compilation of the history of the cell line, characterization of cell banks and testing for adventitious agents. The donor suitability assessment process includes (1) review of the donor's medical records for risk factors and (2) collection of 'informed consent' from the donors prior to the use of the embryo for derivation. However, many of the current hES cell lines listed on the NIH registry do not have a medical history of the donors, but are considered appropriate for clinical use if the cell lines are screened and tested for infectious diseases. The requirement of obtaining donor medical history for future hES cell line derivations is currently being assessed. A review of regulatory issues surrounding hESC-derived therapies has been published (Fink, 2003).

The history of a cell line, beginning with the isolation of the inner cell mass through the creation of a cell bank, should be documented. An example of this would be the sequence of events that was followed during the derivation of the hESC lines in Dr James Thomson's laboratory at University of Wisconsin, which are described in Thomson *et al.* (1998). Briefly, human embryos were cultured to the blastocyst stage followed by inner cell mass isolation. The inner cell mass was then transferred to and cultured on irradiated mouse embryonic feeder (MEF) cells. The resulting outgrowths were sectioned to smaller aggregates and cultured on MEF cells that gave rise to colonies of undifferentiated hESCs. The cells were cultured on MEF cells for 18–20 passages before adapting them to feeder-free conditions at Geron Corporation. Feeder-free conditions include the use of Matrigel™ as an extracellular matrix substrate, and MEF-conditioned medium (CM) supplemented with bFGF as a growth medium (Xu *et al.*, 2001). Research cell banks of four hES cell lines (H1, H7, H9 and H14) have been set up with cells that have been adapted to feeder-free conditions. Cells from the research cell banks of H1, H7, H9 and H14 have been tested and found negative for bacterial, mycoplasma, and human viral (HIV, HTLV, CMV, HBV and HCV) contamination. They were also found to be karyotypically normal as determined by the G-banding method. Furthermore, we have tested two of the four hES cell lines (H1 and H7) at an independent laboratory for infectious agents from murine, porcine and bovine sources (as shown in *Table 15.1*). The results from all the tests showed no detectable pathogens. Establishment of cell banks also allows for future testing of any pathogen that is newly identified (*Table 15.1*).

15.3.2 Raw materials

In the cell manufacturing process, all reagents that come in contact with the cells, directly or indirectly, must be selected, qualified and validated prior to use as per FDA's regulations and guidelines such as Points to Consider (1993), 21 CFR1271 (2001), 21 CFR Part 600 (2002) and 'Source Animal, Product, Preclinical, and Clinical Issues Concerning the Use of Xenotransplantation Products in Humans' (2003). Specifications must be defined and used to release all raw materials with an objective of preventing product impairment or introduction and transmission of communicable disease agents. It is also highly desirable to have all the components

Table 15.1: List of assays used for adventitious agents testing of H1 and H7 cell lines.

Assay
PCR-based Reverse Transcriptase
Human Immunodeficiency Virus (HIV) 1 & 2
Human T-cell Lymphotrophic Virus (HTLV) I/II
Human Cytomegalovirus (CMV)
Hepatitis B and C Viruses (HBV and HCV)
Human Herpes Virus Type 6 (HHV-6), Variant A
Human Herpes Virus Type 6 (HHV-6), Variant B
Epstein–Barr Virus (EBV)
Thin Section Electron Microscopy with Virus Detection & Tabulation
PG4S+L- and XC Plaque Assays
Parvovirus B-19
In Vitro Assay for Adventitious Viruses with 3 Cell Lines
Mouse Antibody Production Test (MAP)
Adventitious Bovine Viruses
Porcine Viruses
In Vivo Assay for Viral Contaminants

of a defined composition lacking animal products. Typical qualification of biologically- (animal or human) derived components will include testing for adventitious agents appropriate for the source material as well as the sources of some of the reagents used in the production of the component. Supporting documents that may be required for acceptance of the reagent include a Certificate of Analysis (COA), Certificate of Origin (for animal sources reagents), Certificate of Suitability, Biochemical Analysis, Manufacturing Documentation, and Quality Control Test Reports including establishment of shelf-life. cGMP audit of vendors may be necessary for critical raw materials such as biologically-derived components.

15.4 Cell production

A proposed scheme for production of cellular products from hESCs is shown in *Figure 15.1*. The fundamental concept of the proposed scheme is that undifferentiated hESCs and, in some cases, differentiated cell progenitors have extensive replicative potential and hence large quantities of these cells can be produced and further matured to yield the target therapeutic population (*Figure 15.1*).

In the proposed scheme, master (1) and working (2) cell banks of the hESC lines are established and characterized according to FDA guidelines. The working cell bank serves as the starting material for production of the differentiated cell types. Optimized, scaled processes are developed to differentiate hESCs into the desired cell type (4) and in some cases, a stable intermediate progenitor cell line (3) is identified. In such instances, cell banks of the intermediate progenitor cell population are established as a starting material for the final phase of production. Quality control tests are performed at different stages during the production process to ensure that process intermediates meet specifications. If the process or the material does not meet specifications, the production process may be halted. In the final step of production, the cells are washed to remove the reagents used

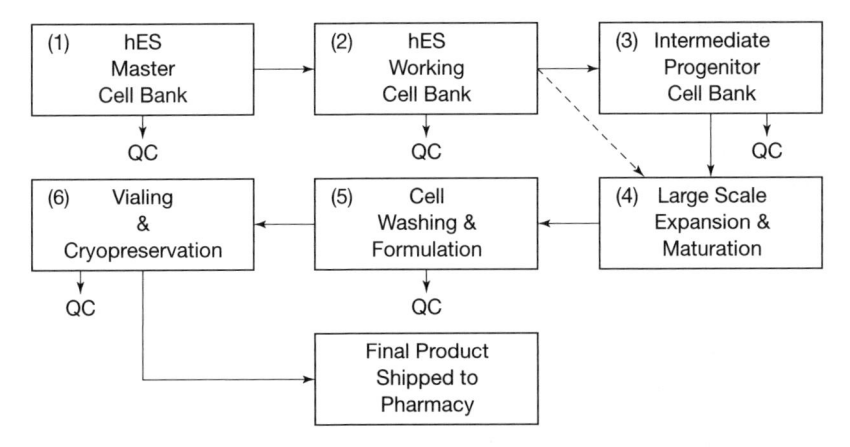

Chapter 15 – hES cell-derived cellular product for therapeutic use

```
(1) hES Master      →   (2) hES Working      ←   (3) Intermediate
    Cell Bank               Cell Bank                Progenitor
                                                     Cell Bank
      ↓                         ↓                         ↓
      QC                        QC                        QC
                                                          ↓
(6) Vialing         ←   (5) Cell             ←   (4) Large Scale
    &                       Washing &                Expansion &
    Cryopreservation        Formulation              Maturation
      ↓                         ↓
      QC                        QC
                                ↓
                        Final Product
                        Shipped to
                        Pharmacy
```

Figure 15.1: Flow diagram of production of glial progenitor cells (GPCs) for clinical use from a master cell bank of undifferentiated hES cells. The arrow and 'QC' beneath an activity indicates that quality control tests are performed at those stages to ensure quality.

during the production process (5), aliquoted into vials, and cryopreserved (6). A statistically representative sample of the cryopreserved product is tested for lot release. If the product does not meet specifications, the particular production lot is rejected. If release specifications are met, the cryopreserved cells are then shipped to pharmacies. As needed, the cryopreserved cells are thawed and suspended in a clinically approved physiological solution prior to administration to patients.

15.4.1 Development of large-scale production process

In the research laboratory, the initial experiments to differentiate hESCs to specific cell types is usually achieved using 'research' scale culture labware with multiple open transfer steps and without culture optimization for function and yield. To develop a cost-effective, scaleable, and optimized process, significant process development activity will be required for each of the cell types. There are a number of parameters such as medium composition, growth factors, attachment factors, inoculum density, perfusion volumes and schedules, duration of culture and oxygenation that need to be optimized for maximizing production while minimizing costs and the number of open transfer steps. Also, the key points to be considered during the development of a large-scale process include: (1) scale-up of critical processes to which the desired cell type is sensitive and (2) selection of appropriate equipment that can implement the optimized process. The final process developed for production of undifferentiated hESCs and the therapeutic cell product will be tested for robustness, reproducibility and reliability prior to process qualification and validation.

Undifferentiated hES cells

Undifferentiated hES cells have been cultured and maintained on mouse embryonic feeder cells in the presence of medium and exogenously added bFGF (Thomson et al., 1998). Although the use of a mouse feeder layer allows for long-term maintenance of hES cultures, this culture system results in cells contaminated with mouse feeder cells and is not preferred for research, development or clinical use. We have developed a feeder-free system that allows culture and maintenance of undifferentiated hESCs on Matrigel™ or laminin coated plates using medium conditioned by mouse embryonic fibroblasts (MEF) and supplemented with 4 ng/ml of bFGF (Xu *et al.*, 2001). hESCs cultured long-term in feeder-free conditions are pluripotent and have been successfully differentiated to specific cell types of all three lineages (Carpenter *et al.*, 2001; Rambhatla *et al.*, 2003; Xu *et al.*, 2002).

The feeder-free hESC culture system described above can be effectively used for generation of cells for research activities such as studying gene and protein expression, understanding development biology, and developing methods to differentiate cells in a controlled and directed manner. However, some components such as conditioned medium and Matrigel™ are murine derived and are not completely defined. It is preferable to use a culture system that is defined and contains human sourced proteins for producing cells for therapeutic applications as it reduces the risk of transmission of zoonotic viruses and other pathogens. We have evaluated medium conditioned by human cells such as dermal fibroblasts (HuF) and supplemented with bFGF (8 ng/ml) for culturing hESCs. The culture was compared with hESCs cultured using MEF-conditioned medium supplemented with bFGF (8 ng/ml). Results from the experiment have shown that HuF-conditioned medium can support expansion of undifferentiated hESCs and maintain expression of undifferentiated hES cell genes, hTERT and Oct3/4 (*Figure 15.2*). Use of a culture system consisting of non-conditioned, defined, serum-free medium

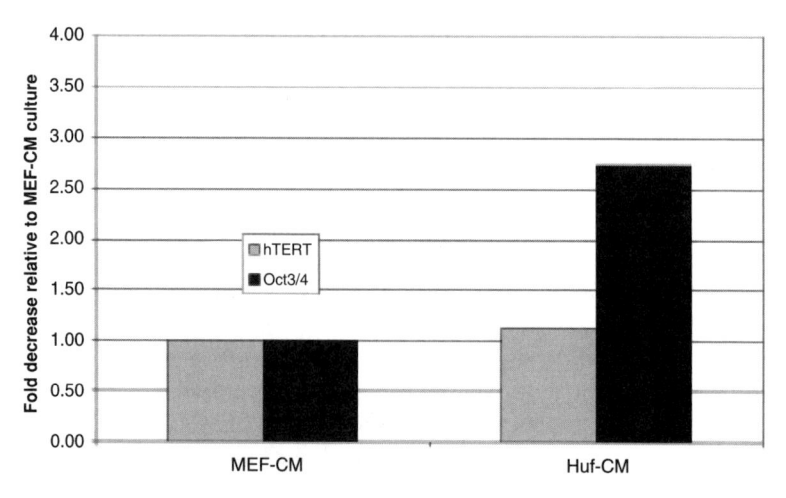

Figure 15.2: hESCs were cultured in MEF Conditioned Medium (CM) and Human Feeder (HuF) Conditioned Medium (CM) (see text for details). Cells were cultured on Matrigel™ and harvested using Collagenase/scraping. hTERT and Oct3/4 expression as measured by TaqMan RT-PCR of cells at passage 10 normalized to MEF CM culture (passage 10) is shown. Value of 1 indicates level of gene expression of hES cell markers of the total population similar to control culture. Values greater or less than 1 indicate the presence of lower or higher expression respectively of hES cell marker genes in the population compared to control (MEF-CM treated) cultures.

containing human or recombinant proteins would further reduce the variability inherent in conditioned medium and aid in standardization (*Figure 15.2*).

Some of the optimization parameters may have a significant salutary effect on productivity while raising material or production costs. For example, a decrease in inoculum density may result in more rapid cell expansion, but a lower overall cell yield, thereby impacting the quantity of media and culture devices required to achieve an equivalent yield. Optimization of such parameters for large-scale production should take operation and cost issues under consideration while maximizing productivity. Production of undifferentiated hESCs in large devices for generation of master/working cell banks or differentiated cells minimizes the required number of vessels thus reducing contamination risks, variability, and cost. The Cell Factory system (Nunc International) is one such vessel that can be used for large-scale production. Approximately $1-2 \times 10^8$ undifferentiated hESCs can be produced in MEF-CM in each 632 cm^2 single-cell layer Cell Factory device. The expression of hES cell-specific genes (hTERT and Oct 3/4) in Cell Factory cultures is equivalent to that observed on cells cultured in standard 6 well plates over four passages (*Figure 15.3*).

Differentiated cells

Undifferentiated hESCs can be manipulated to produce specific functional cell types through multiple steps of differentiation. Usually, the first stage involves directed differentiation commitment by formation of embryoid bodies and subsequent outgrowth under appropriate culture conditions (Xu *et al.*, 2002). Preferential differentiation to a specific lineage may also be achieved by direct exposure of undifferentiated hESCs to certain reagents without formation of embryoid bodies (Carpenter *et al.*, 2001; Rambhatla *et al.*, 2003). Subsequent

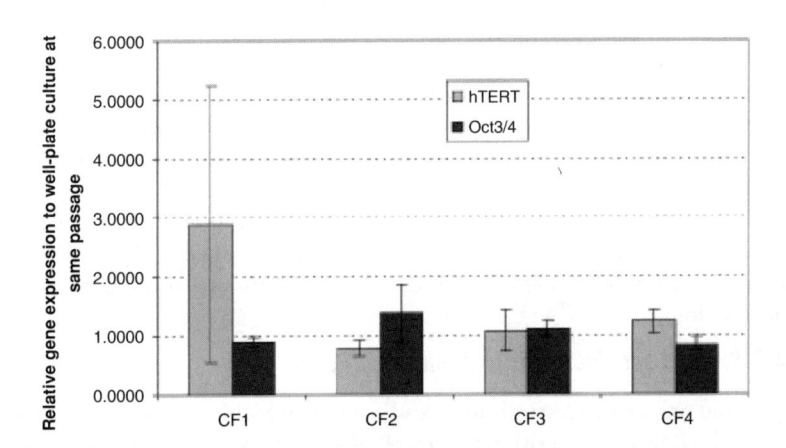

Figure 15.3: Relative expression of hESC genes in Cell Factory (CF) cultures compared with 6-well plate cultures at the same passage number. Results from four consecutive passages from one experiment are shown. A value of 1 denotes equivalent gene expression in CF cultures relative to the well plate control culture. Values greater or less than 1 indicate the presence of higher or lower expression respectively of hESC marker genes in the population compared with control cultures. The error bars represent standard deviation of triplicate analysis of each sample.

exposure to several growth factors, hormones, extracellular matrix factors, and culture media components can preferentially enable maturation to a specific lineage. In many cases, intermediate progenitors with replicative potential may be identified such that a cell bank of progenitors could be established. The progenitor cell bank would then serve as the starting population for the final production process (*Figure 15.1*). A scalable differentiation process must be developed for bulk production of the final product to enable multiple patient doses from each manufactured batch. Parameters that should be addressed to maximize productivity and minimize cost include, but are not limited to, duration of culture, combination and doses of growth factors, medium perfusion schedules and oxygenation. Failure Mode and Effect Analysis (FMEA) should be performed to identify potential failures caused by either process or reagent deficiencies and their effects on the final product. This exercise is critical to identify how manufacturing parameters determine product quality so that process specifications can be defined to minimize failures.

15.4.2 Cryopreservation and formulation

The product concept as described in *Figure 15.1* involves bulk production of the final cell product followed by cryopreservation for storage. At the time of transplantation, the cryopreserved 'off-the-shelf' product can be administered to the patient after minimal processing. Hence, an appropriate formulation and cryopreservation process must be developed that would result in simple, reliable thawing and reconstitution procedure for patient infusion.

The conditions for formulation and cryopreservation are critical for maintaining the function, viability and shelf-life of cells. The parameters that should be considered include freezing medium (non-serum-based medium with high protein content), cryoprotectants, controlled rate freezing conditions and thawing conditions. The optimized cryopreservation/formulation, thawing and reconstitution process will have to be tested for its robustness, reproducibility and maintenance of sterility. One of the key challenges that should be addressed during the development of cryopreservation/thawing process is the final volume of cells for injection. In some instances, a volume concentration step after thawing involving centrifugation of cells, removal of supernatant and re-suspension in specific volume of saline or other appropriate physiological medium that is suitable for the delivery method may be required. A precise protocol that is reproducible and maintains sterility is required for use of this procedure as a routine practice in a clinical setting.

15.4.3 Product specifications and release criteria

Process and product characterization and compliance with release specifications of each lot of cell product is critical to ensure safety and quality. Specifications should be defined for characterization and release of master and working cell banks of undifferentiated hESCs and progenitor cells, for in-process monitoring, and for final product release. Release criteria of cell banks for production could include characterization (identity and purity) and demonstration of ability to produce cells of specified composition and function. In-process monitoring assays may consist of measuring of pH, lactate levels, or a selected metabolite(s) during intermediate production stages.

The product release specifications are defined by product function and requirements. In the case of a cellular product, release criteria will include cell identity and composition (using marker expression), cell viability, functionality specifications, and the lack of adventitious agents. hESCs in their undifferentiated state will form teratomas (tumors) *in vivo* due to their pluripotential and extensive replicative capacity (Amit *et al.*, 2000; Thomson *et al.*, 1998). Hence, appropriate check-points must be employed in the production process such that the final product does not contain undifferentiated hESCs capable of forming teratomas. Undifferentiated hESCs could potentially be removed from the final product by various genetic- or antibody-based methodologies.

Assays used to demonstrate compliance should be tested for sensitivity, reliability and reproducibility in the range desired. Assays could be cell-based (e.g., flow cytometry) or population-based (RT-PCR) or a combination of both as long as they sufficiently characterize the population. In some cases, understanding the limits of detection of an assay becomes very important in defining specifications for release of a product. Although cumbersome, *in vivo* assays may be used in certain instances if appropriate *in vitro* assays are not available.

15.4.4 cGMP production

The production of cells for therapeutic use must be conducted in a validated facility under current Good Manufacturing Practice (cGMP) conditions. Therapeutic products need to be produced via aseptic processes to ensure that the final product is sterile and free from contaminants. This translates to creation of aseptic manufacturing environments with segregated unit operations for activities such as personnel gowning; quarantine and release of raw materials; equipment cleaning and staging; manufacturing; and finished product storage. Where separate areas are not available appropriate control systems could be established and maintained to prevent contamination, cross-contamination and accidental exposure of human cellular products to communicable disease agents.

FDA has published guidelines such as Points to Consider (1993) and regulations (21CFR211, 21CFR 610, 21CFR 820, 21CFR 1270) for development, manufacturing and commercialization of cell-based therapies. In brief, cGMP requires each of the following elements: a quality assurance program including auditing and improvement functions; raw material vendor qualification; documentation management program; equipment qualification, calibration and maintenance; facility maintenance and validation; personnel training; process control development and validation; warehousing, shipping and receiving procedures; implementation of change control processes; and establishment of in-process and final product specification and product release criteria.

15.5 Conclusions

Cellular products derived from human embryonic stem cells have extensive applications in providing cure for some unmet medical needs in the area of regenerative medicine. The potential of the hESCs can be fully realized only if appropriate large-scale manufacturing processes are developed that would enable the product to be safe while functional and cost-effective. The qualification of cell lines and raw materials, reproducible production processes and extensive product characterizations are critical for maintaining product safety and efficacy. Bulk production of hESC-derived cells enables an 'off-the-shelf' product and low cost of production with savings in the areas of cost of goods, quality control testing, documentation and labor. We look forward to the first scaled production of hES-derived product for human clinical trial testing and the realization of the medical potential of this technology.

References

Amit M, Carpenter MK, Inokuma MS, Chiu CP, Harris CP, Waknitz MA, Itskovitz-Eldor J, Thomson JA (2000) Clonally derived human embryonic stem cell lines maintain pluripotency and proliferative potential for prolonged periods of culture. *Dev. Biol.* 227, 271–278.

Carpenter MK, Inokuma MS, Denham J, Mujtaba T, Chiu CP, Rao MS (2001) Enrichment of neurons and neural precursors from human embryonic stem cells. *Exp. Neurol.* 172, 383–397.

Freed CR, Greene PE, Breeze RE, Tsai WY, DuMouchel W, Kao R et al. (2001) Transplantation of embryonic dopamine neurons for severe Parkinson's disease. *N. Engl. J. Med.* 344, 710–719.

Fink DW Jr (2003) Human embryonic stem cells and the Food and Drug Administration. In: *Human Embryonic Stem Cells* (eds A Chiu, MS Rao). Humana Press Inc., Totowa, NJ, pp. 323–343.

Henon PR (1993) Peripheral blood stem cell transplantations: past, present and future. *Stem Cells* 11, 154–172.

Huan SD, Hester J, Spitzer G, Yau JC, Dunphy FR, Wallerstein RO et al. (1992) Influence of mobilized peripheral blood cells on the hematopoietic recovery by autologous marrow and recombinant human granulocyte-macrophage colony-stimulating factor after high-dose cyclophosphamide, etoposide, and cisplatin. *Blood* 79, 3388–3893.

Laughlin MJ, Barker J, Bambach B, Koc ON, Rizzieri DA, Wagner JE et al. (2001) Hematopoietic engraftment and survival in adult recipients of umbilical-cord blood from unrelated donors. *N. Engl. J. Med.* 344, 1815–1822.

Menasche P, Hagege AA, Vilquin JT, Desnos M, Abergel E, Pouzet B et al. (2003) Autologous skeletal myoblast transplantation for severe postinfarction left ventricular dysfunction. *J. Am. Coll. Cardiol.* 41, 1078–1083.

Naughton G (2000) Dermal equivalents. In: *Principles of Tissue Engineering* (eds RP Lanza, R Langer, J Vacanti). Academic Press, San Diego, pp. 891–902.

Parenteau NL, Hardin-Young J, Ross RN (2000) Skin. In: *Principles of Tissue Engineering*, (eds RP Lanza, R Langer, J Vacanti). Academic Press, San Diego, pp. 879–890.

Rambhatla L, Chiu CP, Kundu P, Peng Y, Carpenter MK (2003) Generation of hepatocyte-like cells from human embryonic stem cells. *Cell Transplant.* 12, 1–11.

Rubinstein P, Carrier C, Scaradavou A, Kurtzberg J, Adamson J, Migliaccio AR *et al.* (1998) Outcomes among 562 recipients of placental-blood transplants from unrelated donors. *N. Engl. J. Med.* **339**, 1565–1577.

Shapiro AM, Lakey JR, Ryan EA, Korbutt GS, Toth E, Warnock GL, Kneteman NM, Rajotte RV (2000) Islet transplantation in seven patients with type 1 diabetes mellitus using a glucocorticoid-free immunosuppressive regimen. *N. Engl. J. Med.* **343**, 230–238.

Thomson JA, Itskovitz-Eldor J, Shapiro SS, Waknitz MA, Swiergiel JJ, Marshall VS, Jones JM (1998) Embryonic stem cell lines derived from human blastocysts. *Science* **282**, 1145–1147.

US Food and Drug Administration (1993) 'Points to Consider in the Characteristics of Cell Lines used to Produce Biologics.' US Department of Health and Human Services, Bethesda, MD.

US Food and Drug Administration (1998) 'Guidance for Human Somatic Cell Therapy and Gene Therapy.' US Department of Health and Human Services, Bethesda, MD.

US Food and Drug Administration (2003) 'Source Animal, Product, Preclinical, and Clinical Issues Concerning the Use of Xenotransplantation Products in Humans.' US Department of Health and Human Services, Bethesda, MD.

US Pharmacopoeia, Chapter 1046, 'Cell and Gene Therapy Products'.

Wirth ED 3rd, Reier PJ, Fessler RG, Thompson FJ, Uthman B, Behrman A, Beard J, Vierck CJ, Anderson DK (2001) Feasibility and safety of neural tissue transplantation in patients with syringomyelia. *J. Neurotrauma* **18**, 911–929.

Xu C, Inokuma MS, Denham J, Golds K, Kundu P, Gold JD, Carpenter MK (2001) Feeder-free growth of undifferentiated human embryonic stem cells. *Nat. Biotechnol.* **19**, 971–974.

Xu C, Police S, Rao N, Carpenter MK (2002) Characterization and enrichment of cardiomyocytes derived from human embryonic stem cells. *Circ. Res.* **91**, 501–508.

16. Ethical and policy considerations in embryonic stem cell research

R. Alta Charo

16.1 Federal regulation of embryo research

Federal law governing research using cells and tissues from embryos predates the human embryonic stem cell controversy, and is embedded in policies governing research with human beings, as well as the national debate surrounding abortion.

The core regulations governing research on human beings are codified in the Federal Policy for the Protection of Human Subjects, also known as the Common Rule, because these regulations have been adopted by most federal agencies that sponsor human research (Common Rule, 2003).

The Common Rule comprises subpart A of the HHS regulations, and requires the establishment of Institutional Review Boards (IRBs) to approve all federally funded human subjects research. The Common Rule explicitly outlines the membership of IRBs as well as the criteria for the approval of research. Subpart A covers all kinds of research on human subjects, and makes it subject to IRB approval, if it is subject to federal regulation at all. (Privately funded research may evade federal regulation entirely, though, if it is not regulated by the Food and Drug Administration and if it is not carried out at an institution that has pledged to carry out even otherwise unregulated research in conformity with federal rules.)

Subpart B of the HHS regulations contains specific provisions applicable to certain federal grants and research involving the fetus, pregnant women, and human *in vitro* fertilization (IVF). These regulations primarily address research that may adversely affect living fetuses. The provisions call for additional IRB duties beyond those in subpart A, restrict the use of pregnant women as research subjects, and demand minimal risk standards for therapeutic activities directed towards fetuses *in utero*.

In spite of subpart B's specific applicability to activities involving IVF and fetuses, its scope has been unclear with respect to stem cell research, because the administrative definitions cover fetuses and IVF procedures, but have not encompassed the blastocyst from which embryonic stem cells are derived. Fetus is defined in the regulations as 'the product of conception from the time of implantation ...'. *In vitro* fertilization is defined as any fertilization occurring outside a woman's body. Neither term, however, serves to describe ES cell research. Because it is never implanted into a woman's uterus, a blastocyst created *in vitro* and used to derive ES cells does not meet the administrative definition of fetus. Research on isolated stem cell lines, involving neither human sperm nor egg cells, does not meet the definition of IVF research. Subpart B, while more on point than the Common Rule, still fails adequately to address the complex issues raised by embryonic stem cell research.

In 2002, the Bush administration rewrote the charter for a national advisory body on human subjects research (Weiss, 2002). Under the new charter, the National Human Research Protections Advisory Committee is to examine policies concerning the protection of human embryos used in research even though, as noted below, such research generally cannot be done with federal funding. The effect of this change in the committee's charter is unclear.

16.2 The origins of the *de facto* ban on federal funding for embryo research

Following the emergence of human IVF in the 1970s, the Department of Health, Education and Welfare (HEW – now the Department of Health and Human Services) appointed an Ethics Advisory Board (EAB), to consider and review all applications for federal funding for research involving human IVF. Recognizing both the moral implications and the safety concerns surrounding this new reproductive technology, HEW determined that IVF protocols would first be reviewed for scientific merit, and once assigned a funding priority score, would then be required to undergo a second review by the EAB, to ensure that the work met certain ethical requirements, primarily those regarding the source of gametes and the management of the embryo *in vitro*.

Following the birth of Louise Brown, the first 'test-tube' baby, and increased public interest in the success of IVF in England, the EAB reviewed an application for federal support for an American IVF study. In 1979, the EAB reported that such research was ethically acceptable but subject to several important requirements, including the informed consent of gamete donors. Before federal funding could be released however, the EAB's membership was allowed to lapse. The reasons for this are unclear, but appear to be attributable in large part to a change of administration and a retreat from any commitment to federal support for embryo research. Thus, the requirement for EAB review remained in effect, but the EAB itself was left without members, staff, or physical office space. Because federal regulations required EAB review of all IVF-related studies, the absence of a Board imposed a *de facto* moratorium on IVF research and other studies involving human embryos. The moratorium remained in effect throughout the next decade, that is, during the administrations of Ronald Reagan and George H. W. Bush.

Several related events in the 1980s created a political climate in which the moratorium on embryo-related research was certain to remain in place. In the late 1980s, the Reagan administration asserted its anti-abortion philosophy by taking a stand against the research area of therapeutic fetal tissue transplantation (Charo, 1995a). Although an NIH Advisory Committee by September 1988 unanimously recommended that the moratorium be lifted, the ban remained in place throughout the Bush administration. These events in the 1980s dealt with the area of fetal tissue transplantation research, as opposed to human embryonic stem cell research, which had not yet evolved. Nonetheless, they reflect the spirit in which both administrations approached issues relating to fetal or embryonic research. Both Presidents perpetuated the *de facto* moratorium on research involving human embryos by not appointing EABs to review such research proposals.

16.3 Origins of the *de jure* ban on federal funding for embryo research

The 1990s witnessed a different political approach towards embryo-related research. On his second day in office in January 1993, President Clinton lifted the moratorium on federally funded fetal tissue transplantation research, prompting congressional hearings on its regulation. In March, Congress passed the National Institutes of Health (NIH) Revitalization Act of 1993. The law amended existing federal regulations governing research on human embryos, which required such research to be reviewed by an EAB before such research might proceed. Because prior presidential administrations chose not to appoint an EAB, no funding for such research had in fact been approved. What the new law did was to reverse the conditions for IVF research: it could go forward unless disapproved. Previously it could not go forward unless approved.

Following the passage of the Revitalization Act, the NIH received a number of applications for federal support of research involving human embryos. In response, NIH Director Harold Varmus and HHS Secretary Donna Shalala convened an advisory board known as the Human Embryo Research Panel to establish standards for determining which projects would or would not be considered acceptable for funding (Charo, 1995b). In its report to the Advisory Committee to the Director of the NIH, the Panel identified several research areas considered ethically appropriate for federal support. One such area was the derivation of stem cells from human embryos, as long as the embryos were donated with the fully informed consent of the gamete donors, generally following cessation of an IVF treatment for infertility. The Panel's report sparked controversy, however, in its conclusion that, in some carefully limited situations, it would be appropriate for the federal government to fund research that involved asking individuals to donate their gametes for the purpose of creating embryos by IVF, but solely for research purposes.

The recommendations were formally approved by the Advisory Committee and transmitted to Varmus in December of 1994. On December 2nd – the day after the recommendations were approved by the Advisory Committee, but before there was an official response from the NIH – President Clinton declared that federal funds should not be used to support the creation of human embryos for research purposes, and ordered the NIH not to allocate any resources for such tasks. In light of the presidential declaration, Varmus concluded that the NIH could begin to fund embryonic stem cell research, but only on embryos that had been donated following an abandoned effort at IVF for infertility treatment, and not on those created solely for the purpose of research.

Before any funding decisions were reached, however, Congress attached a rider to that year's HHS appropriations bill that effectively prohibited federal funding of any further human embryo research. The rider to the appropriations bill, the Omnibus Consolidated and Emergency Supplemental Appropriations Act (OCESAA), stated that none of the funds appropriated may be used to support research which involves: (1) creation of a human embryo or embryos for research purposes; or (2) research in which a human embryo or embryos are destroyed, discarded, or knowingly subjected to risk of injury or death greater than that allowed for research on fetuses in utero under Subpart B of the human subjects regulations, described above (45 C.F.R. 46.208(a)(2) 66 and section 498(b) of the Public Health Service Act (42 U.S.C. 289g(b)) 67). The regulations referred to in the rider provide that any risk posed to a fetus be minimal and for the purpose of developing 'important biomedical knowledge' ascertainable by no other means (45 C.F.R. 46.208(a)(2) (1999)) and that the risk standard applied 'be the same for fetuses which are intended to be aborted and fetuses which are intended to be carried to term.'

16.4 Origins of the decision to permit general federal funding of research on embryonic stem cell lines

This portion of the OCESAA Rider, often referred to as the 'Dickey-Wicker Amendment,' unambiguously prohibits research posing any risk to an organism derived by fertilization. Because the derivation of stem cells from a blastocyst destroys the embryo, no federal funds have been allocated to support such research. James Thomson's isolation of human embryonic stem cells from blastocysts, for example, was achieved using private rather than federal funds. The OCESAA rider does not, however, ban the funding of embryo-related research that poses no risk to an embryo. It is this latter category into which much future stem cell studies fall, for research on stem cells like those already isolated by Professor Thomson neither destroys nor poses any risk to an embryo – in this sense it is no different from research on any other human cell or cell line.

Professor Thomson's successful isolation of human embryonic stem cells at the University of Wisconsin and similar accomplishments by John Gearhart at the Johns Hopkins University prompted President Clinton in November 1998 to contact the National Bioethics Advisory Commission (NBAC) to request a thorough review of the medical and ethical issues associated with human stem cell research. In September 1999, after months of scientific, religious, and philosophical research and debate, NBAC provided its report, containing recommendations for responsible federal funding of such research (NBAC, 1999). Shortly thereafter, HHS issued its own interpretation of federal law, holding that funding embryonic stem cell research is permitted.

During deliberations by NBAC and the NIH, the legal debate narrowed to whether federally funded research on stem cells derived from embryos in excess of clinical need would violate the current ban. Contemplating the effect of the Dickey-Wicker Amendment, which clearly prohibited funding of research that destroyed or posed unacceptable risk to an embryo, and other federal restrictions on human embryo research, NIH Director Varmus sought the legal advice of HHS General Counsel Harriet Rabb. Rabb responded with a memorandum to Varmus indicating that federal law did not prevent NIH from funding such research, because stem cells met neither the statutory nor biological definition of a human embryo.

Rabb's memorandum concluded that the statutory prohibition on the use of funds appropriated to HHS for embryonic stem cell research did not apply to research utilizing human pluripotent stem cells because 'such cells are not a human embryo within the statutory definition.' Rabb noted that the term 'human embryo or embryos' was defined in the OCESAA Rider to include any 'organism,' not already protected under HHS regulations, that is derived by any process in which sperm meets egg. She then concluded that pluripotent stem cells are not a human 'organism' as that term is used in the definition of human embryo provided by statute.

Following Rabb's legal opinion, HHS released a Fact Sheet on Stem Cell Research, which stated that 'because pluripotent stem cells do not have the capacity to develop into a human being, they cannot be considered human embryos consistent with the commonly accepted or scientific understanding of that term.' But having concluded that the NIH may fund research using, but not creating, human embryonic stem cells, the NIH nonetheless delayed actual funding until an Ad Hoc Working Group of the Advisory Committee to the Director had been given a chance to develop new guidelines for ethical research, as existing guidelines were then almost 20 years old, dating back to the days of the Carter administration, when embryo research was still funded by HHS with the assistance of the old EAB.

Thus, in an effort to ensure that any research utilizing human embryonic stem cells would be conducted appropriately in light of new developments in scientific and ethical thinking, NIH Director Varmus convened his Working Group to begin developing guidelines for the research. The Working Group consisted of ethicists, scientists, patient advocates, and lawyers, and it considered congressional and public comments, as well as the recommendations in the NBAC Report. On December 2, 1999, the Draft Guidelines were published in the Federal Register, marking the beginning of a 60-day public comment period, later extended to February 22, 2000. The Draft Guidelines proposed specific criteria for informed consent for using stem cells, proposed the establishment of a 'Human Pluripotent Stem Cell Review Group', and listed those areas of research that would and would not be eligible for NIH funding, criteria that remained intact in the final Guidelines.

16.5 The decision to narrow the eligibility requirements for federal funding of research with human embryonic stem cells

Before any funding decisions could be made, however, the 2000 presidential elections intervened, triggering a change in administration policy. Against a backdrop of much speculation and considerable lobbying, President George W. Bush announced that it would endorse the legal interpretation of the Dickey-Wicker amendment that permits federal funding for work using, but not deriving, human embryonic stem cells. He would not, however, authorize such funding except in exceedingly narrow circumstances.

Instead, as he announced on August 9, 2001, the new policy requires researchers to use only cells collected from embryos created for reproductive – rather than research – purposes, and donated without compensation and with informed consent. Most importantly, the embryos must have been destroyed before the President's announcement. His goal was to eliminate not only the remote possibility that future decisions to discard embryos might be influenced by the prospect of federal support for research on stem cells derived from them, but also the appearance that such influence might exist (Press Release, 2001). The President said that approximately 60 lines of the cells existed worldwide at the time of his announcement and directed HHS to issue new guidelines for fundable research (OHRP, 2001).

In addition to announcing that funding would be available only for stem cell lines already in existence at the time of his decision, President Bush announced his opposition to funding research on cell lines derived from embryos that were deliberately and solely made for research purposes, whether by IVF or by somatic cell nuclear transfer (SCNT). His discussion of this point emphasized his opposition to SCNT, whether for reproductive or research purposes, whether done with private or public monies, and whether done here or abroad. This position was taken despite assertions in the scientific community that some SCNT experiments might be done not for reproductive purposes, but solely for the purpose of generating embryos with special characteristics whose stem cells could be used for particular forms of research impossible to do with stem cell lines derived from IVF embryos (NAS, 2001; NAS, 2002; Charo, 2001).

Bans on non-reproductive cloning experiments aimed at producing particular kinds of stem cell lines were passed in the US Congress House of Representatives and introduced in the Senate. The Senate bill, primarily sponsored by Senator Brownback of Kansas, was never passed, resulting in a continuation of the status quo, in which such research is legal but ineligible for federal funding. Competing bills, which would have supplemented the Food and Drug Administration's existing regulatory authority over such research when the aim was to produce transplantable tissue, were introduced in both the House and Senate, but also failed to pass. The change in Senate leadership following the 2002 elections has raised the prospect of renewed efforts to outlaw even privately funded stem cell research that relies on cloned embryos, and as of March 2003, the House of Representatives has once again passed a general prohibition on SCNT research. The Senate is expected to stall on a companion bill, leaving the field once again open to privately funded research, and subject to some FDA oversight where research involves transplantation and cell-based therapy applications of SCNT research that involves derivation of human embryonic stem cells from SCNT embryos.

In light of the federal debate, a number of states have introduced bills either to prohibit or explicitly permit and regulate SCNT research, including SCNT research that involves deriving human embryonic stem cell lines. As of March 2003, for example, prohibitory legislation exists in about a half a dozen states, such as Iowa and Michigan, but California, by contrast, has passed a bill to permit and regulate this work. New Jersey is widely expected to follow California's lead.

16.6 The intersection of embryo research funding and the abortion debate

At the heart of this history of federal regulation is a debate about the moral status of early forms of developing human life, whether embryonic or fetal, a debate that is closely tied to the national debate about abortion.

If an embryo is morally equivalent to a child, it cannot be subjected to harmful experimentation. Unfortunately, there is no easy way to describe what it is about a child that compels us to grant it such a moral status, and therefore it is difficult to determine whether an embryo shares this (or these) key characteristics with a child such that the embryo ought to be viewed as the child's moral equivalent. Therefore, groups such as the Human Embryo Research Panel sought a consensus on those aspects of human existence that implicate a moral status and therefore preclude destructive research. If the embryo embodied these aspects, then it would be treated as the moral equivalent of a live-born baby, and federal regulations governing human research would prohibit funding destructive embryo research, regardless of its scientific potential. If the embryo were viewed as any other human tissue, such as live cells taken from a spleen or liver, then the only concern relating to human research subjects would focus on the adult donors from whom the embryos were to be obtained. If the embryo were viewed as having some sort of intermediate status, then funding of destructive research might be permitted, depending on the balance between the need for the research and the degree of protection to which the embryo was entitled due to its intermediate status.

Unfortunately, every major argument for associating a particular moral status with a particular stage of prenatal development is flawed and leads to paradoxes or requires conclusions clearly at odds with the actual treatment of embryos, as well as actual treatment of trees, pets, and newborns.

16.6.1 Fertilization as the marker of unique personal identity

For those opposing embryo research, the most widely shared analysis of the moral status of the embryo focuses on the moment of fertilization as the key dividing line between unprotected and protected forms of human life. Due in large part to its genetic completeness and uniqueness, many hold that the fertilized egg has now become a full member of humanity.

For example, the 'genetic uniqueness' criterion is used to assert that the fertilized egg now exists in a one-to-one correspondence with a future baby, i.e., that the embryo now represents a single individual. This assertion is especially important for those who claim that fertilization is the moment at which the 'soul' enters the body, but it also resonates strongly with those who merely seek to identify the moment at which the embryo shares a key characteristic with the rest of us who claim an unambiguous moral entitlement to protection.

Reproductive biology, however, reveals that a single fertilized egg can twin (thus creating two babies from one 'unique' embryo); in addition, and perhaps even more conceptually complicating, two different embryos can merge to form a single baby whose body is a combination of the two different genetic patterns embedded in the two original embryos. (Pearson, 2002; Strain 1998) Thus, genetic completeness and uniqueness do not entirely correspond to human individuality.

The genetic completeness criterion is also problematic because it fails to account for other entities that have the same characteristic. Any single cell taken from a human body is genetically complete, but few would argue that each cell in the human body is the moral equivalent of a baby, with a right to be free of experimentation. Genetic uniqueness and completeness, therefore, may be necessary conditions for unambiguous moral claims to protection, but they cannot be sufficient.

16.6.2 The argument of potentiality

Proponents of the genetic completeness argument would respond by noting that a skin cell, though genetically complete, does not have the potential to develop into a baby, whereas the embryo does. Further, while the embryo may not be capable of having any current interest in its own survival because it is not self-aware, it will come to have such an interest in the future. It is this potential that provides the crucial distinction between those entities with a right to life and those without. But this argument, too, suffers from gaps in its logic and discrepancies with common experience.

The simplest objection to this argument is empirical: even under optimal conditions within a woman's womb, nearly 60% of all fertilized eggs will fail to implant or complete their development, with their loss either unnoticed during a menstrual cycle or, in later stages, marked by a miscarriage. Thus, as a British medical society studying embryo manipulation observed: 'It is morally unconvincing to claim absolute inviolability for an organism with which nature itself is so profligate.' But, of course, the fact that few embryos survive under optimal conditions is not necessarily an excuse for affirmatively destroying even more of them.

A more significant objection to the argument of potentiality concerns its reliance on treating acorns as if they were oak trees. That is, an acorn can, under certain conditions, eventually grow into an oak. Therefore, the argument from potentiality goes, the acorn ought to be accorded the same moral status as the oak tree which it could become, and if oak trees are given rights or otherwise protected, then so too should be the acorn, even if it does not at this moment have the attributes of a mature oak tree. In human terms, this argument asks: Assume, for the sake of simplicity, that unambiguous moral status is achieved by being born. Does the *potential* to be born entitle an entity to be treated as if it had already been born?

The potential to be born is present in any embryo left uninterrupted in a woman's womb. But the same potential exists long before fertilization, in any sperm or egg. What is it about fertilization that changes the status of the entity? It must be more than mere genetic completeness and uniqueness. That not only raises the paradoxes of individuation, but it fails to distinguish an embryo from a single sperm and single egg sitting in a petri dish. The contents of that petri dish also represent a genetically complete and unique blueprint for an individual, and, if left alone in the culture, the egg and sperm can combine and begin the processes of cell division. The only difference between the embryo in the womb and the gametes in the petri dish is that the embryo's development can continue with nothing more than 'natural' human intervention (in the form of maternal behavior such as eating) that is largely compelled by other considerations (such as the mother's desire to continue her own existence). The laboratory case requires active, directed, and 'unnatural' (i.e., mechanical transfer into the uterus) intervention by third parties in order to bring about a pregnancy.

But if an embryo cannot be killed because it is a unique, genetically complete entity that will develop to birth if left alone, then embryos outside the body and embryos inside the body would seem to have different moral claims: those inside the body would deserve protection from harm or destruction so they can continue their natural development into a baby; those outside the body, which cannot become babies if they are simply left alone, would be eligible for use in destructive research. While this may seem absurd, abandoning the natural/unnatural criterion leads to a different absurdity. Imagine that human cloning by nuclear transplantation has been perfected. Despite having undergone differentiation at the embryonic stage, any skin cell could now develop into a baby if it were placed into an enucleated egg cell. Surely this would mean every skin cell has a right to life due to its potential for development into a baby, albeit with some artificial assistance? But even those most committed to the argument of potentiality will say that it is ridiculous to think every cell in our body should be protected, even if they do argue that even embryos created through cloning now have a potential that must be protected. Thus, the argument of potentiality must be abandoned. Mere potential is not limited to embryos, nor do all embryos have potential for full development if simply left alone. A strict argument of potentiality would either confer a right-to-life on too many entities (such as sperm, eggs, and skin cells) or restrict itself to too few (by abandoning protection of IVF-created embryos).

16.6.3 Other developmental markers for determining moral status of the embryo or fetus

Others focus on later stages of embryo development, and on attributes that imply the embryo has a current, as opposed to merely potential, interest in continued existence. Developmental markers that have been proposed along these lines include the beginning of brain activity, rudimentary sentience or awareness of surroundings and pain, and sense of self.

Each of these later markers, however, is similarly insufficient to compel the conclusion that such an embryo has the same moral status as a live-born child. 'Brain activity' is a phrase that encompasses the most rudimentary electrical signaling among small clumps of cells (about the sixth or seventh week of development) to the development of a fully functioning nervous system, capable of supporting sentience, awareness, or pain. But rudimentary electrical activity does not indicate a developmental maturity that could encompass a current (as opposed to still potential) interest in continued existence; a current interest requires that an entity has preferences (e.g., a preference for living creates an 'interest' in not being killed). This, in turn, requires some rudimentary awareness in the present or past, so that the preferences can be formed. Thus, for embryos, rudimentary brain activity is nothing more than another kind of potentiality.

On the other hand, brain activity later in development, when it is sufficient to support sentience, pain awareness, and sense of self, would be entirely analogous to the condition of many non-human animals that are used for laboratory experimentation. Use of these later developmental criteria to explain the onset of a human's right to life or a right to be free of experimentation requires an explanation as to why the same attributes in non-human animals do not confer the same rights. Conversely, many newborn babies, who are granted the same moral status as older children and adults by most Americans, would fail to demonstrate some of these characteristics as strongly as some of the non-human animals routinely used in research. Indeed, attitudes toward withholding heroic measures for severely disabled newborns would seem to indicate that there is still some tolerance for infanticide in many modern cultures, including the Western European culture that still dominates the USA, but there is little indication that the public wishes to encourage infanticide in less tragic circumstances.

Thus, embryo research would be permissible if these late developmental markers were adopted as the criteria for protected life, but the implications of the theory are clearly unacceptable in the USA. Further, the very notion of a single criterion by which to measure the moral status of developing life began to seem unrealistic. As one philosopher put it, after having gone through a similar exercise: 'By now, I hope, most readers who followed the convoluted arguments of this section will be feeling that there is something absurd about all these attempts to define a precise moment at which a new human being comes into existence. The absurdity lies in the attempt to force a precise dividing line on something that is a gradual process' (Singer, 1995).

16.6.4 The pluralistic approach to the moral status of the embryo

In recognition of the gradual nature of human development, and in light of the flaws in each single-criterion justification for conferring a particular moral status upon forms of life, some opt for a 'pluralistic' approach: over time, the presence of an increasing number of these individual characteristics leads to an increasingly strong moral claim for protection against treatment that would be unacceptable if directed toward those with an unambiguous moral status. Thus, as an egg fertilizes and develops, it first embodies potentiality, and then genetic completeness and uniqueness. By the time the primitive streak appears and implantation is well underway, the opportunity for twinning or mosaicism has passed, and the embryo goes through individuation and on to cellular differentiation. By now it has also earned significantly better odds of developing to term. As it passes through stages of fetal development and is born, it develops rudimentary neurological tissue and other organs (e.g., the heart), and eventually develops sentience and cognitive ability. Each stage is viewed as developing a stronger claim for recognition as a fully equal member of the human community.

Looking chronologically at prenatal development, early public policy bodies, such as the Warnock Commission in the United Kingdom, concluded that the development of the primitive streak is a verifiable marker that also offers evidence of potentiality, a unique and complete genome, individuation, early differentiation, and the first organic structure associated with the development of a brain that will eventually facilitate awareness and cognition. The primitive streak's appearance is also close in time to the moment when embryos will have implanted in the womb, thus marking the onset of pregnancy. Because many people adopt one of the above markers as the key moment at which the embryo joins the moral community of those already born, the appearance of the primitive streak may represent the outer time limit at which the bulk of the public will tolerate destructive research, as well as a best guess as to when the moral status of the embryo really does become equivalent to that of a baby. The fuzziness of these findings is demonstrated by the Human Embryo Research Panel's willingness to contemplate future federal funding for research up to the 18th day of development in order to do important studies on environmental effects on neural tube formation; the 20th day marks the first appearance of rudimentary heart muscle, and thus creates yet another likely endpoint to public tolerance of destructive research.

What is missing from such an approach is any theory explaining why the presence of a collection of factors, each inadequate in itself, would yield a compelling argument for a particular moral status to be assigned at any particular time. The argument echoes strongly of 'the whole is more than the sum of its parts'. While clearly true in some contexts (the individual body, if reduced to a collection of unrelated cells and fluids, would be incapable of yielding any form of consciousness), it is not self-evident that it is true in the context of assigning a moral status to the embryo. Nor does the existence of a loose consensus about when the embryo's moral status entitles it to protection necessarily mean that this consensus has correctly identified that moment. In fact, the selection of the primitive streak as the limit for nearly all embryo research seems to have little to do with the Panel's stated methodology of determining when the embryo's intrinsic qualities entitle it to protection. Instead, it seems to reflect a concern for the sensibilities of growing numbers of people who would be offended by research on older embryos and a preference for a marker that can visibly signal investigators to stop further work, balanced against the need for the kinds of embryo research that is most likely to yield important public health benefits. And, conveniently, it is also the outer limit on research chosen by national commissions in several countries.

Some view such a 'pluralist' analysis as no more and no less than a purely symbolic respect for the value of embryos: 'Given the proponents' recognition that the embryo is too rudimentary to have interests, their position is best understood in symbolic terms. … Labeling an ethical concern as 'symbolic' is not to denigrate it, but rather to situate it accurately vis-à-vis the interests with which it conflicts. A key point is that symbols do not make moral claims upon us in the same way that persons and living entities do. Because symbolic meanings are so personal and variable, subordinating them to research goals usually violates no moral duties. At issue is a policy choice about what level of costs in lost research are [sic] acceptable to maintain a symbolic commitment to human life' (Robertson, 1995).

This observation does capture the core of the difficulty in balancing the known interests of research beneficiaries against the unknowable interests of embryos. It is also true that some who reject the notion of embryo 'interests' nonetheless accepted the notion of other people having an interest in whether embryos develop or are destroyed. But labeling peoples' interests as symbolic does denigrate their ethical concerns, because it fails to distinguish in a principled way among extremely important, important, and unimportant symbolic values.

When one examines the additional factors that some pluralists cite for a developing claim to equal moral status (such as the degree to which already existing members of the moral community feel a relationship to the developing embryo or the ability of the fetus to survive independent of a woman's body), it appears that one could explain the increasingly strong moral claims of the developing embryo or fetus almost entirely with reference to this 'relational' interest. As the embryo develops, it collects champions whose personal moral principles dictate that a particular developmental marker has now triggered a duty to protect the embryo's life. In other words, it is not the embryo's moral status that changes over time; rather, it is the balance between its champions' demands to have their views accommodated and the resistance of others to curtailing their freedom of action for the embryo's sake. That is, the moral status of prenatal life is not absolute but relative – it is defined by the degree of restraint that we who have been born are willing to tolerate for the benefit of having the prenatal life become a member of our moral community. This, in turn, depends on the nature of the entity, the impact of its mistreatment or suffering on our broadest human interests, and the specific implications its protection has for our liberty.

16.7 Summary

The moral status of the human embryo continues to be a subject of debate among theologians, philosophers, politicians and the general public. Unfortunately, science can only describe the stages of embryonic development. It cannot determine the moral status of developing life at each of these stages. Thus, continued dispute over whether to consider embryos and fetuses as the moral (and legal) equivalent of live-born children will drive public debate and federal regulation of research that uses embryos and embryonic stem cells.

References

Charo RA (2001) Playing God? Or playing Human?. *The Washington Post*, August 12, 2001 at p. B01.
Charo RA (1995a) Le penible valse hesitation: fetal tissue research review, and the use of bioethics commissions in France and the United States. In: *Society's*

Choices: Social and Ethical Decision Making in Biomedicine (eds R Bulger *et al.*). National Academy Press, pp. 477–500.

Charo RA (1995b) The Hunting of the Snark: The moral status of embryos, right-to-lifers, and third world women. *Stanford Law and Policy Review* 6(2), 1–38.

Common Rule (2003) http://www.med.umich.edu/irbmed/FederalDocuments/hhs/HHS45CFR46.html (2003)

(See Bush Says Cloning Is Morally Wrong, Urges Congressional Ban, Bulletin's Frontrunner, Nov. 27, 2001)

National Academy of Sciences (2002) Scientific and Medical Aspects of Human Reproductive Cloning.

National Academy of Sciences (2001) Stem Cells and the Future of Regenerative Medicine.

National Bioethics Advisory Commission (1999) Ethical Issues in Human Stem Cell Research (available online at http://www.georgetown.edu/research/nrcbl/nbac/)

National Institutes of Health Revitalization Act of 1993, Pub. L. No. 103-43, 107 Stat. 122 (codified as amended in scattered sections of 42 U.S.C. 281-89 (1993)

Office of Human Research Protections (OHRP) (2001) DHHS, Guidance for Investigators and Institutional Review Boards Regarding Research Involving Human Embryonic Stem Cells, Germ Cells and Cell-Derived Test Articles (Nov. 16, 2001), http://ohrp.osophs.dhhs.gov/references/HESCGuidance.pdf.)

Pearson H (2002) Human genetics, dual identities. *Nature Science Update* 2002 (available on-line at http://www.nature.com/nsu/020429/020429-13.html

Press Release, Office of the Press Secretary, White House (2001), Remarks by the President on Stem Cell Research (Aug. 9, 2001), available online at http://www.whitehouse.gov/news/releases/2001/08/20010809-2.html.

Robertson JA (1995) Symbolic issues in embryo research. 25 Hastings Center Rep. 37.

Singer P (1995) *Rethinking Life and Death: The Collapse of Our Traditional Ethics*. St. Martin's Press, New York.

Strain L, Dean JCS, Hamilton MPR, Bonthron DT (1998) A true hermaphrodite chimaera resulting from embryo amalgamation after in vitro fertilization. *New Engl. J. Med.* **338**, 166–169.

Weiss R (2002) New status for embryos in research. *Washington Post*, October 30, 2002.

17. Legal framework pertaining to research creating or using human embryonic stem cells

Carl E. Gulbrandsen, Michael Falk, Elizabeth Donley, David Kettner and Lissa Koop

17.1 Introduction

A typical researcher at an American university interested in pursuing study of human embryonic stem cells faces a number of potential legal and regulatory hurdles. At the federal level, legislation may restrict the scope of research allowed in the United States involving human embryonic stem cells. Already, federal funding agency restrictions impose significant limits on the development of new human stem cell lines and on the research uses to which such cells may be put. Issued and pending patents and patent applications owned by private corporations and universities also impact the field of stem cell research. And at a state level, a number of legislative proposals have been brought forward, with varying degrees of success, to limit research done on human embryonic stem cells.

The following sections discuss aspects of a researcher's ability to conduct research on human embryonic stem cells. The first sections of this chapter pertain to US researchers, while the later sections have relevance for researchers in Europe, Asia and Australia. To assist further scientists proposing human embryonic stem cell research, an appendix containing useful resources is provided.

17.2 Federal statute

Presently there is no federal law that prohibits research creating or using human embryonic stem cells. Federal law, at this point, only defines the circumstances under which federal funding can be used to support such research.

Although the enacted statutes in the area of federal regulation on human embryonic stem cells are few, they are arguably the most limiting for researchers. States may attempt to get around federal restrictions by promising state funding for stem cell research, but it is ultimately federal legislation that creates the parameters in which researchers must work. Pertinent legislation comes either directly from the President or through Congressional act, and it is relevant to discuss both in greater detail.

On August 9, 2001, President George W. Bush went before the nation in a televised address in which he addressed the debate surrounding the use and cultivation of human embryonic stem cells. At this time, the debate over use of stem cells was front-page news in the USA and of great public interest. In that speech, the President stated that federal funding would only be used to further research on the 60 cell lines that already existed at that point and 'where the life and death decision [had] already been made.' These restrictions mirrored those laid out in 1994 by President William Clinton, before the most important discoveries had been made in the field of human embryonic stem cell research. President Bush also stressed the need to promote research on umbilical cord, placenta, and adult stem cells and promised \$250 million in federal funds for that research (George W. Bush, Remarks on Stem Cell Research, August 9, 2001).

In the same speech, President Bush also proposed the creation of a President's Council on Bioethics under the direction of Dr Leon Kass of the University of Chicago that would continue the debate and provide recommendations about specific guidelines and regulations in the area of stem cell research. On November 28 of that same year, the President, by executive order, created that council, whose mission includes advising the President 'on bioethical issues that may emerge as a consequence of advances in biomedical science and technology' and studying 'ethical issues connected with specific technological activities, such as embryo and stem cell research, assisted reproduction, cloning … and end of life issues' (November 28, 2001 Executive Order). In July 2002, the council, composed of doctors, ethicists, scientists, lawyers, theologians and political theorists, released its findings in a report entitled 'Human Cloning and Human Dignity: An Ethical Inquiry (http://bioethicsprint.bioethics.gov/reports/cloningreport). The group was divided between two proposals dealing with human cloning, although a majority of ten members advocated one position and seven backed a minority position.

The council made the distinction between cloning for the purpose of producing children and cloning to create embryos for biomedical research. All members agreed that there should be a complete ban on cloning to produce children, but disagreed about the creation of embryos for research purposes. The minority proposal included governmental regulations in that area, but advocated federal support of embryonic research in light of the therapeutic benefits promised by it. The majority proposed a moratorium, or a temporary ban on cloning for research to last four years, during which time scientific evidence could be gathered and the ethical debate over the creation of embryos for research could be continued. Despite this recommendation, the President pressured legislators to pass a 'total ban on human cloning' including cloning to create embryos for stem cell research (President George W. Bush, remarks on Human Cloning Legislation, April 10 2002).

Historically, Congress has certainly felt the President's political pressure as evidenced by the Human Cloning Prohibition Acts of 1998, 1999, 2000, 2001, and 2002; however, none of these bills were successful in passing both houses. In like fashion, the House of Representatives recently passed the Human Cloning

Prohibition Act of 2003 (H.R. 534) by a vote of 241 to 155. It is currently being debated in the Senate, where similar bills have failed in the past three years. The bill would make it illegal 'to perform or attempt to perform human cloning, to participate in an attempt to perform human cloning, to ship or receive for any purpose an embryo produced by human cloning or any product derived from such embryo' and established criminal and civil penalties for such action. Unfortunately, the legislation makes no distinction between cloning for the purpose of producing children and cloning for the purpose of furthering biomedical research.

Although it seems unlikely that these blanket anti-cloning bills will become law, it seems equally unlikely that the efforts to block research in this area will cease. Public perception of cloning and the moral debate surrounding the destruction of embryos ensures a continued tension between those who wish to prevent research on human embryonic stem cells and those who believe in the great life-saving potential of such research.

17.2.1 Funding restrictions imposed by federal agencies

In the USA, no federal laws broadly govern the use of human embryonic stem cells in research, and those laws that do exist only affect federally-funded research, while they do not apply to research funded by private sources. Between 1975 and 1993, no federal funding was available for human embryo research due to a combination of regulatory restrictions and administrative inaction. In 1993, Congress enacted the National Institutes of Health Revitalization Act, providing authority to the National Institutes of Health (NIH) to support human embryo research (Pub. L. No. 103-43 § 492A (1993)). In response, the NIH created the Human Embryo Research Panel to recommend guidelines for reviewing applications for federal research funds intended to support embryological research. In September 1994, the panel endorsed human embryo research, finding that '[t]he promise of human benefit from research is significant, carrying great potential benefit to infertile couples, families with genetic conditions, and individuals and families in need of effective therapies for a variety of diseases' (NIH, Report of the Human Embryo Research Panel, Vol. I, at ix (1994)). In making this endorsement, the panel recommended that federal funding be used to support research involving both spare embryos leftover from *in vitro* fertilization and embryos created specifically for research purposes (Id. at x–xii). In December 1994, the NIH Advisory Committee to the Director accepted the panel's recommended guidelines, but then President William Clinton directed the NIH to forego funding any projects involving the creation of embryos solely for research purposes (John Schwartz and Ann Devroy, Clinton to Ban US Funds for Some Embryo Studies, Wash. Post., Dec. 3, 1994 at A1).

Since that time, two events have occurred that continue to shape today's federal governance of research involving human embryonic stem cells. First, in 1996, Congress enacted legislation prohibiting the use of federal funds in the creation of human research embryos, or embryo research in which embryos are destroyed, discarded, or knowingly subjected to risk of injury (Balance Budget Downpayment Act of 1996, Pub. L. No. 104-99, §128, Stat. 26, 34 (1996)). Then, in August 2001, President George W. Bush announced new federal guidelines limiting the use of federal funds to research conducted using only human embryonic stem cells lines existing as of August 9, 2001, that meet certain defined criteria. Below is a detailed discussion of these regulations and their effect on human stem cell research.

17.2.2 The Dickey Amendment, 1996 federal law

Since 1996, public funding of embryo research has been regulated by federal law commonly referred to as the Dickey Amendment. The Dickey Amendment was first enacted as an attachment to the Balance Budget Downpayment Act of 1996, and has since been continued by way of a rider to the various appropriation bills for the Department of Health and Human Services (DHHS). (See Pub. L. No. 108-07 § 510 (2003); Pub. L. No. 107-116 §510 (2002); Pub. L. No. 106-554 (2000); Pub. L. No. 105-277 (1998); Pub. L. No. 105-78 (1997); Pub. L. No. 104-208 (1996).) This rider provides that appropriated funds shall not be used for any research that involves the creation of human embryos for research purposes, or research 'in which a human embryo or embryos are destroyed, discarded, or knowingly subjected to risk of injury or death greater than that allowed for research on fetuses in utero' (Pub. L. No. 108-07 § 510 (2003)).

In 1999, the General Counsel for the Department of Health and Human Services concluded that this ban on the use of federal funds for human embryo research does not apply to research on human embryonic stem cell lines, but does apply to research in which embryos are actually destroyed (Judith A. Johnson & Brian A. Jackson, Stem Cell Research (Cong. Research Serv. Report No. RS20523 (2000)). In making this determination, the General Counsel concluded that stem cells 'are not a human embryo within the statutory definition' (Id.). Specifically, the statute defines an embryo as an organism, and because human pluripotent stem cells cannot become organisms due to their inability to become a fetus, they cannot be classified as embryos with respect to the law (Id.). In other words, after human embryonic stem cells are derived, or separated from an embryo resulting in destruction of the embryo, the cells no longer constitute an embryo and, thus, are eligible for use in federally funded research (Id.).

By fall 2000, the NIH issued guidelines for conduct of research using human pluripotent stem cells to prevent possible inconsistency with this law (NIH Guidelines for Research Using Human Pluripotent Stem Cells, 65 Fed. Reg. 69951 (November 21, 2000)). These guidelines outline the rules and restrictions governing the use of pluripotent stem cells derived from human embryos and human fetal tissue in research supported by NIH funds (Id.). Initially, these guidelines stated that federal funds could only be used for 'human pluripotent stem cells derived from ... human fetal tissue or ... from human embryos that are the result of in vitro fertilization, are in excess of clinical need, and have not reached the stage at which the mesoderm is formed' (Id.). However, on November 7, 2001, those portions of the Guidelines pertaining to research involving the use of stem cells derived from human embryos were withdrawn by the NIH in view of new criteria announced by President George W. Bush earlier that summer.

17.2.3 Guidelines on stem cell research, August 2001

On August 9, 2001, in a nationally televised address, President Bush announced the federal government would support research involving the use of human embryonic stem cells only for qualifying cell lines that existed at the time of the announcement (President George W. Bush, Remarks on Stem Cell Research (August 9, 2001)). Such existing cell lines must have been, prior to his announcement, already subjected to the derivation process whereby the inner cell mass (ICM) is removed from the intact, complete embryo or blastocyst. By virtue of being derived from the ICM, by definition an 'incomplete' or 'partial' embryo devoid of the part contributing to the placenta, human embryonic stem cells are incapable of developing into a viable, normal human being (Notice of Criteria for Federal Funding of Research on Existing Human Embryonic Stem Cells and Establishment of NIH Human Embryonic Stem Cell Registry, NOT-OD-02-005, November 7, 2001). In addition, the cell lines must have been derived from embryos that were created for fertility treatments but are no longer needed, and the embryos must have come from couples that gave their informed consent free of any financial inducements (Id.). President Bush's guidelines specifically prohibit the use of federal funds for (1) stem cells derived from embryos destroyed after August 9, 2001, (2) for the creation of human embryos for research purposes, and (3) the cloning of embryos (President Bush, Remarks on Stem Cell Research, *supra.*).

The NIH initially released a list of 64 human ES cell lines ('approved human embryonic stem cell lines') that met these criteria. (NIH Update on Existing Human Embryonic Stem Cells (Aug. 27, 2001)). The number of approved stem cell lines has, however, subsequently increased to 78 (as of May 2003) and may be subject to even further upward revision (NIH Human Embryonic Stem Cell Registry, http://escr.nih.gov.). Of these cell lines, several originate from countries other than the USA, and permission may be required for their importation. The primary concern with respect to their importation pertains to the possibility that such cells may serve as carriers for infectious diseases, such as bovine spongiform encephalopathy (BSE – Mad Cow Disease) (NIH: Federal Government Clearances for Receipt of International Shipment of Human Embryonic Stem Cells, NOT-OD-02-013, (November 16, 2001)).

Each source of human embryonic stem cells is aware of the requirements of the US federal government with respect to the importation of their human embryonic stem cell lines (Id.). However, all investigators receiving such cells should consider obtaining a Permit to Import or Transport Controlled Material or Organisms or Vectors (Forms 16-3 and 17-7) from the US Department of Agriculture (USDA) (Id.). In addition, the Centers for Disease Control and Prevention has a separate policy and permit process for the importation of cells and tissues that may harbor agents or organisms of human diseases (Id.). Some investigators may want to consider applying for such a permit from the CDC. The US Food and Drug Administration (FDA) does not presently issue importation permits for cell lines. However, for clinical research applications that involve the use of human embryonic stem cell lines, the agency recommends that investigators contact the FDA regarding policies for the manufacture and administration of biological products (Id.).

17.2.4 Other applicable regulations and laws

In addition to the federal regulations described above, investigators may also wish to consider how their proposed research may be affected by other laws not specific to the use of human embryonic stem cells. For example, clinical research involving biological products such as primary cells, or cell lines, regardless of whether they are genetically manipulated, may be subject to FDA regulations governing investigative new drugs or devices (DHHS Office for Human Research Protections, Guidance for Investigators and Institutional Review Boards Regarding Research Involving Human Embryonic Stem Cells, Germ Cells and Stem Cell-Derived Test Articles (2002); 21 C.F.R. 50, 56, 312 or 812). In addition, although human embryonic stem cells themselves may not qualify as fetal tissue, federally sponsored research involving the use of stem cells derived from fetal tissue is subject to the NIH Guidelines for Research Using Human Pluripotent Stem Cells (65 Fed. Reg. 69951, November 21, 2000), while clinical research involving the transplantation into human recipients of cells or other articles derived from fetal tissue may be subject to Public Law 103-43, 'Research on Transplantation of Fetal Tissue' (42 U.S.C. § 289g-2(a)).

Federally supported research involving the use of human subjects is also subject to the DHHS human subjects protection regulations set forth at Title 45 C.F.R. Part 46, including subpart B, 45 C.F.R. 46.206 (2002) (Guidance for Investigators and Institutional Review Boards Regarding Research Involving Human Embryonic Stem Cells, Germ Cells and Stem Cell-Derived Test Articles, *supra*.). Under these regulations, a human subject is defined as a living individual about whom an investigator (whether professional or student) conducting research obtains (1) data through interaction or intervention with the individual, or (2) identifiable private information (Id.). Research involving neither interactions nor interventions with living individuals or obtaining identifiable private information is not considered human subject research and, thus, not governed by the human subject protection regulations (Id.). Accordingly, investigators should ensure that human embryonic stem cell lines received are free from any information that may identify the donors from which the cell lines are derived.

17.3 Patent rights, licensing programs and agreements

The impact of patent rights on human embryonic stem cell research – and research in general – has been largely dismissed by academic researchers. A recent decision by the chief federal appellate court for patent cases, the Court of Appeals for the Federal Circuit, brings into relief the sparse legal underpinnings of what has come to be known as the 'research exemption' to patent infringement. Any investigator pursuing a course of research involving the development or use of human embryonic stem cells should be cognizant of the potential for the need to secure rights from patent holders in this area in order to practice patented inventions.

Although never codified, the origins of the research exemption date back to the early nineteenth century. The exemption grew out of dicta in an 1813 opinion by the noted jurist Justice Story: 'it could never have been the intention of the legislature to punish a man, who constructed a [patented] machine merely for philosophical experiments, or for the purpose of ascertaining the sufficiency of the machine to produce its desired effects' [*Whittmore v. Cutter, 29 F. Cas. 1120, 1121 (C.C.D. Mass. 1813)*]. Subsequent cases confirmed the doctrine, including dicta from the Federal Circuit in *Roche Products v. Bolar Pharmaceutical Co.*, 733 F.2d 858, 862063 (Fed. Cir. 1984). The decision in *Madey v. Duke* (307 F.3d 1351) by the US Court of Appeals for the Federal Circuit sends a sharp signal to academic researchers that any exemption to the normal rules of patent infringement is narrow indeed.

The decision in *Madey v. Duke* grows out of an unusual set of circumstances. Professor John Madey, an inventor working with free electron lasers, owned two patents related to his work which he developed while a tenured research professor at Stanford University in the mid 1980s. Duke University recruited Madey to their physics department and he left Stanford in 1988 to set up a laboratory at Duke, where he moved his free electron laser research laboratory. This new laboratory included laser equipment employing the technology which he patented and owned. Following a dispute between Duke and Madey related to management of the laboratory and use of the laboratory's equipment, Duke removed Madey as director of the laboratory in 1997 while retaining the equipment in the laboratory and continuing to use it. Madey left Duke in 1998. He subsequently sued Duke for patent infringement of his two patents and brought a variety of other claims. The university argued in its defense that it had no liability for infringement because its use of the patented technology was experimental and in fitting with its academic mission. The lower court thus dismissed Madey's claim based on the experimental use doctrine. Madey appealed, and the Federal Circuit reversed and remanded.

The Federal Circuit's opinion all but eliminates the research use exemption for research universities. Traditionally, as Justice Story's opinion reveals, academic research has been seen as a more selfless cause that puts it beyond the reach of patent law. The opinion of the Court of Appeals, however, rejects this distinction. What must be considered, according to the Court, is the fact that 'these projects unmistakably further the institution's legitimate business objectives, including educating and enlightening students and faculty participating in these projects ... and also serve ... to increase the status of the institution and lure lucrative research grants, students and faculty'. For research to qualify for the experimental use defense the court stresses that it must be 'solely for amusement, to satisfy idle curiosity, or for strictly philosophical inquiry', further emphasizing that the 'non-profit status of the user is not determinative' (*Madey*, 307 F.3d at 11). Thus academic researchers, just like their corporate counterparts, are subject to the patent laws and must be cognizant of obtaining rights to practice patented inventions.

17.3.1 Pertinent patents

In the USA, a person who without authority makes, uses, offers to sell, sells or imports into the USA a patented invention may be liable to the patent owner for infringement under the US patent laws. Actively inducing others to infringe may also be considered infringement (*see* 35 United States Code § 271). Similar laws exist with regard to patents issued by other countries. The jurisdiction of patent laws extends only to the borders of the country granting the patent. However, under US law importing into the USA, offering to sell, selling or using without authority a product made by a patented process may also be considered infringement. Thus, using a process covered by a US patent and then importing into the USA the product made by that process may be considered infringement even though the product itself is not covered by the patent. Laws proscribing importation of a product manufactured by a patented process exist to varying degrees in other countries of the world.

There are two basic patents that cover human embryonic stem cells. The first is US patent 5,843,780 titled Primate Embryonic Stem Cells ('780 patent). This patent was issued to James A. Thomson in 1998 and is owned by the Wisconsin Alumni Research Foundation (WARF) of Madison, Wisconsin. WARF is the patent management organization for the University of Wisconsin-Madison. The '780 patent claims primate embryonic stem cells, including human, that (1) are capable of proliferation in an in vitro culture for over one year, (2) maintain a karyotype in which all the chromosomes characteristic of the primate species are present and not noticeably altered through prolonged culture, (3) maintain the potential to differentiate into derivatives of endoderm, mesoderm and ectoderm tissues throughout the culture, and (4) will not differentiate when cultured on a fibroblast feeder layer. The '780 patent also claims a method of isolating primate embryonic stem cells and the cells isolated by that method. Foreign counterpart applications for '780 patent are currently pending in Canada and Europe as of this writing.

The second basic patent owned by WARF that covers human embryonic stem cells is United States Patent 6,200,806 ('806 patent). This patent is a divisional of the '780 patent. It was issued to James A. Thomson in 2001. The '806 patent claims human embryonic stem cells, a method of isolating human embryonic stem cells and the cells isolated by that method.

The five original human embryonic stem cell lines isolated and cultured by James A. Thomson in 1997 have been assigned to the WiCell Research Institute (WiCell), a non-profit subsidiary of WARF. WiCell was established in 1999 to conduct research on human embryonic stem cells and to distribute human embryonic stem cells to researchers around the world. Human embryonic stem cells are transferred by WiCell to researchers in return for a payment of $5000 ($6000 outside the USA) and agreement to the terms of a material transfer and license agreement. Since its inception, WiCell has transferred cells to over 140 research groups worldwide. The license provided with the transfer is a non-commercial academic research license. Information regarding how a researcher can obtain human embryonic stem cells can be found on the WiCell web site: www.wicell.org.

There are numerous other patents pertinent to scientists conducting or planning to conduct human embryonic stem cell research. Patents exist that pertain to the use of human embryonic stem cells as well as derivative lines. Examples are recent US patents 6,534,052 and 6,506,574 pertaining to improving cardiac function using embryonic stem cells and hepatocyte lineage cells derived from pluripotent stem cells, respectively. Patents covering culture media ingredients and growth factors needed to cause the stem cells to differentiate may be pertinent and number in the hundreds.

17.3.2 Licensing programs

Embryonic stem cell licensing

On September 5, 2001, WiCell Research Institute, Inc. ('WiCell'), a wholly owned subsidiary of the Wisconsin Alumni Research Foundation ('WARF') and the National Institutes of Health ('NIH') reached an agreement to facilitate licensing embryonic stem cell technology to researchers at the Public Health Service. A Memorandum of Understanding ('MOU') between the NIH and WiCell now forms the basis of WiCell's licensing program for academic research institutions working in the embryonic stem cell technology area throughout the world.

WiCell has developed a two-tiered approach to licensing the human embryonic stem cell technology. First, in order to assure that the inventions created at the University of Wisconsin by Professor James A. Thomson in the area of primate and human embryonic stem cells are made widely available to academic researchers around the world, WiCell developed an MOU and Simple Letter Agreement ('SLA'), included as Exhibit A, based on the Memorandum of Understanding between WiCell and the NIH. Secondly, WiCell developed a license for commercial research which returns revenues to the University of Wisconsin (the 'University') for future research in the stem cell area. These two goals, to assure advancement of this important area of science and to return research dollars to the University, reflect the balance of the missions of WiCell and WARF to serve the public good by making the embryonic stem cell materials and patent rights widely available to academic researchers while at the same time observing WARF's responsibility to return licensing revenue to the University.

Academic licensing

As discussed previously, *Madey v. Duke* and other recent cases, have made it clear that academic researchers may not rely on the research exemption to protect them when they use patent rights owned by others in their research. The MOU provides researchers with freedom to conduct non-commercial research, to publish and to file patents on any inventions created by such researchers using the Wisconsin materials and patent rights. The MOU grants a non-exclusive license to conduct non-commercial research under Wisconsin's patent rights which claim a composition of matter as well as method of making both human and primate embryonic stem cells. In addition, the MOU and SLA grant the NIH-funded scientist the rights to use the Wisconsin embryonic stem cell lines which are federally approved embryonic stem cell lines listed on the NIH registry.

The patent rights and materials can be only used for non-commercial research purposes under the MOU. Non-commercial research purposes are defined as research which specifically excludes sponsored research where the sponsor receives the right, whether actual or contingent, to the results of the sponsored research. This provision was included in the MOU to prevent companies from using academic researchers to obtain back door access to the Wisconsin patents and materials without paying reasonable commercial license fees.

This particular provision has also been interpreted to implicate Material Transfer Agreements ('MTA') used by companies when sending materials to research institutions. This is to prevent companies from sending materials to an academic researcher under the terms of an MTA whose terms give the company ownership or a license to the results of the research where the materials are being used in research with human or primate embryonic stem cells. In such cases, WiCell has worked hard to find a compromise appropriate to the situation to allow the research to go on but to avoid allowing companies to obtain rights they would otherwise be required to pay for a license to obtain.

In addition, the NIH MOU provides an implicit right to suppliers of other federally approved cell lines as long as such third-party suppliers of cell lines provide the materials on terms no more onerous than those set forth in the MOU. This provision not only makes Wisconsin cell lines available under the Wisconsin patents but also assures that others who hold federally approved lines can make those lines available to researchers to assure that a diverse group of genetically different stem cells lines are available for research. However, suppliers of lines who charge at a higher fee than the cost reimbursement charged by WiCell, including those who exact a reach-through royalty or attempt to obtain rights whether actual or contingent to the results of the research, must obtain a commercial license from WiCell to be allowed to distribute their cell lines. WiCell's strategy in including this provision was to make stem cells widely available for research but to hold stem cell providers to the terms of the MOU or require them to take a license under the patents to distribute their cell lines.

Finally, the MOU provides a list of experiments that may not be conducted using the Wisconsin materials which were required by the internal review board which originally reviewed Dr Thomson's protocol for deriving human embryonic stem cells at the University of Wisconsin. The prohibited experiments include the mixing of the Wisconsin materials with an intact embryo either human or non-human, the implanting of the Wisconsin materials or products of the materials in a uterus, and attempting to make whole embryos with Wisconsin materials by any method. The researcher and the institution receiving the Wisconsin materials are required to complete an annual certification statement (see Exhibit A) confirming compliance with the restrictions on use of the Wisconsin materials as well as compliance with the terms of the MOU and SLA.

Commercial licensing strategy

The commercial licensing strategy is also a two-tiered approach. The first tier requires that companies obtain a research license to conduct basic research using the embryonic stem cell technology. The commercial Research License incorporates a negative covenant which states that the company will not make, use or sell products that employ or are in any way produced by the practice of, are identified using or arise out of any research involving the inventions claimed in the patents or that would constitute infringement of any claims of the patents without first obtaining a license to make, use and sell products from WiCell.

The Research License sets forth a royalty rate range for products depending on whether they are covered by the licensed patents or produced using the inventions claimed in the licensed patents. The royalty for products that are not covered by the licensed patents but which are produced by the use of the inventions claimed in the licensed patents and the use of the WiCell materials is a quarter percent to two percent. This is a royalty charged for the licensed materials and technique and know-how provided to companies who receive the cells. The royalty rate for products that infringe the licensed patents is 2% to 5% of the selling price of products where the products are covered by the licensed patents or produced using the inventions claimed in the licensed patents. This higher tier royalty rate reflects the fact that not only the materials and technique and know-how are being used, but the claims of the inventions of the licensed patents as well.

Once the company has identified an area in which they are interested in producing a particular product, they can submit a development plan with a timeline for product approval and entry into market to WiCell. If rights are available in the desired area, WiCell will expand the license granted to the company to one which grants a right to make, use, and sell products in a particular licensed field. If the desired rights are not available, the company will be referred to the licensee of those rights to negotiate a sublicense.

Once again the royalty rate will be determined based on the type of products, whether research, diagnostic or therapeutic and whether or not it infringes the claims of the licensed patents or merely uses the license to the materials, technique and know-how provided by WiCell. WiCell used this strategy to make the embryonic stem cell technology widely available not only to academic researchers but to companies interested in developing products in the embryonic stem cell area with as little restriction as possible. However, once a company has identified a path to market with a particular product for which they can set out certain developmental milestones and a plan to market, WiCell will grant a license to make, use and sell such products if those rights are available.

Geron Corporation, Menlo Park California, was an early licensee of WARF in return for their sponsorship of the research that led to the creation of the human embryonic stem cells lines. In return for that sponsorship, Geron obtained an exclusive license to certain fields in the therapeutic and diagnostic areas. Going forward, all licenses to the embryonic stem cell technology will be granted on a non-exclusive basis and the licensees will create their own exclusivity through the patent filings they will inevitably make on inventions they create in their licensed areas.

Through WiCell's two tiered approach to licensing, academic researchers enjoy very little restriction on the research they conduct using the human and primate embryonic stem cell technology. Companies working in stem cell research can also obtain broad rights to conduct research and to commercialize specific products once a development plan has been established. WiCell's approach to licensing the embryonic stem cell technology will allow both academic and commercial scientists to move the research forward while, at the same time, supporting future research at the University of Wisconsin.

17.3.3 Agreements

The PHS agreement with WiCell is an agreement between the US Public Health Service, the parent organization of NIH, and WiCell Research Institute, the University of Wisconsin-Madison research institute charged with the Wisconsin lines of human embryonic stem cells. WiCell is a non-profit institute licensed under the human embryonic stem cell patents owned by WARF.

17.4 State regulation of research involving embryos

In contrast to the paucity of regulations governing embryo research at the federal level, several states have embryo protection laws that can be used to proscribe the creation of human embryonic stem cells. In some states, any research using human embryonic stem cells is also arguably illegal because of their derivation from human embryos. There are strong, disparate movements occurring in several state legislatures that bear mention and watching. For example, the movement that started in California affirmatively to legalize creation and use of human embryonic stem cells may be repeated in several other states. At the same time, a counter movement is underway in a number of states.

The recent explosion of interest in the area of embryonic stem cell research has states scrambling to react legislatively to the ethical issues and economic benefits presented by stem cell therapies. Researchers and research universities must be aware of this ever-evolving area of stem cell regulation on the state level. Varied in language, definitions, and depth, these laws may place additional hurdles in front of scientific advancement. Additionally, these laws present a reaction to federal regulations that limit stem cell research. Watching the activity of state legislatures in this area may provide a harbinger of how the federal government might treat this topic in the future.

Looking at sheer numbers can help convey the importance of this issue for states. According to the National Conference of State Legislatures, 38 states in 2002 considered genetic policy issues in legislation, many pertaining specifically to stem cell research. In the first two months of 2003 alone, 41 bills referencing embryonic and fetal research were introduced in 17 states. Due in large part to the budget crises facing many states, only two state bills relating to embryonic research or human cloning were signed into law in the 2002 cycle, but the public debate surrounding this issue and the current concern with cloning makes stem cell regulation a volatile issue on the state level.

Leading the way in the fight to secure the availability of embryonic stem cells for researchers is the state of California. On 22 September, 2002, then governor, Gray Davis, signed a landmark bill, introduced by Senator Deborah Ortiz (D-Sacramento), which states that 'research involving the derivation and use of human embryonic stem cells, human embryonic germ cells, and human adult stem cells from any source, including somatic cell nuclear transplantation, shall be permitted' and that 'a physician, surgeon, or other health care provider delivering fertility treatment shall provide his or her patient with timely, relevant, and appropriate information to allow the individual to make an informed and voluntary choice regarding the disposition of any human embryos remaining following the fertility treatment'. The bill further stipulates, 'A person may not knowingly ... purchase or sell embryonic or cadaver fetal tissue for research purposes pursuant to this chapter' (SB 253 Art. 5 sec. a, b). Although this bill is limited by federal legislation, the hope of the California legislature, according to UC-San Francisco spokeswoman Jennifer O'Brien, is that the bill will 'tell researchers that if they come to this state ... and have been concerned about the shakiness of the [federal] legislation and whether they are going to be able to get established, they can come here and the state government will support them' (BIOWORLD Today, Volume 13, Issue 200, paragraph 6). Many see the California bill as a direct challenge to federal legislation, which limits federally funded research to the stem cell lines that were in existence as of August 2001. The California bill would make state funds available to be used for the creation of new stem cell lines where federal legislation would not otherwise allow it.

Although California is leading the way in an effort to make their state a leader in stem cell research, there are many other states providing stiff competition. In January 2004, New Jersey became the second state to pass legislation promoting human embryonic stem cell research while specifically outlawing reproductive cloning. The New Jersey legislation 'permit[s] the conduct of research that involves the derivation and use of human embryonic stem cells, human embryonic germ cells and human adult stem cells, including somatic cell nuclear transplantation', and stipulates, like the California bill, that physicians provide information to couples about donating unwanted embryos leftover from fertility treatments (Assembly Bill 2840, paragraphs 1 and 3). In May of 2004, then New Jersey Governor, James E. McGreevey, further advanced his state's efforts by creating the first state-supported stem cell research institute by authorizing the establishment of the Stem Cell Institute of New Jersey, a joint research institute between the University of Medicine and Dentistry of New Jersey (UMDNJ) and Rutgers University, which will be funded through a public–private partnership. Other initiatives are underway in states such as Massachusetts, Minnesota, Illinois and Pennsylvania, while California is considering the issuance of $3 billion in bonds to advance stem cell research in California.

On the other side of the state legislation spectrum are those states that wish severely to limit stem cell research in response to ethical issues raised by the destruction of human embryos. In 2003, Iowa's legislature passed the Human Cloning Prohibition Act, which prohibits both reproductive and therapeutic cloning, defined as 'human asexual reproduction, accomplished by introducing the genetic material of a human somatic cell into a fertilized or unfertilized oocyte whose nucleus has been or will be removed or inactivated, to produce a living organism with a human or predominantly human genetic composition' (Senate File 2118, Sections 2 and 3). Although there is an exception that provides for the embryonic stem cell research permitted by the federal legislation from August 2001, the bill effectively bans research that destroys human embryos, regardless of their source. Proponents of legislation such as this, argue that adult stem cells have proven to be useful and that it is unethical to continue research that destroys human embryos. Some opponents charge that this sort of legislation is unnecessary because federal regulations, and not state laws, should be used as the standard. Still others worry about the chilling effect this sort of legislation will have on research projects at universities around the country (University Wire, Iowa State U). Michigan, South Dakota and Virginia have also passed legislation that prohibits cloning, although it is important to consider carefully the definitions provided in the text of the bills to discern what effect, if any, they will have on embryonic stem cell research.

Obviously, the area of state regulation on stem cell research will continue to be important for researchers, hospitals, universities and biotechnology companies. Undoubtedly, legislative developments at the state level will affect debate in Washington, and the relative value of embryonic stem cell research or the ethical concerns raised by it at both the state and federal levels are likely to evolve over time.

17.5 International legal framework

The international legal framework surrounding the importation and use of human embryonic stem cells deserves mention due to the diversity of legislation originating in Europe, Asia, and Australia. Ranging from the very limiting legislation coming out of Germany to the science-friendly framework set forth by the United Kingdom (UK), these parameters will certainly have an effect on the type and scope of research permitted.

United Kingdom

Dating back to the establishment in 1990 of the Human Fertilization and Embryology Authority (HFEA) which is appointed by the Secretary of State and can grant licenses to perform research on human embryonic stem cells, the UK maintains its position as one of the most legally unrestricted places to practice such research. The 1990 act also lays out specific guidelines that require the consent of the individuals whose gametes were used to create the embryo. In 2001, Parliament took a forward-thinking view and attached the Human Fertilization and Embryology Research Purposes to its pre-existing act, adding three new purposes: 'Increasing knowledge about the development of embryos', 'increasing knowledge about serious disease', and 'enabling any such knowledge to be applied in developing treatments for serious disease'. The HFEA still must approve all licenses and will not allow procedures on embryos that are older than 14 days, but these new purposes have opened the door for more serious research on stem cells. As in the USA, the UK passed the Human Reproductive Cloning Bill in November of 2001, which prevents human cloning for reproductive purposes.

Singapore

Modeling their stem cell research laws on those of the UK, Singapore is currently discussing legislation that would ban reproductive cloning but would allow scientists to take stem cells from aborted fetuses and surplus *in vitro* fertilized embryos less than 14 days old. It is important to note that Singapore is the funding agent for ES Cell International (ESI), which has been a key source of embryos for research in the USA and Europe since 2001 and over 60 countries worldwide since then. Additionally, Professor Ariff Bongso has developed stem cell lines that grow on human fibroblasts, instead of the traditional mouse feeder-cell layers. This advancement may make stem cells more acceptable for transplantation therapies by avoiding contamination by xenogeneic proteins or cells, and inciting less of an inflammatory response. In light of these developments and the potential market for such research, it seems that Singapore will be looking to maintain its position as a science-friendly destination for researchers.

Australia

Australia, like the UK, has adopted legislation regarding the use of human embryonic stem cells that establishes a governing body, the South Australian Council on Reproductive Technology, to oversee research requests. In contrast to the UK, however, Australia's laws are more restrictive. Research on embryos in Australia is limited to those created for (infertility) treatment purposes, and is not permitted on embryos created specifically for research. In addition, the research must not harm the embryo or make the embryo unfit for transfer to a woman. It is interesting to note that the Singapore-funded ES Cell International is located in Australia, signifying the commitment that Australia has to maintaining a good relationship with researchers all over the world. As in the UK, consent of the individuals whose gametes combined to create the embryo is required for any research project.

Germany

On the other end of the research spectrum is Germany, who passed a law in July 2002 that makes it illegal to use human embryonic stem cells derived in Germany or to derive new lines, but that allows for the importation of such cells if they were in existence before January 1, 2002. The first human embryonic stem cells to be imported into Germany arrived in December 2002, after a lengthy wait on the part of researchers. Like all countries that have tackled the issue of human embryonic stem cell legislation, Germany's Budestag was divided for months on the ethical implications of such research and finally came to the import law as a suitable compromise. Opponents of the new law worry that such strict regulation will create a 'brain drain' of the very best scientists, who might move to other countries to perform their research. With the exodus of intellectual capital, so will go the potential revenues generated by a burgeoning biotechnology field.

17.6 Summary

The laws and regulations governing stem cell research are varied; which apply depends a great deal on the type of research being performed. For those scientists endeavoring to derive new cell lines or to study therapeutic cloning procedures, licensing and regulatory hurdles are likely to be greater than for those researchers simply performing *in vitro* manipulations. In the USA, the scientific environment may change rapidly in the near future, as more states undertake debate on therapeutic cloning and stem cell research. As more states in the USA and more countries, such as Japan, pass laws related to stem cell research, the legal framework in which to conduct this research will come into clearer focus.

Exhibit A

Agreement No._____

Memorandum of Understanding

This Memorandum Of Understanding (hereinafter 'Agreement'), effective _____, by and between the _____, having an address at _____ ('Recipient') and the WiCell Research Institute, Inc., a Wisconsin nonprofit corporation having an address at 614 Walnut Street, Madison, Wisconsin 53726 ('WiCell'). Institute and WiCell are referred to herein as the 'Parties'.

WHEREAS, certain technologies and materials concerning primate embryonic stem cells and their cultivation claimed in U.S. Patent 5,843,780, U.S. Patent 6,200,806, U.S. Patent Application 09/522,030 and corresponding U.S. or foreign patent rights and any patents granted on any divisional and continuation applications of any type but only to the extent it claims an invention claimed in a patent application listed herein ('Wisconsin Patent Rights') have usefulness in basic research conducted by Recipient as well as potential utility for commercial applications; and

WHEREAS, specific human embryonic stem cell line materials, their unmodified and undifferentiated progeny or derivatives ('Wisconsin Materials') have been derived consistent with the Presidential Statement of August 9, 2001 from the research efforts of James A. Thomson of the University of Wisconsin – Madison working alone or with other investigators; and

WHEREAS, Wisconsin Materials were made using solely private funds and are the proprietary, tangible property of WiCell and, as such, their ownership is not subject to the rights and obligations granted the Government in the Wisconsin Patent Rights; and

WHEREAS, the Wisconsin Alumni Research Foundation of the University of Wisconsin – Madison ('WARF') and WiCell have a mission to serve the public good and desire to serve the public interest by making the Wisconsin Materials and the Wisconsin Patent Rights widely available to Recipient and other academic researchers; and

WHEREAS, WiCell represents that it has received a license, with the right to grant sublicenses, to Wisconsin Patent Rights from WARF and that WiCell also owns or otherwise has the right to distribute Wisconsin Materials to third parties; and

WHEREAS, WiCell desires to exercise Wisconsin Patent Rights and distribute Wisconsin Materials without placing undue restrictions or burdens upon health research conducted by Recipient;

NOW, THEREFORE, the Parties hereby agree to the following terms and conditions regarding use of Wisconsin Materials or Wisconsin Patent Rights for academic, non-commercial research conducted by Recipient:

(1) The Parties agree that Wisconsin Patent Rights are to be made available without cost for use in the Recipient biomedical research program subject to the following conditions:

(a) Wisconsin Patent Rights may be used in research programs involving Wisconsin Materials only in programs in compliance with all applicable statutes, regulations and guidelines for research of this type. Specifically, Recipient agrees that its research programs will exclude: (i) the mixing of Wisconsin Materials with an intact embryo, either human or non-human; (ii) implanting Wisconsin Materials or products of Materials in a uterus; and (iii) attempting to make whole embryos with Wisconsin Materials by any method. An annual Certification Statement confirming compliance with the restrictions on the use of Wisconsin Materials shall be supplied to WiCell by Recipient and the scientists receiving Wisconsin Materials under the terms of the 'Simple Letter Agreement For The Transfer of Materials.' Recipient agrees that Wisconsin Materials are to be returned to WiCell or destroyed upon a material breach of the terms of the Simple Letter Agreement for the Transfer of Materials Agreement.

(b) Wisconsin Patent Rights may also be used in Recipient research programs involving materials other than Wisconsin Materials that may be within the scope of an issued claim of Wisconsin Patent Rights ('Third Party Materials'). This research may be conducted only in Recipient research programs using Third Party Materials that are derived consistent with the Presidential Statement of August 9, 2001 and in compliance with all applicable statues, regulations and guidelines.

(c) Suppliers of Third Party Materials are granted a limited, revocable, non-commercial, research license by WiCell under the Wisconsin Patent Rights to provide such Third Party Materials to Recipient research programs provided that such Suppliers make such Third Party Materials available on terms no more onerous than those contained in this Agreement. Specifically, but without limitation, Suppliers of Third Party Materials shall not be permitted to directly or indirectly receive rights (either actual or contingent) for themselves or others under agreements or arrangements governing the supply or use of Third Party Materials. The use of Wisconsin Patent Rights in Recipient research programs utilizing Third Party Materials shall be for teaching or non-commercial research purposes only. As used herein, non-commercial research purposes specifically excludes sponsored research wherein the sponsor receives a right whether actual or contingent to the results of the sponsored research, other than a grant for non-commercial research purposes to the sponsor. The Wisconsin Patent Rights may not be used with Third Party Materials for commercial purposes or the direct benefit of research sponsor, except as such research sponsor is permitted to use Wisconsin Patent Rights under a separate written agreement with WiCell or WARF. Specifically, Third Party Materials shall not be used in a Recipient research program where rights (either actual or contingent) have already been granted to a research sponsor who does not have a separate written agreement with WiCell permitting commercial use of Wisconsin Patent Rights.

(d) The Parties recognize that Wisconsin Patent Rights may be used in Recipient research to make patentable discoveries ('Recipient Patent Rights'), which themselves may eventually be the basis of commercial products that benefit public health. Any grant of Wisconsin Patent Rights that may be needed by a third party for commercialization of Recipient Patent Rights shall be done by a separate written agreement with WiCell permitting such use of Wisconsin Patent Rights under terms not less favorable than other similar commercial licenses to the extent such rights are available.

(2) The Parties agree that Wisconsin Materials are to be made available by WiCell for use in Recipient's biomedical research programs. For purposes of transferring Wisconsin Materials to Recipient investigators, WiCell agrees to utilize the Simple Letter Agreement For The Transfer of Materials including the following conditions:

(a) Wisconsin Materials are the property of WiCell and are being made available to investigators at Recipient institution as a service by WiCell. Ownership of Wisconsin Materials shall remain with WiCell.

(b) Wisconsin Materials are not to be used for diagnostic or therapeutic purposes.

(c) Wisconsin Materials may only be used in compliance with all applicable statutes, regulations and guidelines relating to their handling or use. Specifically, Recipient agrees that its research program will exclude: (i) the mixing of Wisconsin Materials with an intact embryo, either human or non-human; (ii) implanting Wisconsin Materials or products of Materials in a uterus; and (iii) attempting to make whole embryos with Wisconsin Materials by any method. An annual Certification Statement confirming compliance with the restrictions on the use of Wisconsin Materials shall be supplied to WiCell by Recipient and the scientists receiving Wisconsin Materials under the terms of the Simple Letter Agreement For The Transfer of Materials. Recipient agrees that Wisconsin Materials are to be returned to WiCell or destroyed upon a material breach of the terms of the Simple Letter Agreement for the Transfer of Materials by Recipient or its investigators.

(d) The use of Wisconsin Materials shall be for teaching or non-commercial research purposes only. As used herein, non-commercial research purposes specifically excludes sponsored research wherein the sponsor receives a right whether actual or contingent to the results of the sponsored research, other than a grant for non-commercial research purposes to the sponsor. The Wisconsin Materials may not be used for commercial purposes or the direct benefit of research sponsor, except as such research sponsor is permitted to use Wisconsin Materials under a separate written agreement with WiCell or WARF. Specifically, Wisconsin Materials shall not be used in a Recipient research program where rights (either actual or contingent) have already been granted to a research sponsor who does not have a separate written agreement with WiCell permitting such commercial use of Wisconsin Materials.

(e) Wisconsin Materials may not be transferred by Recipient to third parties without the written consent of WiCell.

(f) Recipient agrees to acknowledge the source of Wisconsin Materials in any publications or other disclosures reporting their use.

(g) In order to facilitate potential novel collaborative research interactions between Recipient and WiCell that may utilize Wisconsin Materials, Recipient agrees to identify the titles of its planned research in its individual requests for samples of Wisconsin Materials. This information is to be provided to facilitate new inter-disciplinary collaborations among individual scientists at Recipient and WiCell, but not to obligate either Party to a specific program of research utilizing Wisconsin Materials.

(h) The Parties recognize that Wisconsin Materials may be used in the Recipient's research program to make discoveries of different materials ('Recipient Materials') which themselves may eventually be the basis of commercial products that benefit public health. Any grant of rights to Wisconsin Materials or Wisconsin Patent Rights that may be needed by a third party for commercialization of Recipient Materials shall be done by a separate written agreement with WiCell permitting such use of Wisconsin Materials or Wisconsin Patent Rights under terms not less favorable than other similar commercial licenses to the extent such rights are available.

(i) Any Wisconsin Materials delivered pursuant to this Agreement are understood to be experimental in nature and may have hazardous properties. WiCell makes no representations and extends no warranties of any kind, either expressed or implied. There are no express or implied warranties of merchantability for fitness for a particular purpose, or that the use of the Wisconsin Materials will not infringe any patent, copyright, trademark or other proprietary rights. Recipient assumes all liability for claims for damages which may arise from the use, storage, handling or disposal of Wisconsin Materials except that, to the extent permitted by law, WiCell shall be liable to Recipient when the damage is caused by the gross negligence or willful misconduct of WiCell.

(j) A transmittal fee may be requested by WiCell to cover its preparation and distribution costs for samples of Wisconsin Materials requested by Recipient. Such fees will be the responsibility of the requesting Recipient laboratory and are not expected to exceed Five Thousand Dollars ($5,000) to accompany the Recipient Simple Letter Agreement for the Transfer of Materials.

(3) Upon WiCell's written request, Recipient agrees to provide without cost reasonable quantities of any Recipient Materials that it makes in the course of its research program to WiCell for research purposes only at WiCell or the University of Wisconsin after Recipient has publicly disclosed or reasonably characterized such Recipient Materials. Recipient also agrees to grant under the Recipient Patent Rights research licenses to WiCell and the University of Wisconsin.

(4) The provisions of this Agreement and the obligations hereunder with respect to the Wisconsin Patent Rights shall be in effect only during the term of the Wisconsin Patent Rights. However, the provisions of this Agreement and the obligations hereunder with respect to the Wisconsin Materials shall continue as long as Wisconsin Materials, their derivatives or progeny continue to be used by Recipient.

(5) Nothing contained herein shall be considered to be the grant of a commercial license or right under the Wisconsin Patent Rights or to Wisconsin Materials. Furthermore, nothing contained herein shall be construed to be a waiver of WiCell's patent rights under the Wisconsin Patent Rights or WiCell's property rights in Wisconsin Materials.

IN WITNESS WHEREOF, the Parties agree to the foregoing and have caused this Agreement to be executed by their duly authorized representatives.

WiCell Research Institute _____   Recipient_____

By: _____   By: _____
   Carl E. Gulbrandsen, President

Date: _____   Date: _____

   Name: _____

   Title: _____

Reviewed by WiCell's General Counsel:

_____ _____, _____
Elizabeth L.R. Donley, Esq.

(WiCell's attorney shall not be deemed a signatory to this Agreement.)

WiCell Ref: Thomson – P98222US

Agreement No. _____

Simple Letter Agreement for the Transfer of Materials to Recipient Scientists

In response to RECIPIENT's request for MATERIAL (Human Embryonic Stem Cells, WiCell Ref: No. P98222US and its unmodified and undifferentiated progeny or derivatives) for a research program entitled _____
_____WiCell Research Institute, Inc. ('PROVIDER') has entered into a Memorandum of Understanding dated _____ (the 'MOU') with RECIPIENT which is hereby incorporated by reference and asks that the RECIPIENT and the RECIPIENT SCIENTIST agree to the following before the RECIPIENT SCIENTIST receives the MATERIAL:

1. The above MATERIAL is the property of the PROVIDER and is made available as a service to the research community. Ownership of the MATERIAL shall remain with PROVIDER and transfer of the MATERIAL to the RECIPIENT shall not affect PROVIDER's ownership of the MATERIAL.

2. This MATERIAL is not to be used for diagnostic or therapeutic purposes.

3. The MATERIAL will be used for teaching or non-commercial research purposes. As used herein, non-commercial research purposes specifically excludes sponsored research wherein the sponsor receives a right whether actual or contingent to the results of the sponsored research. The MATERIAL may not be used for commercial purposes or the direct benefit of research sponsor, except as such research sponsor is permitted to use MATERIAL under a separate written agreement with PROVIDER. Specifically, MATERIAL shall not be used in a research program where rights (either actual or contingent) have already been granted to a research sponsor who does not have a separate written agreement with PROVIDER permitting such use of MATERIAL.

4. Nothing contained herein shall be considered to be the grant of a commercial license or right under U.S. Patent 5,843,780, U.S. Patent 6,200,806, U.S. Patent Application 09/522,030 and corresponding U.S. or foreign patent rights and any patents granted on any divisional and continuation applications, reissues and reexaminations ('WISCONSIN PATENT RIGHTS') or to the MATERIALS. Furthermore, nothing contained herein shall be construed to be a waiver of PROVIDER's patent rights under the WISCONSIN PATENT RIGHTS or PROVIDER's property rights in the MATERIALS.

5. The MATERIAL will not be further distributed to others without the PROVIDER's written consent. The RECIPIENT shall refer any request for the MATERIAL to the PROVIDER. To the extent supplies are available, the PROVIDER or the PROVIDER SCIENTIST agree to make the MATERIAL available, under a separate Simple Letter Agreement to other scientists for teaching or non-commercial research purposes only.

6. The RECIPIENT agrees to acknowledge the source of the MATERIAL in any publications reporting use of it.

7. Upon PROVIDER's written request, RECIPIENT agrees to provide without cost reasonable quantities of any RECIPIENT MATERIALS that it makes in the course of its research program to PROVIDER for research purposes only at PROVIDER or the University of Wisconsin after RECIPIENT has publicly disclosed or reasonably characterized such RECIPIENT MATERIALS. RECIPIENT also agrees to grant under the RECIPIENT Patent Rights research licenses to PROVIDER and the University of Wisconsin-Madison.

8. Any MATERIAL delivered pursuant to this Agreement is understood to be experimental in nature and may have hazardous properties. THE PROVIDER MAKES NO REPRESENTATIONS AND EXTENDS NO WARRANTIES OF ANY KIND, EITHER EXPRESSED OR IMPLIED. THERE ARE NO EXPRESS OR IMPLIED WARRANTIES OF MERCHANTABILITY OR FITNESS FOR A PARTICULAR PURPOSE, OR THAT THE USE OF THE MATERIAL WILL NOT INFRINGE ANY PATENT, COPYRIGHT, TRADEMARK, OR OTHER PROPRIETARY RIGHTS. Unless prohibited by law, RECIPIENT assumes all liability for claims for damages which may arise from the use, storage, handling or disposal of MATERIAL except that, to the extent permitted by law, PROVIDER shall be liable to the RECIPIENT when the damage is caused by the gross negligence or willful misconduct of the PROVIDER.

9. The RECIPIENT agrees to use the MATERIAL only in compliance with all applicable statutes, regulations and guidelines relating to their handling, use or disposal. Specifically, RECIPIENT agrees that its research program will exclude: (i) the mixing of MATERIAL with an intact embryo, either human or non-human; (ii) implanting MATERIAL or products of MATERIAL in a uterus; and (iii) attempting to make whole embryos with MATERIAL by any method. RECIPIENT shall supply an Annual Certification Statement confirming compliance with the restrictions on the use of MATERIAL supplied by PROVIDER. RECIPIENT agrees that MATERIAL is to be returned to PROVIDER or destroyed upon a material breach of the terms of this Agreement by RECIPIENT.

10. The MATERIAL is provided with a transmittal fee solely to reimburse the PROVIDER for its preparation and distribution costs. The amount of the fee for this transfer of MATERIAL will be $5000.

The PROVIDER, RECIPIENT and RECIPIENT SCIENTIST must sign both copies of this letter and return one signed copy to the PROVIDER. The PROVIDER will then send the MATERIAL.

WICELL RESEARCH INSTITUTE, INC.

By: _____
    Carl E. Gulbrandsen, President

Date: _____

RECIPIENT INFORMATION and AUTHORIZED SIGNATURE

Recipient Scientist: _____

Recipient Organization: _____

Address: _____

Signature of Recipient Scientist: _____

Date: _____

Name of Authorized Official: _____

Title of Authorized Official: _____

Signature of Authorized Official: _____

Date: _____

Reviewed by WiCell's General Counsel:

_____ _____, _____
Elizabeth L.R. Donley, Esq.

(WiCell's attorney shall not be deemed a signatory to this Agreement.)

WiCell Ref: Thomson – P98222US

ANNUAL CERTIFICATION

Annual Certification of Recipient Scientist: I have read and understood the conditions outlined in this Agreement and I agree to abide by them in the receipt and use of the MATERIAL. I further certify that I am not engaged and have not been engaged in commercial research using the MATERIAL or any third party material which requires a license under the WISCONSIN PATENT RIGHTS.

Recipient Scientist: _____

Date: _____

Recipient Scientist: _____

Date: _____

Recipient Scientist: _____

Date: _____

18. Genomic approaches to stem cell biology

Tetsuya S. Tanaka, Mark G. Carter, Kazuhiro Aiba, Saied A. Jaradat, and Minoru S.H. Ko

18.1 Introduction

'Embryogenomics', the systematic analysis of a cohort of genes expressed in specific cell types, is a powerful approach to understand the characteristics of cells and their functions (Ko, 2001). It is now being applied to the study of stem cell biology. The application of such genomic methodologies to stem cell research is also relevant to more general biomedical sciences. For example, early human embryos are not easily accessible for technical and ethical reasons; but human embryonic stem cells (ESC) can provide material to identify new genes and alternatively spliced transcripts that are expressed specifically at early stages. This point, though often overlooked, is very important even in the post-human genome sequence era. The complete catalogue of all genes, a major goal of the Human Genome Project, will indeed provide the fundamental information for the development of new therapies and diagnostic tools. Similarly, the collection of all mouse genes for comparison with their human counterparts is critical, because the laboratory mouse emerges as the premier biomedical research model of the twenty-first century by promising rapid and flexible link between basic and clinical research, and by speeding and enhancing discovery. However, even though nearly complete human (Lander *et al.*, 2001; Venter *et al.*, 2001) and mouse (Waterston *et al.*, 2002) genome sequence are available, some genes have not been identified and properly annotated because the mRNAs are not isolated in the form of cDNA clones.

In this chapter, we first discuss some technical aspects of genomic approaches such as cDNA analyses and expression profiling. We then present two examples of such applications: (1) expression profiling and comparison among various types of stem cells; and (2) expression profiling between cloned mice and normal mice. Following discussion of the large-scale functional analyses of genes, we conclude by outlining some future directions and challenges in stem cell biology.

18.2 Large-scale isolation of new genes from early embryos and stem cells

The fundamental information required for genomics approaches are the sequences and structures of the genes under study. This information is provided by Expressed Sequence Tag (EST) projects (Adams *et al.*, 1991), in which complementary DNA (cDNA) libraries are first made from a variety of tissues and randomly selected individual cDNA clones are sequenced from their 5'- or 3'-ends. These sequences are usually single-pass and very short, up to 500 bp. They therefore often miss protein-coding regions, but can be used as unique identifiers for individual genes. A number of such projects began contemporaneously (Adams *et al.*, 1991; Hoog, 1991; Ko, 1990; Okubo *et al.*, 1992; Sikela and Auffray, 1993), and the work culminated in the publication of EST-based human (Adams *et al.*, 1995) and mouse (Marra *et al.*, 1999) gene indices. However, such indices are likely to miss many genes, because most of the cDNA libraries used were derived from adult or late-stage fetal tissues, and genes that are not expressed in those tissues would thus not be detected. From the outset, the goal of the Mouse cDNA Project we carried out has been to collect all mouse genes, so that many embryonic tissues were included as cDNA sources for library construction (Ko *et al.*, 1998, 2000; Takahashi and Ko, 1994). Several other projects aim specifically at capturing transcripts expressed in mouse preimplantation embryos (Sasaki *et al.*, 1998; Solter *et al.*, 2002) or in early human development (Adjaye *et al.*, 1998; Morozov *et al.*, 1998).

Most early cDNA studies used cDNA libraries with relatively short insert size. They therefore had a low probability of capturing complete coding sequences for most genes. This was particularly problematic in preimplantation embryo cDNA libraries, because the scarcity of materials made library construction difficult. Insert sizes were up to 3 kb, with an average insert size less than 1.5 kb, and often much shorter. Although these cDNA clones were not adapted to the recovery of complete open reading frames, they provided good unique probes for *in situ* hybridization to tissue sections and for cDNA microarrays. As originally proposed (Ko, 1990), short cDNA clones carrying only the 3'-end of transcripts are advantageous for such purposes, because that region is usually the most idiosyncratic part of a transcript, and is thereby relatively specific when spotted on microarrays (e.g., LION Bioscience: http://www.lionbioscience.com/solutions/archived-products/arraytag-arraybase/mouse).

A condensed set of unique cDNA clones is desirable for cDNA microarray construction. The first condensed, non-redundant clone set assembled from our collections was the 'NIA 15K Mouse cDNA Clone Set' that contains 15,247 cDNA clones (Tanaka *et al.*, 2000). These unique cDNA clones were selected from approximately 53,000 3'-ESTs derived from preimplantation stages (unfertilized eggs, 1-cell, 2-cell, 4-cell and 8-cell embryos, morula and blastocyst (Ko *et al.*, 2000)), micro-dissected tissues of embryonic and extra-embryonic parts of E7.5 embryos (Ko *et al.*, 1998), female gonad/mesonephros from E12.5 embryos, and ovary from newborn fetus. About half of the clones represent transcripts with unknown functions (Kargul *et al.*, 2001). Recently, we completed the assembly of the 'NIA 7.4K Mouse cDNA Clone Set (VanBuren et al., 2002)', a non-redundant collection of cDNAs that are not represented in the 15K clone set. The cDNA clones in this set were derived from embryonic tissues (E0.5 to E12.5), as well as the following types of stem cells: ESC, trophoblast stem cells (TSC), mesenchymal stem cells (MSC), neural stem cells (NSC), hematopoietic stem cells (HSC), and embryonic germ cells (EGC). With an average insert size of 2.5 kb, initial analyses indicate that many of these clones contain full-length inserts. Both NIA mouse 15K and 7.4K Clone Sets are available without restriction, and have been distributed to ten academic centers for further distribution to over 200 research centers worldwide (see our web site for details, http://lgsun.grc.nia.nih.gov/cDNA/cDNA.html). Other condensed cDNA clone sets available from a few commercial sources include one from mouse brain tissues by Brain Molecular Anatomy Project (BMAP: http://www.nimh.nih.gov/grants/0006-cbd1.cfm) and those from the IMAGE consortium containing human or mouse cDNA clones (Lennon et al., 1996).

During this period, many cDNA projects have shifted to focus on the collection and sequencing of full-length cDNA clones. Projects that are focusing on such goals include, as major examples, the RIKEN mouse encyclopedia project (Okazaki et al., 2002), the Mammalian Gene Collection (MGC) (Strausberg et al., 2002), the Kazusa cDNA project (Kikuno et al., 2002), the Sugano cDNA project (Suzuki et al., 2002), and the German Cancer Research Center project (Wiemann et al., 2001). Methods used in these projects characteristically require a large amount of starting RNA, and are thus not suited to the study of pre-implantation and early stages of mammalian development. We have recently bypassed this block by developing a novel linker-primer design that, depending on the linker, allows differential amplification of long cDNAs (average 3.0 kb with size ranges of 1–7 kb) or short cDNAs (average 1.5 kb with size ranges of 0.5–3 kb) from a complex mixture (Piao et al., 2001). The method facilitates the generation of cDNA libraries enriched for long transcripts without size selection of insert cDNA. As a result, a significant fraction of these cDNA clones contain complete open reading frames, with many full-length coding regions. To date, the NIA cDNA Project has generated 224,511 ESTs from nearly 50 individual libraries. Individual NIA cDNA clones are currently available from the American Type Culture Collection (ATCC), and information on each cDNA clone, including DNA sequence, is available from our web site (http://lgsun.grc.nia.nih.gov/cDNA/cDNA.html) as well as public databases (e.g., GenBank).

EST projects focused on specific types of stem cells have also been reported. For example, earlier work at Princeton University studied hematopoietic stem cells (Phillips et al., 2000). EST projects on human ESC have also been conducted by various groups. As discussed above, all these projects provide very useful resources for stem cell gene collection and identification.

18.3 Methods for gene expression profiling

There are currently at least five popular methods of expression profiling (*Table 18.1*). These methods can be roughly classified into three groups based on the number of genes that can be monitored.

Table 18.1: Summary of common expression profiling methods.

Method	Description	Advantages	Disadvantages
EST libraries	Thousands of EST clones are derived from each sample, sequenced, and compared to public sequence databases for gene identification. Relative frequencies are calculated for each transcript present (Adams et al., 1991; Okubo et al., 1992). Examples: preimplantation embryos (Ko et al., 2000; Marra et al., 1999; Sasaki et al., 1998), germ cells/oocytes (Choo et al., 2001; Rajkovic et al., 2001; Stanton and Green, 2001a).	– captures unknown genes	– labor-intensive – rare transcript measurement requires very large libraries
SAGE	Hundreds of thousands of sequence tags are cloned, sequenced, and compared with public sequence databases for gene identification. Relative frequencies are calculated for each transcript present (Velculescu et al., 1995). Examples: ES cells (Anisimov et al., 2002), GV oocytes (Stanton et al., 2002).	– measures unknown genes	– labor-intensive – may be problematic for rare transcripts
cDNA microarrays	Thousands of cDNA clones are sequenced and compared with public sequence databases for gene identification. Representative clones are selected and rearrayed for plasmid DNA preparation, PCR amplification of cDNA inserts, and robotic spotting onto nylon membranes or glass slides. RNA samples are labeled and hybridized to the arrays, and relative expression levels for each transcript are calculated from signal measurements for each spot (Schena et al., 1995). Examples: ES/TS cells (Tanaka et al., 2002), HSC (Terskikh et al., 2001). -	– cross-species hybridization possible – standard for in-house production	– array construction is labor-intensive – specificity may be lower than other methods – transcripts must be cloned for measurement
Oligo microarrays	Thousands of oligonucleotides are synthesized and spotted on glass slides. RNA samples are labeled and hybridized to the arrays, and relative expression levels for each transcript are calculated from signal measurements for each spot (Fodor et al., 1991; Hughes and Shoemaker, 2001). Examples: HSC (Ivanova et al., 2002), ES/NSC/HSC (Ramalho-Santos et al., 2002).	– array construction can be less labor-intensive – more transcript-specific – cloned cDNAs not required	– high-quality sequence data required for oligo design
Quantitative real-time RT-PCR	For each transcript, PCR primer pairs are designed and tested. RNA samples are reverse-transcribed to cDNA and amplified using the PCR primers. Signals produced by PCR amplification are normalized to RNA input or housekeeping genes, and relative expression levels are calculated (Higuchi et al., 1993). Examples: ES-osteoblast (Zur Nieden et al., 2003), HSC (Hofmann et al., 2002).	– specificity can be controlled through primer selection – highly quantitative – large dynamic range	– primers must be validated
ATAC-PCR	RNA from each sample is reverse-transcribed to cDNA, digested with restriction enzymes, and ligated to adaptor primers of different lengths. Primers complementary to common sequence in the adaptors are combined with gene-specific primers and used to PCR-amplify transcript fragments of interest. The fragments are separated by gel electrophoresis and quantitated using automated sequencing equipment, and relative expression levels are calculated (Kato, 1997). Examples: cerebellar development (Matoba et al., 2000).	– specificity can be controlled through primer selection – highly quantitative – large dynamic range	– primers must be validated
Differential display	RNA from each sample is reverse-transcribed to cDNA using an anchor primer, then PCR amplified using the same anchor primer combined with an arbitrary primer. Fragments are separated by gel electrophoresis, identified as differentially amplified, isolated, cloned, and sequenced for gene identification (Liang and Pardee, 1992). Examples: HSC (Orelio and Dzierzak, 2002), NSC (Wen et al., 2002).	– uses common molecular techniques	– very labor-intensive – band selection is subjective – high false positive rate

Table 18.1: *continued*.

Method	Description	advantages	disadvantages
Subtraction (SSH, RDA)	cDNA pools from two samples are hybridized with one in excess, and hybrids are removed, leaving non-hybridized (differential) cDNAs behind to be cloned and sequenced. Examples: ES cells (Du et al., 2001).	– uses common molecular techniques – good at detecting differences in highly-expressed transcripts	– difficult/labor-intensive – unidirectional comparisons only – high redundancy of non-relevant clones – may require multiple rounds of selection
Massive parallel signature sequencing (MPSS)	cDNAs are cloned onto microbeads, which are then packed in a movement-restricting flow cell. Repeated cycles of 'encoded' adaptor ligation, hybridization of labeled 'decoder' oligos, image capture, and type II restriction endonuclease cleavage produce short (20bp) sequence signatures for millions of templates (Brenner et al., 2000).	– >10^5 tags generated in a single run	– technically challenging – tags are short – significant failure rate at microbead level

Abbreviations: GV, germinal vesicle; ATAC, adaptoc tagged competitive; SSH, suppression subtractive hybridization; RDA, representational difference analysis.

- The first group is able to monitor a large number of genes, potentially all genes. Although their ease and cost varies, methods included in this group include microarrays (Hughes *et al.*, 2001; Lipshutz *et al.*, 1999; Pease *et al.*, 1994; Schena *et al.*, 1995), serial analysis of gene expression (SAGE; Velculescu *et al.*, 2000), massively parallel signature sequencing (MPSS; Brenner *et al.*, 2000), and EST frequency measurement.

- The second group monitors the expression levels of a limited number of genes. These methods include quantitative-PCR (Q-PCR) and its variations, as well as Northern blotting (which also directly provides information about transcript size that can be compared to the cDNA clone size). Ready-made TaqMan primers for all human transcripts are becoming available from Applied Biosystems and other firms.

- A third group of methods falls between other two. It can monitor a few hundred to one thousand genes, though the labor and cost significantly increases with the number of genes. Differential Display (Liang and Pardee, 1992) and ATAC-PCR (Kato, 1997) are included in the third group.

Each method described here, as well as those not mentioned, has advantages and disadvantages – the nature of tissues, the amount of available materials, biological processes, and the number and types of genes under study will determine which approach or combination of approaches is appropriate and feasible. For example, tissues and cells from early embryos and stem cells are often very limited, so that very sensitive detection methods such as Q-PCR must be used to detect rare transcripts, or cDNA amplification methods are required to amplify the labeled probe (Kacharmina *et al.*, 1999).

The number and complexity of technical issues related to expression profiling not only make these studies challenging to perform, but also make evaluation of the data they produce very complex. Critical reading of the literature requires an understanding of the strengths and pitfalls of each technique.

In general, experimental goals require the monitoring of a large number of genes. Thus, microarray, MPSS, or SAGE would be methods of choice in the beginning phase of the work. However, the recent documentation of high variability in expression measurements from one sample to another demonstrates the importance of repeated measurements using different batches of biological samples to achieve statistical validity (Yang and Speed, 2002). To perform the necessary repeated measurements, assays must be quick, easy and relatively inexpensive. These requirements position microarrays as the method of choice, with more sensitive or specific methods used for validation and downstream experimentation.

Two forms of microarrays are currently available: (1) cDNA microarrays made by spotting PCR-amplified cDNA inserts onto nylon membranes or glass slides (Schena et al., 1995); (2) oligonucleotide microarrays, with probes of much shorter (25–70 nt) DNA sequences from each cDNA, synthesized and covalently attached to glass slides (Hughes et al., 2001; Lipshutz et al., 1999; Pease et al., 1994). cDNA arrays require large numbers of properly-identified cDNA clones, and cross hybridization of longer cDNA probes detects not only the transcript from which they were derived, but also splice variants and closely related transcripts. In some cases, cross-species hybridization has been possible using cDNA microarrays. Oligo arrays, on the other hand, require high-quality sequence data for probe design, rather than physical clones. Oligos can be synthesized individually and applied to the slides later, but newer technologies have allowed oligonucleotide synthesis to be performed directly on a glass substrate (Hughes et al., 2001; Lipshutz et al., 1999; Pease et al., 1994). When designed properly, the shorter probe sequences are highly specific for individual transcripts, and can distinguish the expression of splice variants and gene family members. To provide standard mouse microarrays in this format, we have recently developed 60-mer oligonucleotide arrays, which cover a wide variety of genes expressed in early embryos and stem cells (Carter et al., 2003).

18.4 Data analysis and bioinformatics

Large-scale genomic methodologies differ from traditional biological studies in the scope and the volume of the experimental data, i.e., information on not one or a few genes, but thousands. This feature is both a challenge and an asset – on the one hand, such volumes of data can be overwhelming to the researcher; on the other hand, such compendium information will provide much deeper levels of understanding that are not obtained from gene-by-gene analysis. Various bioinformatics and statistical analysis tools have been developed and utilized for the analysis of expression profiles of a large number of genes (*Table 18.2*). In particular, hierarchical clustering (Eisen et al., 1998), self-organizing map (SOM; Tamayo et al., 1999), and k-means clustering analyses have been widely used. In our own experience, the k-means clustering algorithm usually produces the most satisfactory results (Chen et al., 2002). A number of excellent reviews and perspectives have been published (Churchill, 2002; Quackenbush, 2002; Yang and Speed, 2002), and the summary of available tools, though not comprehensive, is listed in *Table 18.2*. Some examples of actual data analysis will be shown later in this chapter.

Table 18.2: Bioinformatical tools available through internet.

Software

http://abs.cit.nih.gov/	Analytical Biostatistics Section at NIH
http://linus.nci.nih.gov/BRB-ArrayTools.html	BRB Array Tools
http://arrayanalysis.nih.gov/	CIT Microarray Data Analysis at NIH
http://rana.lbl.gov/index.htm?software/	Eisen Lab at LBL
http://www.cse.ucsc.edu/research/compbio/genex/	Support Vector Machines
http://www.mged.org/Workgroups/MIAME/miame_software.html	MIAME Compliant Softwares
http://web.wi.mit.edu/bio/pub/biopage5tools.html	Tools at Whitehead Institute
http://genome-www.stanford.edu/~sherlock/	Sherlock Lab at Stanford University
http://www.systemsbiology.org/research/software.html	Institute for Systems Biology
http://pevsnerlab.kennedykrieger.org/snomadinput.html	Pevsner Lab at Johns Hopkins Univ.
http://pga.tigr.org/tigr-scripts/magic/r1.pl	TIGR RESOURCERER
http://www.tigr.org/software/	TIGR Softwares
http://lgsun.grc.nia.nih.gov/ANOVA/	NIA Array Tools

Protocol

http://cmgm.stanford.edu/pbrown/	Brown Lab at Stanford University
http://derisilab.ucsf.edu/	DeRisi Lab at UCSF
http://cmgm.stanford.edu/~kimlab/wmdirectorybig.html	Kim Lab at Stanford University
http://research.nhgri.nih.gov/microarray/index.html	NHGRI Microarray Project at NIH
http://web.wi.mit.edu/young/	Young Lab at Whitehead Institute

Statistical discussion

http://aretha.jax.org/jax-cgi/churchill/index.cgi	Churchill Lab at the Jackson Laboratory
http://www.stat.berkeley.edu/users/terry/zarray/html/index.html	Speed Lab at UC Berkeley

Information/database for microarray

http://www.biochipnet.de/	BiochipNet
http://www.biologie.ens.fr/en/genetiqu/puces/bddeng.html	Databases for Microarray
http://www.gene-chips.com/	DNA Microarray (Genome Chip)
http://www.deathstarinc.com/science/biology/chips.html	DNA Microarrays
http://zenith.berkeley.edu/genearrays/	Gene-array mailing list
http://www.hugeindex.org/index.html	HuGEIndex.org
http://www.microarrays.org/index.html	Microarray.org
http://ihome.cuhk.edu.hk/~b400559/arraysoft.html	My Microarray Software Comparison
http://genome-www5.stanford.edu/MicroArray/SMD/restech.html	Stanford Microarray Database
http://www.pubgene.uio.no/	The PubGene (Gene Interactions)

Mouse genomics

http://www.ensembl.org/	EBI
http://www.bcgsc.bc.ca/	Genome Science Center at Canada
http://www.ncbi.nlm.nih.gov/genome/guide/mouse/index.html	NCBI Mouse Genome Resource
http://lgsun.grc.nia.nih.gov/cDNA/cDNA.html	NIA Mouse cDNA Project
http://genome.gsc.riken.go.jp/index.html	RIKEN Mouse Encyclopedia Project
http://www.jax.org/ges/index.html	The Jackson Lab
http://www.tigr.org/search/	TIGR
http://genome.ucsc.edu/	UCSC Genome Bioinformatics

Another important aspect of bioinformatics is the data mining of existing data in the public database. Although most of such data has been generated to ask specific biological questions, once incorporated into the public database, they become building blocks for much larger-scale analysis. For example, pre-implantation EST libraries were analyzed to obtain the expression repertoire and patterns of pre-implantation embryos (Ko *et al.*, 2000; Marra *et al.*, 1999; Sasaki *et al.*, 1998; Solter *et al.*, 2002), and were re-analyzed as a part of larger data sets to identify transcripts that may be specific to germ cells, stem cells, oocytes, and early embryos (Choo *et al.*, 2001; Rajkovic *et al.*, 2001; Stanton and Green, 2001a). Genes that are likely to be specifically expressed in germ cells, for example, can be identified by comparing the digital expression profiles of primordial germ cells, oocytes, and ovaries to those of other tissues (Rajkovic *et al.*, 2002; Stanton and Green, 2001b; Suzumori *et al.*, 2002). At a more global level, researchers have generated 'digital expression profiles' – calculated estimates of relative transcript abundance based on EST frequencies, e.g., NCBI Digital Differential Display (http://www.ncbi.nlm.nih.gov/UniGene/ddd.cgi?ORG=Hs).

There have been extensive efforts to incorporate data sets generated by microarray experiments into a single database so that the data obtained from different experiments and laboratories can be compared. Two main standards are Gene Expression Omnibus (GEO; Edgar *et al.*, 2002) and Minimum Information About a Microarray Experiment (MIAME; Brazma *et al.*, 2001). GEO is intended to be a public repository for microarray data, and is designed to accept a variety of data sets in a very flexible format (Edgar *et al.*, 2002). On the other hand, MIAME is intended to be a repository for more standardized and controlled data sets by requiring detailed information about experimental conditions and array platforms (Brazma *et al.*, 2001). Such vast amounts of expression data have the potential to yield incisive new information as new and different analysis techniques are applied. However, as illustrated later in this chapter, many variables in experimental design, microarray platforms, and data reproducibility impose obstacles to the widespread exchange and integration of microarray data. As platforms mature and microarrays and methods increase reliability, international consortia are working to standardize expression profile data sets and incorporate them into increasingly referenced, standardized public databases.

18.5 Expression profiling of stem cells

As techniques become more powerful and reliable, the general goals of gene expression profiling in stem cells become more accessible. They include (1) to obtain a comprehensive repertoire of genes expressed in stem cells; (2) to obtain comprehensive views of absolute expression levels of individual genes; (3) to identify genes differentially expressed among various types of stem cells; and (4) to provide a first step toward the determination of genetic pathways that maintain pluripotency or lead to the differentiation of stem cells into specific cell types. As a subgoal (5), groups are attempting to define 'stemness,' i.e., the combination of gene expression and environmental cues that are required for 'generalized' stem cell character.

Straightforward expression profiling of stem cells by ESTs ((Bain *et al.*, 2000); NIA Mouse cDNA Project: http://lgsun.grc.nia.nih.gov/cDNA/cDNA.html) and SAGE (Anisimov *et al.*, 2002) can provide useful information. However, more comprehensive gene expression profiles from a variety of multipotent cell types, including HSC and precursors (Park *et al.*, 2002; Terskikh *et al.*, 2001), NSC (Terskikh *et al.*, 2001; Zhou *et al.*, 2001), TSC (Tanaka *et al.*, 2002; Xu *et al.*, 2002), and ESC (Kelly and Rizzino, 2000; Tanaka *et al.*, 2002; Xu *et al.*, 2002), have been analyzed by cDNA microarrays using statistical significance tests and pattern-finding algorithms to identify distinct 'clusters' of genes with similar expression patterns across many stem cell types. Most of these studies have focused on the comparison between stem cells and their differentiated cell types, addressing issues such as (1)–(4); 'stemness' in molecular terms (5) has just begun to be addressed.

For example, Terskikh *et al.* (2001) compared cDNA microarray expression profiles between HSC and NSC. They found that many genes expressed in HSC were also expressed in NSC, suggesting that the cell types share part of their genetic programs. Ramalho-Santos *et al.* (2002) compared the expression profiles of larger numbers of genes between ESC, HSC, and NSC, using Affymetrix oligonucleotide GeneChip arrays. Common genes among stem cells were identified, with 216 genes potentially involved in properties of stem cells per se. They proposed several essential attributes of stemness, including activation of signaling systems such as JAK/STAT and Notch signaling, and up-regulation of genes involved in DNA repair. Interestingly, some of the putative genes for stemness are apparently clustered in the proximal region of mouse chromosome 17 – the region known as the t-complex, in which mutations cause a variety of early embryonic lethal phenotypes. This suggests that a fraction of stemness genes might be co-regulated at the level of large genomic segments. One of those genes, *D17Ertd197e*, was previously mapped as a gene expressed during preimplantation mouse development (Ko *et al.*, 2000). This again illustrates the relevance of preimplantation genomics projects to stem cell biology. Furthermore, Ramalho-Santos *et al.* compared gene expression profiles to ask whether stem cells are more similar to one another or to their differentiated counterparts. Indeed, NSC turn out to be more similar to ESC than to cells from the lateral ventricle of the brain (a source of NSC) or HSC.

Ivanova *et al.* (2002) also reported expression profiles from ESC, HSC, and NSC using Affymetrix oligonucleotide GeneChip arrays. They have also compared these expression profiles with mouse fetal and adult HSC to find HSC-specific genes. Their analysis showed that ~70% of genes are common among them. Mouse and human HSC profiles were also compared to obtain insights into evolutionary conservation of HSC-specific genetic programs; at least 40% of expressed genes are shared between species. Finally, Ivanova *et al.* identified 283 genes shared by the three types of stem cells they investigated.

Expression profiles of another combination of stem cells have also been reported (Tanaka *et al.*, 2002). In that work, we focused on the first cell differentiation event in mouse embryogenesis, which gives rise to pluripotent inner cell mass (ICM) and lineage-committed trophectoderm (TE). The ICM will differentiate into embryo proper with germ cells, whereas the TE will differentiate solely into the cells of trophoblast lineage like trophoblast giant cells and spongiotrophoblast in placenta (*Figure 18.1*). When cultured *in vitro*, the ICM will turn into pluripotent ESC (Edwards, 2001; Evans and Kaufman, 1981; Martin, 1981), whereas the TE will turn into multipotent TSC (Tanaka *et al.*, 1998). By taking advantage of this cell culture system, we performed gene expression profiling of pluripotent ESC and multipotent TSC using the NIA 15k mouse cDNA microarray (Tanaka *et al.*, 2000). We used two different TSC lines, TS3.5 and TS6.5, which were derived from different developmental stages, as well as mouse embryonic fibroblasts (MEF) as an example of terminally differentiated primary culture cells.

348 Human Embryonic Stem Cells

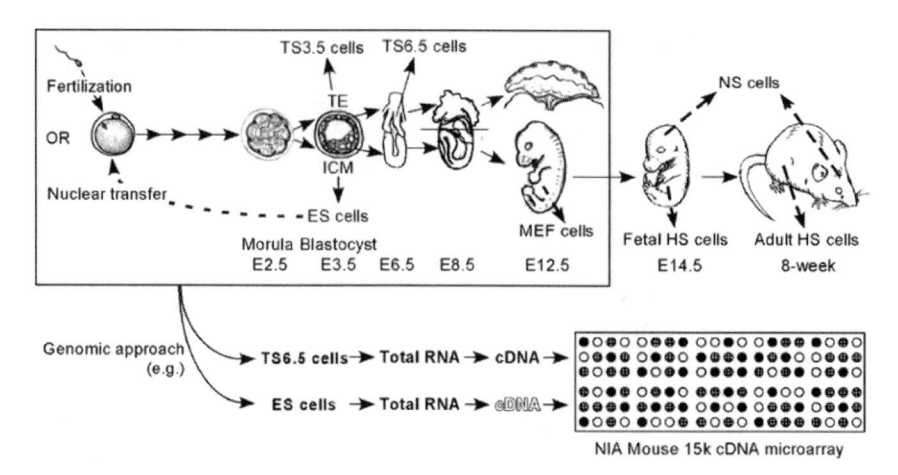

Figure 18.1: Schematic representation of the topic discussed in this chapter. Mouse embryogenesis will be initiated by either fertilization between a sperm and an egg, or nuclear transfer from a somatic cell (in this example, from an ESC) into an enucleated unfertilized egg. During the pre- and peri-implantation stages, two stem cell types have been identified; ESC from pluripotent inner cell mass (ICM) of blastocysts, and trophoblast stem (TS) cells from either trophectoderm (TE) of blastocysts (TS3.5) or extraembryonic ectoderm from E6.5 embryos (TS6.5). In addition to these early embryo-derived stem cells, two more stem cell populations have been identified; haematopoietic stem (HS) cells from either the liver at E14.5 embryos or the bone marrow from the femora of adult mice, and neural stem (NS) cells from either the lateral ventricle striatum of E14.5 embryos or the subventricular zone of adult mouse brains. The rectangle indicates the cell/tissue types described deeply in this chapter, including mouse embryo fibroblast (MEF) cells. In the bottom, cDNA microarray hybridization experiment was summarized schematically as an example of the genomic approach to mouse early developmental study. Total RNA was extracted from cell/tissue types of interest (in this case from ESC and TS6.5 cells) and used as a template for the first strand cDNA synthesis, during which fluorescent or radioactive labels were incorporated into cDNAs. Labeled cDNAs (bold or outline) were hybridized onto a cDNA microarray, where specific transcripts in either cell/tissue types are represented as black or white circles, and transcripts expressed at approximately equal levels in both cell/tissue types are shown as hatched circles.

MEF also serve as a negative control to find any genes related to the adaptation of cells to *in vitro* culture systems.

Clustering analysis of triplicate data sets identified 346 signature genes: 124 ESC-specific genes, 94 TSC-specific genes, and 77 MEF-specific genes. Fifty-one genes were expressed in both ESC and TSC, but not in MEF cells, and are therefore designated as potential 'stemness' genes. The majority (67%) of ESC signature genes are uncharacterized, whereas only 49% of TSC signature genes are known and include genes with well-defined roles in placental differentiation. Examples of such genes with expression patterns in other *in vivo* cell types, including morula, blastocyst, E8.5 embryo and placenta, and E12.5 embryo and placenta, are shown in *Figure 18.2*, in which many TSC signature genes are expressed highly in placenta, whereas many ESC signature genes are underrepresented in differentiated tissues. Thus, there is a clear distinction between pluripotent ESC and lineage-committed multipotent TSC in terms of global gene expression patterns.

Although both Ramalho-Santos *et al.* and Ivanova *et al.* independently identified >200 'stemness' genes shared by three types of stem cells, the comparison

Chapter 18 – Genomic approaches to stem cell biology 349

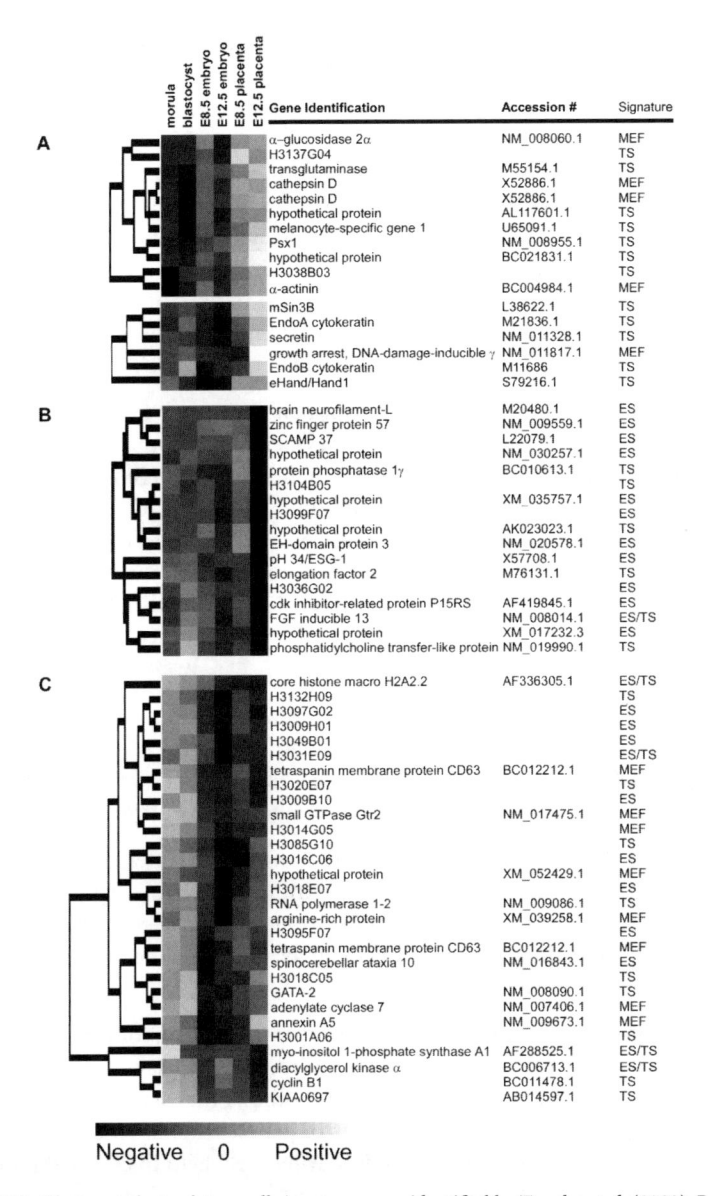

	morula	blastocyst	E8.5 embryo	E12.5 embryo	E8.5 placenta	E12.5 placenta	Gene Identification	Accession #	Signature
A							α-glucosidase 2α	NM_008060.1	MEF
							H3137G04		TS
							transglutaminase	M55154.1	TS
							cathepsin D	X52886.1	MEF
							cathepsin D	X52886.1	MEF
							hypothetical protein	AL117601.1	TS
							melanocyte-specific gene 1	U65091.1	TS
							Psx1	NM_008955.1	TS
							hypothetical protein	BC021831.1	TS
							H3038B03		TS
							α-actinin	BC004984.1	MEF
							mSin3B	L38622.1	TS
							EndoA cytokeratin	M21836.1	TS
							secretin	NM_011328.1	TS
							growth arrest, DNA-damage-inducible γ	NM_011817.1	MEF
							EndoB cytokeratin	M11686	TS
							eHand/Hand1	S79216.1	TS
B							brain neurofilament-L	M20480.1	ES
							zinc finger protein 57	NM_009559.1	ES
							SCAMP 37	L22079.1	ES
							hypothetical protein	NM_030257.1	ES
							protein phosphatase 1γ	BC010613.1	TS
							H3104B05		TS
							hypothetical protein	XM_035757.1	ES
							H3099F07		ES
							hypothetical protein	AK023023.1	TS
							EH-domain protein 3	NM_020578.1	ES
							pH 34/ESG-1	X57708.1	ES
							elongation factor 2	M76131.1	TS
							H3036G02		ES
							cdk inhibitor-related protein P15RS	AF419845.1	ES
							FGF inducible 13	NM_008014.1	ES/TS
							hypothetical protein	XM_017232.3	ES
							phosphatidylcholine transfer-like protein	NM_019990.1	TS
C							core histone macro H2A2.2	AF336305.1	ES/TS
							H3132H09		TS
							H3097G02		ES
							H3009H01		ES
							H3049B01		ES
							H3031E09		ES/TS
							tetraspanin membrane protein CD63	BC012212.1	MEF
							H3020E07		TS
							H3009B10		ES
							small GTPase Gtr2	NM_017475.1	MEF
							H3014G05		MEF
							H3085G10		TS
							H3016C06		ES
							hypothetical protein	XM_052429.1	MEF
							H3018E07		ES
							RNA polymerase 1-2	NM_009086.1	TS
							arginine-rich protein	XM_039258.1	MEF
							H3095F07		ES
							tetraspanin membrane protein CD63	BC012212.1	MEF
							spinocerebellar ataxia 10	NM_016843.1	ES
							H3018C05		TS
							GATA-2	NM_008090.1	TS
							adenylate cyclase 7	NM_007406.1	MEF
							annexin A5	NM_009673.1	MEF
							H3001A06		TS
							myo-inositol 1-phosphate synthase A1	AF288525.1	ES/TS
							diacylglycerol kinase α	BC006713.1	ES/TS
							cyclin B1	BC011478.1	TS
							KIAA0697	AB014597.1	TS

Negative 0 Positive

Figure 18.2: Cluster analysis of stem cell signature genes identified by Tanaka *et al.* (2002). Based on the cDNA microarray analysis comparing ES, TS3.5, TS6.5 and MEF cells, 'signature' genes highly expressed in the different cell types, and expressed commonly between ES and TS (ES/TS) cells were identified. These signature genes were further grouped by hierarchical clustering methods (Eisen *et al.*, 1998), based on the similarity of expression patterns among the separate cDNA microarray analyses from morula, blastocyst, embryonic parts at E8.5 and E12.5 and placental parts at E8.5 and E12.5. The relative expression levels are in gray-scale. The nodes on the left show how similar the individual patterns of gene expressions are. (A) Many TS signature genes were found in the two clusters showing high expression levels in placenta. (B) Many ES signature genes were underrepresented in other cell/tissue types examined. (C) This cluster showed high expression levels in morula and blastocyst consisting of many kinds of signature genes. This suggests that the signature genes do not represent embryonic stages.

between their gene lists revealed only six genes in common, even though they used identical microarray chips and the same cell types. The comparison of these gene lists with the one by Tanaka *et al.* is not straightforward, because different microarray platforms were used. Fortunately, however, RESOURCERER, a tool developed at TIGR (Tsai *et al.*, 2001), can find the corresponding genes based on Genbank accession numbers of cDNA sequences. The comparison revealed that none of the 51 'stemness' genes identified by Tanaka *et al.* matched any of the 283 'stemness' genes identified by Ivanova *et al.* or the 216 'stemness' genes identified by Ramalho-Santos *et al.*

It can be instructive to ask why the lists of 'stemness' genes do not significantly overlap. Among the factors that can influence the results are:

- First, there are differences in the types of SCs used for the experiments. Although it is still under debate, there is evidence that both NSC and HSC have the ability to differentiate into cells of different lineages (Bjornson *et al.*, 1999; Mezey *et al.*, 2000), and it is likely that both NSC and HSC share a set of genes with ESC that allow them to differentiate into cells of different lineages. On the other hand, it has been shown that TSC can differentiate into only extraembryonic lineages, but not into other cell lineages (Tanaka *et al.*, 1998), and ESC lack the ability to differentiate into trophoblast cell lineages (Beddington and Robertson, 1989).

- Second, there are differences in the microarray platforms. Based on NCBI's UniGene IDs, Affymetrix chip, 'Mouse Genome U74v2 (A, B and C)' contains about 36,000 genes, whereas NIA mouse 15k clone set contain 15,000 genes. But, only 7033 genes are common between these two platforms.

- Third, ESC, TSC, and NSC are cultured *in vitro* with different growth factors, and thus, the cell culture conditions may influence overall expression patterns of genes significantly. Particularly in the work by Tanaka *et al.*, the comparison with MEF, the primary cultured cells, may have excluded genes commonly expressed in cultured cells.

Thus, although studies until now provided important clues about the genes critical for the function of stem cells, it is obviously too early to say that the 'stemness' genes have been defined. In fact, the null hypothesis that stemness can be maintained by several routes, with alternate sets of genes, remains to be proven. More extensive and rigorous expression profiling will surely help to clarify the situation further.

18.6 Follow-up study of cDNA microarrays

It has become standard practice to validate data from EST frequencies and microarray expression profiles using additional methods (Chuaqui *et al.*, 2002). This is done both to refine results and to verify them. Because large-scale genomics methods test so many genes, and each test is in fact a small experiment, even very low error rates can cause significant numbers of false-positive results. Furthermore, clone-handling errors during cDNA microarray construction can cause the misidentification of genes. RNA blotting, RT-PCR, and Quantitative real-time RT-PCR, or Q-PCR (Higuchi et al., 1993) require only small amounts of RNA and can be applied to many genes. Q-PCR also measures small expression differences reliably, and can distinguish between transcripts with regions of high sequence similarity. Correlative work can further focus on interesting genes by examining expression levels in tissues and cell types not studied by genome-wide screening.

Although expression profiling provides a global snapshot of gene expression patterns, deeper understanding requires studies of pathways and function. As a typical example, we discuss a gene, called *Embryonal stem cell-specific gene 1* (*Esg1*), which was identified by its high and unique expression in ESC by the microarray analysis mentioned above (Tanaka et al., 2002). Originally, *Esg1* was identified as down-regulated when embryonal carcinoma cells (ECC) were induced to differentiate (Astigiano et al., 1991) and highly expressed in preimplantation stage mouse embryos (Bierbaum et al., 1994). In terms of its relation to pluripotency, we further found that the *Esg1* expression pattern in pre-implantation mouse embryos was slightly different from that of *Oct3/4*, a well-known transcription factor necessary for pluripotency of ESC or ECC. In contrast, in genetically manipulated ESC, the expression pattern of *Esg1* was tightly correlated with that of *Oct3/4* and *Lif-Stat3*. In other words, *Esg1* seems to be located downstream in the pathway of *Oct3/4* and/or *Lif-Stat3*. *In vitro*, tight connection of *Oct3/4* and *Lif-Stat3* pathways, which might be bridged by *Esg1*, could speculatively be necessary for ESC to maintain indefinite pluripotency. A next step in the analysis will be to examine what happens if *Oct3/4* and/or *Stat3* are downregulated by siRNA in *Esg1*-overexpressing ESC.

Another sample gene, H3001A06, which is expressed throughout extraembryonic cell lineages such as trophoblast giant cells and spongiotrophoblasts (Tanaka et al., 2002), is a member of the Polycomb group (PcG). It contains one MBT (malignant brain tumor) domain, which is also found in Sex comb on midleg (Scm) in *Drosophila*, another PcG member. Our current working hypothesis is that *Esg1* and this novel trophoblast-specific Scm-like gene may be involved in the divergence of the first two cell types in the embryo, ICM and TE.

18.7 cDNA microarray analysis of cloned animals

The successful generation of cloned sheep (i.e., 'Dolly') from adult somatic cells has raised the possibility (or spectre) of cloning humans, and aroused continuing controversy in the scientific community and society at large (Wilmut et al., 1997). One possible future therapeutic application of such technology, 'therapeutic cloning,' generates blastocysts by the transfer of nuclei from a donor's ('patient's') somatic cells (Lanza et al., 1999). Human ESC can then in principle be derived from those blastocysts and directed to differentiate into various cell types (Odorico et al., 2001). Such cells would be genetically identical to the patient, and might thus be transplanted back into the patient without rejection or other complications. The study of nuclear transplantation (NT) or 'cloning' thus becomes an important part of stem cell biology. Without considering the ethical issues currently debated, such an application of NT technology will be difficult and is at least currently unjustifiable, because of the serious limitations of NT technology at present. These include extremely low successful birth rates as well as abnormalities that are associated with NT offspring, such as placentomegaly (Hochedlinger and Jaenisch, 2002; Solter, 2000; Wilmut et al., 2002; Yanagimachi, 2002).

Several attempts to understand the mechanisms causing such abnormalities include a study that utilized Affymetrix GeneChip arrays to examine expression differences of ~10,000 genes in livers and placentas after normal fertilization and nuclear transplantation (Humpherys et al., 2002). Both ESC and cumulus cells were used as donor cells. Up to 4% of genes showed abnormal expression patterns in cloned placentas, but most of these dysregulations show variability between individuals. Another study utilized the NIA mouse 15K cDNA microarrays and examined the expression differences between placentas produced by NT of one-cell embryos as a control and placentas produced by NT in ESC (Suemizu et al., 2003). The use of this control allowed us to investigate the differences caused by the donor nucleus (because the control also went through the NT procedure).

Five principal aberrant events have been reported: (1) inappropriate expression of imprinted genes; (2) altered expression of genes involved in global regulation of gene expression, such as DNA methyltransferase and histone acetyltransferase; (3) increased expression of oncogenes and growth-promoting genes; (4) overexpression of genes involved in placental growth, such as *Plac1*; and (5) identification of many novel genes overexpressed in ESC-derived NT mouse placentas, including *Pitrm1*, a new member of the metalloprotease family (Suemizu et al., 2003). The results indicate that placentomegaly in ESC-derived NT mice is associated with large-scale dysregulation of normal gene expression patterns.

Humpherys et al. (2002) reported that the expressions of several hundred genes were dysregulated in cloned placenta. Among those, they selected 74 genes as most significantly altered (t-test, $P<0.05$). Suemizu et al. used slightly different statistical criteria (FDR = 0.05 and greater than twofold difference), and found 753 genes altered. Of 74 genes specified by Humpherys et al., 45 genes were present on the microarrays used by Suemizu et al. Surprisingly, only four genes of 45 showed positive correlation between these studies, whereas two genes showed anticorrelation! As in the case of 'stemness,' various differences in the experimental conditions may be involved, including the array platforms, biological materials, etc. The four genes that showed positive correlation are *Plac1*, Prolactin-like protein G, Glypican 1, and the 1700012A18Rik gene. These genes have indeed become primary candidates for dysregulation in the placentas of cloned animals. Interestingly, *Plac1*, which was suggested as one of the critical genes based on its map location, biological function, etc. in our study (Cocchia et al., 2000; Suemizu et al., 2003), was also shown to be differentially expressed in the other study (Humpherys et al., 2002). The actual role of *Plac1* in this setting remains unknown, but it will be interesting to learn whether the forced downregulation of *Plac1* might be an effective intervention to prevent the development of abnormally large placentas in cloned mice.

18.8 Large-scale functional studies of genes

While the expression patterns of genes are reliably assessed and many genes of interest have been identified, the functions of genes, including protein–protein interactions and phenotypes, are open for study. However, compared with gene identification and expression profiling, the functional analysis of individual genes is time-consuming and difficult. For example, even the simplest functional analyses of genes by the generation of loss-of-function knockout mice and gain-of-function transgenic mice are slow. Generation of such mice can take several months to a year, and detailed phenotypic and molecular analyses take longer. Based on a one-gene-at-a-time paradigm, such studies are hard to adapt to high-throughput genomics approaches. Similarly, ENU-mutagenesis programs (Brown and Balling, 2001) have limitations, because mutations occur at random sites in the genome, and it is thus not a simple matter to link the genes identified by the large-scale expression profiling to a specific mutant phenotype. Other methods that alter specific regions of the mouse genome, such as chromosomal engineering by Cre/loxP site-specific recombination systems (Yu and Bradley, 2001) and radiation-induced deletion of specific genomic regions (Goodwin et al., 2001), may be more efficient, but suffer from similar limitations.

Large-scale gene-trapping methods may be the most suitable method at present for two reasons. First, short flanking sequences of trapped genes are normally associated with mutant ESC, and can thereby be easily linked to genes with cDNA clone information and expression levels. As has been proposed by the gene-trap mutagenesis consortium, once a large collection of such gene-trapped ESC is generated and archived, a specific ESC line carrying a mutation in a gene of interest can be easily identified, and the phenotype of mice can be observed relatively rapidly (Stanford et al., 2001). Secondly, because these gene traps were made in ESC, differentiation of such ESC in culture to specific cell lineages, such as hematopoietic cells, neural cells, muscle cells, etc. (Stanford et al., 2001), can be used to gain insight into the function of a gene of interest in specific cell lineages. This combination of technologies and biological platforms is thus highly favorable for one of the goals of the application of embryogenomics to stem cell biology: finding the genes that specifically direct the differentiation of stem cells to specific lineages (for the web addresses of these programs, see Beckers and Hrabe de Angelis, 2002).

Another direct way to analyze loss-of-function phenotypes for a large number of genes identified by expression profiling is to ablate their expression in a dominant-negative manner. Unlike methods targeting genomic DNAs, both alleles are knocked down simultaneously, requiring no further manipulations (like breeding mice to homozygosity) to produce loss-of-function phenotypes. Antisense oligonucleotides have been traditionally used, and relatively new compounds that incorporate Morpholino (Heasman, 2002) and 2'-methoxyethoxy modifications (Kimber et al., in press) show some potential. RNA interference (RNAi) and small interfering RNA (siRNA), which have been discovered and developed recently, seem to have rapidly become a method of choice for several reasons (McManus and Sharp, 2002; Scherr et al., 2003). They are effective, work for a wide range of genes, and can be easily adapted to large-scale applications. Perhaps the application of this technology to ESC will provide rapid functional assays with a throughput that compares with microarray analysis of various stem cells.

18.9 Future perspectives

Stem cell biology holds much promise for biomedical science. Clinical applications, such as the treatment of patients with dysfunctional and aging organs, are important and ultimate goals for stem cell biology. However, understanding the nature of stem cells is a prerequisite for such goals. There are important questions to address: involving the outcome of experimental manipulations, the pathways involved in pluripotency, and how they interact. Conversion of differentiated cells such as fibroblasts into embryonic stem cells may perhaps be an ultimately ideal solution for stem cell therapy.

In summary, embryogenomic approaches will continue to serve two purposes in stem cell biology. First, they provide tools for molecular biology – for example, the use of cDNA microarrays to identify genes differentially expressed between different stem cell types, or between stem cells and their cognate differentiated cells. The new approaches provide more comprehensive and thorough screening than previous molecular biological approaches, such as differential screening etc. Secondly, they can be used to investigate the global nature of stem cells. For example, the differentiation potential of stem cells should eventually be defined – not by the presence or absence of a few gene markers, but by global expression patterns of many genes with the details of their interactions and functions. This is the fruit of application of high-throughput genomics: dramatic increases in information about gene action that cannot be attained by the study of one or a few genes at a time.

Acknowledgments

The authors would like to thank Dr David Schlessinger for critical reading of the manuscript, and the members of the Developmental Genomics and Aging Section for their discussion on this topic. T.S.T. and M.G.C. were supported by fellowships from the Japan Society for the Promotion of Science (JSPS) and the National Institute of General Medicine (NIGMS) PRAT program, respectively. K.A. and S.A.J. were supported by fellowships from the National Institute on Aging (NIA).

References

Adams MD, Kelley JM, Gocayne JD, Dubnick M, Polymeropoulos MH, Xiao H, Merril CR, Wu A, Olde B, Moreno RF *et al.* (1991) Complementary DNA sequencing: expressed seqeunce tags and human genome project. *Science* 252, 1651–1656.

Adams MD, Kerlavage AR, Fleischmann RD, Fuldner RA, Bult CJ, Lee NH, Kirkness EF, Weinstock KG, Gocayne JD, White O *et al.* (1995) Initial assessment of human gene diversity and expression patterns based upon 83 million nucleotides of cDNA sequence. *Nature* 377, 3–174.

Adjaye J, Daniels R, Monk M (1998) The construction of cDNA libraries from human single preimplantation embryos and their use in the study of gene expression during development. *J Assist. Reprod. Genet.* 15, 344–348.

Anisimov SV, Tarasov KV, Tweedie D, Stern MD, Wobus AM, Boheler KR (2002) SAGE identification of gene transcripts with profiles unique to pluripotent mouse R1 embryonic stem cells. *Genomics* 79, 169–176.

Astigiano S, Barkai U, Abarzua P, Tan SC, Harper MI, Sherman MI (1991) Changes in gene expression following exposure of nulli-SCC1 murine embryonal carcinoma cells to inducers of differentiation: characterization of a down-regulated mRNA. *Differentiation* 46, 61–67.

Bain G, Mansergh FC, Wride MA, Hance JE, Isogawa A, Rancourt SL, Ray WJ, Yoshimura Y, Tsuzuki T, Gottlieb DI *et al.* (2000) ES cell neural differentiation reveals a substantial number of novel ESTs. *Funct. Integr. Genomics* 1, 127–139.

Beckers J, Hrabe de Angelis M (2002) Large-scale mutational analysis for the annotation of the mouse genome. *Curr. Opin. Chem. Biol.* 6, 17–23.

Beddington RS, Robertson EJ (1989) An assessment of the developmental potential of embryonic stem cells in the midgestation mouse embryo. *Development* 105, 733–737.

Bierbaum P, MacLean-Hunter S, Ehlert F, Moroy T, Muller R (1994) Cloning of embryonal stem cell-specific genes: characterization of the transcriptionally controlled gene esg-1. *Cell Growth Differ.* 5, 37–46.

Bjornson CR, Rietze RL, Reynolds BA, Magli MC, Vescovi AL (1999) Turning brain into blood: a hematopoietic fate adopted by adult neural stem cells in vivo. *Science* 283, 534–537.

Brazma A, Hingamp P, Quackenbush J, Sherlock G, Spellman P, Stoeckert C, Aach J, Ansorge W, Ball CA, Causton HC *et al.* (2001) Minimum information about a microarray experiment (MIAME) – toward standards for microarray data. *Natl. Genet.* 29, 365–371.

Brenner S, Johnson M, Bridgham J, Golda G, Lloyd DH, Johnson D, Luo S, McCurdy S, Foy M, Ewan M *et al.* (2000) Gene expression analysis by massively parallel signature sequencing (MPSS) on microbead arrays. *Nat. Biotechnol.* 18, 630–634.

Brown SD, Balling R (2001) Systematic approaches to mouse mutagenesis. *Curr. Opin. Genet. Dev* 11, 268–273.

Carter MG, Hamatani T, Sharov AA, Carmack CE, Qian Y, Aiba K, Dudekula DB, Brzoska PM, Hwang SS, Ko NT (2003) In situ-synthesized novel microarray optimized for mouse stem cell and early developmental expression profiling. *Genome Res.* 13, 1011–1012.

Chen GX, Jaradat SA, Banerjee N, Tanaka TS, Ko MSH, Zhang MQ (2002) Evaluation and comparison of clustering algorithms in analyzing es cell gene expression data. *STATISTICA SINICA* 12, 241–262.

Choo KB, Chen HH, Cheng WT, Chang HS, Wang M (2001) In silico mining of EST databases for novel pre-implantation embryo-specific zinc finger protein genes. *Mol. Reprod. Dev.* 59, 249–255.

Chuaqui RF, Bonner RF, Best CJ, Gillespie JW, Flaig MJ, Hewitt SM *et al.* (2002)

Post-analysis follow-up and validation of microarray experiments. *Nat. Genet.* **32 (Suppl 2)**, 509–514.

Churchill GA (2002) Fundamentals of experimental design for cDNA microarrays. *Nat. Genet.* **32**, 490–495.

Cocchia M, Huber R, Pantano S, Chen EY, Ma P, Forabosco A, Ko MSH, Schlessinger D (2000) PLAC1, an Xq26 gene with placenta-specific expression. *Genomics* **68**, 305–312.

Du Z, Cong H, Yao Z (2001) Identification of putative downstream genes of Oct-4 by suppression-subtractive hybridization. *Biochem. Biophys. Res. Commun.* **282**, 701–706.

Edgar R, Domrachev M, Lash AE (2002) Gene Expression Omnibus: NCBI gene expression and hybridization array data repository. *Nucleic Acids Res.* **30**, 207–210.

Edwards RG (2001) IVF and the history of stem cells. *Nature* **413**, 349–351.

Eisen MB, Spellman PT, Brown PO, Botstein D (1998) Cluster analysis and display of genome-wide expression patterns. *Proc. Natl Acad. Sci. USA* **95**, 14863–14868.

Evans MJ, Kaufman MH (1981) Establishment in culture of pluripotential cells from mouse embryos. *Nature* **292**, 154–156.

Fodor SP, Read JL, Pirrung MC, Stryer L, Lu AT, Solas D (1991) Light-directed, spatially addressable parallel chemical synthesis. *Science* **251**, 767–773.

Goodwin NC, Ishida Y, Hartford S, Wnek C, Bergstrom RA, Leder P, Schimenti JC (2001) DelBank: a mouse ES-cell resource for generating deletions. *Nat. Genet.* **28**, 310–311.

Heasman J (2002) Morpholino oligos: making sense of antisense? *Dev. Biol.* **243**, 209–214.

Higuchi R, Fockler C, Dollinger G, Watson R (1993) Kinetic PCR analysis: real-time monitoring of DNA amplification reactions. *Biotechnology (NY)* **11**, 1026–1030.

Hochedlinger K, Jaenisch R (2002) Nuclear transplantation: lessons from frogs and mice. *Curr. Opin. Cell Biol.* **14**, 741–748.

Hofmann WK, de Vos S, Komor M, Hoelzer D, Wachsman W, Koeffler HP (2002) Characterization of gene expression of CD34+ cells from normal and myelodysplastic bone marrow. *Blood* **100**, 3553–3560.

Hoog C (1991) Isolation of a large number of novel mammalian genes by a differential cDNA library screening strategy. *Nucleic Acids Res.* **19**, 6123–6127.

Hughes TR, Shoemaker DD (2001) DNA microarrays for expression profiling. *Curr. Opin. Chem. Biol.* **5**, 21–25.

Hughes TR, Mao M, Jones AR, Burchard J, Marton MJ, Shannon KW *et al.* (2001) Expression profiling using microarrays fabricated by an ink-jet oligonucleotide synthesizer. *Nat. Biotechnol.* **19**, 342–347.

Humpherys D, Eggan K, Akutsu H, Friedman A, Hochedlinger K, Yanagimachi R, Lander ES, Golub TR, Jaenisch R (2002) Abnormal gene expression in cloned mice derived from embryonic stem cell and cumulus cell nuclei. *Proc. Natl Acad. Sci. USA* **99**, 12889–12894.

Ivanova NB, Dimos JT, Schaniel C, Hackney JA, Moore KA, Lemischka IR (2002) A stem cell molecular signature. *Science* **298**, 601–604.

Kacharmina JE, Crino PB, Eberwine J (1999) Preparation of cDNA from single cells and subcellular regions. *Methods Enzymol.* **303**, 3–18.

Kargul GJ, Dudekula DB, Qian Y, Lim MK, Jaradat SA, Tanaka TS, Carter MG, Ko MSH (2001) Verification and initial annotation of the NIA mouse 15K cDNA clone set. *Nat. Genet.* **28**, 17–18.

Kato K (1997) Adaptor-tagged competitive PCR: a novel method for measuring relative gene expression. *Nucleic Acids Res.* **25**, 4694–4696.

Kelly DL, Rizzino A (2000) DNA microarray analyses of genes regulated during the differentiation of embryonic stem cells. *Mol. Reprod. Dev.* **56**, 113–123.

Kikuno R, Nagase T, Waki M, Ohara O (2002) HUGE: a database for human large proteins identified in the Kazusa cDNA sequencing project. *Nucleic Acids Res.* **30**, 166–168.

Kimber WL, Puri N, Borgmeyer C, Ritter D, Sharov A, Seidman M, Ko MSH (2003) Efficacy of 2-methoxyethoxy (2′-MOE)-modified antisense oligonucleotide for the study of mouse preimplantation development. *Reprod. Biomed. Online* **6**, 318–322.

Ko MSH (1990) An 'equalized cDNA library' by the reassociation of short double-stranded cDNAs. *Nucleic Acids Res.* **18**, 5705–5711.

Ko MSH (2001) Embryogenomics: developmental biology meets genomics. *Trends Biotechnol.* **19**, 511–518.

Ko MSH, Threat TA, Wang X, Horton JH, Cui Y, Pryor E, Paris J, Wells-Smith J, Kitchen JR, Rowe LB *et al.* (1998) Genome-wide mapping of unselected transcripts from extraembryonic tissue of 7.5-day mouse embryos reveals enrichment in the t-complex and under-representation on the X chromosome. *Hum. Mol. Genet.* **7**, 1967–1978.

Ko MSH, Kitchen JR, Wang X, Threat TA, Hasegawa A, Sun T *et al.* (2000) Large-scale cDNA analysis reveals phased gene expression patterns during pre-implantation mouse development. *Development* **127**, 1737–1749.

Lander ES, Linton LM, Birren B, Nusbaum C, Zody MC, Baldwin J *et al.* (2001) Initial sequencing and analysis of the human genome. *Nature* **409**, 860–921.

Lanza RP, Cibelli JB, West MD (1999) Prospects for the use of nuclear transfer in human transplantation. *Nat. Biotechnol.* **17**, 1171–1174.

Lennon G, Auffray C, Polymeropoulos M, Soares MB (1996) The I.M.A.G.E. Consortium: an integrated molecular analysis of genomes and their expression. *Genomics* **33**, 151–152.

Liang P, Pardee AB (1992) Differential display of eukaryotic messenger RNA by means of the polymerase chain reaction. *Science* **257**, 967–971.

Lipshutz RJ, Fodor SP, Gingeras TR, Lockhart DJ (1999) High density synthetic oligonucleotide arrays. *Nat. Genet.* **21**, 20–24.

Marra M, Hillier L, Kucaba T, Allen M, Barstead R, Beck C *et al.* (1999) An encyclopedia of mouse genes. *Nat. Genet.* **21**, 191–194.

Martin GR (1981) Isolation of a pluripotent cell line from early mouse embryos cultured in medium conditioned by teratocarcinoma stem cells. *Proc. Natl Acad. Sci. USA* **78**, 7634–7638.

Matoba R, Saito S, Ueno N, Maruyama C, Matsubara K, Kato K (2000) Gene expression profiling of mouse postnatal cerebellar development. *Physiol. Genom.* **4**, 155–164.

McManus MT, Sharp PA (2002) Gene silencing in mammals by small interfering RNAs. *Nat. Rev. Genet.* **3**, 737–747.

Mezey E, Chandross KJ, Harta G, Maki RA, McKercher SR (2000) Turning blood into brain: cells bearing neuronal antigens generated in vivo from bone marrow. *Science* **290**, 1779–1782.

Morozov G, Verlinsky O, Rechitsky S, Kukharenko V, Goltsman E, Ivakhnenko V, Gindilis V, Strom C, Kuliev A, Verlinsky Y (1998) Sequence analysis of libraries from individual human blastocysts. *J. Assist. Reprod. Genet.* **15**, 338–343.

Odorico JS, Kaufman DS, Thomson JA (2001) Multilineage differentiation from human embryonic stem cell lines. *Stem Cells* **19**, 193–204.

Okazaki Y, Furuno M, Kasukawa T, Adachi J, Bono H, Kondo S *et al.* (2002) Analysis of the mouse transcriptome based on functional annotation of 60,770 full-length cDNAs. *Nature* **420**, 563–573.

Okubo K, Hori N, Matoba R, Niiyama T, Fukushima A, Kojima Y, Matsubara K (1992) Large scale cDNA sequencing for gene expression analysis: quantitative and qualitative aspects of gene expression in a liver cell line, Hep G2. *Nature Genet.* **2**, 173–179.

Orelio C, Dzierzak E (2002) Identification of two novel genes developmentally regulated in the mouse aorta-gonad-mesonephros region. *Blood* **101**, 2246–2249.

Park IK, He Y, Lin F, Laerum OD, Tian Q, Bumgarner R *et al.* (2002) Differential gene expression profiling of adult murine hematopoietic stem cells. *Blood* **99**, 488–498.

Pease AC, Solas D, Sullivan EJ, Cronin MT, Holmes CP, Fodor SP (1994) Light-generated oligonucleotide arrays for rapid DNA sequence analysis. *Proc. Natl Acad. Sci. USA* **91**, 5022–5026.

Phillips RL, Ernst RE, Brunk B, Ivanova N, Mahan MA, Deanehan JK, Moore KA, Overton GC, Lemischka IR (2000) The genetic program of hematopoietic stem cells. *Science* **288**, 1635–1640.

Piao Y, Ko NT, Lim MK, Ko MSH (2001) Construction of long-transcript enriched cDNA libraries from submicrogram amounts of total RNAs by a universal PCR amplification method. *Genome Res.* **11**, 1553–1558.

Quackenbush J (2002) Microarray data normalization and transformation. *Nat. Genet.* **32**, 496–501.

Rajkovic A, Yan MSC, Klysik M, Matzuk M (2001) Discovery of germ cell-specific transcripts by expressed sequence tag database analysis. *Fertil. Steril.* **76**, 550–554.

Rajkovic A, Yan C, Yan W, Klysik M, Matzuk MM (2002) Obox, a family of homeobox genes preferentially expressed in germ cells. *Genomics* **79**, 711–717.

Ramalho-Santos M, Yoon S, Matsuzaki Y, Mulligan RC, Melton DA (2002) 'Stemness': transcriptional profiling of embryonic and adult stem cells. *Science* **298**, 597–600.

Sasaki N, Nagaoka S, Itoh M, Izawa M, Konno H, Carninci P *et al.* (1998) Characterization of gene expression in mouse blastocyst using single-pass sequencing of 3995 clones. *Genomics* **49**, 167–179.

Schena M, Shalon D, Davis RW, Brown PO (1995) Quantitative monitoring of gene expression patterns with a complementary DNA microarray [see comments]. *Science* 270, 467–470.

Scherr M, Morgan MA, Eder M (2003) Gene silencing mediated by small interfering RNAs in mammalian cells. *Curr. Med. Chem.* 10, 245–256.

Sikela JM, Auffray C (1993) Finding new genes faster than ever. *Nature Genet.* 3, 189–191.

Solter D (2000) Mammalian cloning: advances and limitations. *Nat. Rev. Genet.* 1, 199–207.

Solter D, de Vries WN, Evsikov AV, Peaston AE, Chen FH, Knowles BB (2002) Fertilization and activation of the embryonic genome. In: *Mouse Development: Patterning, Morphogenesis, and Organogenesis* (eds J Rossant, PPL Tam). Academic Press, San Diego, pp. 5–19.

Stanford WL, Cohn JB, Cordes SP (2001) Gene-trap mutagenesis: past, present and beyond. *Nat. Rev. Genet.* 2, 756–768.

Stanton JL, Green DP (2001a) Meta-analysis of gene expression in mouse preimplantation embryo development. *Mol. Hum. Reprod.* 7, 545–552.

Stanton JL, Green DP (2001b) A set of 840 mouse oocyte genes with well-matched human homologues. *Mol. Hum. Reprod.* 7, 521–543.

Stanton JL, Bascand M, Fisher L, Quinn M, Macgregor A, Green DP (2002) Gene expression profiling of human GV oocytes: an analysis of a profile obtained by Serial Analysis of Gene Expression (SAGE). *J. Reprod. Immunol.* 53, 193–201.

Strausberg RL, Feingold EA, Grouse LH, Derge JG, Klausner RD, Collins FS *et al.* (2002) Generation and initial analysis of more than 15,000 full-length human and mouse cDNA sequences. *Proc. Natl Acad. Sci. USA* 99, 16899–16903.

Suemizu H, Aiba K, Yoshikawa T, Sharov AA, Shimozawa N, Tamaoki N, Ko MS (2003) Expression profiling of placentomegaly associated with nuclear transplantation of mouse ES cells. *Dev. Biol.* 253, 36–53.

Suzuki Y, Yamashita R, Nakai K, Sugano S (2002) DBTSS: DataBase of human Transcriptional Start Sites and full-length cDNAs. *Nucleic Acids Res.* 30, 328–331.

Suzumori N, Yan C, Matzuk MM, Rajkovic A (2002) Nobox is a homeobox-encoding gene preferentially expressed in primordial and growing oocytes. *Mech. Dev.* 111, 137–141.

Takahashi N, Ko MSH (1994) Toward a whole cDNA catalog: construction of an equalized cDNA library from mouse embryos. *Genomics* 23, 202–210.

Tamayo P, Slonim D, Mesirov J, Zhu Q, Kitareewan S, Dmitrovsky E, Lander ES, Golub TR (1999) Interpreting patterns of gene expression with self-organizing maps: methods and application to hematopoietic differentiation. *Proc. Natl Acad. Sci. USA* 96, 2907–2912.

Tanaka S, Kunath T, Hadjantonakis AK, Nagy A, Rossant J (1998) Promotion of trophoblast stem cell proliferation by FGF4. *Science* 282, 2072–2075.

Tanaka TS, Jaradat SA, Lim MK, Kargul GJ, Wang X, Grahovac MJ *et al.* (2000) Genome-wide expression profiling of mid-gestation placenta and embryo using a 15,000 mouse developmental cDNA microarray. *Proc. Natl Acad. Sci. USA* **97**, 9127–9132.

Tanaka TS, Kunath T, Kimber WL, Jaradat SA, Stagg CA, Usuda M, Yokota T, Niwa H, Rossant J, Ko MS (2002) Gene expression profiling of embryo-derived stem cells reveals candidate genes associated with pluripotency and lineage specificity. *Genome Res.* **12**, 1921–1928.

Terskikh AV, Easterday MC, Li L, Hood L, Kornblum HI, Geschwind DH, Weissman IL (2001) From hematopoiesis to neuropoiesis: evidence of overlapping genetic programs. *Proc. Natl Acad. Sci. USA* **98**, 7934–7939.

Tsai J, Sultana R, Lee Y, Pertea G, Karamycheva K, Antonescu V, Cho J, Parvizi P, Cheung F, Quackenbush J (2001) RESOURCERER: a database for annotating and linking microarray resources within and across species. *Genome Biol.* **2**, software0002.1–0002.4.

VanBuren V, Piao Y, Dudekula DB, Qian Y, Carter MG, Martin PR *et al.* (2002) Assembly, verification, and initial annotation of the NIA mouse 7.4K cDNA clone set. *Genome Res.* **12**, 1999–2003.

Velculescu VE, Zhang L, Vogelstein B, Kinzler KW (1995) Serial analysis of gene expression. *Science* **270**, 484–487.

Velculescu VE, Vogelstein B, Kinzler KW (2000) Analysing uncharted transcriptomes with SAGE. *Trends Genet.* **16**, 423–425.

Venter JC, Adams MD, Myers EW, Li PW, Mural RJ, Sutton GG *et al.* (2001) The sequence of the human genome. *Science* **291**, 1304–1351.

Waterston RH, Lindblad-Toh K, Birney E, Rogers J, Abril JF, Agarwal P *et al.* (2002) Initial sequencing and comparative analysis of the mouse genome. *Nature* **420**, 520–562.

Wen T, Gu P, Chen F (2002) Discovery of two novel functional genes from differentiation of neural stem cells in the striatum of the fetal rat. *Neurosci. Lett.* **329**, 101–105.

Wiemann S, Weil B, Wellenreuther R, Gassenhuber J, Glassl S, Ansorge W *et al.* (2001) Toward a catalog of human genes and proteins: sequencing and analysis of 500 novel complete protein coding human cDNAs. *Genome Res.* **11**, 422–435.

Wilmut I, Schnieke AE, McWhir J, Kind AJ, Campbell KH (1997) Viable offspring derived from fetal and adult mammalian cells. *Nature* **385**, 810–813.

Wilmut I, Beaujean N, de Sousa PA, Dinnyes A, King TJ, Paterson LA, Wells DN, Young LE (2002) Somatic cell nuclear transfer. *Nature* **419**, 583–586.

Xu RH, Chen X, Li DS, Li R, Addicks GC, Glennon C, Zwaka TP, Thomson JA (2002) BMP4 initiates human embryonic stem cell differentiation to trophoblast. *Nat. Biotechnol.* **20**, 1261–1264.

Yanagimachi R (2002) Cloning: experience from the mouse and other animals. *Mol. Cell Endocrinol.* **187**, 241–248.

Yang YH, Speed T (2002) Design issues for cDNA microarray experiments. *Nat. Rev. Genet.* **3**, 579–588.

Yu Y, Bradley A (2001) Engineering chromosomal rearrangements in mice. *Nat. Rev. Genet.* **2**, 780–790.

Zhou FC, Duguid JR, Edenberg HJ, McClintick J, Young P, Nelson P (2001) DNA microarray analysis of differential gene expression of 6-year-old rat neural striatal progenitor cells during early differentiation. *Restor. Neurol. Neurosci.* **18**, 95–104.

Zur Nieden NI, Kempka G, Ahr HJ (2003) In vitro differentiation of embryonic stem cells into mineralized osteoblasts. *Differentiation* **71**, 18–27.

19. Proteomics and embryonic stem cells

Michael R. Sussman, Adrian D. Hegeman, Amy C. Harms and Clark J. Nelson

19.1 Introduction

The typical multicellular advanced eukaryote, either human, animal or plant, contains approximately 30,000 different genes. DNA 'chips' have allowed the measurement of all 30,000 gene products (mRNAs) in a single experiment. In molecular biology jargon, this means that we can do 30,000 Northern blots simultaneously. In less than a decade, we have gone from taking a week to do a one-gene Northern blot to doing a genome's worth of Northern blots in one afternoon. This is a breathtaking increase in throughput and data gathering that, unfortunately, is not yet possible in proteomic studies. For whole genome protein analysis (proteomics), the best technology available either uses two-dimensional gel electrophoresis coupled to a simple mass spectrometry measurement (usually, MALDI-TOF, which stands for matrix-assisted laser desorption ionization/time of flight), or a high performance liquid chromatography (HPLC) based method coupled with ESI (electrospray ionization)-tandem mass spectrometry. In either technique, one is capable of detecting the presence of less than a thousand of the most abundant proteins, rather than all 30,000. Despite these limitations, since it is the proteins rather than the mRNA, which really do all the biological 'work' in a cell, there is a great deal of interest in performing high-throughput, massively parallel 'proteomic' measurements.

While there are no published proteomics studies of embryonic stem cells to date, these techniques with little doubt will be brought to bear in future analyses. This review is intended as a general primer for the biological reader not yet versed in state of the art of proteomics, and will discuss this technological field from the vantage point of the human embryonic stem cells. Proteomics techniques rely on mass spectrometry (MS) as a general detection technique that allows amino acid sequencing through analysis of peptide fragmentation. Most universities now have mass spectrometry facilities capable of performing proteomic experimentation and despite the fact that an instrument typically costs in the range of $200,000–

$400,000, many biologists are increasingly obtaining access to the technology. The authors' laboratory is biased towards the use of HPLC and ESI rather than two-dimensional gel electrophoresis. Hence, this chapter will focus on such techniques using HPLC to separate complex peptide mixtures, followed by tandem mass spectrometry for sequence analysis. We hope that it helps you get started in this exciting and rewarding new area.

19.2 Making elephants fly

Mass spectrometry (MS) relies on manipulation and measurement of ions in the gas phase. Until the development of matrix-assisted laser desorption/ionization (MALDI) and electrospray ionization (ESI) (Karas and Hillenkamp, 1988; Tanaka et al., 1988; Wong et al., 1988), this proved a challenge for proteins since they are difficult to get out of a liquid solution and into the gas phase. The significance of MALDI and ESI is underscored by the awarding of the 2002 Chemistry Nobel Prize to Koichi Tanaka and John B. Fenn for research leading to the development of these technologies (Vestling, 2003). When asked by journalists why he was awarded the Nobel Prize, John Fenn reportedly answered 'for making elephants fly'. This statement alludes to the fact that proteins are generally large polymers (>10,000 Da) formed from the 20 different monomeric amino acids. Both ESI and MALDI are 'soft' ionization methods, in that they produce intact molecular ions of proteins.

19.2.1 Ionization mechanisms

In MALDI, the analyte is dissolved in a molar excess of matrix molecules, which are aromatic compounds that absorb UV radiation. After combining analyte and matrix, the mixture is spotted onto a metal target plate, dried, and placed into the mass spectrometer. To achieve ionization, a UV laser is pulsed at the dried sample on the target. The matrix absorbs photons, ionizes and vaporizes from the surface carrying analyte with it into the gas phase. Following vaporization from the target, a proton is transferred from matrix to analyte via a poorly understood mechanism (Gluckmann et al., 2002; Land and Kinsel, 2001; Papantonakis et al., 2002).

Like any technique, MALDI possesses strengths and weaknesses. One such strength is that analyte predominantly exists as a singly charged species, making spectrum interpretation easy. However, peptide sequencing is less effective with singly charged peptides than with doubly or triply charged peptides, which tend to produce better fragmentation patterns. MALDI is fairly tolerant of impurities; therefore, it requires less clean-up and is more suitable for mixtures. Other advantages of MALDI include low sample consumption, with hundreds of attomoles of analyte or less potentially sufficient for identification (Ekstrom et al., 2001; Caprioli et al., 2001) and suitability for rapid automated analysis via robotics. Real time analysis of separations is difficult with MALDI, which can be viewed as a disadvantage.

ESI, the other commonly used ionization method, involves spraying a liquid from a capillary through a small exit hole at high voltage (1–5 kV), resulting in charging of the effluent. Liquid is sprayed from the capillary as tiny charged droplets that break up into smaller and smaller droplets, until coloumbic repulsion exceeds surface tension of the droplet resulting in vaporization of the solvent and entry of analyte molecules into the gas phase for MS analysis. While MALDI is a pulsed ion source, ESI is a continuous source of ionization.

As with MALDI, there are advantages and disadvantages with ESI. One drawback is that though ESI can be automated, it is slower than analysis by MALDI. Typically an analyte has multiple charge states (+2 and +3 are most common with peptides, and +20 or more are possible with proteins), which can be advantageous for peptide sequencing but can prove challenging for spectral deconvolution when more than one species is present. Another weakness is ESI's intolerance of impurities such as salts, detergents and polymers. A strong point of ESI is its ease of utilization coupled to HPLC as a front end (LC/MS), a common separation method in proteomics, allowing real-time analysis of separations.

19.2.2 Sample fractionation

Despite its cost, a mass spectrometer has a limited capacity for analysis; so, for complex (i.e., biological) mixtures of hundreds or thousands of proteins it becomes necessary to fractionate the sample into reasonable reasonably sized samples for analysis. The traditional method has been separation of whole proteins via two-dimensional gel electrophoresis (2-DGE). This method separates proteins by their isoelectric point (pI) and molecular weight. Resolved proteins may be stained for visualization and quantitation, cut from the gel, digested, and analyzed by ESI or MALDI. Though a venerable tried and true technique, 2-DGE has several limitations when dealing with proteins that are hydrophobic, extreme in pI or size, or low in abundance. For those interested in 2-DGE, see the following references (Gygi *et al.*, 1999a, 2000; Rabilloud, 2000).

An alternative approach, which has proven quite powerful in recent years, relies on fractionation of peptides from a proteolytic digest of a protein mixture rather than analysis of intact proteins. This is referred to as 'bottom-up' or 'shotgun' proteomics. Peptide fractionation is accomplished using multiple dimensions of liquid chromatography (LC/LC) with concomitant MS analysis. After each fraction is processed, the data are pooled and analyzed to identify the proteins present. Different types of chromatography have been used: size exclusion, affinity, ion exchange, and reverse-phase (RP) (Ficarro *et al.*, 2002; Han *et al.*, 2001; Liu *et al.*, 2002; Wolters *et al.*, 2001). The most commonly used combination of chromatographies are reverse-phase (RP) and strong cation exchange (SCX) since they give sharp peaks and rely on different properties of the peptide. The first dimension (SCX) of chromatography can be performed off-line or in-line with the mass spectrometer with advantages and drawbacks to either (Peng *et al.*, 2002; Wolters *et al.*, 2001). Another variable is the choice of ionization, with both ESI and MALDI used (Peng *et al.*, 2002; Wolters *et al.*, 2001; Griffin *et al.*, 2001). Sample analysis by both MALDI and ESI will increase the protein coverage due to the complimentary nature of these two ionization mechanisms (Stevens *et al.*, 2002). As reported at the 2002 American Society for Mass Spectrometry conference, ESI is more successful with smaller peptides while MALDI is more successful with larger peptides, resulting in different protein identifications (Juhasz *et al.*, 2002).

One of the more familiar versions of LC/LC/MS relies on packing both resins into a biphasic column (*Figure 19.1*), and is referred to as the multidimensional protein identification technology (MudPIT). A capillary with a pulled tip is first packed with a C18 RP resin using a pneumatic packing chamber followed by packing with a SCX resin. When the column is loaded with sample, the peptides are retained on the SCX resin, which they encounter first. Next, a small population of peptides is eluted from the SCX resin onto the RP resin using a low concentration of salt such as ammonium acetate. Once the salt has been washed away, an analysis is performed using RP chromatography and ESI-MS. Following completion of the RP analytical run, another small fraction is eluted from the SCX phase onto the RP phase using a slightly higher concentration of salt; again the salt is washed away and another RP run conducted. This cycle can be repeated as many times as required based on sample complexity. Sample loss, which occurs in gels or other intermediate handling steps, is minimized with this approach. Sample quantitation and visualization, however, become more challenging. These analyses are routinely performed using 75 or 100-micron columns. With smaller columns, lower flow rates are used resulting in lower elution volumes. Lower elution volumes mean higher relative concentrations of peptide species and since ESI is a concentration dependent process, sensitivity is increased (Davis *et al.*, 1995). With higher sensitivity, less material is required, which is valuable for scarce samples. On the down side, smaller columns are more prone to plugging, and as a result more care in sample preparation and solvent handling is required.

There are alternative technologies to the MudPIT method just described. One such method relies on resolving the sample via denaturing sodium-dodecyl sulfate polyacrylamide gel electrophoresis (SDS-PAGE), with separation based on

Source Modifications for 2D LC MS/MS

4.5kV
Pt wire

Inlet From HPLC and Autosampler

Mixing Cross

Reverse Phase(C18) Medium

Strong Cation Exchange

Q-TOF MS/MS

Divert Valve
- closed on load
- open for LC run

~90% Flow to waste

○ C18 and SCX chromatographic media are packed into a "pulled" fused silcia capillary (I.D. 100 μm, O.D. 362 μm)

○ Peptides are resolved and eluted directly into the Mass Spectrometer

○ Sensitivity is increased at lower flow rates (~100 nL/min)

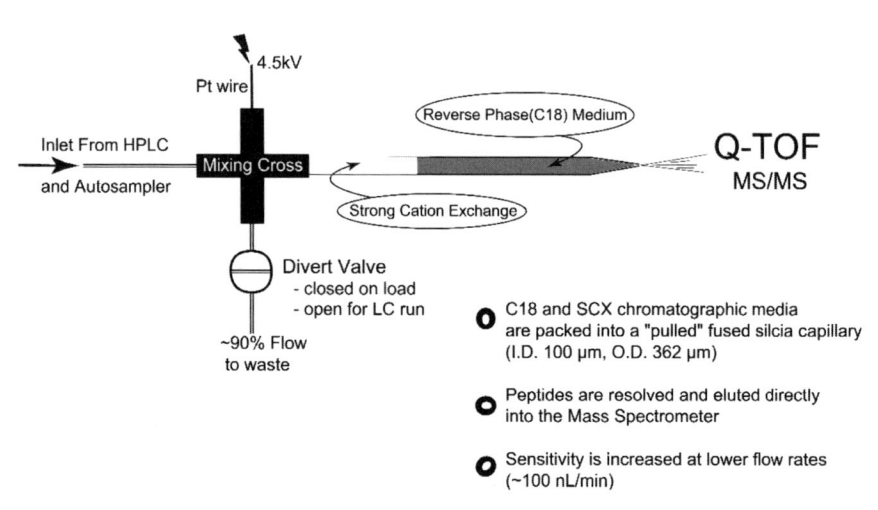

Figure 19.1: Multidimensional chromatography with a direct MS interface for analysis of complex mixtures of peptides.

molecular weight. The gel is cut into equal sized bands, in-gel proteolytic digests performed, and one-dimensional LC/MS analysis conducted (Eden *et al.*, 2002; Lasonder *et al.*, 2002). Another method relies on multiple one-dimensional RP analysis of the sample using successive narrow, bracketed mass ranges for picking candidate peptides for sequencing (VerBerkmoes *et al.*, 2002; Blonder *et al.*, 2002; Spahr *et al.*, 2001).

19.3 Mass spectrometry instrumentation

19.3.1 Mass analyzers

Once proteins or peptides have been successfully fractionated and introduced into the gas phase, several types of analyzers can be used to separate these species according to their mass-to-charge ratio *(m/z)*. Commonly used analyzers include the time of flight mass analyzers, quadrupole mass filters, quadrupole ion traps, and Fourier transform ion cyclotron resonance (FT-ICR) mass spectrometers (Yates *et al.*, 1995a, b; Eng *et al.*, 1994).

Time-of flight (TOF) mass analyzers separate ions by the different times required to reach the detector. Ions are accelerated down a flight tube and ions with lower masses will have higher velocities and a shorter flight time. Based on the TOF, the instrument assigns a m/z to the various products detected following ion extraction. Modern TOF instruments have high resolution, and all masses are measured in each ion pulse giving this analyzer high sensitivity. Because of the pulsed nature of both the MALDI source and the TOF analyzer, the MALDI-TOF is a common configuration for mass spectrometers.

Quadrupole mass filters consist of four parallel rods to which direct current and radio frequency (RF) voltages are applied. Only ions with a selected *m/z* will have a stable trajectory through the analyzer. Systematically changing the electric fields will cause different *m/z* to be passed through allowing a range of *m/z* values to be scanned. Most quadrupoles have a limited upper *m/z* range of 3-4,000 making ESI the ionization source of choice as the multiple charging allows for the analysis of higher MW compounds. Single quadrupoles can be used alone to generate MS data, but are more powerful when used as a component of a tandem mass spectrometer. One such configuration is the triple quadrupole. In this instrument two mass separating quadrupole sections are separated by a middle quadrupole, which serves as a collision cell. The first quadrupole selects an ion of interest, energetic collisions with neutral gas molecules cause fragmentation in the second quadrupole, and the third quadrupole is scanned to detect the product ions. This type of experiment gives MS/MS data, which is used for determining peptide sequence. Programming the independent quadrupole sections in other ways gives additional scanning options. A recent instrument design, which has proven useful, replaces the third quadrupole with a time of flight to create a quadrupole time-of-flight instrument (Q-TOF). The high accuracy and resolution of the TOF analyzer results in unambiguous determinations of charge-state, and an increased chance of success in data interpretation.

Quadrupole ion traps use RF voltages to collect and store ions between electrodes. Ions can be manipulated in this space, energy can be added to induce fragmentation, and ions can be systematically ejected on the basis of their m/z values and detected. Ion traps are able to do MS-MS experiments by first trapping all ions from the source, isolating the selected ion of interest by ejecting ions that do not have the desired m/z, inducing fragmentation of the ion of interest, and sequentially ejecting fragment ions for detection. An additional 'n' number of rounds of isolation and fragmentation can be added for more detailed structural studies and this is why this method is sometimes called MS^n.

Another trapping instrument is the Fourier transform ion cyclotron resonance (FT-ICR) mass spectrometer. In this instrument, ions are trapped in a strong magnetic field. Sensitivity and mass accuracy can be outstanding, but due to the complexity of the instrumentation its use for proteomic applications is not as widespread as other analyzers.

19.4 Protein and peptide chemistry

19.4.1 Whole protein or peptide fragment analysis

Mass spectrometric analysis of complex mixtures of proteins may be performed using intact proteins, sometimes referred to as the 'top-down approach', or by using peptide fragments of the proteins, called the 'bottom-up' approach (Kelleher et al., 1999). While considerable progress has been made developing top-down techniques (Jensen et al., 1999), the bottom-up approach is currently more widely used, in part because the instrumentation needed for this approach is more accessible. In the bottom-up approach, proteins are identified using either sequence data (from MS/MS spectra of individual peptides) and/or the observation of predicted peptide fragment masses (Wolters et al., 2001). Post-translational modifications make the top-down approach less effective for identifying proteins in complex mixtures as the measured mass may differ considerably from the predicted value (Mann et al., 2001). In cases where the protein identity is already known the top-down approach is excellent for characterizing these modifications. A combination of the two approaches has been employed that required specialized instrumentation, but ultimately, capitalized on the strengths of both techniques (VerBerkmoes et al., 2002).

19.4.2 Peptide sequencing

Bottom-up approaches take advantage of the relative ease of obtaining sequence data from peptides rather than intact proteins. When peptide ions are dissociated in a tandem mass spectrometer several bonds along the backbone can be broken. The most common fragmentation is at the amide bond with charge retention on either the amino (N-) or carboxy (C-) terminal fragments and these are commonly referred to as b and y ions, as shown in *Figure 19.2*. It is possible to interpret the fragmentation ladders to get the peptide sequence. However, problems can arise since b and y fragments are indistinguishable, not all fragments are present at detectable levels, and additional observed ions that are not b or y ions complicate *de novo* interpretation. A key advance in automated interpretation of peptide fragmentation is the development of algorithms directly correlating MS-MS data with

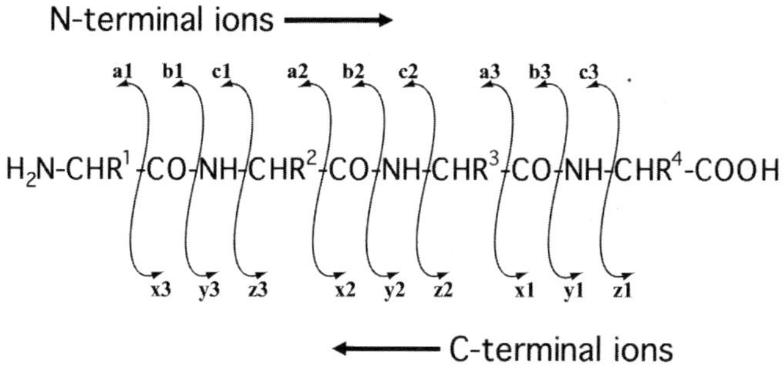

Figure 19.2: Peptide fragmentation and ion nomenclature. Fragmentation of a peptide can occur anywhere along the backbone. Peptide bond cleavage results in b- and y- ions, nitrogen $C\alpha$ bond cleavage produces c- and z- ions, while carbonyl $C\alpha$ bond cleavage generates a- and x- ions. Typical collisionally induced fragmentation in tandem mass spectrometry results in a series of y- and/or b-ions that can be used to deduce the amino acid sequence. Higher energy fragmentation techniques may typically produce other types of fragmentation including bond cleavage in the amino acid side chains.

peptide sequences in databases. This automation combined with good separation techniques and the appropriate mass analyzer allows for a very large number of proteins to be identified in a single run.

19.4.3 Proteolysis

The proteolytic digestion of proteins can be accomplished easily by using any of a number of commercially available sequencing-grade proteases. These enzymes catalyze the hydrolysis of protein amide bonds adjacent to specific amino acid residues producing predictable peptide fragments, which is key for the computer-based peptide-mass prediction routines. Commonly used enzymes include trypsin, Lys-C, Glu-C, Asp-N; the specificities for these and other enzymes are described in detail by the *ExPASy* web site program *peptidecutter* (http://us.expasy.org/tools/peptidecutter). Sequencing grade trypsin is often used as it generates peptides that yield informative MS/MS fragmentation series resulting in high quality sequence data. As both the N-terminal and the C-terminal residues (Arg or Lys) carry a positive charge, complete series of positively charged b- and y-ions may be observed. Chemical cleavage reagents can also be used in some cases, and some of the most useful reagents are also summarized in the *peptidecutter* program (see above link).

19.4.4 Coverage

A small fraction of the total sequence of any particular protein is observed in bottom-up proteomics experiments. A single protein identification may be made based on one or two unique peptides if the sequence data is of high quality. The fraction of the protein sequence that corresponds to observed peptide sequence is called the 'coverage.' The concept of coverage is important for understanding limitations in detecting post-translational modifications. Use of multiple proteases in separate analyses has been shown to enhance coverage of individual proteins as well as increase the number of proteins identified from a complex protein mixture (Choudhary et al., 2002).

19.4.5 Post-translational modifications

Post-translational modifications (PTMs) are ubiquitous, diverse and frequently physiologically important chemical alterations of proteins. Some PTMs of mammalian proteins include: acetylation, deamidation, farnesylation, glycosylation, geranyl-geranylation, hydrolysis (like signal peptide cleavage or N-terminal methionine removal), hydroxylation, lipoylation, methylation, myristoylation, palmitoylation, phosphorylation, sulphation, and sulphoxide formation (Wilkins et al., 1999). PTMs are important in the regulation of protein function, and their transient appearance is a common mechanistic motif in signal transduction. Proteomic applications to PTM identification have been somewhat successful with 2-DGE (Dwek et al., 2001; Sickmann et al., 2001).

Despite the limitations of 2-DGE listed earlier in this chapter, it may be used to provide information on the PTM state of abundant soluble proteins by comparing 2-DGE gels of, say glycosidase treated and untreated samples, rapidly to identify protein spots that migrate differently on the two gels. If reasonable coverage can be obtained for the untreated protein spot, it may be possible to localize the site of glycosylation within the protein. The 2-DGE approach does not provide access to as high a percentage of the proteome as bottom-up LC-ESI approaches.

While greater than 300 PTMs were reported using MudPIT on rat-liver Golgi membrane protein at the 2002 ASMS conference (Wu et al., 2002), comparable peer-reviewed reports have yet to materialize. As of yet, bottom-up approaches have been less successful in identification of PTMs than for large lists of proteins. This is largely because identification of proteins may be accomplished in bottom-up techniques with marginal coverage, while more thorough coverage is necessary to ensure that at least some fraction of the protein is unmodified. Unfortunately, PTMs often interfere with the ability to detect or assign peptides to a protein as the chemical properties and masses of the peptides are altered. The combined top-down/bottom-up approach gives good proteome coverage and allows identification of PTMs by analyzing the intact protein (VerBerkmoes et al., 2002). Although the localization of the sites of PTMs still depends on a high level of protein sequence coverage, improvements in methodology, instrumentation and search algorithms may allow bottom-up approaches to surpass 2-DGE and top-down approaches in the extent of proteome covered, and the identity and localization of PTMs.

As mentioned above, PTMs frequently alter the chemical properties of peptides. For example, a lipoylated peptide will not elute from reverse phase medium as it would if it were unmodified. Often, PTMs cause more insidious problems. Phosphorylated peptides, for example, do not ionize efficiently in positive mode in the presence of un-phosphorylated peptides. This is one example of a general phenomenon referred to as 'ion suppression', which makes chromatographic-MS essential for phospho-proteomics (Mann et al., 2002). Often, the unique chemical properties of a PTM can be exploited to facilitate its localization on the protein. This can be accomplished by detection of unique fragmentation properties of the PTM-peptide. For phosphopeptides containing phosphoserine, phosphothreonine and phosphotyrosine a loss of either H_3PO_4 (–98 Da) and/or HPO_3 (–80 Da) may be observed under mild fragmentation conditions. This 'neutral loss' (there is no net change in charge) can be utilized by data mining software to select phospho-serine/threonine/tyrosine peptides from large quantities of MS/MS data (Ficarro et al., 2002). The observation of a neutral loss can also be used as a trigger for 'data dependent switching' so that only the phosphopeptides are sequenced (Schlosser et al., 2001). 'Precursor ion scanning' has been effectively used for identification of peptides containing phosphotyrosine (Steen et al., 2001). This technique looks for the phosphotyrosine derived immonium ion (216.043 Da) in MS/MS data.

19.4.6 Phosphopeptide enrichment strategies

The chemical properties of some PTMs, like phosphorylation, lend themselves to the implementation of various enrichment strategies. Immobilized metal affinity chromatography (IMAC) with Fe^{3+} or Ga^{3+} (Andersson and Porath, 1986; Posewitz and Tempst, 1999) has been used to enrich phosphopeptides in complex mixtures (Vener et al., 2001). Ficarro and coworkers (2002) were able to identify 383 phosphorylation sites in yeast by using Fe^{3+} IMAC and by blocking acidic groups with methyl esterification to reduce background and increase enrichment. PTM specific antibodies can also be used to pull down proteins that contain PTMs such as phosphotyrosine (Pandey et al., 2000). Other strategies directly modify the PTM to make it easier to detect or enrich (Zhou et al., 2001). Phosphates can be eliminated from peptides containing phosphoserine/threonine under alkaline conditions to generate an α/β-unsaturated system, which can then be covalently modified by nucleophiles. This has been utilized to attach affinity tags, and other useful functional groups to phosphopeptides for facile purification and differential labeling (Goshe et al., 2001; Li et al., 2002; Weckwerth et al., 2000).

19.5 Better proteomics through chemistry

19.5.1 Quantitation

Some of the most interesting biological questions for proteomics concern the dynamics of proteins. Changes in protein concentrations, cellular location and PTM status can be monitored by comparing proteomic analyses of proteins from biological test and control samples. Unfortunately, any given peptide will experience ionization and detection at different efficiencies that are a function of its unique chemical properties and its ionization milieu. Every species measured will have its own extinction coefficient for any given ionization condition. Thus, absolute quantitation can only be obtained by constructing calibration curves under the appropriate solvent/co-analyte conditions. Several approaches have been developed to allow quantitation of interesting proteomic changes that rely on relative quantitation rather than absolute quantitation. One general strategy, summarized in *Figure 19.3*, involves derivatizing peptides prior to analysis with a reagent that contains either a stable heavy isotope label (^2H, ^{13}C, ^{15}N or ^{18}O), or

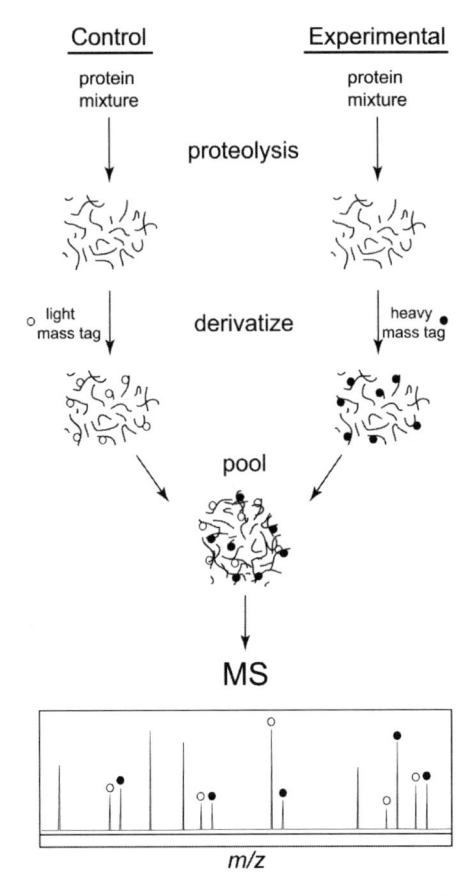

Figure 19.3: Isotope tag strategy for comparative proteomics. Experimental and control protein batches are processed in parallel and then combined for the MS analysis. Peptides in each population can be distinguished by differences in mass resulting from the use of either natural abundance or heavy atom labeled mass tags during processing. Tags can be attached using any number of chemical modification strategies. Comparison of chemically identical mass-tagged peptides in the same analysis sidesteps problems with changes in ionization efficiency and detection from run to run.

natural abundance, making one reagent a specific number of mass units heavier than the other. In this way peptides from control and experimental samples can be modified with either the heavy or light reagent, and combined for analysis. The integrated areas under heavy and light peaks can be compared quantitatively, as the two species are chemically identical, and they ionize at the same time and under the same conditions. A relative ratio of abundance can then be established.

19.5.2 ICAT reagent

The most commonly used reagent for comparative quantitation is the isotope coded affinity tag (ICAT) reagent developed in the Aebersold laboratory (Gygi *et al.*, 1999b). This compound selectively modifies the sulfhydryl moiety of cysteine residues effectively restricting the pool of modified peptides. The reagent also incorporates a biotin moiety as an affinity tag so that modified peptides can be purified from the unmodified peptides. This strategy allows low abundance proteins to be analyzed in complex mixtures, and is particularly effective in combination with multidimensional LC-MS (Gygi et al., 2002). ICAT reagents have been used effectively for studies of interesting biological questions, such as a comparison of proteins in a rat cell line with or without the oncogene *myc* (Shiio et al., 2002). However, implementation of the multiple chemical steps involved in the use of this reagent is far from trivial for laboratories that do not perform peptide modification chemistry on a semi-regular basis. ICAT reagents are commercially available and, generally, it is best to use the newest generation of reagents, as these use $9 \times {}^{13}C$ atoms for the isotope tag rather than $8 \times {}^{2}H$. The larger mass difference is used to prevent confusion with oxidized methionine residues in doubly charged peptides, and the use of ${}^{13}C$ eliminates the significant deuterium-isotope effects, which can result in heavy and light tagged peptides having slightly different retention times on reverse phase C18 HPLC columns (Zhang and Regnier, 2001). The new ICAT reagent also has an acid labile linkage to the biotin moiety, allowing its facile removal, improving MS/MS data quality.

Other isotope tag protocols have been developed that utilize a variety of tagging strategies, and modify different peptide functional groups. Primary amines, including the peptide N-termini and lysine ϵ-NH_2 may be modified using $[{}^{13}C_4/{}^{12}C_4]$succinic anhydride (Zhang and Regnier, 2001) or 1-($[{}^{2}H_4/{}^{1}H_4]$nicotinoyloxy)succinimide ester (Münchbach et al., 2000). Other strategies incorporate ${}^{18}O$ (vs. ${}^{16}O$ from natural abundance water) into peptides by performing the proteolytic digests in ${}^{18}O$ enriched water (Reynolds et al., 2002; Liu and Regnier, 2002). Richard Smith's laboratory has developed a technique for isolating and quantitatively comparing phosphopeptides called the phosphoprotein isotope-coded affinity tag (PhIAT) approach (Goshe et al., 2001). PhIAT is chemically more complex than the ICAT approach, but has tremendous possibilities for phosphoproteome analyses. Lastly, it should be noted that while MS techniques are generally used for mass measurements, rather than analyte quantitation, there is a fairly good linear correlation between protein concentration and peak areas via LC-MS (Chelius and Bondarenko, 2002). If ionization conditions are stable, and some sort of normalization procedure is used, it is reasonable to expect that comparative quantitative information can be obtained via multiple LC-MS replicate experiments.

19.6 Baby steps

To our knowledge, no one has published the application of any of the proteomics techniques described in this document to the study of embryonic stem cells. Two peripheral studies have been reported using 2-DGE. Lian and coworkers (2002) using tandem mass spectrometry, identified 123 protein species resolved by 2-DGE (out of 220 protein gel spots) from murine myeloid progenitor (MPRO) cell lines during induced differentiation. A rough measure of abundance for a few of the proteins, based on gel spot staining intensity, was used to test whether mRNA abundance correlated with protein abundance. This study, and more extensive reports using yeast, indicate that for many genes, especially those expressed at low levels, changes in mRNA level during differentiation are not necessarily reliable indicators of changes in protein abundance (Gygi et al, 1999a). Another study also using 2-DGE identifies 132 protein components of mouse embryonic fibroblast feeder layers used for human embryonic stem cell propagation (Lim and Bodnar, 2002). These reports represent baby steps in an emerging field and hardly represent comprehensive proteomic analyses as the many hundreds of thousands of proteins that are likely expressed in these samples have yet to be observed or temporally characterized.

In conclusion, we are in the midst of the birth of a new generation of proteomic technologies. Although the gestation period is over, these technologies currently fall short of the degree of throughput and reduced cost that is ultimately needed to perform comprehensive genome-wide analyses of proteins in cell extracts on a routine basis.

References

Andersson L, Porath J (1986) Isolation of phosphoproteins by immobilized (Fe^{3+}) affinity chromatography. *Anal. Biochem.* **154**, 250–254.

Blonder J, Goshe M B, Moore RJ, Pasa-Tolic L, Masselon CD, Lipton MS, Smith RD (2002) Enrichment of integral membrane proteins for proteomic analysis using liquid chromatography-tandem mass spectrometry. *J. Proteome Res.* **1**, 351–360.

Caprioli RM, Farmer TB, Gile J (2001) Molecular imaging of biological samples: localization of peptides and proteins using MALDI-TOF MS. *Anal. Chem.* **69**, 4751–4760.

Chelius D, Bondarenko PV (2002) Quantitative profiling of proteins in complex mixtures using liquid chromatography and mass spectrometry. *J. Proteome Res.* **1**, 317–323.

Choudhary G, Wu S-L, Shieh P, Hancock WS (2002) Multiple enzymatic digestion for enhanced sequence coverage of proteins in complex proteomic mixtures using capillary LC with ion trap MS/MS. *J. Proteome Res.* **2**, 59–67.

Davis MT, Stahl DC, Hefta SA, Lee TD (1995) A microscale electrospray interface for online capillary liquid chromatography/tandem mass spectrometry of complex peptide mixtures. *Anal. Chem.* **67**, 4549–4556.

Dwek MV, Ross HA, Leathem AJC (2001) Proteome and glycosylation mapping identifies post-translational modifications associated with aggressive breast cancer. *Proteomics* **1**, 756–762.

Eden S, Rohatgi R, Podtelejnikov AV, Mann M, Kirschner MW (2002) Mechanism of regulation of WAVE1-induced actin nucleation by Rac1 and Nck. *Nature* **418**, 790–793.

Ekstrom S, Ericsson D, Onnerfjord P, Bengtsson M, Nilsson J, Marko-Varga G, Laurell T (2001) Signal amplification using 'spot on-a-chip' technology for the identification of proteins via MALDI-TOF MS. *Anal. Chem.* **73**, 214–219.

Eng JK, McCormack AL, Yates JR 3rd (1994) An approach to correlate tandem mass spectral data of peptides with amino acid sequences in a protein database. *J. Am. Soc. Mass Spectrom.* **5**, 976–989.

Ficarro SB, McCleland ML, Stukenberg PT, Burke DJ, Ross MM, Shabanowitz, J, Hunt DF, White FM (2002) Phosphoproteome analysis by mass spectrometry and its application to *Saccharomyces cerevisiae*. *Nature Biotechnol.* 20, 301–305.

Gluckmann M, Pfenninger A, Kruger R, Thierolf M, Karas M, Horneffer V, Hillenkamp F, Strupat K (2002) Mechanisms in MALDI analysis: surface interaction or incorporation of analytes? *Int. J. Mass Spectrom.* 210, 121–132.

Goshe MB, Conrads TP, Panisko E.A, Angell NH, Veenstra TD, Smith RD (2001) Phosphoprotein isotope-coded affinity tag approach for isolating and quantitating phosphopeptides in proteome-wide analyses. *Anal. Chem.* 73, 2578–2586.

Griffin TJ, Gygi SP, Rist B, Aebersold R, Loboda A, Jilkine A, Ens W, Standing KG (2001) Quantitative proteomic analysis using a MALDI quadrupole time-of-flight mass spectrometer. *Anal. Chem.* 73, 978–986.

Gygi SP, Rochon Y, Franza RB, Aebersold R (1999a) Correlation between protein and mRNA abundance in yeast. *Mol. Cell Biol.* 19, 1720–1730.

Gygi SP, Rist B, Gerber SA, Turecek F, Gelb MH, Aebersold R (1999b) Quantitative analysis of complex protein mixtures using isotope-coded affinity tags. *Nat. Biotechnol.* 17, 994–999.

Gygi SP, Corthnls GL, Zhang Y, Rochon Y, Aebersold R (2000) Evaluation of two-dimensional gel electrophoresis-based proteome analysis technology. *Proc. Natl Acad. Sci. USA* 97, 9390–9395.

Gygi SP, Rist B, Griffin TJ, Eng J, Aebersold R (2002) Proteome analysis of low-abundance proteins using multidimensional chromatography and isotope-coded affinity tags. *J. Proteome Res.* 1, 47–54.

Han DK, Eng J, Zhou HL, Aebersold R (2001) Quantitative profiling of differentiation-induced microsomal proteins using isotope-coded affinity tags and mass spectrometry. *Nat. Biotechnol.* 19, 946–951.

Jensen PK, Pasa-Tolic L, Anderson GA, Horner JA, Lipton MS, Bruce JE, Smith RD (1999) Probing proteomes using capillary isoelectric focusing-electrospray ionization fourier transform ion cyclotron resonance mass spectrometry *Anal. Chem.* 71, 2076–2084.

Juhasz P, Falick A, Graber A, Hattan S, Khainovski N, Marchese J *et al*. (2002) ESI and MALDI LC/MS-MS approaches for large scale protein identification and quantification: are they equivalent? *Proceedings of the 50th American Society for Mass Spectrometry (ASMS) Conference on Mass Spectrometry and Allied Topics, Orlando, Florida, June 2-6.*

Karas M, Hillenkamp F (1988) Laser desorption ionization of proteins with molecular masses exceeding 10,000 daltons. *Anal. Chem.* 60, 2299–2301.

Kelleher NL, Lin HY, Valaskovic GA, Aaserud DJ, Fridriksson EK (1999) Top down versus bottom up protein characterization by tandem high-resolution mass spectrometry. *J. Am. Chem. Soc.* 121, 806–812.

Lasonder E, Ishihama Y, Andersen JS, Vermunt AMW, Pain A, Sauerwein RW *et al*. (2002) Analysis of the *Plasmodium falciparum* proteome by high-accuracy mass spectrometry. *Nature* 419, 537–542.

Land CM, Kinsel GR (2001) The mechanism of matrix to analyte proton transfer in clusters of 2,5-dihydroxybenzoic acid and the tripeptide VPL. *J. Am. Soc. Mass Spectrom.* 12, 726–731.

Li W, Boykins RA, Backlund PS, Wang G, Chen H-C (2002) Identification of phosphoserine and phosphothreonine as cysteic acid and β-methylcysteic acid residues in peptides by tandem mass spectrometric sequencing. *Anal. Chem.* 74, 5701–5710.

Lian Z, Kluger Y, Greenbaum DS, Tuck D, Gerstein M, Berliner N, Weissman SM, Newberger PE (2002) Genomic and proteomic analysis of the myeloid differentiation program: global analysis of gene expression during induced differentiation in the MPRO cell line. *Blood* 100, 3209–3220.

Lim JW, Bodnar A (2002) Proteome analysis of conditioned medium from mouse embryonic fibroblast feeder layers which support the growth of human embryonic stem cells. *Proteomics* 2, 1187–1203.

Liu HB, Lin DY, Yates JR (2002) Multidimensional separations for protein/peptide analysis in the post-genomic era. *BioTechniques* 32, 898–911.

Liu P, Regnier FE (2002) An isotope coding strategy for proteomics involving both amine and carboxyl group labeling. *J. Proteome Res.* 1, 443–450.

Mann M, Hendrickson RC, Pandey A (2001) Analysis of proteins and proteomes by mass spectrometry. *Annu. Rev. Biochem.* 70, 437–473.

Mann M, Ong S.-E, Grønborg M, Steen H, Jensen ON, Pandey A (2002) Analysis of protein phosphorylation using mass spectrometry: deciphering the phosphoproteome. *Trends Biotechnol.* 20, 261–268.

Münchbach M, Quadroni M, Miotto G, James P (2000) Quantitation and facilitated de novo sequencing of proteins by isotopic N-terminal labeling of peptides with a fragmentation-directing moiety. *Anal. Chem.* 72, 4047–4057.

Pandey A, Podtelejnikov AV, Blagoev B, Bustelo XR, Mann M, Lodish HF (2000) Analysis of receptor signaling pathways by mass spectrometry: identification of Vav-2 as a substrate of the epidermal and platelet-derived growth factor receptors. *Proc. Natl Acad. Sci. USA* 97, 179–184.

Papantonakis MR, Kim J, Hess WP, Haglund RF (2002) What do matrix-assisted laser desorption/ionization mass spectra reveal about ionization mechanisms? *J. Mass Spectrom.* 37, 639–647.

Peng J, Elias JE, Thoreen CC, Licklider LJ, Gygi SP (2002) Evaluation of multi-dimensional chromatography coupled with tandem mass spectrometry (LC/LC-MS/MS) for large scale protein analysis: the yeast proteome. *J. Proteome Res.* 2, 43–50.

Posewitz, MC, Tempst P (1999) Immobilized Gallium(III) affinity chromatography of phosphopeptides. *Anal. Chem.* 71, 2883–2892.

Rabilloud T (2000) Detecting proteins separated by 2D gel electrophoresis. *Anal. Chem.* 72, 48A–55A.

Reynolds KJ, Yao X, Fenselau C (2002) Proteolytic ^{18}O labeling for comparative proteomics: evaluation of endoprotease Glu-C as the catalytic agent. *J. Proteome Res.* 1, 27–33.[AQ2]

Schlosser A, Pipkorn R, Bossemeyer D, Lehmann WD (2001) Analysis of protein phosphorylation by combination of elastase digestion and neutral loss tandem mass spectrometry. *Anal. Chem.* 73, 170–176.

Shiio Y, Donohoe S, Yi EC, Goodlett DR, Aebersold R, Eisenman RN(2002)

Quantitative proteomic analysis of Myc oncoprotein function. *EMBO J.* 21, 5088–5096.

Sickmann A, Marcus K, Schäfer H, Butt-Dörje E, Lehr S, Herkner A, Suer S, Baher I, Meyer HE (2001) Identification of post-translationally modifies proteins in proteome studies. *Electrophoresis* 22, 1669–1677.

Spahr CS, Davis MT, McGinley MD, Robinson JH, Bures EJ, Beierle J et al. (2001) Towards defining the urinary proteome using liquid chromatography-tandem mass spectrometry I. Profiling an unfractionated tryptic digest. *Proteomics* 1, 93–107.

Steen H, Küster B, Fernandez, M, Pandey A, Mann M (2001) Detection of tyrosine phosphorylated peptides by precursor ion scanning quadrupole TOF mass spectrometry in positive ion mode. *Anal. Chem.* 73, 1440–1448.

Stevens SM, Kem WR, Prokai L (2002) Investigation of cytolysin variants by peptide mapping: enhanced protein characterization using complementary ionization and mass spectrometric techniques. *Rapid Commun. Mass Spectrom.* 16, 2094–2101.

Tanaka K, Waiki H, Ido Y, Akita S, Yoshida Y, Yoshida T (1988) Protein and polymer analysis up to m/z 100,00 by laser ionization time-of-flight mass spectrometry. *Rapid Commun. Mass Spectrom.* 2, 151–153.

Vener AV, Harms A, Sussman MR, Vierstra RD (2001) Mass spectrometric resolution of reversible protein phosphorylation in photosynthetic membranes of *Arabidopsis thaliana*. *J. Biol. Chem* 276, 6959–6966.

VerBerkmoes NC, Bundy JL, Hauser L, Asano KG, Razumovskaya J, Larimer F, Hettich RL, Stephenson JL (2002) Integrating 'top-down' and 'bottom-up' mass spectrometric approaches for proteomic analysis of *Shewanella oneidensis*. *J. Proteome Res.* 1, 239–252.

Vestling MM (2003) Using mass spectrometry for proteins. *J. Chem. Educ.* 80, 122–124.

Weckwerth W, Willmitzer L, Fiehn O (2000) Comparative quantification and identification of phosphoproteins using stable isotope labeling and liquid chromatography/mass spectrometry. *Rapid Commun. Mass Spectrom.* 14, 1677–1681.

Wilkins MR, Gasteiger E, Gooley AA, Herbert BR, Molloy MP, Binz, P-A et al. (1999) High-throughput mass spectrometric discovery of protein post-translational modifications. *J. Mol. Biol.* 289, 645–657.

Wolters DA, Washburn MP, Yates JR (2001) An automated multidimensional protein identification technology for shotgun proteomics. *Anal. Chem.* 73, 5683–5690.

Wong SF, Meng CK, Fenn JB (1988) Multiple charging in electrospray ionization of poly(ethylene glycols). *J. Phys. Chem.* 92, 546–550.

Wu CC, MacCoss MJ, Howell KE, Yates JRIII (2002) Proteomic analysis of golgi membrane proteins: identification, topology, and covalent modifications. *Proceedings of the 50th American Society for Mass Spectrometry (ASMS) Conference on Mass Spectrometry and Allied Topics, Orlando, Florida, June 2-6.*

Yates JR, Eng JK, McCormack AL (1995a) Mining genomes: correlating tandem mass spectra of modified and unmodified peptides to sequence in nucleotide databases. *Anal. Chem.* **67**, 3202–3210.

Yates JR, Eng JK, McCormack AL, Schieltz, D (1995b) Method to correlate tandem mass spectra of modified peptides to amino acid sequences in the protein database. *Anal. Chem.* **67**, 1426–1436.

Zhang R, Regnier FE (2001) Minimizing resolution of isotopically coded peptides in comparative proteomics. *J. Proteome Res.* **1**, 139–147.

Zhou H, Watts JD, Aebersold R (2001) A systematic approach to analysis of protein phosphorylation. *Nat. Biotechnol.* **19**, 375–378.

Appendix
Human embryonic stem cell resources

Cheryl Scadlock

Every effort has been made to ensure accurate web site information; however, web sites evolve, and some details may have changed.

United States Government Stem Cell Links

National Institutes of Health Human Embryonic Stem Cell Registry
Includes cell lines which meet the eligibility criteria for federally funded research
http://stemcells.nih.gov/research/registry

National Institutes of Health Stem Cell Basics
Updated September 2002
http://stemcells.nih.gov/info/basics

Federal Funding Announcement by President George W. Bush
August 9, 2001 announcement of federal funding for stem cell lines
www.whitehouse.gov/news/releases/2001/08

National Institutes of Health Guidelines for Research Using Human Pluripotent Stem Cells
Federal Register 65 FR 51976 August 25, 2000, corrected November 21, 2000 65 FR 69951.

The following web site is searchable by volume, issue date or text.
http://www.gpoaccess.gov/fr/index.html

National Institutes of Health Funding and Grant Opportunities
http://grants1.nih.gov/grants/oer.htm

National Institutes of Health Stem Cell Task Force
http://stemcells.nih.gov/policy.taskForce

Stem Cell Material Transfer Agreements and Memorandums of Understanding
http://stemcells.nih.gov/research/registry

United Kingdom Embryo Research Link

Human Fertilization and Embryology Authority (HFEA)

The statutory body which regulates licenses and collects data on human embryo research:
http://www.hfea.gov.uk/home

Australian Parliament and Individual Australian States to Reproductive Technology

Contains links to House of Representatives, Senate and bills.
http://www.aph.gov.au/

The Commonwealth Government of Australia cannot universally legislate for reproductive technology (including stem cell research), so each state and territory is responsible for designing and implementing separate legislation. The following web site contains links to the other Australian states.
http://www.sa.gov.au/government/other

Stem Cell Training Programs

ES Cell International (Melbourne, Australia and Singapore)
http://www.escellinternational.com/products/trainingsupport.html
The course is designed so trainees observe every procedure undertaken in the production and maintenance of hES cells, in a sequential fashion.

Jackson Laboratory (Bar Harbor, Maine)
http://www.jax.org/courses
The workshop provides hands-on training for investigators learning how to culture, manipulate, and differentiate ES cells from humans *in vitro*.

University of California at San Francisco (San Francisco, California)
http://escells.ucsf.edu/Training/Trng.asp
Training sessions include expert instruction, hands on experience with each procedure demonstrated, all cells and supplies required for training, and the UCSF Human Embryonic Stem Cell Culture Protocol Handbook.

Wisconsin Alumni Research Foundation WARF
(WiCell Research Institute, Madison, Wisconsin)
http://www.wicell.org/learn
Training program includes basic training in human embryonic stem cell culture techniques with one-on-one assistance throughout the course. Class participants receive additional support materials, including a CD of protocols.

The **National Institutes of Health** is funding stem cell mini-courses at universities around the United States. Each 7- to 14-day session will allow scientists to practice techniques with various mediums to keep human embryonic stem cell colonies alive and undifferentiated. For details go to http://stemcells.nih.gov/research/training or call the Science Policy and Planning Branch of the NIH 301.402.2313.

Permanent Federal Government Links

The White House, News and Policies
Text and date searchable.
http://www.whitehouse.gov/news/

Thomas, US Congress on the Internet
Legislation is searchable by Bill Number or Word/Phrase.
thomas.loc.gov

Senate and House of Representatives
http://www.house.gov/
http://www.senate.gov/

State Legislation Links

National Conference on State Legislatures
Searchable by keyword, major topic, date or author.
http://www.ncsl.org/

Law Librarians' Society of Washington, D.C.
http://www.llsdc.org/sourcebook
(Then click on state legislatures, state laws, and state regulations.)

Publicly Available Searchable Database for Scientific Articles

National Library of Medicine's PubMed Database
http://www.ncbi.nlm.nih.gov/entrez/query.fcgi?db=PubMed
Advanced search capabilities including MeSh (controlled vocabulary) searching with headings stem cells, erythroid progenitor cells, tumor stem cells, totipotent stem cells, myeloid progenitor cells, hematopoietic stem cells, pluripotent stem cells, and multipotent stem cells.

Society Devoted to Stem Cells

International Society for Stem Cell Research (ISSCR)
www.isscr.org
Promotes and fosters the exchange and dissemination of information and ideas related to stem cells, encourages the general field of research involving stem cells, and promotes professional and public education in all areas of stem cell research and applications.

Publications Focusing on Stem Cells

Publisher contacts are provided for informational purposes only. No endorsement is intended.

'Stem Cell Research News' (electronic journal)
'Stem Cell Business News' (electronic journal)
'Guide to Stem Cell Research Companies' (electronic directory)
'Who's Who in Stem Cell Research' (electronic directory)
DataTrends Publications Inc.
P.O. Box 4460
Leesburg VA 20177-8541
Phone: 703.779.0574
info@stemcellresearchnews.com

'Stem Cell Week'
NewsRx
P.O. Box 5528
Atlanta GA 31107-0528
Phone: 800.726.4550
www.newsrx.com

'Cloning and Stem Cells'
Mary Ann Liebert, Inc.
2 Madison Avenue
Larchmont NY 10538
Phone: 914-834.3100
www.liebertpub.com

'Journal of Hematotherapy and Stem Cell Research'
Mary Ann Liebert, Inc.
2 Madison Avenue
Larchmont NY 10538
Phone: 914-834.3100
www.liebertpub.com

'Stem Cells; the International Journal of Cell Differentiation and Proliferation'
AlphaMed Press, Inc.
One Prestige Place, Ste 290
Miamisburg OH 453542-3758
Phone: 937.291.2355
www.stemcells.com

Index

A2B5, 163, 164
ABO blood group antigens, 233
Abortion debate, 307–312
Adipogenic differentiation, 84, 88, 89
Adult stem cells, 29–30, 199
   human mesenchymal stem cells, 83–100
   MAPCs, 50–51, 87–88
   plasticity, 45–60
   potential uses, 55–56
Alkaline phosphatase, 3, 33
Alloantigen recognition pathways, 235–236
Allografts, 232–233
   strategies to prevent rejection, 245–250
Aneuploidy, 3, 33, 68
Angioblasts, 15, 137–138
Angiogenesis, 146, 147
Animal models of development and disease, 15–16, 263–265
Animal pathogens, exposure to, 65, 267
Anterior visceral endoderm, 178
Anti-angiogenic gene therapy, 147
Anti-apoptotic molecules, 244
Antigen presenting cells (APC), 235–236, 248
Antiproliferative agents, 247
Antisense constructs, 223–224
Applications of ES cell technology, 15–17
   see also Clinical applications
ATAC-PCR, 343
Australia, legal framework for research, 329
Azathioprine, 247

Bare lymphocyte syndrome, 243
Basic fibroblast growth factor (bFGF), 63–64, 138, 260

Basic helix-loop-helix transcription factors, 106–107, 108
Bcr/abl, 124, 276
β1 integrin 207
β cells, 176, 186–187
   identification, 188, 189
   in mouse and human, 70–71, 182
   and nestin, 177
   and transcription factors, 178, 180
β-globin, 129, 131
Biological agents, 247, 248
Biology of ES cells, 1–28
Blastema formation, 54–55
Blastocysts, 2, 30, 301–302
Blood group antigens, 233
Blood islands, 137–138
Blood transfusion, 123, 276
Blood vessels, tissue engineered, 145–146
Bone marrow cells
   plasticity of hematopoietic, 47–49
   see also Human mesenchymal stem cells
Bone marrow transplantation, 56, 121, 239
Bone morphogenetic proteins (BMPs), 64, 109, 114, 201
   anti-BMP signaling, 155, 158
   BMP2, 89
   BMP4, 15, 109, 110, 111–112, 128, 261
   BMP6, 180
Bone tissue engineering, 279
Brain activity, and moral status of embryo, 309–310
Brn4, 178

Calcineurin antagonists, 246–247

California, state regulation of research, 326–327
Capillary tube formation, 143
Cardiac arrhythmias, 209
Cardiac development, 72, 261
  early signals in, 200–201
Cardiac grafts, bioartificial, 206
Cardiac valves, tissue-engineered, 146
Cardiomyocytes, 72, 262
  differentiation in hES cell progeny, 199–213
  electrophysiological studies, 203
  in vitro differentiation of mouse and human ES cells to, 201–204
  and MSCs, 50, 89–90
  and myocardial regeneration, 204–209, 279
Catheter-based delivery strategies, 263
Cationic lipid reagents, 216–217
CD31: 139, 142
CD34: 124, 126, 128, 129, 139, 140, 147
CD44: 86
CD49b: 86
CD90: 86
CD105: 86
Cdx2, 105, 107
Cell banks, 237, 238, 266, 291, 292, 296
Cell-based therapies, *see* Clinical applications
Cell cycle of ES cells, 9–10
Cell Factory system, 295
Cell fusion, 54, 84–85
Cell holder, 263
Cellular cardiomyoplasty, 205
CFU-f (colony-forming unit-fibroblastic), 83
Chondrogenic differentiation, 84, 88–89
Chorionic gonadotrophin (CG), 109, 111, 113–114
Chorionic villi formation, 101, 102–103
Chromosomal abnormalities, 3, 68–69
Clinical applications, 16–17, 121, 257–287
  and adult stem cells, 55–56
  cellular transplantation with gene therapy for inherited disorders, 270–271
  continuous delivery of bioactive molecules by *ex vivo* gene therapy, 271–273
  and endothelial cell progenitors, 144–148
  engraftability of hES cell-derived neural precursors, 166
  future prospects, 280
  and genetic engineering, 224–226
  goals for bringing hES cells to, 258–269
    avoidance of immune rejection, 265–267
    clinical trials, 268–269
    efficacy in animal models of disease, 263–265
    efficient and safe cell delivery approaches, 263
    isolation of well-characterized donor cell population, 259–262

Clinical applications – *contd*
  goals for bringing hES cells to – *contd*
    safety, 267–268
    testing functional properties of cells/tissues *in vitro*, 262–263
  myocardial regeneration, 204–209
  preimplantation genetic diagnosis, 129–132
  production of cellular product for, 289–299
  prospects in disease-specific therapies, 275–280
  safety issues, 168, 209, 263, 267–268, 290
  and somatic cell nuclear transfer, 273–275
  tissue engineering, 16–17, 145–146, 206, 263, 269–270
Cloned animals, cDNA microarray analysis of, 351–352
Colony-forming cells (CFCs), hematopoietic, 125–127, 128
Common Rule, 301
Complementary DNA (cDNA) libraries, 340, 341
Complementary DNA (cDNA) microarrays, 340, 344, 350–352
Congestive heart failure, 204–205, 278
Connexin 45: 204
Corticosteroids, 247
Cryopreservation, 296
Cyclin A, 9–10
Cyclin E, 9–10
Cyclosporine, 246–247
Cytomegalovirus infection, 248
Cytotrophoblasts
  extravillous, 104
  villous, 103

Dedifferentiation, 54–55
Dendritic cells, 235, 236
Developmental potential, 29
Diabetes, 17, 173, 185, 270, 272, 278
Diapause, 6
Dickey Amendment (OCESAA rider), 304, 318
Differential display, 343
Differentiation of ES cells, 10–15, 33
  for clinical applications, 260–262
  default pathway, 13
  directed differentiation in culture, 12–13
  enrichment for differentiated cell populations, 11–12
  in vitro and *in vivo* in mouse and human ES cells, 66–68
  and large-scale production, 295–296
  and LIF signaling via ERKs, 8
  lineage induction and positional information, 13–14
  mechanistic vs. technical differences in mouse and human ES cells, 74–75

Differentiation of ES cells – *contd*
  similar principles in mouse and human ES cells, 73–74
  spontaneous, 10–11
  as tool for discovery, 75
  towards mesenchymal lineages, lessons from hMSCs, 88–91
Digital expression profiles, 346
Dlx-3, 107
Donor cells, tracking of, 264
Donor suitability assessment, 291
Dopaminergic neurons, 12, 14, 17, 73, 74, 157, 272, 277
Drugs, evaluation of, 17, 75
Dystrophin, 47–48

E1A-like activity, 39
E2F, 10
Early primitive ectoderm-like (EPL) cells, 12, 13
  differentiation as EBs (EPLEBs), 12–13
Early response to neural induction gene, 154–155
eGFP, 221, 225
Electroporation, 216
Electrospray ionization (ESI), 364–365
Embryogenesis, mammalian
  ES cells as *in vitro* model, 69–73
  understanding through ES cell modeling, 14–15
Embryogenomics, 339
Embryoid bodies (EBs), 10–12
  and cardiomyocyte differentiation, 202–203, 204
  in EC-derived teratocarcinomas, 32–33
  and endothelial cell differentiation, 138
  and hematopoietic differentiation, 127–128
  and islet differentiation, 185
  and isolation of donor cells for clinical applications, 260–261
  and neural differentiation, 156, 161
  and trophoblast differentiation, 112–114
Embryoid body-derived (EBD) cells, 37–38
Embryonal carcinoma (EC) cells, 30–33
Embryonal stem cell-specific gene 1 (*Esg1*), 9, 351
Embryonic germ (EG) cells, 34–37, 62, 66
Embryonic sources of stem cells, 30
Embryonic stem cell research
  ethical and policy issues, 301–313
  legal framework, 315–337
Embryonic stem cell test, 17
Embryonic stem cells
  biology, 1–28
  characteristics, 33–34
Encapsulation devices, 272–273
Endocrine deficiencies, 278

Endoderm
  endodermal epithelial cells, 48
  endodermal origin of islets, 176–177
  formation of and pancreatic morphogenesis, 174–176
  patterning of and inductive tissue interactions, 178–180
  primitive endoderm and cardiomyocyte induction, 201, 206–207
Endogenous stem cells, 29–30
Endothelial cells
  applications for gene therapy, 147–148
  characterization techniques for isolated cells, 142–144
  derivation from hES cells, 137–152
  development of progenitors, 137–140
  differentiation and vascularization in ES cells, 138–140
  and embryonic vasculogenesis, 137–138
  expression of markers, 142
  and *in vivo* vessel formation, 143–144
  and ischemia, 146–147
  isolation of human endothelial cells and progenitors, 141
  isolation of murine endothelial cells and progenitors, 140–141
  and LDL incorporation, 143
  and restenosis, 147
  separation using selectable marker, 141–142
  separation using surface receptors, 140–141
  therapeutic applications of progenitors, 144–148
  and therapeutic neovascularization, 146–147
  and tissue engineered blood vessels and other constructs, 145–146
  and tube formation on matrigel/collagen, 143
Eomes, 105
Epiblast, 30, 174
ERKs, 8, 89
ERRβ 105
ES Cell International, 328
ES cell renewal factor (ESRF), 8
Ethical and policy considerations in research, 301–313
Ethics Advisory Board (EAB), 302, 303
ExGen 500 reagent, 217
Expressed Sequence Tag (EST) projects, 340, 341

Failure Mode and Effect Analysis, 296
Fas ligand, 244
Federal regulations and guidelines
  and embryo research, 301–307, 315–320
  and production of cells for therapeutic use, 290–291, 297

Feeder layer-free conditions, 65, 267–268, 291, 294
Feeder layers, 33, 208
 human, 65, 267, 294, 328
 and large-scale production, 293–295
 novel, 64–65
Fertilization, as marker of unique personal identity, 307–308
Fibroblast growth factor (FGF)
 FGF2: 157–158, 161, 164, 165–166
 FGF4: 5, 39, 105
 FGF10: 180
 FGFR1: 64
 FGFR2: 105
 FGF signaling, 63–64, 154–155, 158
Flk-1, 14–15, 139
Flt-1, 139
Fluorescence-activated cell sorting, 262
Fourier transform ion cyclotron resonance (FT-ICR) mass spectrometer, 368
FoxA3, 177
Foxd3, 5, 39, 66
FTY720, 247–248
Fumarylacetoacetate hydrolase (FAH), 48–49
Functional assays, 262–263

GATA-2, 139
GATA4, 201, 203
Gcm1, 107–108
Gene expression
 and abnormalities in cloned animals, 352
 alteration, 220–224
 expression profiling methods, 341–344
 expression profiling of pluripotent cells, 9
 expression profiling of stem cells, 346–350
 of marker genes, 221–222
 over-expression, 221–222
 silencing, 222–224, 225
Gene Expression Omnibus, 346
Gene targeting, 1, 15–16, 222–223
Gene therapy
 anti-angiogenic, 147–148
 combined with cell therapy, 124, 270–271
Genetic completeness criterion, 308
Genetic disorders, 16, 129–132, 226, 270–271
Genetic diversity of ES cells, 231–232
Genetic engineering, 15–16, 215–229, 271
 alteration of gene expression, 220–224
 for cellular therapy, 224–225
 gene targeting, 222–223
 and immune matching, 241–244
 infection, 217–219
 introduction of DNA into hES cells, 216–219
 in nuclear transplantation therapy, 225–226
 over-expression of genes, 221–222

Genetic engineering – *contd*
 potential clinical applications, 224–226
 silencing gene expression, 222–224, 225
 transfection, 216–217, 219
 transient vs. stable integration, 219
Genetic mutations, ES cell lines with defined mutations, 130–131
Genetic testing, 129–132
Genetic uniqueness criterion, 307–308
Gene trapping, 223, 353
Genome replacement approach, 239–241
Genomic approaches, 339–361
 cDNA microarray analysis of cloned animals, 351–352
 data analysis and bioinformatics, 344–346
 expression profiling, 346–350
 follow-up study of cDNA microarrays, 350–351
 future perspectives, 354
 large-scale functional studies of genes, 353–354
 large-scale isolation of new genes, 340–341
 methods for expression profiling, 341–344
Germany, legal framework for research, 329
Germ cells, 30, 33, 34
Germ layers, 30, 257
 challenges to concept, 90
Glial cell line-derived neurotrophic factor, 272
Glial cells, 167
Glucagon, 186
Good Manufacturing Practice (cGMP) production, 297
Gp130: 6, 8, 63
Green fluorescent protein, 264
Growth factors, 11, 138, 141, 206

H3001A06 gene, 351
Hand1: 107, 108
Hash2, 108
Heart disease, 204–205, 278–279
Hedgehog signaling, 138
Helix-loop-helix transcription factors, 106–107, 108
Hemangioblasts, 14–15
Hematologic disorders, 276–277
Hematopoiesis, 14–15, 121–135
 definitive, 276–277
 from hES cells, 124–129
 lessons from mouse ES cells, 123–124
 in mouse and human, 71–72
 and preimplantation genetic diagnosis, 129–132
 reasons for research on hES cells, 122–123
Hematopoietic cell transplantation (HCT), 121–122, 129

Hematopoietic stem cells (HSCs), 46, 71–72
  and adult stem cell plasticity, 47–49, 52–53
  and clinical applications, 276–277
  and expression profiling, 347, 350
  functional assays, 263
  and mixed hematopoietic chimerism, 248–250
Hepatocytes, 48–49, 53, 54
HepG2 conditioned medium, 12, 13
Herpes simplex thymidine kinase gene, 225
Heterokaryon technique, 54
High performance liquid chromatography (HPLC), 363, 364
Histocompatibility antigens, 233–244
Homeobox factors, 107–108
Homologous recombination, 15–16, 104, 163–164, 223
HoxB4, 12, 72, 124, 276–277
Human and murine ES cell lines
  chromosomal alterations, 68–69
  derivation, growth and morphology, 62–69
  differences and similarities, 34, 61–81, 139–140, 159–160
  differentiation *in vitro* and *in vivo*, 66–68
  and endothelial cells, 139–140
  and ES cells as renewable source of functional cells, 73–75
  hES cells as model to study human development, 72–73
  and *in vitro* model of early mammalian development, 69–73
  markers of undifferentiated state, 66
  mechanistic vs. technical differences in differentiation, 74–75
  novel feeder layers, 64–65
  self-renewal factors, 63–64
  similar principles governing differentiation, 73–74
Human embryo, moral status of, 307–312
Human embryonic germ cells, 35, 62
Human Embryo Research Panel, 303, 311, 317
Human feeder layers, 65, 267, 294, 328
Human leukocyte antigens (HLA), 234–235, 237–244
  and hematopoietic cell transplantation, 122
  HLA-A, 234, 237–238
  HLA-B, 234, 237–238
  HLA-DR, 234, 237–238
  HLA expression by hES cells and derivatives, 237
  HLA-G, 237, 244
  HLA matching, 237–239, 245
  homozygous HLA haplotypes, 240–241
Human mesenchymal stem cells (hMSCs), 83–100
  adherence to tissue culture plastic, 83, 85

Human mesenchymal stem cells (hMSCs) – *contd*
  cell surface markers, 86–87
  in culture, 83–88
  and discovery of MAPCs, 87–88
  in vitro differentiation towards mesenchymal lineages, 88–90
  isolation techniques, 85–86
  mesenchymal potential and *in vivo* experiments, 90
  multiple sources, 87
  non-mesodermal lineages from, 90–91
  standard culture conditions, 86–87
  synergy of hMSC and hES research, 91–92
Hypoblast, 30
Hypoxanthine guanine phosphoribosyltransferase gene, 223

Id proteins, 106–107
Immune barrier to transplantation, 231–256
  and bringing cell-based therapy to clinical practice, 265–267
  and genetic manipulation, 225, 241–244
  immune profile, 232–237
  and myocardial regeneration, 209
  strategies for matching donor and recipient, 237–244
  strategies to prevent allograft rejection, 245–250
  and therapeutic cloning, 225–226, 239–241, 275
Immunosuppressive therapy, 245–248, 265
Importation of hES cell lines, 319
Indian hedgehog signaling, 138
Individualized cell production, 289
Indoleamine 2,3-dioxygenase, 244
Infection techniques, 217–219
Infectious agents, testing for, 291
Inner cell mass (ICM), 3, 30, 199–200, 347
Insulin-producing cells, 183–185, 278
Integrin–fibronectin interactions 138
Interferon-γ 235, 237
Interleukin 6 family cytokines, 1–2, 6
In vitro fertilization, 301, 302
Ionization mechanisms, 364–365
Ion suppression, 370
Ischemia, 146–147
Isl1, 178
Islets of Langerhans
  development in vertebrates, 174–182
  differentiation from ES cells, 182–187
  endodermal origin, 176–177
  lineage differentiation from mouse and human ES cells, 185–187
  modeling development through ES cell differentiation, 173–198

Islets of Langerhans – *contd*
  recapitulation of developmental pathways of differentiation in ES cells, 187–188
  transplantation, 263, 278
Isotope coded affinity tag (ICAT) reagent, 372–373

JAK/STAT3 pathway, 6–8

Kidney transplantation, 237–238

Lateral plate mesoderm, 179
Legal framework for research, 301–307, 315–337
Lentiviral vectors, 218–219
Leukemia inhibitory factor (LIF), 2, 31–32, 63, 260
  LIF-independent signaling, 8
  LIF signaling, 6
  receptor β, 6
  signaling via ERKs, 8
  signaling via JAK/STAT3 pathway, 6–8
Licensing programs, and research, 323–325, 330–337
Liver engineering, 146
Low density lipoprotein, acetylated, 143

Major histocompatibility complex (MHC) antigens, 233–236, 265–266
  MHC class I molecules, 225, 234, 235, 241–242, 243–244, 266
  MHC class II molecules, 234, 235, 236, 242, 266
Markers of endothelial cells, 139, 141–142
Markers of ES cells, 3
  expression of marker genes, 221–222
  markers of pluripotency, 39–40
  markers of undifferentiated mouse and human ES cells, 66
Markers of MSCs, 86–87
Mash-2, 107, 108
Mass spectrometry (MS), 363–364, 367–368
Material Transfer Agreements, 323–324
Matrix-assisted laser desorption/ionization (MALDI), 363, 364, 365, 367
Mesenchymal stem cells
  and bone tissue engineering, 279
  plasticity, 49–50
  see also Human mesenchymal stem cells
Mesenchymal-to-epithelial signaling, 179, 188
Mesodermal progenitors, 12–13
Mesoderm lineages, and hMSCs, 83–100
Metaplasia, 46
Microarrays, 340, 343, 344, 350–352
Minimum Information About a Microarray Experiment, 346

Minor histocompatibility antigens, 235, 236, 275
Mixed hematopoietic chimerism, 248–250
Monoclonal antibodies, 85–86, 247
Moral status of embryo, 307–312
Motor neurons, 14, 74, 158–159, 277
Mouse embryonic fibroblasts (MEF), 31, 33, 65, 208, 294, 347–348
Mouse ES cells
  derivation and definition, 1–3
  see also Human and murine ES cell lines
Msx1: 55
Multidimensional protein identification technology (MudPIT), 366
Multipotent adult progenitor cells (MAPCs), 50–51, 87–88
Muscular dystrophy, 47–48
Mycophenolate mofetil, 247
Myocardial infarction, 278, 279
Myocardial regeneration, 204–209, 279
  cardiomyocyte purification, 207–208
  directing cardiomyocyte differentiation, 206–207
  in vivo transplantation and anti-rejection strategies, 208–209
  scale-up needed, 208
Myocardin, 201

Nanog, 39
National Bioethics Advisory Commission, 304–305
National Institutes of Health Revitalization Act 1993: 303, 317
Natural killer (NK) cells, 236, 237, 243
Natural/unnatural criterion, 309
Neovascularization, therapeutic, 146–147
Nestin, 176–177
Neural induction, 13, 69–70, 153
  'default' model, 9, 158
  and neural specification from human ES cells, 160–163
  in vertebrates, 154–155
Neural patterning, 167
Neural rosettes, 161, 162, 163, 165
Neural specification from human ES cells, 153–172
  application of neural induction principles, 160–163
  directed differentiation in culture, 13
  direction of hES cells to glia/neurons with regional identities, 167
  engraftability of hES cell-derived neural precursors, 166
  ES cells and mammalian neural development, 155–156

Neural specification from human ES cells – *contd*
  ES cells and modeling neural lineage development, 157–159
  functionality of *in vitro* generated neurons, 167–168
  functional properties of hES cell-derived neurons, 165–166
  identity of hES cell-generated neural cells, 164–165
  isolation of ES-derived neural cells, 159
  isolation of neural cells from differentiated hES cell population, 163–164
  neural differentiation from hES cells, 159–166
  neural differentiation from mouse ES cells, 73, 156–159
  neural induction in vertebrates, 154–155
  safety of ES cell-derived neural cells for cell therapy, 168
  specification of neural fate from hES cells, 166–167
Neural stem cells
  gene expression profiling, 9, 347, 350
  plasticity, 52
Neuroectoderm, 13, 49–50, 70, 153, 162
Neuroepithelial cells, 154, 157, 161, 162, 163
Neurogenesis, 69–70
Neurogenin 3 (*Ngn3*), 177–178
Neurological diseases, 277
Neurospheres, 163
Nkx2.2, 178
Nkx2.5, 201, 203
Nkx6.1, 178, 180
Nodal, 174, 188
Nuclear transfer, *see* Somatic cell nuclear transfer
Nurr1, 12

Oct4 (Oct3/4), 3–5, 9, 39, 55, 66, 88, 105, 223, 351
Oligodendroglia, 167
Oligonucleotide microarrays, 344
Oligospheres, 157
Oocyte manipulation, and immune matching, 232, 239–241
Organizer, 154
Osteogenic differentiation, 83, 84, 85, 88, 89, 92

P48/Ptf1a, 177, 180, 186
PA6 cells, 13, 14, 157
Pancreatic duodenal homeobox 1 (*pdx1*), 177–178, 179, 180, 181, 186–187, 188
Pancreaticogenesis, 173–174
  and endoderm formation, 174–176
  human compared with mouse, 70–71, 181–182

Pancreaticogenesis – *contd*
  transcription factors involved, 177–178
  in vertebrates, 174–182
Parkinson's disease, 17, 74, 166, 272, 277
Parthenogenesis, 232, 241
Pax4, 12, 178
Pax6, 178
Pem, 39
Peptide sequencing, 368–369
Phosphopeptide enrichment strategies, 371
Phosphoprotein isotope-coded affinity tag approach, 373
Plac1, 352
Placental development, 68, 69, 101–119
  bridging mouse–human gap in placental biology, 108–114
  human ES cells as model for morphogenesis, 112–114
  initiation and trophectoderm formation, 105–106
  and lineage determination, 101
  morphogenesis in mouse placenta, 108
  in primates, 102–105
  and trophoblast differentiation, 101
  and villous morphogenesis, 101
Plasmid transfection, 219
Plasticity of adult stem cells, 45–60
  hematopoietic bone marrow cells, 47–49
  mechanisms, 52–55
  mesenchymal stem cells, 49–50
  neural cells, 52
  skeletal muscle cells, 51
Platelet-derived growth factor, 15, 138
Pluralistic approach, to moral status of embryo, 310–312
Pluripotency of ES cells, 2, 55
  and cell cycle, 9–10
  and gene profiling, 9
  and LIF signaling, 6–8
  markers, 39–40
  molecular basis, 3–10
  and transcriptional regulators, 3–5
Polymer scaffolds, 269–270
Poly-sialylated neural cell adhesion molecule (PSA-NCAM), 160, 163, 164
Positional information, 13–14, 73, 158
Positive negative selection, 223
Post-transplant lymphoproliferative disease, 248
Potentiality argument, 308–309
Precursor cells, 37
Precursor ion scanning, 371
Pregnancy, immunological tolerance in, 244
Preimplantation genetic diagnosis, 129–132
President's Council on Bioethics, 316
Primate Embryonic Stem Cells patent, 322

Primitive streak, 174, 311
Primordial germ cells (PGCs), 34, 35
Privileged sites, and transplant immunology, 232–233, 244
Production of hES cell-derived cellular product for therapeutic use, 289–299
　cell production scheme, 292–297
　cGMP production, 297
　cryopreservation and formulation, 296
　development of large-scale process, 293–296
　differentiated hES cells, 295–296
　individualized, 289
　product specifications and release criteria, 296–297
　properties required for cell therapy, 290
　qualification of hESCs and raw materials, 290–292
　undifferentiated hES cells, 293–295
Progenitor cells, 37
　and clinical applications, 75, 260, 263
　endothelial 137–142, 144–148
　MAPCs, 50–51, 87–88
Proteolysis, 369
Proteomics, 363–378
　coverage, 369–370
　ICAT reagent, 372–373
　ionization mechanisms, 364–365
　mass analyzers, 367–368
　mass spectrometry instrumentation, 367–368
　peptide sequencing, 368–369
　phosphopeptide enrichment strategies, 371
　post-translational modifications, 370–371
　protein and peptide chemistry, 368–371
　proteolysis, 369
　quantitation, 371–372
　sample fractionation, 365–367
　whole protein or peptide fragment analysis, 368
Pulmonary disease, 280

Quadrupole ion traps, 367–368
Quadrupole mass filters, 367
Quantitative-PCR, 343

Research
　academic licensing, 323–324
　agreements, 325–326
　commercial licensing strategy, 324–325
　decision to narrow eligibility for federal funding, 306–307
　ethical and policy considerations, 301–313
　federal regulation, 301–302, 315–320
　funding restrictions imposed by federal agencies, 302–304, 317–318
　guidelines, August 2001: 318–319

Research – contd
　international legal framework, 328–329
　intersection of funding with abortion debate, 307–312
　legal framework, 301–307, 315–337
　licensing, 323–325, 330–337
　origins of decision to permit general federal funding, 304–305
　origins of *de facto* ban on federal funding, 302–303
　origins of *de jure* ban on federal funding, 303–304
　patent rights, 320–322
　'research exemption' to patent infringement, 320–321
　state regulation, 326–328
Restenosis, 147
Retinoic acid (RA), 156–157, 158–159, 160, 165–166
Retroviral vectors, 217–218, 219
RNA interference (RNAi) method, 224

S17 cells, 125, 126, 127
Safety issues, and clinical applications, 168, 209, 263, 267–268, 290
Scl, 14–15
Self-renewal of ES cells, 2–3, 63–64, 260
Serial analysis of gene expression (SAGE), 343
Serum-free conditions, 65
Short inhibitory RNAs (siRNAs), 224
Shp2/Ras-dependent pathway, 6, 8
Sickle cell anemia, 129–130
Singapore, legal framework for research, 328–329
Sirolimus, 247
Skeletal muscle cells, plasticity of, 51
Skin cancer, 248
Skin engineering, 146, 269
Smooth muscle cell differentiation, 138
Somatic cell nuclear transfer, 273–275
　cDNA microarray analysis of cloned animals, 351–352
　ethical and policy issues, 306–307
　for genetic disorders, 17, 225–226
　and immune matching, 232, 239–241, 266–267, 275
　and stem cell plasticity, 54, 55
Sonic hedgehog (*Shh*), 138, 158, 179, 188
Sox2, 5, 39, 66, 222
Sox17, 177, 188
Stage-specific embryonic antigens, 33, 66
STAT3: 6–8, 9, 10, 63
State regulation, and research, 326–328
Stem cell factor, 34, 64

Stem cells
  definition, 45–46
  and developmental potential, 29
  sources, 29–37
'Stemness'
  definition, 84, 346
  genes for, 347–350
Stromal cell-derived neural inducing activity (SDIA), 13
Stromal cells from bone marrow, 83–100
Suicide genes, 225
Survival promoting factors, 14
SXY module, 243
Syncytiotrophoblasts, 102, 103–104, 111

Tacrolimus, 246–247
T cells, 234, 236, 246, 275
Telomerase, 3, 55, 88, 91
Teratocarcinomas, 30–33
Teratomas, 66, 68, 112, 168, 225
Therapeutic cloning, 225–226, 239–241, 273–275
Tie-1, 139
Tie-2, 139
Time-of flight (TOF) mass analyzers, 367
Tissue engineering, 16–17, 145–146, 206, 263, 269–270, 279
Totipotency of ES cells, 2, 29
Toxins, evaluation of, 17
TRA antigens, 33, 66
Transcriptional regulators, 3–5
Transdifferentiation, 54–55
Transfection techniques, 216–217, 219
Transforming growth factor beta, 64, 138, 154
Transgenic mice, 216, 242
Transplantation, immune barrier to, 231–256

Transwell cultures, 127
Trisomy-8 ES cells, 3
Trophectoderm, 101, 105–106, 347
Trophoblast differentiation from ES cells, 69, 101–119
  comparison of human and mouse, 68, 104–106, 110–111
  from human ES cells, 108–112
  homeobox and other factors, 107–108
  molecular control, 106–108
  in primates, 102–105
Tumor formation, avoidance of, 91, 268
Two-dimensional gel electrophoresis (2-DGE), 365, 370, 373–374

Unipotent cells, 29
United Kingdom, legal framework for research, 328
'Universal' donor cell phenotype, 266

Vascular endothelial growth factor (VEGF), 15, 138
Vasculogenesis, embryonic, 137–138
VE-cadherin, 139
Visceral endoderm, 11, 12, 14
vWF, 142

WiCell Research Institute, 322, 323, 324–326, 330–337
Wisconsin Alumni Research Foundation (WARF), 322, 323
Wnt signaling, 64, 70, 155, 201

Yolk sac, 137–138

Zoonotic pathogens, 65, 267